D1553473

AQUATIC INSECT ECOLOGY

1. Biology and Habitat

AQUATIC INSECT ECOLOGY

1. Biology and Habitat

J. V. Ward

Professor of Biology
Colorado State University
Fort Collins, Colorado

John Wiley & Sons, Inc.

New York / Chichester / Brisbane / Toronto / Singapore

In recognition of the importance of preserving what has been written, it is a policy of John Wiley & Sons, Inc., to have books of enduring value published in the United States printed on acid-free paper, and we exert our best efforts to that end.

Library of Congress Cataloging-in-Publication Data:

Ward, J. V.

Aquatic insect ecology / by J. V. Ward.

p. cm.

Includes bibliographical references and index.

Contents: v. 1. Biology and habitat.

ISBN 0-471-55007-8 (v. 1)

1. Aquatic insects—Ecology. I. Title.

QL472.W37 1992

595.7092—dc20 91-22057

CIP

Printed in the United States of America

10 9 8 7 6 5 4 3 2 1

To my mentor
Robert W. Pennak

Errata for

AQUATIC INSECT ECOLOGY
1. Biology and Habitat

J.V. Ward

The figure on page 45 should appear as follows:

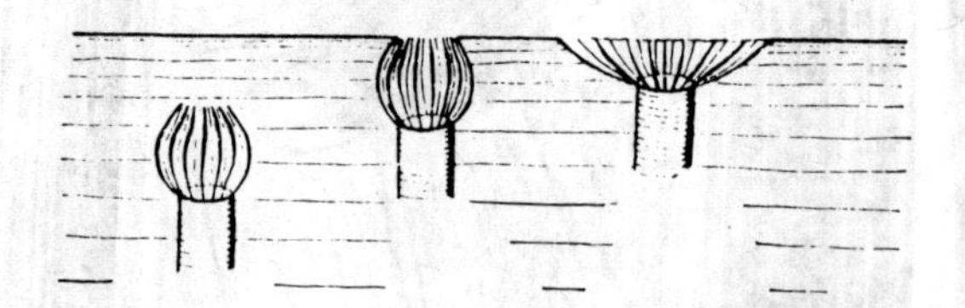

ISBN 0-471-55007-8

PREFACE

It is my aim to provide a comprehensive treatment of the ecology of aquatic insects to be used as a textbook at the advanced undergraduate and beginning graduate levels, and as a reference work by those studying aquatic habitats and their inhabitants.

This first volume begins with a brief overview of the insect orders with aquatic representatives. This is followed by evolutionary considerations; habitat occurrences (lentic, lotic, marine, saline) of aquatic insect communities (pleuston, plankton, nekton, benthos, hyporheos) and finally the relationship of aquatic insects to environmental variables (temperature, substrate, flow and water level, suspended sediment, light, oxygen, and other chemical factors).

Volume II will begin with the feeding ecology of aquatic insects. This will be followed by predator-prey and competitive interactions, and other symbiotic relationships; biodiversity patterns; life stages and production; spatial patterns (from microdistribution to zoogeography); and will end with the influence of man on aquatic insects.

Although the published literature is extensive and diverse, there are surprisingly few major comprehensive works dealing exclusively with aquatic insects, and most are badly out-of-date (Miall 1895, Rousseau 1921, Karny 1934, Wesenberg-Lund 1943, Bertrand 1954). Available books are concerned primarily with taxonomy and identification. An exception is the book edited by Resh and Rosenberg (1984) that deals with selected topics of aquatic insect ecology.

It is, of course, not possible in one volume to even cursorily deal with the entire literature on aquatic insect ecology. Even if one were to include only papers written in English, the task would be an overwhelming, as well as futile exercise. I have selected citations according to my best judgment, with an emphasis on European and North American literature. Many important works of a largely taxonomic or morphological nature have been purposely omitted.

A book dealing with any aspect of ecology must include information on habitat variables as they relate to the organisms under consideration. Rather than merely referring the reader to textbooks in limnology, stream ecology, or marine biology, a certain amount of basic data on environmental conditions and phenomena influencing aquatic insects has been included. This has been done with the express knowledge that such information will already be known by the experienced investigator. The emphasis on inland waters reflects the restriction of most aquatic insects to freshwater habitats.

A challenging and intellectually stimulating treatment of aquatic insect ecology requires using a rather extensive technical vocabulary. In the words of Lavoisier:

> It is impossible to dissociate language from science or science from language, because every natural science always involves three things: the sequence of phenomena on which the science is based; the abstract concepts which call these phenomena to mind; and the words in which the concepts are expressed. To call forth a concept a word is needed; to portray a phenomenon, a concept is needed. All three mirror one and the same reality.
>
> Antoine Laurent Lavoisier, 1789

Many of my colleagues have assisted me in preparing this volume. I am especially grateful to the late G. Evelyn Hutchinson, Sterling Professor of Zoology Emeritus, Yale University, for providing me with sections of his manuscript being prepared for Volume IV (Benthos), *A Treatise on Limnology*. Professor Hutchinson has graciously permitted me to cite from his manuscript in the form "Hutchinson MS." I am also grateful to my mentor, Professor Emeritus Robert W. Pennak, who suggested that I write a book on aquatic insect ecology and to whom this volume is dedicated. Professor Pennak critically reviewed all chapters as they were completed. Others who provided guidance include Drs. N. H. Anderson, L. Botosaneanu, J. E. Brittain, H. E. Evans, W. D. Fronk, C. A. S. Hall, B. C. Kondratieff, B. L. Peckarsky, V. H. Resh, S. J. Saltveit, B. P. Smith, J. A. Stanford, and P. Zwick. Drs. B. L. Peckarsky and M. J. Tauber kindly hosted my sabbatical leave at Cornell University, enabling me to focus my attention on writing and library research with few distractions. Original drawings were skillfully executed by L. O'Keefe (frontispiece), A. Blackstone, J. Bodenham, D. Carlson, A. Dixon, R. Hite, and T. Sechrist. Mrs. Nadine Kuehl not only skillfully typed the manuscript, she exhibited great patience as the developing manuscript was twice transferred to updated word processing systems. Finally, I thank the professional staff of Wiley's New York office, especially Senior Editor Dr. Philip C. Manor and Associate Managing Editor Ms. Allison Ort.

J. V. Ward

Fort Collins, Colorado

CONTENTS

I

INTRODUCTION

Nature's secrets are of unfathomable depth, but it is granted to us ... to look into them more and more; and the very fact that she remains unfathomable at last perpetually charms us to approach here again and again, and ever to seek for new lights and new discoveries.

—Johann Wolfgang von Goethe

1

INSECTS OF AQUATIC ENVIRONMENTS

Although less than 3% of all species of insects have aquatic stages (Daly 1984), in some freshwater biotopes insects may comprise over 95% of the total individuals or species of macroinvertebrates. Aquatic insects exhibit a vast array of morphological, physiological, and behavioral adaptations enabling inhabitation of virtually all bodies of water. Aquatic stages of insects occur in hot and cold springs, intertidal pools, temporary and aestival ponds, water-filled tree holes, intermittent streams, and saline lakes as well as in less severe running and standing-water habitats. Rarely are conditions in natural or even polluted waters so extreme as to exclude insects. The virtual absence of insects in the open sea, a topic dealt with in Chapter 5, provides an interesting exception to the ubiquity of the group.

HABITAT OCCURRENCES

Thirteen orders of insects contain species with aquatic or semiaquatic stages (Tables 1.1 and 1.2). In five of these orders (Ephemeroptera, Odonata, Plecoptera, Megaloptera, and Trichoptera), aquatic stages are possessed by all species (rare exceptions are discussed in the text). The remaining eight orders contain terrestrial as well as aquatic or semiaquatic representatives. Semiaquatic species live in damp marginal habitats (e.g., some hemipterans), or are associated with the upper surface of the air-water interface (e.g., some collembolans), or normally live above the water surface only submerging temporarily, perhaps for concealment (e.g., some orthopterans). With the exception of a few rare and

Table 1.1 Aquatic Association and habitats of insect orders containing aquatic or semiaquatic representatives[a,b]

Order	Aquatic Association	Habitats of Life Stages			
		Eggs	Larvae	Pupae	Imagos
Collembola	2	A,T	A,S	—	S
Ephemeroptera	1	A	A	—	T
Odonata	1	A,T	A	—	T
Hemiptera	2	A,T	A,S,T	—	A,S,T
Orthoptera	3	T	S	—	S
Plecoptera	1	A	A	—	T
Coleoptera	2	A,T	A,T	T	A,S,T
Diptera	2	A,T	A	A,T	T
Hymenoptera	2	P	P	P	A,T
Lepidoptera	2	A	A	A	T
Megaloptera	1	T	A	T	T
Neuroptera	2	T	A	T	T
Trichoptera	1	A,T	A	A	T

[a]Habitats refer to those of only the aquatic or semiaquatic species of those orders that also contain terrestrial species. Rare exceptions are not indicated in the table, but are discussed in the text.
[b]Key: 1 = all species are aquatic; 2 = primarily terrestrial, some truly aquatic species; 3 = primarily terrestrial, some semiaquatic species, no truly aquatic species; A = aquatic; T = terrestrial; S = semiaquatic; P = parasitic on an underwater host.

Table 1.2 Occurrence of life stages in major habitat types for aquatic and semiaquatic representatives of insect orders[a]

Order	Terrestrial	Freshwater	Estuarine	Marine
Collembola	A,L	A,L	A,L	A,L
Ephemeroptera	A	L	L	—
Odonata	A	L	L	—
Hemiptera	A	A,L	A,L	A,L
Orthoptera	A,L	A,L	—	—
Plecoptera	A	L	L	—
Coleoptera	A,L,P	A,L,P[b]	A,L	A,L
Diptera	A,P	L,P	L,P	L,P
Hymenoptera	A	A,L,P	—	A,L,P
Lepidoptera	A	L,P	L,P	—
Megaloptera	A,P	L	—	—
Neuroptera	A,P	L	—	—
Trichoptera	A	L,P	L,P	L,P

[a]Key: A - adult; L - larvae and nymphs; P - pupae.
[b]A few Psephenidae and the African species of Torridincolidae have truly aquatic pupae (see text).

interesting examples, only aquatic beetles and bugs contain species in which both adult and immature stages occur under water; aerial adults characterize the other aquatic orders that are, therefore, amphibiotic. Isolated cases of aquatic or semiaquatic species in a few other orders of otherwise terrestrial insects are briefly referred to at the end of this chapter.

The pupae of some holometabolous taxa remain within the aquatic habitat, whereas in others the last instar larvae move onto land to pupate. The pupal stage thus provides the transition from aquatic larva to terrestrial adult in neuropterans, megalopterans, and some dipterans, but most aquatic coleopterans move back into the water following pupation on land.

Some parasitic wasps (Hymenoptera) have aquatic hosts. The female wasp crawls underwater to oviposit on the eggs or immature stages of aquatic insects and other invertebrates.

ORDERS WITH AQUATIC REPRESENTATIVES

The following brief synopsis of the orders with aquatic and semiaquatic representatives is presented as a prelude to the remainder of this book, which is structured according to ecological rather than taxonomic categories.

Collembola—Primitively Wingless Hexapods Collembolans or springtails are small ametabolous apterygotes. The ordinal name is derived from collophore, based on the presumed adhesive function of the ventral tube, a structure on the first abdominal segment. The primary function of the ventral tube, however, appears to be osmoregulatory [see refs. cited in Joossee (1976)]. The common name relates to the possession by primitive forms of a springing organ or furcula attached to the fourth abdominal segment. Whether these primitive hexapods are truly insects is subject to contention (Mackerras 1970).

Members of the order are primarily terrestrial, commonly inhabiting soil and other moist habitats. The majority of species associated with aquatic habitats are accidentals, often as temporary inhabitants of the water surface. Dense masses of collembolans appear as a film of powder on the surfaces of quiet backwaters, especially in early spring. When disturbed the water surface appears to "come alive" as individuals release the springing organ. Thirty species in four families are associated with European waters (Gisin 1978). Waltz and McCafferty (1979) consider 50 species found on or near North American freshwaters under three headings. Primary aquatic associates (10 species) are essentially restricted to aquatic habitats and exhibit a high degree of specialization for an aquatic existence (primarily morphological adaptations facilitating mobility on the water surface); secondary aquatic associates (5 species) are commonly found in and near aquatic habitats; and tertiary aquatic associates (35 species) may be regarded as transitory. Virtually all collembolans, because of their requirement for high humidity, have an affinity for areas marginal to aquatic habitats.

Marine collembolans occur primarily in the intertidal zone (Joossee 1976).

They cannot tolerate permanent submergence and when the tide comes in refuge is sought in rock crevices or other microhabitats with high humidity. Species may exhibit zonal distribution patterns within the intertidal zone. Some species tend to be associated with rocky shores, whereas a specialized collembolan fauna occurs in the littoral subsoil of marine beaches. Several species are associated with salt marshes, river estuaries, and other brackish waters.

In the common freshwater species *Podura aquatica,* and perhaps other species, the male performs a courtship dance on the water surface (Waltz and McCafferty 1979). Spermatophores deposited by the male are taken up by the female through the genital pore. The eggs develop under water. Although the young may remain submerged for a time, once they break the surface film the hydrophobic cuticle prevents future submergence. North temperate populations of *Anurida maritima,* the most common intertidal species, deposit eggs in rock fissures in August (Joossee 1976). The eggs overwinter and hatch during spring. The young grow larger, but otherwise change little as they pass through the immature instars.

Relatively little is known regarding collembolan food habits. Many marine and freshwater species are apparently primarily scavengers or detritivores, although those of the rocky intertidal tend to be carnivorous.

Although few ecological data are available, literature dealing with collembolans associated with marine (Joossee 1976) and freshwater (Waltz and McCafferty 1979) habitats has been summarized.

Ephemeroptera—Mayflies Mayflies are hemimetabolous insects, the nymphs of which are aquatic, the adults aerial. The ephemeral nature of the adult stage, generally of 2–3 days' duration or less, accounts for the Latin name of the order and the common names "day flies" and "Eintagsfliegen." "Mayflies" apparently refers to the time of year when adults first appear in large numbers at certain locales.

The world fauna consists of slightly over 2000 extant species in about 310 genera and 22 families (Hubbard 1990). There are about 650 species in North America. The European fauna contains 217 species.

The first scientific study of a mayfly was published over 300 years ago (Swammerdam 1675) based on the burrowing species now known as *Palingenia longicauda.* Delightful accounts, quotations, and drawings from this historical work are available in English (Miall 1895), German (Illies 1968), and French (Arvy and Peters 1975). Swammerdam vividly describes the emergence of this species from rivers in Holland. Apparently each adult lives less than 1 day and the entire emergence is completed in a 3-day period each year.

With the exception of a semiterrestrial South American baetid (Riek 1973), all mayfly nymphs are strictly aquatic. Except for a very few species which venture into brackish areas, Ephemeroptera occur exclusively in freshwaters. Various species occupy a wide array of lotic and lentic biotopes. The most diverse ephemeropteran fauna generally occurs in warm lotic habitats (Wiggins and Mackay 1978, Ward and Berner 1980).

Nymphal morphology more distinctly reflects microhabitat occurrence than perhaps in any order of aquatic insects. Pictet (1843) divided mayfly nymphs into burrowing, flattened, swimming, and creeping forms. Needham et al. (1935) grouped characteristic genera into three lentic types: (1) climbers amid vegetation, (2) sprawlers on the bottom, (3) burrowers in the bottom; and three lotic types: (1) agile free-ranging, streamlined forms, (2) close-clinging limpet forms found under stones, (3) stiff-legged trash and slit inhabiting forms.

Using essentially the same categories as Pictet, but altering the sequence to reflect an evolutionary perspective, Hutchinson (MS) groups immature mayflies into four life forms as follows:

1. Swimming nymphs. Characteristic families: Siphlonuridae, Ametropodidae, Metropodidae, Baetidae (in part).
2. Creeping and climbing nymphs. Characteristic families: Baetidae (in part), Ephemerellidae, Tricorythidae, Caenidae, Neoephemeridae.
3. Flattened, streamlined nymphs. Characteristic families: Heptageniidae, Baetiscidae, Prosopistomatidae.
4. Burrowing nymphs. Characteristic families: Ephemeridae, Palingeniidae, Polymitarcyidae.

A most aberrant nymphal morphology is exhibited by *Prosopistoma* (Prosopistomatidae), an Old World genus classified with the Crustacea for over a century (Hubbard 1979). The closest Nearctic relative is *Baetisca* (Baetiscidae), which is restricted to North America. Both genera have an expanded thoracic shield forming a carapace partly covering the abdomen and forming a protective chamber for the gills (Edmunds et al. 1976). Morphological adaptations of mayflies and other aquatic insects will be considered in some detail in the following chapters.

Most mayfly nymphs are herbivorous, feeding on fine organic detritus or algae. A few species are specialized filter-feeders or carnivores. Adults of extant mayflies have poorly sclerotized vestigal mouthparts and do not feed.

Ephemeroptera are unique in having two winged stages. The initial winged stage, the subimago, which emerges from the last nymphal instar, is usually not sexually mature. Most species undergo a final terrestrial ecdysis before transformation to the imago. Most species form exclusively male swarms in the imago stage. Individual females briefly enter the swarm; almost immediately the female and an accompanying male leave the swarm and mate in flight. The female generally oviposits on the water surface, although in a few species the females crawl under water to deposit eggs.

Many of the classic dry fly patterns of anglers are meant to be imitations of mayfly subimagos (duns) or imagos (spinners).

Mayfly life cycles range from two or more generations per year (multivoltine) to one generation every 2 years [semivoltine—some semivoltine populations also have 3-year cohorts (Brittain 1982)]. In temperate areas univoltinism is the most common life-cycle strategy of mayflies.

Palingenia longicauda, the burrowing nymphs so abundant in rivers in Holland during the seventeenth century, is now extinct in the Netherlands and western Europe (Puthz 1978). *Ephoron virgo* once emerged in such great numbers from the Vltava River in Prague that the masses of dead bodies were collected and fed to caged birds (Edmunds 1973). Mass emergences of burrowing species in North America, that resulted in bodies of dead and dying adults one meter deep on bridges, became less common and more geographically restricted but recently have exhibited a recurrence (Fremling and Johnson 1990).

Ecological data are included in the major works on mayflies by Needham et al. (1935), Verrier (1956), and Edmunds et al. (1976). The proceedings of the international conferences on Ephemeroptera (Peters and Peters 1973, Pasternak and Sowa 1979, Flannagan and Marshall 1980, Landa et al. 1984, Campbell 1990, Alba-Tercedor and Sanchez-Ortega 1991) contain important ecological information. Illies (1968) and Brittain (1982) provide comprehensive reviews of mayfly biology.

Odonata—Dragonflies Dragonflies are hemimetabolous insects, the nymphs of which are aquatic and very different in appearance from the adults. The aerial adults are conspicuous, often brightly colored, and highly vagile compared to the terrestrial stages of many aquatic insects. Odonata, along with Ephemeroptera, are paleopterous insects since they lack the musculature and articulation necessary to place the wings in a backward folded position when at rest. Dragonflies may serve as the common name for the entire order (Corbet 1980), or may refer to the suborder Anisoptera (Gr. = unequal wings), with damselflies the vernacular for Zygoptera (Gr. = yoke-wings). A third suborder, Anisozygoptera, contains only two known species, which occur in Japan and Nepal. A variety of other names such as mosquito hawks, darning needles, and horse stingers have been applied to the order based on real or imagined adult characteristics.

About 5500 extant species comprise the world odonate fauna. In North America at least 650 species are known; Europe has 127 species.

The adults of all known species are terrestrial. The immatures are aquatic with a very few exceptions. Species with terrestrial nymphs living in damp leaf litter are known from the South Pacific Islands of Hawaii and New Caledonia, and the subtropical rainforest of Australia (Corbet 1962, 1980). The immatures of the vast majority of Odonata occur in freshwater, although a few species are occasionally found in brackish estuaries. Although Odonata have not been reported from truly marine habitats, nymphs may occupy inland waters in which the salinity exceeds that of seawater. The most diverse odonate faunas occur in lentic environments, although nymphs occupy a vast array of running and standing inland waters including hot springs, water-filled tree holes, and temporary water bodies.

Newly-hatched young are enclosed in a membranous sheath and are called "pronymphs." The pronymphal sheath is shed within a few minutes. There generally are 11–14 nymphal instars. The young of some species characteristically are found climbing in beds of aquatic plants or in masses of organic detritus. Others slowly move along the bottom, whereas yet others burrow into soft substrate. Odonata are exclusively carnivorous in nymphal and adult stages. Nymphs have a uniquely structured prehensile and extensible labium (which may be extended one-fourth the body length) used to capture prey. Some stalk their prey, whereas others (e.g., burrowers) are "sit and wait" predators. Food consists of other invertebrates. Cannibalism, so common among specimens held under crowded conditions, is rare in natural habitats. Dragonfly (Anisoptera) nymphs are generally more robust than damselflies (Zygoptera). Whereas damselflies have three leaf-like caudal tracheal gills, dragonfly nymphs have numerous small gills in the form of folds on the inner lining of the rectal chamber. Pulsations of the abdomen move water in and out of the anus, providing a continuous source of oxygenated water for the rectal gills. Rapid contraction of the abdomen expells the contained water and results in a form of "jet propulsion." A very few zygopterans have paired tracheal gills along the sides of the abdomen (Snodgrass 1954).

The last nymphal instar leaves the water before undergoing the final ecdysis to the imago stage. After a short (2 days to 2 weeks in temperate species) prereproductive period of maturation, the adult enters a reproductive period of one to several weeks' duration. Adult Odonata are conspicuous, often brilliantly colored, strong flying insects. Damselflies have four wings of equal size that are held vertically in repose. In dragonflies, the fore wings differ from the hind wings, and both pairs are extended horizontally at rest. Adult Odonata have large compound eyes and a head that can be rotated. Adult body length of extant species ranges from <20 to >150 mm, although some fossil dragonflies are much larger (see Chapter 2). Although lacking the highly modified labium of the larvae, adults are voracious predators, feeding predominantly on flying insects. Zygopterans often attack immobile prey, but anisopterans almost invariably capture insects in flight. Furneaux (1897, pp. 249–250) offers the following description of predatory behavior: "Now a butterfly flits across its sacred track; and the Dragon Fly, with one steady but rapid dart, seizes the creature in its powerful jaws, and continues its flight as before. Then down falls a wing, which has been clipped from the butterfly's body, like a faded leaf on the moss below. Then another and another, till, in the space of a few seconds, the Dragon Fly is feasting in the air on the wingless trunk."

Reproductive behavior of Odonata has received considerable study. Conspecific males spatially or temporally partition rendezvous sites where mating often takes place. Copulation may occur in flight or on a perch. Oviposition may occur in the absence of the male, or the pair may be linked in tandem, with the male clasping the head or thorax of the female. Eggs of endophytic species (those ovipositing within plant tissues) may be deposited above or below the water surface. Some species descend under water (individual females or tandem pairs) to oviposit. Females of such species commonly remain submerged for a few minutes, although continued submergence for over an hour has been reported. Other species oviposit by inserting only the abdomen under water, or by touching the tip of the abdomen to the water surface while in flight, or by dropping eggs while flying a short distance above the water.

Species of tropical lowland habitats complete one or more generations per year. Temperate species are univoltine or semivoltine, and at high latitudes, up to 6 years may be required for each generation. Populations of *Ischnura elegans* vary from trivoltine to univoltine to semivoltine along a latitudinal gradient from 43 to 58°N (Corbet 1980).

Although adults are highly vagile and thus of limited value as pollution indicators, the nymphs as a group are highly sensitive to habitat degradation. Their potential as environmental monitoring agents, however, is limited by insufficient knowledge of the ecology and taxonomy of the immatures. Odonata have little direct economic value to humans. Although feeding on mosquitoes and other pest species, they are opportunistic, catholic predators, which limits their value in reducing populations of individual prey species. However, an assemblage of several species of dragonflies apparently exerts some biological control over red locusts in Africa (Stortenbeker 1967). Adults of some species may prey on bees, and nymphs of the larger species reportedly will occasionally attack trout fry.

The biology of Odonata has been reviewed in book form (Tillyard 1917, Walker 1953, Corbet 1962) and in review papers (St. Quentin and Beier 1968, Corbet 1980). Ethological studies of adults comprise a significant portion of published literature.

Hemiptera: Heteroptera—Water Bugs The vast majority of the true bugs are terrestrial. The suborder Heteroptera contains all the truly aquatic species. The suborder Homoptera contains a few semiaquatic species associated with vegetation of the intertidal zone or freshwater margins (Foster and Treherne 1976, Polhemus 1984). Hemiptera, meaning "half wing," refers to the division of the forewings of most heteropterans into a coriaceous (leathery) base and a membranous apical portion. A few groups (such as *Halobates,* the marine water strider) are consistently wingless. Others exhibit alary (wing) polymorphism; the entire range from aptery to macroptery may be exhibited within a single species population. The nymphs of these hemimetabolous insects have wing pads in the last two or three instars (in winged forms). Aquatic hemipterans, along with coleopterans, include species which are truly aquatic both as imagos and immatures, although fully winged adults may leave the aqueous medium to fly to another water body. Attraction to lights during dispersal flights of some species accounts for the appellation "electric light bugs."

The world fauna of aquatic and semiaquatic Heteroptera is comprised of over 3300 species in 16 families. There are about 400 species in North America; Europe has 129 species (excluding Saldidae). World patterns of distribution of aquatic and semiaquatic hemipterans are analyzed by Hungerford (1958) and Jaczewski and Kostrowicki (1969).

Hemipterans occur in freshwaters, brackish waters, and the sea. The marine water strider *Halobates* may be found on the sea surface hundreds of kilometers from land. Most species living on the water surface or submerged within aquatic habitats are associated with ponds, slow streams, or the littoral zone of lakes. Rapid streams have few species, those present usually occurring on the water surface along the edges. The broad-shouldered water strider *Rhagovelia,* however, is adapted to dart about on the riffles of rapidly moving water. In Table 1.3 the families of Heteroptera are placed under semiaquatic, water surface, and submerged habitat categories. The seven subcategories are roughly organized in order of increasing adaptation to an aquatic existence. Members of subcategory 1 (Table 1.3), which inhabit the shore zone, generally occur on or in the water only accidentally. The three families residing on the water surface are collectively known as *water striders*. Members of the truly aquatic families (subcategories 5-7, Table 1.3) leave the water only during migratory dispersal flights. None of the hemipterans have developed tracheal gills. All truly aquatic bugs, except a few species with plastron respiration (see Chapter 2), must periodically surface to renew their air supply.

Hemipteran mouthparts are modified to form a piercing beak. Whereas most terrestrial bugs are phytophagus, using the beak to pierce plant tissue, aquatic hemipterans (including many Corixidae) inject enzymes to liquefy the tissues of animal prey that are sucked up through a food channel in the beak. Prey normally consists of small insects and crustaceans, although there are accounts of giant water bugs (Belostomatidae:Lethocerinae) attacking and subduing frogs, fish, and water snakes. The largest aquatic insects are members of the genus *Lethocerus*.

Table 1.3 A generalized habit and habitat classification of the families of Heteroptera

Semiaquatic	Water Surface	Submerged
1. Wet margins: Saldidae, Ochteridae,[a] Gelastocoridae,[a] Hebridae	3. Near shore: Hydrometridae, Veliidae (in part)	5. Poor swimmers: Nepidae
	4. Open water surface: Veliidae (in part), Gerridae	6. Swimmers and climbers: Belostomatidae (in part) Pleidae Naucoridae Helotrephidae
2. Shore and water surface: Macroveliidae, Mesoveliidae		7. Strong swimmers: Belostomatidae (in part) Corixidae Notonectidae

[a]Ochteridae and Gelastocoridae are apparently secondarily partly terrestrial (Parsons 1966) and are classified with the underwater bugs despite their present semiaquatic mode of life.

The general hemipteran life cycle in the temperate zone involves egg laying in the spring, nymphal development during summer and autumn, and adult overwintering. Exceptions to this univoltine pattern occur, and in tropical regions some species exhibit nearly continuous reproduction. Eggs of aquatic species are most commonly inserted in plant tissues or attached to a solid surface (often an aquatic plant). In some water bugs (e.g., *Belostoma*), the eggs are glued to the back of the male, who exhibits a variety of brooding behaviors that enhance egg survival (Smith 1976).

There are five nymphal instars except in some species of *Nepa, Microvelia,* and *Mesovelia,* which have four. Nymphs reside in the same habitat and exhibit the same general habits and structure as the adults.

Scent glands are normally present on the metasternum, which may account for a general avoidance of hemipterans by potential fish predators. However, corixids, which are reportedly the one group heavily preyed on by fishes, also have well-developed scent glands.

Sound production by aquatic hemipterans has been known for some time. Hutchinson (MS) provides a detailed account of stridulation in corixids and notonectids. Sound is produced by toothed structures on the abdomen (strigil) or front legs (pars stridens), which are rubbed against a sharp edge. The palar pegs on the front legs of the males of some corixids, formerly thought to be stridulatory organs, apparently function to grasp the female during copulation (Popham 1961).

Although some terrestrial hemipterans are of considerable economic concern, aquatic species have little direct influence on humans. The giant water bugs may become pests in fish hatcheries and along with other biting forms are sometimes nuisances in swimming pools. The common names "toe-biters" or "toe-stabbers" vividly describe the annoyance that may be inflicted upon humans wading in certain aquatic habitats. In Southeast Asia, giant water bugs are collected and prepared as human fare (Polhemus 1984). In Mexico, dried corixids and their eggs are used as human food and are also fed to pet birds, fish, and turtles. Aquatic hemipterans offer potential as agents of biological control of mosquitoes (Ellis and Borden 1970), and as indicators of water quality (Jansson 1977).

Although considerable ecological data are available for aquatic hemipterans, most works are limited to certain geographical regions or taxonomic groups. The polyphyletic origins of aquatic Hemiptera, with each family having distinctive morphology and habits, likely accounts for the paucity of general works on aquatic bugs. The classic publication by Hungerford (1919) remains the major comprehensive treatment of the biology and ecology of water bugs. Hungerford also provides a review of aquatic hemipteran literature back to the seventeenth century. Andersen's (1982) book on semiaquatic bugs includes information on adaptations and distribution patterns.

Semiaquatic Orthoptera The grasshoppers, locusts, and crickets constitute an almost exclusively terrestrial order. Semiaquatic species occur in the following families: Tetrigidae (grouse locusts), Tridactylidae (pygmy mole crickets), Acrididae (short-horned grasshoppers), Tettigoniidae (katydids), Gyrllidae (crickets), and Gryllotalpidae (mole crickets).

All the semiaquatic species frequent wet margins of freshwater bodies or are found on emergent littoral vegetation. A few of these hydrophilous species have structural adaptations for swimming or for support on the surface film. In addition to the accidental incursions to the water surface, some species dive into the water when disturbed (Cantrall 1984). The pygmy mole cricket *Tridactylus minutus* has blade-like calcaria (swimming plates) on the ends of the hind tibia; tarsi are absent from the hind legs (La Rivers 1956). Pygmy mole crickets were commonly encountered on lichen-covered emergent boulders in a forested stream in Malaysia (Bishop 1973a). When disturbed they moved into the water and burrowed into the gravel. A pygmy locust in Australia, *Bermiella acula,* has several structural adaptations for a semiaquatic life, including dense hair patches on the distal abdominal sterna, and an air chamber over the first abdominal spiracles formed by the costal area of the forewings (Key 1970).

Plecoptera—Stoneflies Stoneflies are hemimetabolous insects, the nymphs of which are aquatic, the adults terrestrial. The ordinal name, meaning folded wing, refers to the fact that most species fold the anal fan of each hind wing under the remainder of that wing when at rest (Burmeister 1839). The common name refers to the characteristic nymphal habitat—stony streams. The Russian name, Vesnyanka, meaning "spring insect," relates to the emergence period of many species in north temperate latitudes (Zhadin and Gerd 1961).

Most extant families of Plecoptera are confined to the temperate zone of either the Southern or Northern Hemispheres (Illies 1965, Zwick 1980). The world fauna consists of about 2000 recent species in 15 families. North America has approximately 540 species; there are 387 species in the European fauna.

There are a very few exceptions to the aquatic occurrence of nymphs and the terrestrial habits of the adult. In cool, moist habitats in the Southern Hemisphere (Patagonia, New Zealand, and two subantarctic islands), the nymphs of a few genera live a semiterrestrial existence (Illies 1969). Aquatic adults of *Capnia lacustra* have been collected at some depth in Lake Tahoe (California-Nevada), but nothing is known of their biology or ecology (Jewett 1963).

Plecoptera nymphs characteristically occur in cold running waters. Lotic species may sometimes be collected on wave-swept shores of lakes. Lentic records are generally confined to cold oligotrophic bodies of water. The only estuarine record of stoneflies (Müller and Mendl 1979) is for the upper Baltic Sea, where the water is only slightly brackish (4.2‰ salinity). No marine stoneflies are known.

The nymphs of most families are herbivores, feeding primarily on plant detritus (Hynes 1976). Some groups are largely carnivorous, but the early instars of all species feed on fine detritus. Submerged wood is a major food of some Southern Hemisphere genera, although the major source of nutrition is probably derived from associated microorganisms. Although there have been reports of nymphs feeding on salmonid eggs, some evidence indicates that only dead eggs are consumed, thus reducing fungal infestations to the live eggs in the redd (Claire and Phillips 1968, Nicola 1968).

Although some stoneflies do not feed as adults, most do, and all drink water. Some species must feed following emergence to produce eggs. Many forage on lichens growing on tree trunks and branches. *Taenionema pacificum* has been reported feeding on the foliage, buds, and fruit of orchard trees in northwestern North American (Newcomer 1918).

Mature stonefly nymphs range in length from a few millimeters in some capniids to >5 cm for the Nearctic *Pteronarcys* and the large eustheniids of the Southern Hemisphere. Nymphs and adults are generally quite similar in appearance, even compared to other hemimetabolous groups. In north temperate latitudes emergence of a given species is usually confined to a relatively short period, usually spring or early summer before water temperature maxima are attained. "Winter stoneflies" may emerge while the stream banks are still snow-covered. The few species that occur at low latitudes may emerge year round; in thermally constant temperate springbrooks, adults of some, but not all, species may be present during most of the year. Species that do not feed as adults have a terrestrial life of only a few days; other species may live for several months as adults.

Oviposition generally occurs on or slightly above the water surface. Apterous or brachypterous females may run across the water surface while ovipositing. The Southern Hemisphere family Notonemouridae inserts eggs in crevices at the water's edge. Females of some genera of Southern Hemisphere stoneflies crawl under water and oviposit on the undersides of stones.

Adult stoneflies are poor fliers; some are apterous or short-winged. Wing length is apparently controlled by environmental factors (often water temperature). High-altitude populations may be short-winged in species which exhibit normal wing length at lower elevations. Genetic factors, however, are also operative in some cases. Often it is the males that have short wings, although only part of the population may exhibit this feature. The low vagility of even fully winged stoneflies has made them especially interesting subjects of zoogeographic studies.

Northern, but — strangely — not Southern, Hemisphere Plecoptera exhibit an interesting mating behavior called "drumming" in which the adults beat their abdomens against the substrate, thus transmitting a signal that is species-specific. This behavior is thought to facilitate mating and serves as a mechanism to isolate species.

Because of the often highly specific environmental requirements of the nymphs (e.g., Baumann 1979), stoneflies are particularly good water-quality indicators, especially where oxygen-demand pollutants are concerned.

The known distribution of the world fauna has been catalogued by Claassen (1940), Illies (1966), and Zwick (1973). Reviews have treated the phylogeny and zoogeography (Illies 1965, Hynes 1988) and the biology (Hynes 1976) of Plecoptera. Comprehensive works containing ecological data include Frison

(1935), Hynes (1941), Brinck (1949), Needham and Claassen (1925), Zwick (1980), Lillehammer (1988), Stewart and Stark (1988), and an important series of papers on the ecology of Norwegian stoneflies (Lillehammer 1974–1976). The recent international symposia on Plecoptera have published proceedings that include papers dealing with ecological aspects (Berthélemy 1984, Campbell 1990, Alba-Tercedor and Sanchez-Ortega 1991).

Coleoptera—Water Beetles There are more kinds of beetles than all the other insects combined. This highly diverse holometabolous order is, however, primarily terrestrial. The world fauna contains about 6000 aquatic and semiaquatic beetles (Hutchinson MS) in at least 27 families. In North America 1073 species are known; Europe has 1072 species.

Both adult and larval stages of most aquatic beetles reside under water. Last instar larvae generally move onto land to pupate, and most species return to the water following transformation to the adult stage. Three genera of water pennies (Psephenidae) and the African species of Torridincolidae contain species in which the pupae have spiracular gills and are truly aquatic (Hinton 1955, 1969). At least five other genera in three families also pupate under water, but the pupae are in air-filled cocoons (Leech and Chandler 1956, Resh and Solem 1984). Species in which terrestrial adults have aquatic larvae occur in six families (Psephenidae, Helodidae, Ptilodactylidae, Curculionidae, Lampyridae, and Chrysomelidae), but the imagines of most aquatic coleopterans leave the water only for dispersal flights. The combination of terrestrial larvae and aquatic adults in dryopid and some hydraenid beetles is most exceptional (Hinton 1955). Nearly all species are aquatic as both imagines and larvae in those families, such as Elmidae (= Elminthidae), where adults are specialized for an aquatic existence (Leech and Sanderson 1959).

The name *Coleoptera*, meaning sheath-winged, refers to the coriaceous anterior wings or elytra, which form a hardened cover over the membranous second pair of wings of adult beetles. The elytra are held forward during flight, which is performed by the membranous pair of wings. Some species, however, are flightless, lacking functional wing muscles.

Beetles inhabit freshwater, brackish, and marine environments. The most diverse and abundant fauna occurs in well-vegetated freshwater habitats. Members of a few families, however, reside primarily in rocky-bottomed rapid streams. For example, in high mountain streams in the Colorado Rockies, the elmid *Heterlimnius corpulentus* is usually the only beetle present. Most aquatic beetles are closely associated with the substrate. Virtually all marine species are highly cryptic and some construct burrows (Doyen 1976). Only the adults of some primarily freshwater forms freely swim in open water.

The larvae of some truly aquatic species are gilled or obtain sufficient oxygen through the general body surface, but others must periodically visit the surface film. Larvae and pupae of some groups obtain air by tapping the aerenchyma of aquatic plants. Adult whirlygig beetles (Gyrinidae) swim partly submerged in the surface film (each eye is completely divided so that aerial and underwater vision are separated). The spiracles of adult beetles open into a chamber under the elytra (probably as an adaptation to reduce desiccation in the terrestrial environment). The aquatic imagines of most species hold an air supply in the subelytral chamber that can be renewed at the air–water interface. Adult elmid beetles, which characteristically reside in rapid well-oxygenated waters, have evolved plastron respiration, which eliminates the necessity of a surface visit.

Aquatic coleopterans exhibit diverse feeding habits. Many families are predaceous as both larvae and adults. Large dytiscids are especially voracious predators, attacking and devouring even small fishes and tadpoles. There is a documented case of a dytiscid larva attacking and killing a garter snake (Drummond and Wolfe 1981). Other families consist primarily of herbivores or detritivores. Those few aquatic species with terrestrial adults generally have a short imaginal stage during which feeding may not occur.

Most truly aquatic species oviposit underwater. Females of some hydrophilids carry the eggs on the underside of the abdomen. Semiaquatic beetles, and a few truly aquatic forms, deposit eggs in damp marginal areas. There are generally three to eight larval instars. The last-instar larva generally leaves the water to construct a pupal chamber. Nearly all aquatic species of the Temperate Zone are univoltine.

Some adult beetles stridulate, apparently as part of the mating behavior. The "screech beetle" *(Hygrobia hermanni)* makes a squeaking sound when captured in a dipnet (Clegg 1959). The sound is produced by rubbing the roughened surface on the underside of the wing case against the edge of the abdomen.

Aquatic beetles are of minor economic importance. Larvae and adult rice weevils *(Lissorhoptrus* spp.) may damage cultivated rice, and some predaceous species, especially large dytiscids, are pests in fish hatcheries. Some large dytiscids and hydrophilids are used as food in China, and dried elmid beetles serve as a seasoning called *chupe de chiche* in Peru (Brown 1987).

The larvae of a few species of the predominately terrestrial lampyrids (fireflies) occur in ponds and mountain streams of Asia (Okada 1928). These beetles, which are bioluminescent in larval, pupal, and adult stages, are the only known luminous aquatic insects.

At least 27 of the more than 250 families of beetles have aquatic or semiaquatic species (Table 1.4). Most truly marine beetles are specialized species of primarily terrestrial families (e.g., Staphylinidae, Carabidae). Conversely, families such as Elmidae, which are common and widespread in freshwater, generally do not have marine representatives (Doyen 1976). With the exception of Chrysomelidae, a primarily terrestrial family, beetles of brackish waters are from families comprised of primarily freshwater species.

Bertrand (1972) treats the immatures of the world fauna of aquatic beetles. Doyen (1976) reviews the ecology and distribution of marine Coleoptera. Crowson's (1981) monograph contains some ecological data on aquatic beetles.

Diptera—True Flies The true flies include a highly diverse assemblage of holometabolous insects characterized by only one pair of functional wings in the

Table 1.4 Classification of Coleoptera families containing aquatic and semiaquatic species[a]

1. Primarily or exclusively freshwater families
   Amphizoidae—running waters
   Dytiscidae—standing and slow-running waters
   Elmidae—running waters
   Gyrinidae—standing and slow-running waters
   Haliplidae—standing and running waters
   Hydraenidae—running and standing waters
   Hydrophilidae—standing and slow-running waters
   Hydroscaphidae—running waters
   Hygrobiidae—standing waters
   Noteridae — standing and slow-running waters
   Torridincolidae—running waters
2. Larvae terrestrial, adults primarily in freshwater
   Dryopidae—running and standing waters
3. Adults terrestrial, larvae primarily in freshwater
   Helodidae—standing waters, including phytotelmata
   Psephenidae—running waters
   Ptilodactylidae—running waters
4. Primarily terrestrial families with some freshwater species
   Chrysomelidae—standing waters
   Lampyridae—standing and running waters
5. Primarily terrestrial families with a few marine species
   Salpingiidae—rocky intertidal
   Staphylinidae—rocky and sandy intertidal
6. Primarily terrestrial families with some semiaquatic species
   Carabidae—rocky intertidal
   Curculionidae—marine and freshwaters
   Melyridae—marine
   Ptiliidae—marine and freshwaters
7. Primarily semiaquatic families
   Georyssidae—freshwaters
   Sphaeriidae—freshwaters
   Limnichidae—marine and freshwaters
   Heteroceridae—marine and freshwaters

[a]Classification into seven groups based on the predominant occurrence of adults and larvae.

adult stage. Although the vast majority of species in this large order are terrestrial, those with aquatic immatures may be the predominant insects in many freshwater habitats. Over half of all known species of aquatic insects are dipterans; the chironomids constitute the largest family of freshwater insects, rivaled only by the beetle family Dytiscidae. Many dipterans which are familiar as adults, such as mosquitoes, horse flies, and crane flies, may be aquatic as larvae or pupae. Virtually all aquatic insects that are of major economic importance or which serve as vectors of human disease are dipterans.

Except for the important economic species, aquatic dipterans have been relatively well studied only in certain parts of Europe and North America. In many areas of the world little is known, especially regarding the immature stages.

At least 30 familes have aquatic or semiaquatic representatives (Table 1.5). In North America, somewhat less than 5000 species are known. In Europe there are about 4000 species.

Dipterans are found in virtually every conceivable aquatic environment. Congeneric species may occupy terrestrial, semiaquatic, and aquatic habitats. Preimaginal stages occur in such diverse water bodies as hot springs, saline lakes, aestival ponds, glacial brooks, pitcher plants, and the profundal zone of eutrophic lakes. Dipterans are the most successful insect colonizers of the marine environment. Members of this order are often the only insects in freshwater habitats with extreme environmental conditions.

Dipterous adults are invariably terrestrial. The larval stage of aquatic species occurs in water. The pupae of some groups (e.g., mosquitoes and midges)

Table 1.5 Broad ecological classification[a] of Diptera families containing aquatic or semiaquatic species

I. Primarily terrestrial families
  A. Marine species: Dryomyzidae
  B. Freshwater and marine species: Tipulidae
  C. Freshwater species
    1. Primarily running waters: Athericidae, Empididae
    2. Primarily standing waters:[b] Calliphoridae, Phoridae, Sarcophagidae, Scatophagidae, Stratiomyidae, Syrphidae
    3. Standing and running waters: Muscidae
    4. Parasitic: Tachinidae[c]

II. Primarily aquatic families[d]
  A. Primarily semiaquatic: Dolichopodidae, Ephydridae
  B. Parasitic: Sciomyzidae[e]
  C. Primarily standing waters:[b] Ceratopogonidae, Tabanidae
  D. Standing and running waters: Chironomidae, Psychodidae

III. Exclusively aquatic families
  A. Primarily marine: Canaceidae
  B. Primarily standing waters:[b] Chaoboridae, Culicidae
  C. Primarily running waters: Blephariceridae, Deuterophlebiidae, Dixidae, Nymphomyiidae, Simuliidae, Tanyderidae, Thaumaleidae
  D. Standing and running waters: Ptychopteridae

[a]Classification based on the predominant habitats of larvae. "Freshwater" denotes inland waters, irrespective of salinity.
[b]Some may also occupy slowly flowing depositional areas of streams.
[c]Parasitic on terrestrial and aquatic insects.
[d]All have marine or brackish water representatives.
[e]Parasitic on terrestrial and freshwater mollusks.

remain within the aquatic realm, whereas in others (e.g., most Tipulidae), the last instar larvae leave the water to pupate on land.

Although most dipterans are terrestrial in all life stages, many of those families that have aquatic representatives are largely or exclusively composed of species that inhabit water as larvae (Table 1.5). It should be emphasized, however, that our knowledge of preimaginal stages is far from complete. As new data on larval habits and habitats become available, the placement of families in the categories indicated in Table 1.5 may require emendation.

Nearly all of the primarily terrestrial and primarily aquatic families contain marine representatives. In contrast, of those exclusively aquatic families (Table 1.5), only the beach flies (Canaceidae) and a relatively few species of mosquitoes (Culicidae) are marine. None of the families occurring primarily in running waters have marine representatives.

Many aquatic dipterans (e.g., chironomids) lack functional spiracles as larvae (apneustic), and respiratory exchange occurs through gills or the general body surface. Invertebrate hemoglobin is possessed by a few species. Those with functional spiracles may have only the last pair open (metapneustic), or may have functional thoracic and terminal spiracles (amphipneustic). Respiratory tubes or siphons which pierce the surface film are often present to facilitate respiratory exchange with the atmosphere. Some species have spiracular structures modified for piercing the aerenchyma of aquatic plants.

All major types of food habits are exhibited by larval dipterans, sometimes within a single family. The food habits of adults vary from those that do not feed to those groups in which females require a blood meal to produce viable eggs.

Oviposition habits vary widely among aquatic species. Eggs are deposited on objects above the water surface, on the water surface, or under water. Some mosquitoes deposit eggs in dry depressions that later fill with water. There are normally three or four larval instars. The larval are distinguished from all other free-living aquatic insects (except a few coleopterans) by the absence of jointed legs. Fleshy prolegs may be present on one or more segments. Pupation occurs on land or in the water. Most pupae are inactive and some are enclosed in a hardened puparium. Culicid and chironomid pupae are, however, capable of considerable movement. The pattern of voltinism in aquatic dipterans varies from species that produce several generations per year under favorable conditions to those that may require several years to complete one generation.

Some dipteran adults are serious pests of humans and their domestic plants and animals. Swarms of flies make certain parts of the world nearly uninhabitable during certain seasons. Females of most black flies suck blood and are serious pests of humans and wild as well as domestic animals, in some regions. Some species of mosquitoes and other dipterans serve as vectors of diseases such as malaria and yellow fever.

Dipterans are, however, important components of aquatic and terrestrial food webs. Their value as processors of organic matter is especially evident in aquatic systems receiving organic wastes where "blood worms" (chironomids containing invertebrate hemoglobin) may attain densities of many thousands of individuals per square meter. Aquatic dipterans have been widely used as pollution indicators, and chironomids and chaoborids are important components in lake typology schemes (see Chapter 3).

Aquatic dipterans exhibit far too many fascinating ecological phenomena to even mention in this brief overview of the order. Many remarkable relationships are, however, referred to in subsequent chapters.

There are few comprehensive works dealing with the biology and ecology of aquatic dipterans. Works of a broad scope are generally primarily taxonomic treatments at the family or subfamily level, with data on biology and ecology presented anecdotally. Johannsen's memoirs (1934–1937) are an exception. Although primarily taxonomic, the four-part treatise deals with the preimaginal stages of the whole of aquatic Diptera. Hennig's *Die larvenformen der Dipteren* (1948–1952) treats both aquatic and terrestrial species, as do the *Manual of Nearctic Diptera* (McAlpine et al. 1981, 1987) and the *Catalogue of Palaearctic Diptera* (Soós and Papp 1984 et seq.). Some of the most comprehensive studies of families or subfamilies deal with economically important groups such as mosquitoes (e.g., Horsfall 1955, Laird 1988) and black flies (e.g., Grenier 1949, Rubtsov 1956, Carlsson 1962, Kim and Merritt 1987). Thienemann's (1954) tome on chironomids remains the definitive work on the biology of what is probably the most abundant and diverse family of aquatic insects.

Hymenoptera—Aquatic Wasps Sir John Lubbock (1862) reported the aquatic occurrence of two tiny species of hymenopterous insects, one of which swam with its wings, the other by means of its legs. Prior to that time, the order Hymenoptera, which includes bees, ants, wasps, and related insects, was thought to be an exclusively terrestrial group. The ordinal name refers to the highly membranous wings characterizing many species.

In North America 51 species of aquatic Hymenoptera have been reported from nine families (Table 1.6); in Europe there are 74 species.

Adults enter the water to parasitize or obtain hosts, usually the aquatic stages of other insects. The specific habitat of aquatic Hymenoptera is dictated by the habitat of the hosts that occur in both running and standing freshwaters. Hymenoptera are known to parasitize members of all orders of aquatic and semiaquatic insects except Collembola, Ephemeroptera, Orthoptera, and Plecoptera. However, since most aquatic wasps parasitize hosts which are associated with higher plants (eggs attached to or embedded in plant tissues; leaf miners), densely vegetated freshwaters (ponds, slow streams, the littoral zone of lakes) provide the most common habitats. A proctotrupid wasp, observed along with marine copepods under rocks on the coast of France (Moniez 1894) and a scelionid wasp that parasitizes the eggs of a spider living in the intertidal zone of South Africa (Masner 1968), apparently are the only recorded marine occurrences of hymenopterans.

Species range from those that are only marginally aquatic to those in which all life stages occur under water. For example, the chalcid *Mestocharis bimacularis* does not enter the water and drowns rapidly if immersed (Jackson

Table 1.6 Families and hosts of North American aquatic Hymenoptera

	Host	
Hymenoptera Family	Taxon	Stage Parasitized
Braconidae	Diptera	Egg, larva, pupa
	Lepidoptera	Larva
Diapriidae	Diptera	Pupa
	Coleoptera	Prepupa, pupa
Eulophidae	Coleoptera	Egg, prepupa, pupa
Ichneumonidae	Lepidoptera	Larva, pupa
	Coleoptera	Larva
	Diptera	Pupa
Mymaridae	Hemiptera	Egg
	Odonata	Egg
	Coleoptera	Egg
Pompilidae	*Dolomedes* (spider)	Adult
Pteromalidae	Diptera	Larva, pupa
	Coleoptera	Larva
	Neuroptera	Larva, pupa
Scelionidae	Hemiptera	Egg
	Lepidoptera	Egg
Trichogrammatidae	Odonata	Egg
	Hemiptera	Egg
	Coleoptera	Egg
	Megaloptera	Egg
	Hydracarina	Egg

Source: Modified from Hagen (1984).

1964). Able to walk on the surface film, this wasp parasitizes the eggs of dytiscid water beetles. As the water level of the pond drops, some of the dytiscid eggs, which are embedded in aquatic plants, become available to the wasp. If the host egg remains above water, development is rapid and adult wasps emerge from it in several weeks. If the host egg is submerged following oviposition, *Mestocharis* develop to the mature larval stage, but further development depends on exposure to the atmosphere.

The European ichneumon *Agriotypus armatus* is an ectoparasite on prepupae and pupa of caddisflies in the families Goeridae and Odontoceridae (Elliott 1982). The female enters the water in search of a suitable host. Submerged females are surrounded by a film of air and remain underwater for several hours (Grenier 1970, Elliott 1983). Early larval instars of the wasps have no special respiratory adaptations; apparently the current of water moving through the case (produced by undulations of the caddis pupa) supplies sufficient oxygen to the larval cuticle. By the time the host dies, and thus current through the case ceases, the parasite has produced a silk ribbon about 3 cm long (Clausen 1950) that protrudes from the case and functions as a plastron (see Chapter 2 for an explanation of plastron respiration). The ribbon serves the respiratory needs of the mature larva, the pupa, and the overwintering adult. Removal of the ribbon results in death of the parasite. Other species of *Agriotypus* with similar habits occurs in Japan and India (Elliott 1982).

Caraphractus cinctus, a fairyfly (Mymaridae), parasitizes dytiscid eggs (Jackson 1958). This is the minute species (1.5 mm in length) observed by Sir John Lubbock swimming under water with its wings. Essentially the entire life cycle occurs under water. Mating may occur underwater or on the surface film.

Most aquatic wasps are internal parasites (parasitoids) that do not exhibit any specializations, either as adults or immatures, for an aquatic existence (Hagen 1984). Ichneumon wasps of the family Agriotypidae are the only external parasites of underwater hosts, and the only aquatic wasps with a special morphological adaptation (the respiratory ribbon) for an underwater existence. The extremely small size of most aquatic hymenopterans provides a surface/volume ratio favorable for respiratory exchange.

Henriksen (1918) provided an early treatment of the biology of the aquatic Hymenoptera of Europe. Hagen (1984) presents a brief general review of aquatic hymenopterans, followed by further details on the taxonomy, hosts, and distribution of the North American fauna.

Lepidoptera—Aquatic Moths Butterflies and moths are conspicuous, primarily terrestrial and typically phytophagous insects. The ordinal name refers to the overlapping scales on the wings of the adults. Several families of moths contain aquatic or semiaquatic representatives. However, even the family Pyralidae, within which most of the truly aquatic species occur, is a primarily terrestrial group containing such well-known insects as the European corn borer and sod webworms.

In addition to Pyralidae, aquatic or semiaquatic moths occur in the families Arctiidae, Nepticulidae, Cosmopterygidae, Tortricidae, Olethreutidae, Noctuidae, Cossidae, and Sphingidae. Lange (1984) lists 279 species in five families for North America. There are 148 North American species of aquatic or semiaquatic pyralids in 21 genera; the European fauna consists of five species in four genera.

With few exceptions, aquatic and semiaquatic Lepidoptera are intimately associated with aquatic vascular plants, and thus most species occur in ponds, the littoral of lakes, and in slow-flowing depositional areas of running waters. Most are restricted to freshwaters, but at least one species *(Acentropus niveus)* also ventures into brackish estuaries (Green 1968). Pyralids have been collected from saline lakes in Australia (Williams 1981c) and North America (Galat et al. 1981). All preimaginal stages (eggs, larvae, pupae) of aquatic pyralids occur in the water. Adults are terrestrial; however, the female brachypterous form of *Acentropus niveus* is truly aquatic (Berg 1941). Copulation with the terrestrial male occurs at the surface film during the night.

Larvae of most aquatic species feed on the various tissues of higher plants. Some are leaf miners, or bore within the stems or roots. Yet others construct cases of plant fragments. A few species of *Petrophila* (= *Parargyractis*) reside in rapid streams in North and South America (Lloyd 1914). The larvae spin silken tents within which they scrape algae from the rock surface. Larvae of some species of aquatic moths are gilled; others are not.

Pupae of the aquatic species occur underwater within silken cases. Larvae of some species cut an "escape slit" in the cocoon immediately prior to pupation (Lange 1984). Adults emerge from the pupal cocoon and either float or swim to the surface. In Britain, the adults of aquatic moths are called "china mark moths" because the wing patterns resemble the potters' marks inscribed on the bottom of fine china (Clegg 1959).

Oviposition by plant-dwelling aquatic species occurs without submergence of the female. Eggs are generally attached to the undersides of floating leaves of hydrophytes, although some species deposit the eggs within leaves or flowers. Females of rock-dwelling species enter the water to oviposit. The middle and hind legs possess well-developed swimming hairs. Eggs are deposited on rocks, sometimes at considerable depths in quite rapid water.

Life-cycle data are available for relatively few aquatic Lepidoptera. Lavery and Costa (1976) reported a long overwintering generation and a short summer generation for *Petrophila canadensis* in a stream in New York.

Larvae occasionally occur in such numbers as to depredate the plants on which they feed. They may be a nuisance in cultivated water lily ponds, and are injurious to rice crops in several regions of the world. *Nymphula depunctalis* is called the "rice caseworm" in India (Berg 1950). Pyralid larvae have been suggested as potential biological control agents for nuisance aquatic plants (Balciunas and Center 1981).

The first detailed description of the larval ecology of an aquatic moth (*Nymphula nymphaeata*) is contained in Réaumur's (1736) treatise. To date there is no single comprehensive treatment of aquatic Lepidoptera. Lange (1984) provides distributional data for all known aquatic and semiaquatic species in North America. Although a considerable European literature exists, publications are generally limited to one or a few species in a limited area, as are the few works on aquatic moths from regions outside Europe or North America.

Megaloptera—Alderflies, Dobsonflies, and Fishflies This small order of holometabolous insects includes the alderflies (Sialidae), and the dobsonflies and fishflies (Corydalidae), all of which are aquatic as larvae and terrestrial as eggs, pupae, and adults. Not only are the wings large (in proportion to the body) as indicated by the ordinal name, but the group contains some of the largest species of aquatic insects. Mature dobsonfly larvae may reach 8 cm in length.

Corydalids are widely distributed in temperate and tropical waters, but are absent from Europe. Sialids are widely distributed in the temperate zones of both hemispheres. The megalopteran world fauna contains over 250 species. North America has 43 species; there are 6 species of *Sialis* in Europe.

Larval megalopterans have seven or eight pairs of cylindrical abdominal gill filaments. Functional abdominal spiracles are also present so that larvae may survive several days out of water if kept in a moist environment. Some species are occasionally found along the edges of water bodies, under stones, or in wet debris or mud. Larvae normally occur in ponds or the littoral of lakes, and in running-water habitats. In certain river habitats dobsonflies may be underrepresented in collections because they reside under large rocks at least by day. Alderflies are more commonly associated with soft-bottom habitats.

Larval megalopterans are voracious predators with well-developed mandibles. They feed on a variety of aquatic organisms, but insects make up the majority of the diet in most habitats. The imagos live for only a few days and do not feed, despite the huge hypertrophied mandibles of some dobsonflies. Despite the relatively large wings, megalopterans are weak fliers.

Eggs are laid on vegetation or other objects above the water surface so that larvae normally enter the water on hatching. Members of the family Sialidae are univoltine or semivoltine with up to 10 larval instars. The larger corydalids have 10–11 larval instars and life cycles of 2–5 years. Corydalid larvae are sometimes called *hellgrammites*. Members of both families leave the water prior to pupation and pupate in chambers in the soil or in decaying logs. *Archichauliodes diversus* of New Zealand is unique in that each larval instar leaves the water to molt, but returns to the aquatic realm following ecdysis (Hamilton 1940).

No single comprehensive treatment of Megaloptera has been published since Weele's monographic revision of the order (Weele 1910).

Neuroptera—Spongillaflies This is a primarily terrestrial order that includes lacewings and ant lions, the so-called nerve-winged insects. Only one small family, the Sisyridae or spongillaflies, contains members that are truly aquatic. As the common name implies, spongillaflies are associated with freshwater sponges. Here is an example of a single freshwater family of an otherwise terrestrial group of insects depending for its existence on a small freshwater family (Spongillidae) of an otherwise exclusively marine phylum (Porifera).

There are four genera and about 45 species of spongillaflies distributed throughout the world. The genera *Sisyra* and *Climacea,* each have three North American species. In Europe there are five species of the cosmopolitan genus *Sisyra,* three of which occur in the British Isles. The closely related neuropteran families Osmylidae and Neurorthidae have semiaquatic larvae that live in wet margins along freshwater bodies.

Only the larvae of sisyrids are aquatic; the eggs, pupae, and adults occur in the terrestrial realm. Eggs are deposited in small groups on emergent vegetation or other overhanging objects and are enclosed in a web of silk. On eclosion, the larvae fall into the water. There are three larval instars. Although first instars lack gills, seven pairs of tracheal gills are attached to the ventral abdominal segments of mature larvae. Final instar larvae are 4–8 mm long. The larvae live on the surfaces and within the cavities of several species of freshwater sponges without evident host specificity (Pennak 1978). Mouth parts are modified to form long hollow tubes used to pierce the sponge tissue and withdraw fluids.

Apparently the sponge is not seriously damaged by the feeding activities of spongillaflies. As with most terrestrial neuropterans, the alimentary canal is closed at the end of the larval midgut so that no feces are voided prior to pupation. Larvae occur in both running and standing freshwaters. Final instar larvae crawl out of the water just before pupation and construct hemispherical double-walled cocoons of silk spun from anal glands. Some species complete several generations per year. Overwintering apparently occurs as prepupae within the pupal cocoon. Adults are small typical neuropterans. Few data are available on imaginal biology or ecology.

Relatively little is known regarding the ecology of sisyrids. Only Navas (1935) and Parfin and Gurney (1956) have provided relatively comprehensive treatments of the family.

Trichoptera—Caddisflies Caddisflies are holometabolous insects with aquatic larvae and pupae and terrestrial adults. The derivation of the common name may relate to the "silk" secreted by the larval, since "cadace" refers to a commercial grade of silk. A fanciful resemblance between cased Trichoptera and cadice men, vendors who attached their wares (ribbons, pieces of braid, etc.) to their coats, provides another possible explanation for the common name of the group (Hickin 1967). Trichoptera means "hair-wing," referring to the hair-like setae that clothe the wings of adults.

The world fauna consists of about 10,000 species. Over 1200 species of Trichoptera have been reported for North America; Europe has 895 species.

Exceptions to the amphibiotic condition occur in some species residing in temporary waters that are adapted to survive the dry phase in preimaginal stages (Wiggins et al. 1980). Four wholly terrestrial caddisflies are known, the European *Enoicyla pusilla* (Limnephilidae), *Philocasca demita* and *Cryptochia pilosa* (Limnephilidae) from western North America, and *Caloca saneva* (Calocidae) from Tasmania (Erman 1981, Wisseman and Anderson 1987). *Desmona bethula*, the larvae of which repeatedly leave and return to the water, may represent a transitional stage in the secondary invasion of terrestrial habitats (Erman 1981). Adult caddisflies are terrestrial winged insects, although flightless species have adapted to live in the open waters of Lakes Tanganyika (Africa), Baikal (Siberia), and Titicaca (South America) (Marlier 1962).

Species of Trichoptera occur on all continents except Antarctica, and are found in freshwaters, brackish waters, and the sea. Most occur in inland waters where they have adapted to conditions in a variety of permanent and temporary running and standing-water biotopes. The most diverse fauna is found in cool lotic habitats. The high ecological diversity of caddisflies has been attributed to silk production, enabling the evolution of respiratory, feeding, hydrodynamic, and bouyancy adaptations (Wiggins and Mackay 1978). Silk (a salivary gland secretion) production makes possible the construction of cases possessed by many larvae and all caddisfly pupae.

A most flamboyant account of caddisflies, from Réaumur's *History of Insects* (1736), is presented by Miall (1895). Réaumur compares the myriad types of caddis cases to the diversity of human attire. He further indicates that caddis larvae may construct their cases from the shells of snails and small bivalves (presumably Sphaeriidae) that may still be alive, which (continuing the analogy) he compares with "a savage who should cover his body, not with furs, but with living musk-rats and moles" [cited in Miall (1895)]. Caddisfly cases have held a fascination for amateur naturalists and entomologists alike. Cases may be constructed solely of silk, or of a silk lining covered with bits of plant, animal, or mineral matter. Some cases are very distinctive at the generic level, but many species may use a variety of materials for case construction. Following the eruption of Mt. St. Helens (Washington State, USA) in 1980, pumice became a common case-building material for populations of the limnephilid *Dicosmoecus,* which resided in streams receiving the ashfall. Silk is also used for net construction in those Trichoptera that spin nets (almost exclusively lotic species). In Table 1.7 caddisflies are grouped into five categories according to Wiggins' (1984) typology, based on net and case construction.

Trichoptera larvae exhibit a vast array of food habits and feeding mechanisms. Many species show little selectivity in food ingested, although feeding mechanisms are often highly specialized. For example, the design and mesh size

Table 1.7 Categories of Trichoptera based on net and case construction

Category	Description	Families[a]
Free-living forms	Pupal case only	Rhyacophilidae Hydrobiosidae
Saddle-case makers	Portable larval cases having the shape of tortoise shells	Glossosomatidae
Purse-case makers	Free-living until final larval instar; purse- or barrel-shaped, mainly portable cases	Hydroptilidae
Net-spinners or retreat makers	Fixed larval retreats, most with affixed capture nets	Philopotamidae, Psychomyiidae, Xiphocentronidae, Polycentropodidae, Hydropsychidae
Tube-case makers	Tubular, portable larval cases	Phryganeidae, Brachycentridae, Limnephilidae, Uenoidae, Lepidostomatidae, Beraeidae, Sericostomatidae, Odontoceridae, Molannidae, Helicopsychidae, Calamoceratidae, Leptoceridae

[a]North America.

Source: Modified from Wiggins (1984).

of capture nets of filter-feeders may be highly species specific, but a variety of living and dead plant and animal material entrapped by the net may be ingested. The free-living *Rhyacophila* are, however, largely predaceous, and some limnephilids feed primarily on coarse plant detritus, at least in later instars. The saddle-case makers, such as *Glossosoma,* scrape periphyton from the upper surfaces of rocks in lotic waters. Adult Trichoptera have nonfunctional mandibles and feed very little, although water and other liquids such as nectar may be ingested.

Adult caddisflies superficially resemble moths (Lepidoptera). They range in length from <2 mm in some of the microcaddisflies (Hydroptilidae) to >40 mm in large limnephilids. The duration of the adult stage is normally 30 days or less, although some species inhabiting temporary pools have an extended adult phase as an adaptation to the special habitat.

During the day imagines generally are quiescent, usually concealed in vegetation. The adults of many species are most active at dusk and during the night. Oviposition occurs underwater in some species. The females crawl or dive underwater and attach the eggs to rocks or submerged plants. A female *Hydropsyche* was observed to remain submerged for 37 min during oviposition (Badcock 1953). Others deposit their eggs on the surface film. Some limnephilids attach eggs on vegetation above the water surface.

Most caddisflies have five larval instars. The final larval instar constructs a pupal case (free-living forms), or seals itself within the larval case after anchoring it to the substrate with silk. Hydroptilids are free-living in the first four instars, constructing a case at the beginning of the fifth instar. The pupal stage lasts 2–3 weeks in the majority of species, although a larval diapause of up to 6 months' duration (within the sealed case) may precede pupation. The pupa has large sclerotized mandibles, which it uses to extricate itself from the pupal case. The pupa swims to the water surface aided by dense hair fringes on the middle tarsal segments. Eclosion to the adult stage occurs shortly thereafter. In middle latitudes, a univoltine life cycle is the rule, although bivoltinism and semivoltinism have been reported for some species.

Because of their considerable habitat diversification, caddisflies play an important ecological role in most freshwaters. Larval caddis are generally intolerant of pollution and thus serve as indicators of water quality. Some species of hydropsychids are, however, quite tolerant of organic pollution. In some regions mass emergence may reach nuisance proportions. Instances of adult trichopterans serving as inhalent allergens have been reported; the wing hairs are apparently the major causative agent (Henson 1966).

The known distribution of the world fauna has been delineated in the *Trichopterorum Catalogus* (Fischer 1960--1973). Hickin (1967) and Malicky (1973) reviewed the biology of Trichoptera. Wiggins' (1977) generic treatment of Nearctic larvae includes valuable ecological data. The proceedings of the international symposia on Trichoptera (Malicky 1976a, Crichton 1978, Moretti 1981, Morse 1984, Bournaud and Tachet 1987) contain important ecological papers. Trichoptera larvae and pupae of Russia are treated by Lepneva (1964, 1966).

OTHER ORDERS WITH AQUATIC SPECIES

In addition to the insect orders previously listed, there are isolated records of aquatic or semiaquatic species in other groups. While the order Mecoptera (scorpionflies) consists largely of terrestrial species, members of the Nannochoristidae, a family restricted to Australia, New Zealand, and South America, are truly aquatic (Williams 1980). The Order Blattodea (cockroaches) contains several semiaquatic species which frequent streams in India and eastern Asia (Hutchinson MS). Bishop (1973a) reported three species of semiaquatic epilamprid cockroaches in a Malayan stream. They primarily inhabited the shore areas, but were collected under submerged stones and in detritus accumulations in the stream. Semiaquatic dermapterans (earwigs) are known from Sumatra and Bali (Günther 1934). Drs. G. F. Edmunds, Jr., W. M. Beck, Jr., and W. L. Peters, while collecting aquatic insects from a stream in Borneo, found what appear to be truly aquatic earwigs. It seems likely that further such examples remain to be discovered, especially in tropical and subtropical regions.

2

EVOLUTION AND ADAPTATION

This chapter deals with the origin of insects and the evolutionary colonization of the aquatic environment. Following a brief synopsis of insect evolution in general [see Manton (1977), Boudreaux (1979), Gupta (1979), and Hennig (1981) for comprehensive accounts of this topic], attention will be directed toward salient features of respiratory and osmoregulatory adaptations of aquatic insects. An evolutionary perspective is essential to a comprehensive understanding of the ecology of the extant aquatic insect fauna.

Adaptations associated with swimming and other types of locomotion, and the mechanisms that have enabled insects to adapt to low-oxygen waters and the special conditions in rapid streams, temporary ponds, the intertidal zone, and other biotopes will be presented in subsequent chapters.

THE EVOLUTION OF INSECTS

Insects evolved in the Paleozoic, apparently from a myriopod (Protosymphyla) ancestral line (Tiegs and Manton 1958). Although until recently the earliest insect fossils known were from the Upper Carboniferous some 280 million years ago, the main evolutionary lines already present in such strata suggested that the first insects began their evolutionary development in the Lower Carboniferous or Devonian (Carpenter 1977). Indeed, discoveries of fossil insects of early Devonian age (Kukalová-Peck 1987, Labandeira et al. 1988) substantiate that suggestion. By the early Mesozoic most higher orders of insects were present.

Carpenter (1953) recognized four major stages in the evolution of insects.

The first stage is that of the primitively wingless Apterygota. The development of the Pterygota [generally accepted as of monophyletic origin; see Kristensen (1981), Kukalová-Peck (1985)] greatly enhanced the fossil record since most insect fossils consist primarily of wings. The second stage in the evolution of insects consisted of the Paleoptera, the winged insects lacking the alar articulation necessary to flex the wings back over the body when at rest. Ephemeroptera and Odonata are the extant representatives of the Paleoptera. Flight may have evolved in response to predation pressures, especially since the aerial realm apparently remained unoccupied, except by insects for about 50 million years (Carpenter 1953). Morphological modifications enabling wing flexing, the development of the Neoptera, represent the third stage of insect evolution in Carpenter's scheme. Insects that could fold their wings over the abdomen in repose could more easily assume cryptic behavior (well exemplified by adult stoneflies), and thus reduce predation by hiding in crevices, under stones, or in other protected places. Although paleopterous forms once predominated, 97% of the extant species of insects are neopterous (Carpenter 1953). Plecoptera, Orthoptera, and Hemiptera are orders of Exopterygota (hemimetabolous) Neoptera.

The fourth major stage in insect evolution involved development of holometabolism resulting in the Endopterygota Neoptera. In contrast to the exopterygotes, which have external wing pads in immature stages, the developing wings of the endopterygotes invaginate into the larval body and the last larval instar becomes the pupal stage. In the early Permian only about 5% of the known species of insects were endopterygotes, whereas nearly 90% of the present fauna has complete metamorphosis (Hinton 1977). Holometabolism, in which the immature stages may occupy different habitats and eat different foods than the adults, is thus a form of intraspecific niche segregation.

The lines of insect evolution in Figure 2.1 illustrate a hypothesized sequence of these four stages. However, as Boudreaux (1979) points out, whereas the group Pterygota is phylogenetically meaningful, the Apterygota is an artificial assemblage of several evolutionary lines.

Development of the tracheal system and a relatively impermeable cuticle were requisite for inhabiting dry terrestrial biotopes (Hinton 1977). Much of the early evolutionary history of insects involved adaptations to a dry terrestrial environment. A comprehensive understanding of the evolutionary ecology of aquatic insects is predicated on an appreciation of the earlier phases of insect evolution.

The Origin and Evolution of Wings A diverse array of hypotheses have been set forth to explain the origin of insect wings and the selective values incurred in the early stages of their development. Proposed selective advantages conferred by the possession of pro-wings in early insects include thermoregulation (Douglas 1981), predator avoidance (Hamilton 1971), enhancement of passive dispersal (Wigglesworth 1963), and sexual attraction (Alexander and Brown 1963). An explanation for the evolutionary development of wings in which the

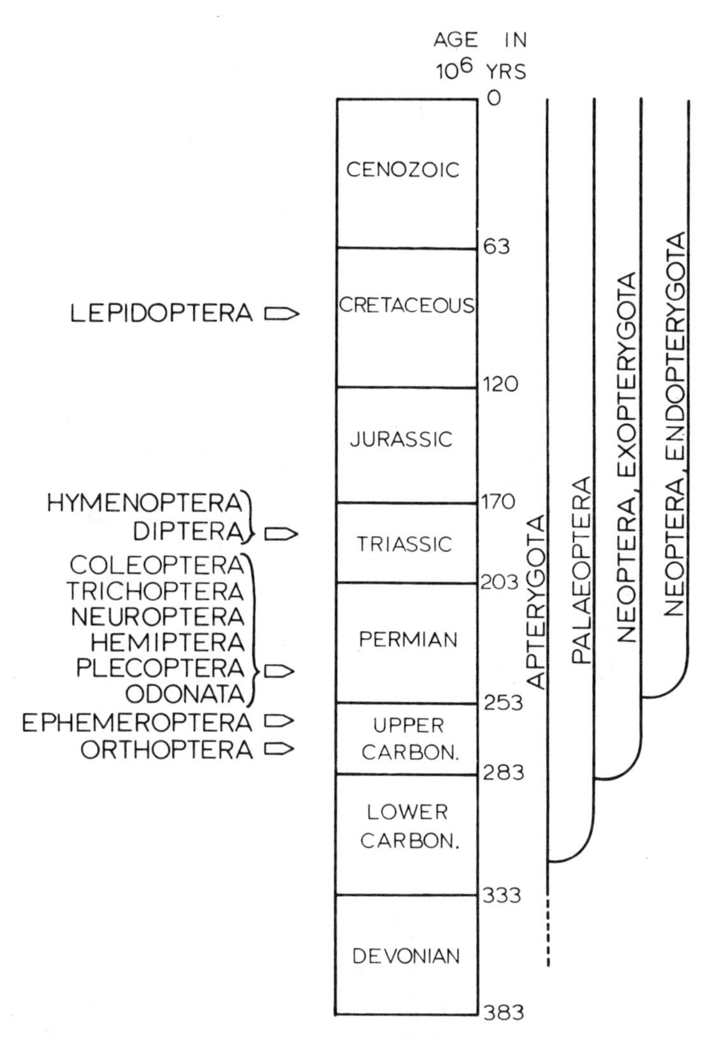

Figure 2.1 Hypothetical diagram of the major lines of insect evolution and the first appearance of fossils for the pterygote orders with extant aquatic representatives. All known Permian Hemiptera are in the suborder Homoptera. Neuroptera includes Megaloptera. [Modified from Carpenter (1953, 1977).]

aquatic environment plays a major role will be presented in the next section of this chapter. Much of the following discussion is based on the excellent review of this topic by Kukalová-Peck (1978), who also presents convincing arguments refuting several widely held viewpoints while developing a new interpretation of the evolution of insect wings (Fig. 2.2).

One of the oldest and most widely held concepts of wing origin in insects is the paranotal theory, the proponents of which believe that wings developed from lateral expansions of thoracic terga (Wootton 1976, Hinton 1977). According to

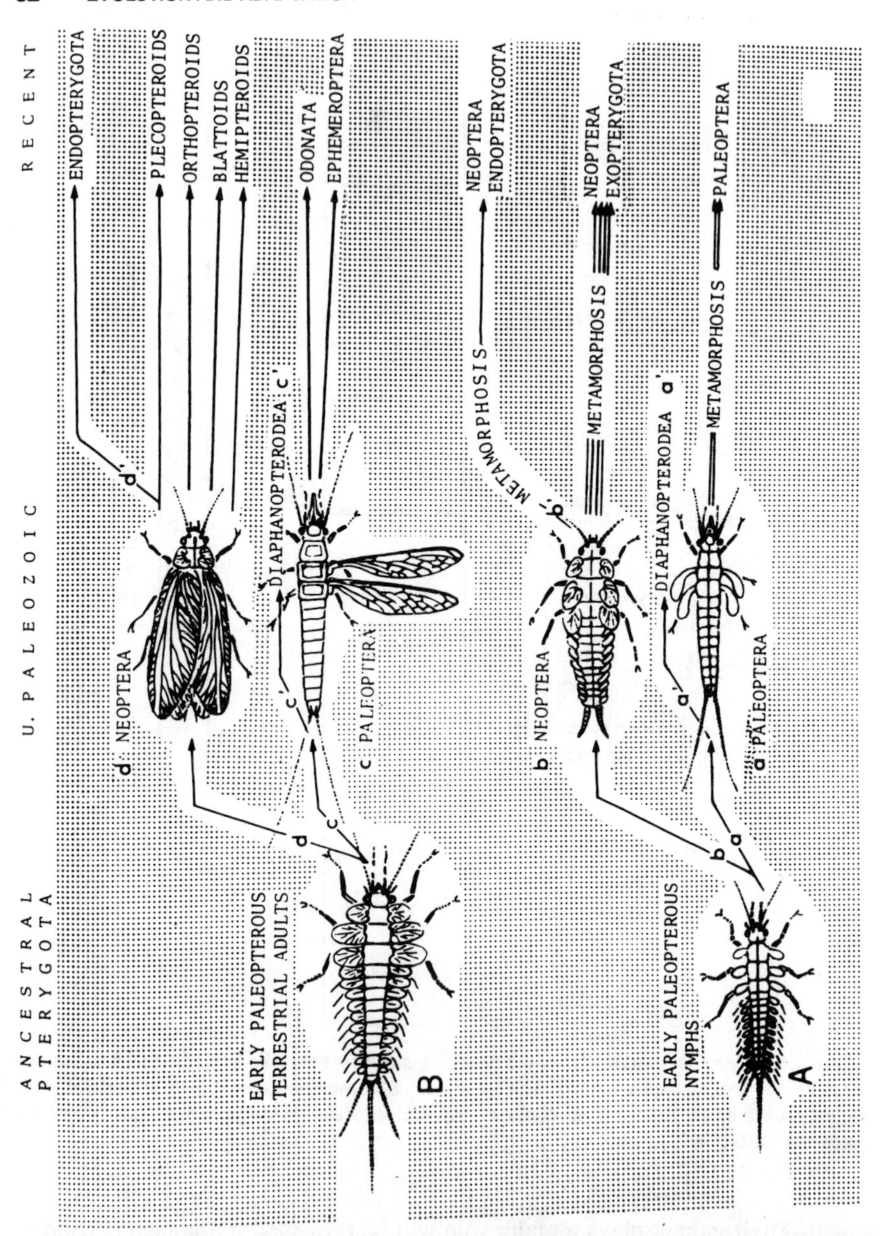

Figure 2.2 A conceptual interpretation of the major steps in the evolution of nymphal (*A*) and adult (*B*) pterygotes, sponsored by the development of juvenile wings. Pterygota were derived monophyletically from Apterygota through acquiring a paleopterous type of pro-wings on all nymphal body segments. In the paleopterous lineage, the meso/metathoracic wings in nymphs increased in size and became curved backward (*a*), or flexed backward (*a*′) along the fold distal to the hinge line. In the neopteran lineage, which originated monophyletically from the early paleopteran stock by developing a pivoting third axillary, the nymphal meso- and metathoracic wings became streamlined by actively flexing backward (*b*) along the fold within the hinge line. In both Paleoptera and Neoptera, the prothoracic and abdominal wings became more or less

this theory the enlarged paranotal lobes progressively evolved from primitive gliding structures, to steering vanes, to flapping wings capable of sustained flight. Many fossil insects from the Carboniferous and Permian periods had prothoracic flaps in addition to the functional wings of the meso- and metathorax. The premises upon which the seemingly attractive paranotal theory are based are systematically refuted by Kukalová-Peck (1978). She presents fossil evidence and data from comparative morphology to support the view that the wings of pterygotes developed from freely articulated lateral appendages, not as tergal outgrowths. She believes that the immobilized wing pads fused to the tergum in modern nymphs are derived structures rather than providing an ontogenetic recapitulation of phylogeny (thus supporting the paranotal theory). Indeed, studies of the ontogenetic development of modern Odonata, Ephemeroptera, and other insects demonstrate that wings begin as free lateral structures that fuse with the tergum in later nymphal instars (Kukalová-Peck 1978). Paleozoic Ephemeroptera (Kukalová 1968) and most primitive Neoptera had articulated and movable, although incipient thoracic wings as *nymphs.*

Wigglesworth (1973, 1976) suggests that wings originated from the coxal styli of apterygotes. He also proposes that thoracic wings are homologous with the abdominal gills of Ephemeroptera nymphs and with the abdominal styli of the Apterygota. Kukalová-Peck (1978) generally disputes these views, but believes that the thoracic wings of mayfly adults are homologous with the abdominal gill lamellae of the nymphs (which are quite unlike the filamentous gills of the immatures of other aquatic insects). The serial homology of wings and gill plates is clearly demonstrated in the Permian mayfly nymph *Kukalova americana,* the abdominal gills of which increase in size from posterior to anterior segments (Hutchinson MS). Those on anterior abdominal segments appear intermediate between gill plates and thoracic wing buds. Paleozoic mayflies possessed gill plates on all nine abdominal segments (Kukalová-Peck 1985). Vestiges of true venation often are present on the gill lamellae of modern mayflies; in Paleozoic nymphs the venation of the gills serially repeat a simplified pattern of the thoracic wing venation.

Terrestrial plants of considerable height were present by the end of the

---

reduced. The endopterygote lineage probably originated monophyletically from the juvenile plecopteroid Protorthoptera through the withdrawal of deemphasized wings (*b′*) into subcuticular pockets. Toward the Recent (era), in all (*a*, *b*, and *b′*) lineages juvenile wings independently and parallelly decreased in size and/or became fused with the terga, and as a consequence developed a metamorphic instar. Outspread prowings in paleopterous adults of ancestral pterygotes were flapping and were in elastic equilibrium for effortless soaring. In the paleopteran lineage, flapping forward flight occurred first (*c*). In the neopteran lineage elastic equilibrium was lost as a result of wing flexing (*d*) and indirect flight musculature became involved in flight (*d*, *d′*). Toward the Recent, the wings often tended to become smaller and the wing beat frequency increased by auxiliary structures. Hypothetical early Pterygota: (*A*) older nymph; (*B*) adult. Examples of typical Paleozoic insects: (*a*) Megasecoptera, nymph; (*b*) Protorthoptera, nymph; (*c*) Megasecoptera, adult; (*d*) Protorthoptera, adult. [From Kukalová-Peck (1978), Alan R. Liss, Inc., Publisher and Copyright Holder.]

Devonian. Insects climbing among such vegetation that possessed pro-wings would be preadapted to escape from predators, survive falls without injury, enhance mating success, and to engage in short-distance dispersal. It is not difficult to envision the selective pressures leading to sustained flight once well-developed movable pro-wings were present. A more difficult question concerns the origin of such movable structures, and the selective advantages conferred by their continued evolutionary development into pro-wings.

An Aquatic Origin of Wings and Flight? There is evidence that the development of wings had an aquatic origin, or at least that conditions in the aquatic habitat provided selective pressures for the continuous evolution of pro-wings of increasing size and the development of the musculature and articulation necessary for future flight. In the following text, an explanation for the origin of wings and flight is presented (using Ephemeroptera as a model) in which the aquatic environment is presumed to have played a major role. This hypothetical interpretation is based in large part on Wigglesworth (1973, 1976) and Kukalová-Peck (1978, 1985).

It is assumed that winged insects arose from a secondarily aquatic ancestor, as suggested by Wigglesworth (1973), and that the evolution of pterygotes was a monophyletic event (Kukalová-Peck 1985). Originally it may have been only the smallest species or earliest instars that were able to reside in the aqueous medium. Larger species and later instars would require special structures to increase the surface for respiratory exchange. Perhaps spiracular flaps, developed for controlling respiratory water loss in the terrestrial environment, were the precursors of gills. Spiracles were apparently primitively present on all body segments. Spiracular flaps originally may have allowed insects to cover the spiracles on entering the water. Paleozoic mayflies likely possessed gills and spiracles on all body segments. The musculature and articulation necessary for flapping flight may have been originally developed as gills became adapted for propulsion (underwater flying) and to ventilate respiratory surfaces. Gill plates of modern mayflies function not so much as sites of respiratory exchange, as to create currents over the fibrillar portion of the gills or tracheated areas of the abdominal surface (Wigglesworth 1973, Hutchinson MS). The greater density of water (compared to air) may have speeded the evolution of gill mobility. Once such mobility had evolved, the gills of insects were preadapted for the development of aerial flight. Dispersal of aquatic insects from drying or temporary habitats may have provided the selective pressures for the development of wings. Although "wings" were primitively present on all body segments, only the meso- and metathoracic wings are suitably positioned for efficient flying (Bradley 1942). The prothoracic wings have been suppressed or modified for protection or streamlining in modern insects, and abdominal wings (gill plates) are retained (although reduced in number) only in mayfly nymphs where they are still used for locomotion by some species. Ontogenetic development of primitive pterygotes apparently involved a continuous enlargement of the "wings" from the first instar to the adult (Kukalová-Peck 1978). The wings remained movable throughout development, without becoming immobilized and fused to the terga, as occurs in later instars of modern nymphs.

As larger thoracic wings evolved, they became cumbersome in both terrestrial and aquatic environments. The evolution of wing flexing, the neopterous condition, is believed by Kukalová-Peck to have begun in immatures as an adaptation for streamlining that eventually resulted in the backwardly projected, protective wing pads fused to the terga of modern Paleoptera and exopterygote Neoptera. The wings of endopterygote Neoptera (Holometabola) are present only as invaginated imaginal disks in the larval stage. Suppression of wings and other structures during the larval stage of endopterygotes necessitated an additional metamorphic stage (prepupae and pupae) for the rapid development of adult structures.

The Evolutionary Invasion of Freshwater The evidence is overwhelming that insects that now occupy aquatic habitats during all or part of their life cycles, evolved from terrestrial insects and are thus secondarily aquatic (Hinton 1953, 1977; Hutchinson 1967; Wootton 1972; Wigglesworth 1976; Kukalová-Peck 1978; but see Riek 1971). The success of insects in dry terrestrial environments was made possible by tracheal respiration combined with a cuticle relatively impermeable to water (and therefore oxygen). For such organisms to successfully invade aquatic habitats, it was necessary to (1) continue to breathe atmospheric air or to tap the air stores of underwater plants, (2) reduce cuticular impermeability, or (3) develop special structures for underwater respiration. The respiratory adaptations involved in the invasion of freshwater by insects are discussed in some detail later in this chapter.

The rise of angiosperms during the Mesozoic and their secondary invasion of freshwater undoubtedly played a role in the evolutionary history and adaptive radiation of aquatic insects, especially in lentic habitats (Hutchinson 1981). Insects have evolved highly specialized respiratory adaptations for tapping the underwater air stores of vascular plants, as will subsequently be described. Coexistence of insects with major predators such as fishes was made possible partly by the refugia (spatial heterogeneity) provided by higher aquatic plants. Although aquatic angiosperms are little utilized as a direct food source, at least while alive (Hutchinson 1975), their surfaces are often densely colonized by epiphytes, which, with associated detritus, form a rich food supply [the "universal pabulum" of Hutchinson (1981)].

Terrestrial vegetation has also shaped aquatic insect communities through direct and indirect mechanisms. H. H. Ross presented evidence to indicate "that the freshwater fauna has had essentially the same evolutionary history as the terrestrial components of the climax community of the biome" (Ross 1963, p. 242).

Running waters appear to have been the ancestral habitat of many major groups of aquatic insects, some of which colonized lentic water bodies later in their evolutionary histories (Hynes 1970a, b, Wootton 1972, Wiggins et al. 1980, Ward and Stanford 1982, Wiggins and Wichard 1989). Major groups apparently primarily associated with lotic habitats throughout their evolutionary histories include Plecoptera, Megaloptera, and several families of Diptera (e.g., Simuliidae). Other groups, such as Trichoptera and Chironomidae, which evolved in running-water biotopes, have since adapted to a wide array of habitat types. The ancestral habitats of Odonata and Ephemeroptera are less certain, but Wootton (1972) believes that these orders also may originally have been running water forms. The branchial structure of the fossil mayfly nymph *Kukalova americana* suggests that it inhabited cool lotic waters (Hutchinson MS). Some Lower Permian Protorthoptera are believed to have resided in lotic habitats, whereas others are associated with lacustrine deposits. The major evolutionary lines of aquatic Heteroptera, Coleoptera, and some Diptera (e.g., Culicidae) are more closely associated with standing waters than most other aquatic insects. For additional information on the evolution and phylogeny of major groups, the reader is referred to Fraser (1957) for Odonata; Rohdendorf (1964) for Diptera; Illies (1965) for Plecoptera; China (1955) and Popov (1971) for Hemiptera; Crowson (1960, 1981) for Coleoptera; Ross (1956, 1967) and Wiggins and Wichard (1989) for Trichoptera; and Tshernova (1970), Edmunds (1972), and McCafferty and Edmunds (1979) for Ephemeroptera.

EVOLUTIONARY ADAPTATIONS OF EXTANT AQUATIC INSECTS

Ecological conditions prevailing in aquatic habitats differ in several major respects from conditions in the terrestrial realm. Many evolutionary adaptations exhibited by aquatic insects are attributable to the special characteristics of the aqueous medium they inhabit.

Water as a substance has several unusual, if not unique, properties (Henderson 1913, Barnes and Jahn 1934, Clarke 1954, Hutchinson 1957a). Several of these special characteristics relate to the thermal behavior of water. The specific gravity of ice at 0°C is 8.5% less than that of liquid water at the same temperature. The fact that water contracts on melting, a characteristic exhibited by few other substances, means that ice floats. Because ice floats and has high insulation qualities, only the shallowest water bodies freeze to the substrate. Imagine how different conditions on this planet would be if ice expanded, rather than contracted, on melting. The temperature of maximum density of water is 4°C, another anomaly with myriad ecological implications for aquatic habitats and inhabitants. The high specific heat of water and the large amounts of energy required to induce phase changes, account for the relative thermal stability of aquatic environments.

The high surface tension of liquid water, exceeded only by mercury, has allowed the evolutionary development of organisms such as water striders that are supported by the surface film. The high density of water, 775 times greater than the density of air, has provided a medium sufficiently buoyant for the evolution of planktonic organisms. Because of the great density of water, hydrostatic pressure rapidly increases with depth (1 atm for each 10-m depth).

Some aquatic insects are equipped with depth (pressure) receptors. See Menzies and Selvakumaran (1974) for a review of effects of hydrostatic pressure on aquatic organisms.

Because even clear water is relatively opaque to solar radiation, the euphotic zone of most inland waters is limited to the uppermost few meters, and most heating (and cooling) is restricted to surface strata. This influences the depth distributions of aquatic plants, oxygen, and other chemical constituents, and ultimately aquatic insects.

In contrast to the relative thermal constancy of water, chemical conditions exhibit a great deal more spatial and temporal variation in aquatic habitats than characteristically occur in terrestrial environments. For example, whereas the composition of atmospheric gases is essentially constant, dissolved oxygen is a modifiable factor in aquatic habitats, with concentrations ranging from anaerobic to supersaturated.

Respiratory Adaptations

The evolutionary invasion of water from the terrestrial environment has resulted in a remarkable array of respiratory adaptations among aquatic insects. The development of a relatively impermeable cuticle and a tracheal system in which the respiratory epithelia are located deep within the body, while requisite for successful colonization of land by insects, posed major problems in the secondary invasion of water.

Air contains over 200,000 ppm O_2 (parts per million of oxygen), whereas oxygen-saturated waters do not normally exceed 15 ppm O_2 (Eriksen et al. 1984); if water is saturated with oxygen, the partial pressure is equivalent to that of air and the propensity of oxygen to cross membranes is also theoretically equivalent. However, local oxygen depletions, virtually unknown in the terrestrial environment, commonly occur in aquatic biotopes because of the low oxygen reserve and because of density properties of water that impede circulation. The high density of water precludes (if only from an energetic standpoint) respiratory systems dependent on moving the fluid medium to respiratory epithelia contained within the body, as is so common among terrestial animals.

For an insect to become fully aquatic (i.e., to exclusively utilize dissolved oxygen to satisfy respiratory requirements) requires a greater respiratory surface, externally located respiratory epithelia, and modification of the tracheal system. Although variously modified, as will be subsequently discussed, the tracheal system was not lost as insects invaded aquatic habitats and further adapted to life in the water. The probable reason for its retention, as Hutchinson (1967) indicates, is that diffusion within the gas-filled tracheal network provides underwater insects with an efficient distribution system. [Gaseous diffusion is several hundred thousand times faster than liquid diffusion (Miller 1974).] Because of the high density and viscosity of water, oxygen diffusion gradients tend to build up on respiratory surfaces. Various mechanisms function to maintain currents across the respiratory epithelia of aquatic insects. Of course, many so-called aquatic insects breathe atmospheric air or otherwise satisfy respiratory needs without extracting oxygen from solution.

When discussing respiratory adaptations of aquatic insects it is convenient, if somewhat artificial, to distinguish between open and closed tracheal systems, based on the presence or absence of functional spiracles. Ontogenetic development of the tracheal system is completed prior to the appearance of air within it (Wigglesworth 1972). Normally, at about the time of hatching from the egg, the fluid is absorbed from the tracheal system by surrounding tissues, and the tracheae fill with gas. Some aquatic insects, such as mosquitoes, with open tracheal systems expose their functional spiracles to atmospheric air shortly after eclosion from the egg, but often the tracheae fill with gas while the insect remains submerged. In the latter situation, apparently the gases that fill the tracheal system are liberated from solution from the tissue fluids. However, in a few aquatic (and endoparasitic) insects the tracheae remain filled with fluid. For example, the tracheal systems of the dipterans *Chironomus* and *Simulium*, and the lepidopteran *Acentropus* do not become gas-filled until the second larval instar (Wigglesworth 1972).

Cutaneous Respiration Virtually all species of aquatic insects utilize exchange through the general body surface to satisfy a portion of their respiratory needs (Table 2.1). A surprising number of species rely entirely on cutaneous respiration during the aquatic phase of their life cycle. The evolutionary invasion of aquatic habitats generally involved loss of the wax layer, and often the cement layer, of the epicuticle to render the insect body surface more permeable to oxygen.

Many insects of small size (favorable surface/volume ratio for respiratory exchange) are devoid of any special respiratory structures; even fairly large active species of cool, well-aerated habitats (e.g., some Plecoptera) may depend entirely on diffusion through the general body surface. In some cases (e.g., *Chironomus*) a uniformly thin cuticle covers a richly tracheated body wall (Mill 1974). Larval *Atrichopogon trifasciatus* (Ceratopogonidae) possess eight elliptical "respiratory organs" on the dorsal surface, regions where the cuticle is very thin and the underlying hypodermis is richly supplied with tracheoles.

Tracheal Gills Tracheal gills are tracheated evaginations of the body wall. They may have developed as outpocketings of specialized regions of the body wall, such as the "respiratory organs" of *Atrichopogon*. It is intriguing to envision the evolutionary development of closed respiratory systems in aquatic insects as occurring in the following three stages: (1) a uniformly thin cuticle overlying a richly tracheated body wall, (2) a concentration of tracheoles and cuticular permeability into delineated regions of the body wall, and (3) evagination of such regions to form tracheal gills.

Tracheal gills are possessed by many taxonomically diverse groups of aquatic insects (Table 2.2), although they have not evolved in adult insects; nor do they occur in pupae, with the exception of some species of Trichoptera. Tracheal gills

Table 2.1 Respiratory mechanisms of underwater stages of aquatic Pterygota

Respiratory Mechanisms	Orders and Life Stages[a]
Cutaneous[b]	Some Plecoptera (L), some Diptera (L), some Trichoptera (L,P), Hymenoptera (endoparasitic L), very few Hemiptera (L,A), early instars of many aquatic orders
Tracheal gills	Ephemeroptera (L), Odonata (L), Megaloptera (L), Neuroptera (L), many Plecoptera (L), some Lepidoptera (L), a few Copeoptera (L), some Diptera (L), many Trichoptera (L, a few P)
Excursions to surface	Most Hemiptera (L,A), some Megaloptera (L), a few Lepidoptera (L), many Coleoptera (L,A), some Diptera (L,P)
Extensions to surface	A few Diptera (L)
Tap plant aerenchyma	A few Coleoptera (L,P,A), a few Diptera (L,P), a few Lepidoptera (L)
Physical gills	Most Hemiptera (L,A), many Coleoptera (L,A), a few Lepidoptera (L)
Hair plastron	A very few Hemiptera (A), some Coleoptera (A), Lepidoptera (wingless *Acentropus niveus*)
Spiracular gills	A few Coleoptera (L,P), some Diptera (P)

[a] Key: L = larvae (s.l.), P = pupae, A = adults
[b] Virtually all aquatic insects utilize cutaneous respiration. Those listed herein depend on cutaneous respiration exclusively (i.e., apneustic without tracheal gills).

are unknown in any life stages of Hemiptera, Hymenoptera, or Orthoptera, but are universally present in larval Ephemeroptera, Odonata, Megaloptera, and sisyrid Neuroptera. Many Plecoptera lack gills of any kind, but others, especially larger species, have well-developed tracheal gills. [Shepard and Stewart (1983) present detailed gill terminology for Plecoptera.] Some larval aquatic Lepidoptera have tracheal gills, others do not. Many larval Trichoptera and a few pupae possess tracheal gills. Among aquatic Coleoptera, tracheal gills are possessed by all or most larvae of Elmidae, Gyrinidae, Hygrobiidae, and Ptilodactylidae; by some Psephenidae and Dryopidae; and by a few genera of Haliplidae, Hydrophilidae, and Dytiscidae. Tracheal gills do not occur in most families of Diptera. In a few families, such as Tipulidae, tracheal gills occur rarely, being restricted to a few species in one or two genera. Among Diptera, only larvae of the net-winged midges (Blephariceridae) characteristically possess tracheal gills.

There are two basic types of tracheal gills (Table 2.2). Lamellate (plate-like) tracheal gills occur as lateral abdominal structures on mayfly nymphs, or as caudal projections on zygopteran dragonflies. Ephemeropterans may also possess filamentous tracheal gills at the bases of the maxillae (Oligoneuriidae) or coxae (e.g., *Isonychia* and *Dactylobaetis*). The gill lamellae of some species of mayflies have a fibrilliform tuft attached at the base. Two families of Zygoptera

Table 2.2 Occurrence and locations of lamellate and filamentous tracheal gills on larvae (L) and pupae (P) of pterygote orders of aquatic insects[a]

	Lamellate		Filamentous					
	Abdominal	Caudal	Lateral Abdominal	Legs and Thoracic	Head	Cervical	Caudal	Rectal
Ephemeroptera	L		L	(L)	(L)			
Odonata		L	(L)					L
Plecoptera			(L)	L	(L)	(L)	(L)	
Megaloptera			L					
Neuroptera			L					
Lepidoptera			L	L				
Coleoptera			(L)	(L)			(L)	
Diptera			(L)	(L)				
Trichoptera			L(P)	(L)				

[a]A designation does not necessarily signify the universal occurrence of gills at that location. Less than common occurrences are indicated by parentheses.

(Epallagidae, Polythoridae) possess filamentous abdominal gills in addition to caudal lamellae. Snodgrass (1954) postulates that abdominal gills were primitively present on Odonata nymphs, although Boudreaux (1979) suggests that the lateral abdominal gills of extant zygopterans are new structures that evolved as an adaptation to low-oxygen waters. Filamentous abdominal gills also occur in Plecoptera (Eustheniidae, Pteronarcidae), Megaloptera, sisyrid Neuroptera, some aquatic Lepidoptera, many Trichoptera (including pupae in a few species), and several families of Coleoptera and Diptera. In addition to the mayflies already referred to, filamentous tracheal gills occur on the head region of some Plecoptera. Some species of stoneflies also have gills in the cervical (neck) region. Filamentous tracheal gills are attached to the thorax in many Plecoptera, some aquatic Lepidoptera, a few Coleoptera larvae in two families (Haliplidae, Hygrobiidae), and a very few Diptera and Trichoptera. A few Ephemeroptera (already mentioned) and some Plecoptera have filamentous tracheal gills attached to the bases of the coxae. A very vew Plecoptera and Coleoptera larvae possess caudal (anal) filamentous tracheal gills (not to be confused with the blood gills found in many dipterans and some other aquatic insects). Rectal gills are restricted to nymphs of anisopteran dragonflies. The rectum has become enlarged to form a branchial chamber, the walls of which are lined with tracheal gill folds. Water is pumped in and out of the rectum through the anus. This is achieved by the action of two sets of valves operating in conjunction with the contraction and expansion of the abdomen. Some zygopteran nymphs appear to have supplementary rectal respiration, which is exemplified by the movement of water in and out of the anus of early instars and nymphs from which the caudal gills have been removed (Tillyard 1917). Snodgrass (1954) suggests that rectal respiration, originally more widespread among the Odonata, was perfected (with a secondary locomotor function) in the Anisoptera, whereas the caudal lamellae of the Zygoptera (originally developed for locomotion) later took on a primary respiratory role. However, Pennak and McColl (1944) failed to find a correlation between ambient oxygen and the frequency of rectal ventilation in zygopteran nymphs. They suggested an osmoregulatory role for the zygopteran rectum, which has subsequently been confirmed (Komnick 1977).

The number of tracheal gills possessed by aquatic insects may increase as larvae grow, presumably to maintain a favorable surface/volume ratio for respiratory exchange. For example, gills are absent from first instar larvae of the aquatic moth *Nymphula maculalis,* whereas the final instar possesses over 400 gills (Welch 1916). The number of gills increases from zero to seven pairs (the full complement) during the first seven instars of the mayfly *Leptophlebia cupida* (Clifford et al. 1979). The caudal gills of some zygopteran (Odonata) nymphs exhibit protrusive growth in which the distal zone, which has a thinner cuticle than the proximal zone, increases in length as the nymph matures (MacNeill 1960). Larvae of *Dubiraphia,* one of the few elmid beetles to reside in still waters, has especially long gills (Hutchinson MS).

In addition to a direct respiratory function, tracheal gills may serve to ventilate respiratory surfaces, thus reducing the oxygen diffusion gradient. Gills are also used for swimming and feeding, or may be modified as an adaptation to current. Tracheal gills also function in osmoregulation, as discussed in the next section. Osmoregulation appears to be the primary function of blood gills, the hemolymph-filled extensions of the body wall possessed by many dipteran larvae and a few other aquatic insects, although a respiratory role may be important in some circumstances.

The uptake of oxygen by tracheal gills requires that the walls of the tracheae are rigid so that volume changes do not occur. Rigidity is established by spiral bands of chitinized fibers (taenidia) in tracheal walls, already well developed in terrestrial insects. The restriction of most aquatic insects to near-surface depths may relate to the limitations of tracheal wall rigidity given the hydrostatic pressures prevailing in aquatic habitats. Only species with highly modified tracheal structures, such as the dipteran *Chaoborus,* occur in deep waters.

Gaseous exchange between the tracheoles and the aqueous medium apparently occurs by simple diffusion (Wigglesworth 1972). As oxygen is utilized by the tissues, pressure in the uncollapsible tracheal tubes is lowered and oxygen diffuses in from the water to restore the equilibrium pressure. Removal of the full complement of tracheal gills may or may not produce immediately discernible adverse effects on aquatic insects. Removal of all 60 gills from the hydropsychid caddisfly *Macrostemum zebratum* had little or no effect on oxygen consumption, although the elimination of carbon dioxide was significantly reduced (Morgan and O'Neil 1931). Specimens without tracheal gills lived as larvae for 8 months prior to undergoing apparently normal transformations to the pupal stage. In contrast, removal of gills from nymphs of the burrowing mayfly *Litobrancha recurvata,* although not immediately lethal during winter, was fatal in a short time following amputation under springtime water temperatures (Morgan and Grierson 1932). Oxygen consumption of normal nymphs was twice that of individuals from which gills had been removed, and normal nymphs eliminated carbon dioxide over twice as rapidly as those without gills.

Dodds and Hisaw (1924) reported a negative correlation between the total surface area of gills (per unit of total body weight) of different species of mayflies and the oxygen content of the habitat. Species with the proportionately largest gills occurred in standing waters or in the dead-water zones under stones in streams. Exceptions were explicable in terms of behavioral factors or enlargement of the gills to form a friction device as an adaptation to high current velocity. However, as mentioned earlier in this chapter, the lamellate gills of many mayflies are no more effective in respiratory exchange than is the general body surface. The larger gills of species living in biotopes where low oxygen levels occur may function primarily by creating ventilatory currents. For example, nymphs of the mayfly *Cloeon* increase the frequency of gill movements as oxygen levels decline. Nymphs from which the gills have been removed exhibit a rapid decline in oxygen uptake at a much higher level of environmental oxygen than normal nymphs when in still water, but not if a current is created (Wingfield 1939). It is apparent that the tracheal gills of some aquatic insects play a major role in gaseous exchange only at low oxygen tensions, when they become directly important as respiratory sites or increase oxygen uptake by creating ventilatory currents that reduce the diffusion gradient (Mill 1974).

Whereas many insects that possess tracheal gills are apneustic, some also have functional spiracles during aquatic stages. Hinton (1947) reviewed the ontogenetic and evolutionary occurrence of functional versus nonfunctional spiracles in aquatic insects, and developed some interesting ecological generalizations. Hinton points out that generally the extent to which the functional spiracles are reduced evince the degree to which the larvae of a given group of aquatic insects have adapted to life in the water. For example, since all mayfly nymphs are apneustic (lacking functional spiracles) in all instars, thereby depending exclusively on oxygen in solution for satisfying respiratory requirements, ephemeropterans are an example of an entire order in which immature stages have completely adapted to an aquatic mode of existence. In contrast, aquatic stages of many Holometabola have open tracheal systems and are dependent on the oxygen in the terrestrial environment.

Ontogenetic changes in the number of functional spiracles (which may increase, but never decrease as larvae develop) generally reflect changes in potential modes of respiration (Hinton 1947). Increases in the number of functional spiracles often occurs in the final larval instar. For example, although abdominal spiracles of Odonata remain nonfunctional in all nymphal instars, thoracic spiracles become functional in the final instar or even before. Nymphs of Odonata leave the water prior to final ecdysis and thus are dependent on atmospheric oxygen for that period. Some Anisoptera nymphs may even leave the water during the night (Tillyard 1917), and mature nymphs of many species expose their thoracic spiracles to atmospheric air at the water surface if oxygen levels in the water drop (Hinton 1947). Ephemeroptera, which are apneustic in all nymphal instars, undergo ecdysis at the water surface or while submerged. The early instars of corydalid megalopterans are apneustic and have lateral abdominal gills. The last larval instar is peripneustic (possessing functional spiracles on most of the body segments), although the tracheal gills are retained (the functional abdominal spiracles open dorsal to the base of the gills). The functional spiracles are needed because the final larval instar leaves the water to construct a pupal chamber on land. *Archichauliodes diversus*, a primitive New Zealand corydalid, is unique in that the larvae move from the water onto land to moult at the end of each instar (Hamilton 1940). It is, therefore, not surprising that early as well as late instars of this species are peripneustic.

Excursions to the Water Surface Some aquatic insects make periodic excursions to the water surface, or remain more or less permanently at the air–water interface, as they are totally reliant on atmospheric oxygen. Such a method of respiratory exchange may represent an initial stage in the evolutionary invasion of water by aquatic insects with open tracheal systems. Further adaptations may have resulted in the evolutionary development of (1) extensions to reach the water surface, (2) mechanisms to tap the underwater air stores of plants, (3) structures to carry air on the body thereby reducing the frequency of surface visits, or (4) highly evolved cuticular structures that eliminate excursions to the surface. These adaptations are subjects of the subsequent sections in this chapter.

It is tempting to postulate that the secondary invasion of water occurred via two major evolutionary pathways in different lineages of aquatic insects. The levels of adaptation described in the preceding paragraph, beginning with complete reliance on atmospheric oxygen, may represent one pathway. The other possible series begins with sole reliance on cutaneous respiration and proceeds to the development of tracheal gills as evaginations of highly tracheated regions of the body wall. Stated in another way, independence of atmospheric oxygen appears to have been achieved separately in species with open and closed tracheal systems. The imagines of many Heteroptera and Coleoptera have become highly adapted to aquatic life, yet they retain an open tracheal system. Hutchinson (1981) queries as to the reasons why no adult aquatic insect respires by means of tracheal gills when surfacing increases the danger of being preyed upon, is energetically expensive, and limits the depths that may be colonized. However, an adult insect depending on tracheal gills for respiratory exchange would relinquish the possibility of aerial dispersal flights, the maladaptive nature of which should be apparent given the generally transient nature of inland waters. Hutchinson goes on to point out that the only apneustic adult aquatic insects known, two genera of tiny hemipterans *(Paskia* and *Idiocoris),* are endemic to Lake Tanganyika, an extant ancient lake. However, as we shall see, retention of an open tracheal system does not preclude the evolution of respiratory adaptations that free aquatic insects from dependence on atmospheric oxygen, at least not in well-oxygenated biotopes.

Aquatic insects dependent on obtaining oxygen at the water surface require hydrofuge structures to break through the surface film, and indeed such structures may be required by underwater insects with functional spiracles to prevent water from flooding the tracheal system. Hydrofuge structures have a greater affinity for air than for water, as for example, unwettable hairs. The cohesive forces between water molecules are greater than the adhesive forces between water and the hydrofuge surface. Hydrofuge structures thus repell water so that when, for instance, a spiracle surrounded by hydrofuge hairs reaches the surface film, a direct and immediate communication is established between the atmosphere and the air within the tracheal system (Fig. 2.3). On submergence the hydrofuge hairs retain a film of air around the spiracular opening. The surface of the cuticle around spiracles may be hydrofuge, apparently as a result of oily secretions from perispiracular glands (Wigglesworth 1972). In some cases perispiracular valves close the spiracles when the insect submerges (e.g., the respiratory siphon of mosquitoes).

Extensions to the Water Surface Several aquatic insects have special respiratory siphons for piercing the surface film. For example, mosquito (Culicidae) larvae and pupae possess short respiratory tubes for that purpose. However, only a few aquatic insects have developed tubular extensions that allow them to breathe atmospheric air without making excursions to the water surface. Such

Figure 2.3 Spiracle surrounded by a crown of hydrofuge hairs shown in three positions in relationship to the water surface. [Slightly modified from Figure 244(C, D, E) in V. B. Wigglesworth (1972).]

species are able to remain on the bottom of shallow waters yet maintain contact with the water surface. The so-called rat-tailed maggot is a syrphid fly, the larvae of which have a highly extensible caudal respiratory tube (Fig. 2.4). The telescopic siphon may be extended up to 6 times the length of the body proper. The virtually impervious cuticle (Krogh 1943), coupled with utilization of atmospheric oxygen, enable syrphid larvae to inhabit waters unsuitable for most aquatic insects. Larvae of Ptychopteridae, the phantom crane flies, also have long, extensible caudal respiratory tubes that enable them to lie concealed in shallow water among accumulations of organic matter with their respiratory tubes projecting to the water surface.

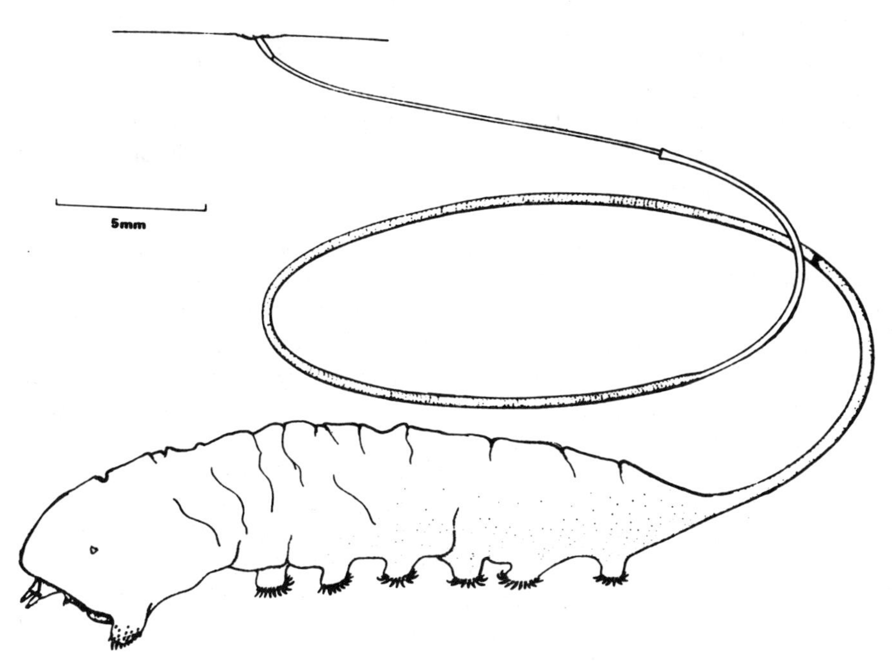

Figure 2.4 Larva of *Myathropa florea* (Syrphidae), an inhabitat of rot-holes in beech trees, showing the telescopic respiratory tube. [From Roberts (1970). Reprinted with permission of Pergammon Press plc.]

Tapping Air Stores of Aquatic Plants Utilization of the oxygen contained within the aerenchyma of underwater parts of plants has enabled a few species of Coleoptera, Diptera, and Lepidoptera to become independent of the water surface. At least one beetle *(Elmis)* and an aquatic moth *(Hydrocampa)* bite into the air spaces of plants to obtain oxygen. The evolutionary development of special respiratory siphons adapted to pierce plant tissues has apparently evolved independently in several families of Diptera and Coleoptera (Varley 1937, 1939).

The chrysomelid beetle *Donacia simplex* resides in the mud surrounding the roots of aquatic plants where anoxic conditions prevail (Houlihan 1969). The larvae have the functional spiracles on the eighth abdominal segment modified as long pointed structures that pierce the aerenchyma of the plant on which they are feeding (Fig. 2.5). There is an uninterrupted pathway for oxygen between the air spaces of the plant and the tracheal system of the larvae. The amount of oxygen within the root tissues is sufficient to meet the respiratory needs of both larva and plant. The final instar larva bites a hole in the root epidermis over which a cocoon is constructed (Houlihan 1970). The pupa and teneral adult depend on oxygen within the plant which diffuses into the cocoon through the root scar (Fig. 2.5). The adult overwinters within the cocoon in a state of diapause and is able to survive very low oxygen partial pressures, but not anoxia.

Larvae of other species of *Donacia* feed on plant leaves where ambient oxygen levels are relatively high. During the winter leaf-piercing species descend to the rhizomes, where they encounter low-oxygen conditions similar to *D. simplex.* The weevil *Lissorhoptrus* (Curculionidae) and the noterid (*Noterus*) pierce the air stores of plants during larval stages. A few genera of mosquitoes (e.g., *Mansonia, Coquillettidia*) tap the oxygen stores of underwater plants, as do *Notiphila* (Ephydridae) and some species of *Chrysogaster* (Syrphidae), and the tipulid *Erioptera.*

Physical Gills Many insects dependent on atmospheric oxygen have reduced the frequency of surface visits by carrying air bubbles under water. Such physical gills or gas gills are normally in communication with the functional spiracles, and thus provide air stores on which the insect may draw when submerged. Air stores are associated with pubescent regions of the body surface (Fig. 2.6), and in coleopterans and hemipterans may also be contained within subelytral chambers.

It was first suggested by Comstock (1887) that underwater air stores carried on the bodies of insects may in fact function as physical gills in that oxygen from the surrounding water could diffuse into the bubble. Ege (1915) demonstrated that a "gill effect" does indeed occur and is most effective at low temperatures. The physical gill of *Naucoris* (Naucoridae) supplies 96% of the insect's total oxygen uptake at 10°C (December), 54% at 15°C (May), and 28% at 20°C (August) (Wolvekamp 1955).

In describing the operation of physical gills only the partial pressures of nitrogen and oxygen need be considered (Mill 1974). Carbon dioxide diffusion

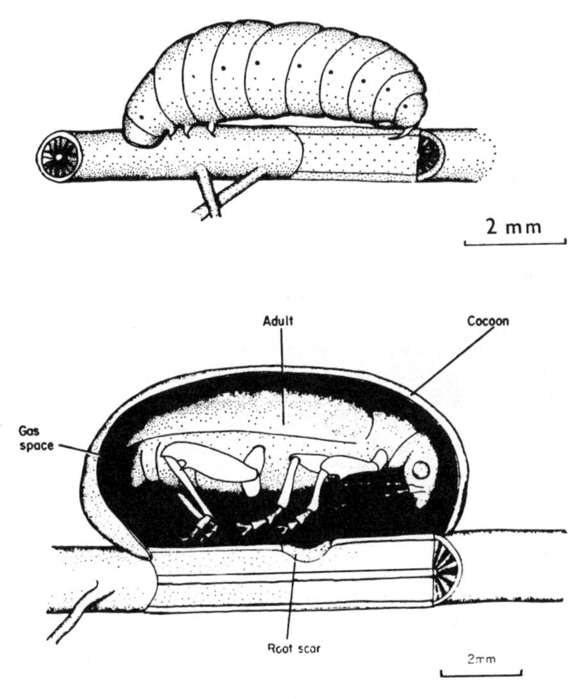

Figure 2.5 *Top*: The larva of the chrysomelid beetle *Donacia simplex* tapping the air spaces of *Typha latifolia* while feeding on the root of the plant. A portion of the plant epidermis has been removed to illustrate the modified larval spiracle piercing the air spaces. *Bottom*: The teneral adult of *D. simplex* overwinters within the air-filled cocoon constructed by the final instar larva prior to pupation. Oxygen diffuses into the cocoon through the root scar formed over the hole excavated in the root by the larval mouthparts. [Top is from Houlihan (1969); bottom is from Houlihan (1970). Reprinted with permission of Pergammon Press plc.]

within animals is much greater than that of oxygen, and carbon dioxide is highly soluble in water so that it diffuses into the surrounding environment as rapidly as it is produced.

As the submerged insect consumes oxygen, the oxygen tension within the physical gill decreases. In air-saturated water a disequilibrium is immediately established between the oxygen tension in the bubble and that in the surrounding water, and oxygen diffuses into the physical gill. The consumption of oxygen also creates a disequilibrium between the nitrogen tension within and surrounding the bubble, resulting in the outward diffusion of nitrogen and a gradual decrease in the volume of the physical gill. If the diffusion rates of nitrogen (outward) and oxygen (inward) were equal, the "gill effect" would increase the lifetime of the physical gill about 4 times. However, because oxygen diffuses inward 3.2 times more rapidly than nitrogen diffuses outward, the insect can theoretically extract nearly 13 times as much oxygen from the surrounding water as that originally contained in the physical gill (Ege 1915). The actual gill

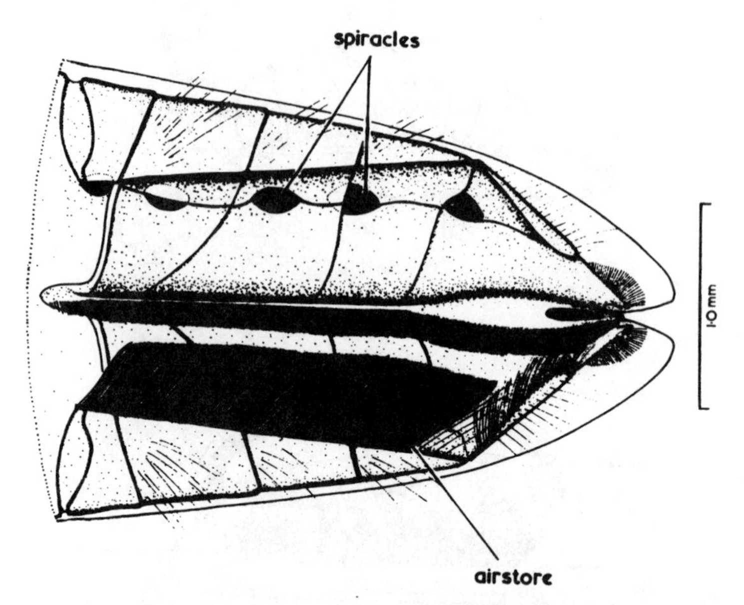

Figure 2.6 A ventral view of the abdomen of the backswimmer *Anisops pellucens* showing the hairs that hold the air store. The hairs of the left side have been removed revealing the spiracles. [From Miller (1964), reproduced by permission of the Royal Entomological Society.]

lifetime is somewhat less (see Mill 1974). Eventually, however, the gas volume is reduced to the extent that the insect must surface to replenish the air store. Actually it is the nitrogen that requires replacement. *Notonecta* denied access to the surface survived for 7 hours in air-saturated water, but only 35 minutes in water saturated with oxygen (Ege 1915, Thorpe 1950). A system that prevented loss of nitrogen from the air film would result in a more or less permanent physical gill. A very few aquatic insects in fact do possess remarkable adaptations for retaining nitrogen gas as will be seen in the next section of this chapter.

The larvae of at least two genera of aquatic arctiid moths carry air stores in lateral hair piles (Myers 1935), but otherwise physical gills are primarily restricted to Coleoptera and Hemiptera. Insects that enter the water to oviposit and semiaquatic species typically have hydrofuge hairs that carry air bubbles during submergence; such air stores may also function as physical gills.

In addition to the film of air carried on various pubescent regions of the body surface, the subelytral air stores of bugs and beetles apparently play an important respiratory role. With the exception of one or two genera of naucorid bugs that have highly specialized respiratory adaptations detailed in the next section, essentially all submerged coleopterans and heteropterans have well-developed elytra or hemelytra (Hutchinson MS). This is in contrast to terrestrial or surface inhabiting hemipterans in which striking reductions of the hemelytra are exhibited by members of all major families.

Some aquatic beetles and bugs merely expose hydrofuge surfaces to atmospheric air at the surface film, but others have highly specialized structures and behavior for renewing air stores. Water scorpions (Nepidae) possess a long, flexible respiratory siphon on the end of the abdomen (Fig. 2.7). The respiratory siphon of the adult is formed by two terminal filaments, the concave inner surface of each forming half a tube, held together by hairs (Menke 1979). When replenishing the subhemelytral air stores the siphon is thrust through the surface film. Air moves down the siphon and enters longitudinal tracheal trunks. From the tracheal trunks air moves through the first abdominal spiracles and into the subhemelytral space. Nepids have pressure receptors on the abdomen and always remain close to the water surface (Thorpe and Crisp 1947c, China 1955). The nymphal filaments contain a ventral longitudinal channel rather than forming a tube. There are two spaces along the abdomen of nymphs, formed by ventrally folded lateral extensions of the terga, where air stores are held by long hairs. The "airstraps" of belostomatid bugs are homologous with the respiratory siphon of nepids, being also derived as extensions of the eighth abdominal segment and bearing spiracles at their base. The airstraps are retracted when the

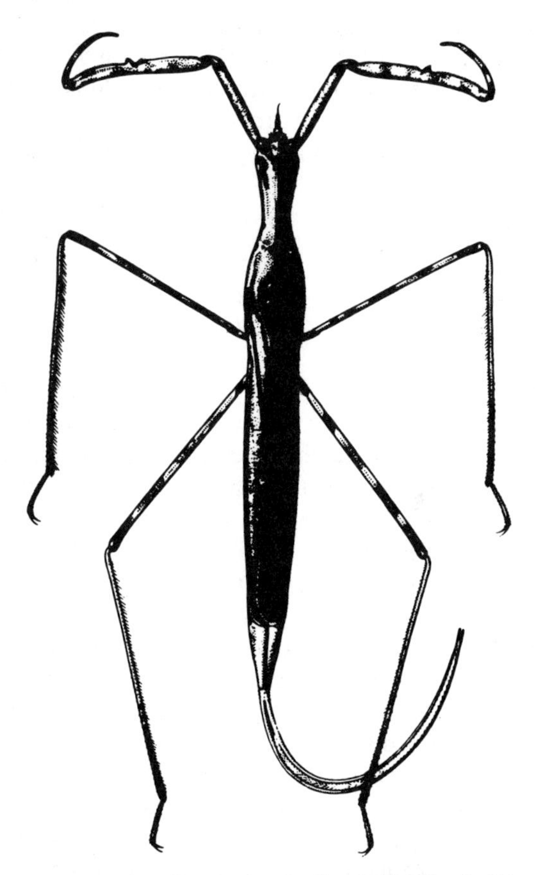

Figure 2.7 A male water scorpion, *Ranatra brevicollis*, showing the flexible respiratory siphon. (From Robert Usinger, *Aquatic Insects of California, with Keys to North American Genera and California Species*. © 1956, The Regents of the University of California.)

bug is not replenishing the air stores. Belostomatid nymphs lack airstraps and air is renewed merely by penetrating the surface film with the tip of the abdomen, a method common to adults of many aquatic bugs and beetles. Water boatmen (Corixidae), however, break the surface film head first and atmospheric air enters via spaces between the head and prothorax.

Hydrophilid beetles break the surface film with the antennae, the terminal clubbed segments of which are clothed with hydrofuge hairs. The antennae are so positioned that a connection is formed between the atmospheric air and the air stores of the insect (Hrbâĉek 1950). Adult haliplid beetles are characterized by greatly expanded hind coxae that form plates covering much of the abdominal sternum. An air store is carried in the space between the coxal plates and the ventral abdominal surface.

Because cold water holds more oxygen and metabolic requirements are less than in warm water, physical gills may provide most or all of small aquatic insect's respiratory needs during times of low temperature. The pigmy backswimmer *Neoplea striola* overwinters as a dormant adult on the bottom of ponds (Gittelman 1975), whereas some other backswimmers must periodically replenish the air in the physical gill or spend the winter in the egg stage. Gittelman (1975) determined mean survival times for the adults of three species of backswimmers that were prevented from surfacing (25°C), and the average duration of dives in minutes when allowed free access to the surface:

	Survival Time	Dive Duration
Neoplea striola	585	39
Buenoa confusa	22	2
B. margaritacea	27	4

The efficiency of the physical gill of *Neoplea* undoubtedly relates to their small body size. Adult *Neoplea* are able to reduce the size of their air stores as water temperatures decline in autumn, enabling them to attain negative buoyancy while satisfying respiratory needs. The greater size of *Buenoa* and *Anisops* necessitates periodic surfacing by adults, although buoyancy control is facilitated by a special use of hemoglobin (Miller 1964, 1966).

Hutchinson (MS) points out that the relationships between gill efficiency, water temperature, and body size may explain the predominance of Micronectinae among water boatmen in tropical waters, whereas the larger Corixinae characteristically occur in temperate latitudes, as they are essentially restricted to high elevations near the equator. The respiratory function of physical gills appears to be limited above 30°C, although little or no research has been conducted on tropical species normally living at high temperatures.

Underwater air stores also function as hydrostatic organs. The possession of gas bubbles normally ensures that the insect rises to the surface when not swimming, if it is not holding on to the substrate. A greater or lesser degree of buoyancy control is exhibited by aquatic insects which is apparently achieved primarily by controlling the volume of the air store (e.g., Thorpe and Crisp 1949).

The notonectids *Anisops* and *Buenoa* use hemoglobin for buoyancy control. Within the abdomens of these bugs are hemoglobin-filled cells, each of which is invaded by a tracheolar mesh (Fig. 2.8). The strikingly low oxygen affinity of notonectid hemoglobin permits dissociation of the pigment during dives, even in well-oxygenated waters, enabling these bugs to carry only small air stores (Miller 1966, Wells et al. 1981). The air stores provide about 25% of the oxygen used by *Anisops pellucens* during a normal dive, whereas 75% comes from the deoxygenation of hemoglobin. (The gill effect appears to be insignificant in supplying oxygen during a dive.) The notonectids possessing hemoglobin thus have the unusual ability to remain poised in midwater without leg strokes. In contrast, *Enithares sobrin*, a notonectid without hemoglobin that may move to middepths when disturbed, requires extremely rapid leg strokes (10–15 per second) to maintain a fixed position (Miller 1964). Deoxygenation of hemoglobin has enabled these bugs to attain neutral buoyancy and has, therefore, made it energetically feasible for notonectids possessing hemoglobin to exploit the midwater niche where they are protected from both benthic and surface predators. (*Anisops* and *Buenoa* typically inhabit water bodies devoid of fishes.) Hemoglobin has thus allowed the Anisopinae to exploit the midwater zone, a

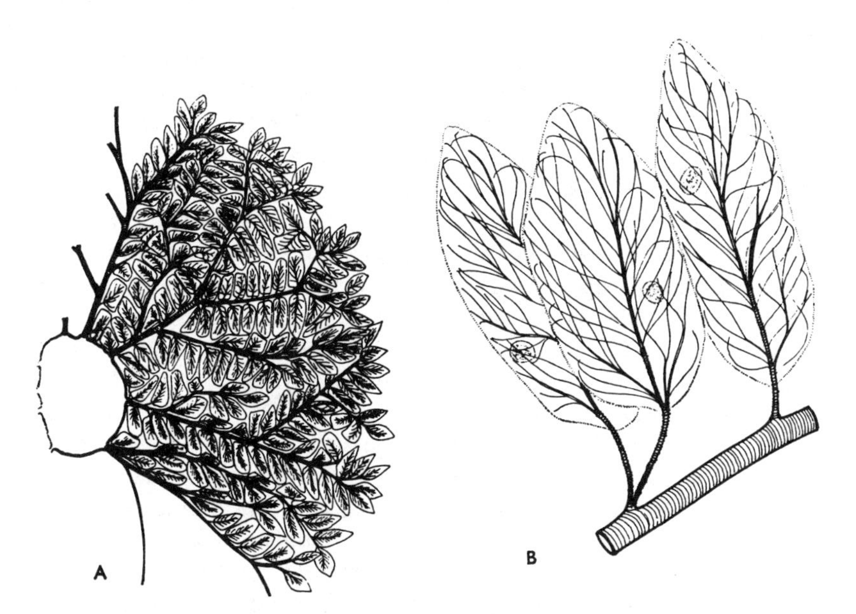

Figure 2.8 (*A*) Portion of a "tree" of hemoglobin-filled cells from the abdomen of the notonectid *Anisops pellucens*; (*B*) three hemoglobin-filled cells (each 60–100 μm long) enlarged to show the tracheolar mesh. [From Miller (1966), with permission of The Company of Biologists Ltd.]

region essentially unoccupied by insects with the exception of chaoborid dipterans. However, unlike chaoborids, the anisopin bugs apparently do not exert active control over their specific gravity (Miller 1964). They are, therefore, less fully adapted to the midwater zone than, for example, *Chaoborus* (= *Corethra*), in which the closed tracheal system has been virtually reduced to two pairs of air sacs (hydrostatic organs). Chaoborus larvae control their buoyancy by inducing volume changes in the hydrostatic organs.

Plastron Respiration A very few aquatic insects with open tracheal systems possess a permanent physical gill, called a *plastron*, which obviates the necessity of surfacing just as tracheal gills allow continuous submergence for insects with closed tracheal systems. With few exceptions, insects utilizing plastron respiration reside in aquatic habitats such as rapid streams, where the difficulties and dangers of surfacing would be considerable and where oxygen depletions do not normally occur.

A plastron is a gas store that resists wetting (the ingress of water) and volume changes against some degree of hydrostatic pressure (Thorpe 1950, Hinton 1976b). The plastron functions on the same general principles as a physical gill with one major exception—the volume of the plastron remains essentially constant. Because the plastron resists volume changes, nitrogen is not lost, and the insect need not periodically replenish the air store. A plastron may thus be envisioned as an incompressible physical gill.

The evolutionary development of plastron respiration was contingent on modifications of the body surface to maintain the structural integrity of the plastron (i.e., surficial structures resistant to volume changes). Plastrons are divided into two types based on the nature of the supporting structures involved. "Hair plastrons," as the name implies, derive rigidity from specialized hydrofuge hairs; "spiracular gills" contain cuticular modifications that produce minute labyrinthine networks to hold the plastron.

THE HAIR PLASTRON The first definitive studies of plastron respiration were reported in a series of papers by Thorpe and Crisp (1947a–c) which dealt with *Aphelocheirus*, an Old World water bug (Aphelocheiridae) that inhabits rapid stretches of rivers. The wings of *A. aestivalis montandoni* are mere vestiges suitable neither for flying nor retaining a subhemelytral air store. This remarkable insect has the entire ventral surface and portions of the dorsum covered with an extremely dense pile of fine hydrofuge hairs that effectively retain the plastron. The spiracles are highly modified rosette-like structures, the arms of which are tiny channels in the exocuticle that terminate in minute pores, the spiracular openings (Fig. 2.9, top). A hair pile contiguous with the surficial hair pile lines these channels (Fig. 2.9, middle). The plastron hairs on the surface are 5-6 μm long and have a density of 2.5×10^6 mm^{-2} according to Thorpe and Crisp (1947a), although Hinton (1976b), using stereoscan electron microscopy, estimated about 4 million hairs per square millimeter. The tip of each hair is bent at right angles to the shaft and derives support from adjacent hairs (Fig. 2.10). There are numerous projections 0.10–0.18 μm in length that keep the individual hairs separated when the plastron is subjected to pressure. Thorpe and Crisp (1947a) calculated that the plastron of *Aphelocheirus* would be completely wetted by an excess pressure of three atmospheres (equivalent to 30 m of water). According to Hinton (1976b), the combined strength of the overlapping hairs is such that at least 40 atm is required to cause the plastron hairs to buckle.

Only the adult of *Aphelocheirus* has a plastron. In stark contrast, the final nymphal instar has a closed tracheal system and depends entirely on cutaneous respiration (Thorpe 1950). The hair plastron is formed on the adult cuticle while beneath the cuticle of the fifth nymphal instar. Two or three days prior to the final ecdysis, which occurs underwater, a gas layer forms within the plastron of the adult and is visible as a sheen through the translucent nymphal cuticle. The adult thus already has a functional plastron when liberated from the nymphal exuviae.

There is evidence suggesting that true plastron respiration may also occur in some New World naucorid bugs such as species of *Cryphocricos* and *Cataractocoris* that reside in turbulent streams or rapid sections of rivers (Parsons and Hewson 1975, Hinton 1976b). Two species of helotrephid bugs inhabiting rapid streams, *Neotrephes usingeri* in Brazil and *Paratrephes hintoni* in French Guiana, apparently possess hair plastrons (Hinton 1976b). Some intertidal saldid bugs utilize plastron respiration during submergence (Polhemus 1976).

Several beetles exhibit hair plastron respiration as adults (Thorpe and Crisp 1949, Hinton 1976b). The adults of many elmid beetles utilize plastron respiration. Adaptations involved with plastron respiration appear incompatible with flight and plastron-bearing elmids have lost the ability to fly (Thorpe and Crisp 1947a). There may be an initial dispersal flight after adults emerge from the terrestrial pupal chamber, but once adults enter the water the indirect flight muscles undergo an apparently irreversible degeneration (Hinton 1976b). Most species studied in detail show a somewhat less efficient plastron than *Aphelocheirus* and withstand a pressure increase of only slightly less than 1 atm (Thorpe 1950). Riffle beetles primarily inhabit shallow streams and do not require the margin of safety exhibited by *Aphelocheirus,* which resides in rivers. The chrysomelid beetle *Macroplea* (= *Haemonia*) *mutica* has a hair plastron covering the antennae and most of the body surface except the elytra. Although Hinton (1976b) lists numerous aquatic and semiaquatic weevils (Curculionidae) as possessing plastron respiration, all except one of these species are considered "transitional forms" by Thorpe and Crisp (1949). Only *Eubrychius* (= *Phytobius*) *velatus* among the weevils is known to have a true plastron. *Eubrychius velatus* is an active swimmer living among beds of aquatic plants. The plastron is held by overlapping cuticular scales clothed with plastron hairs (Fig. 2.9, bottom). It may be that the scales or similar cuticular structures characteristic of many weevils preadapts this group for the evolutionary development of plastron respiration (Thorpe 1950). The elmid beetle *Cylloepus barberi* also has minute plastron hairs borne on overlapping scales, a structure very similar to that of *E. velatus.*

The beetles and bugs associated with aquatic habitats are placed in four

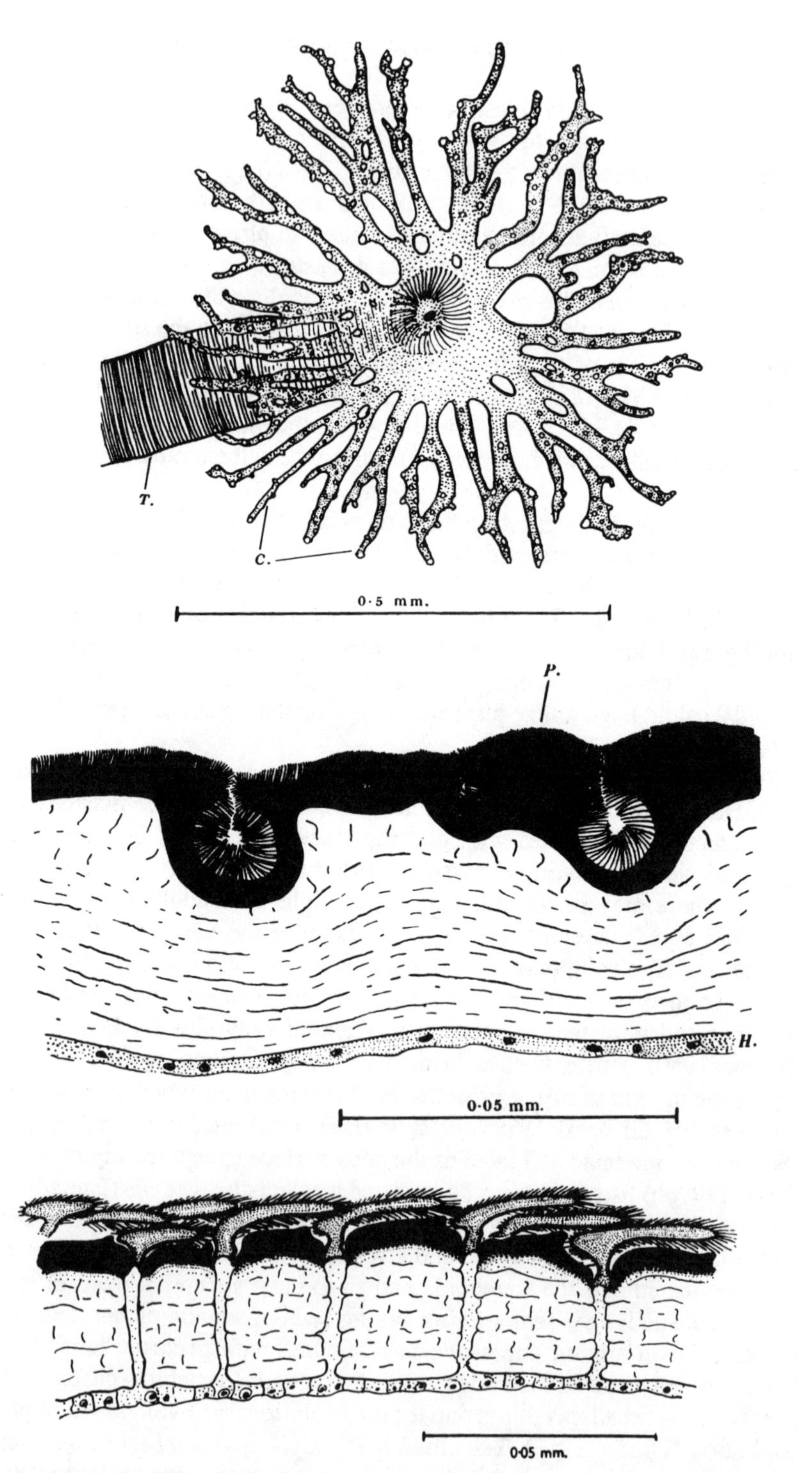

Figure 2.9 *Top*: Spiracular "rosette" of third abdominal segment of the heteropteran *Aphelocheirus aestivalis* (C = channels in the exocuticle forming the arms of the rosette; T =

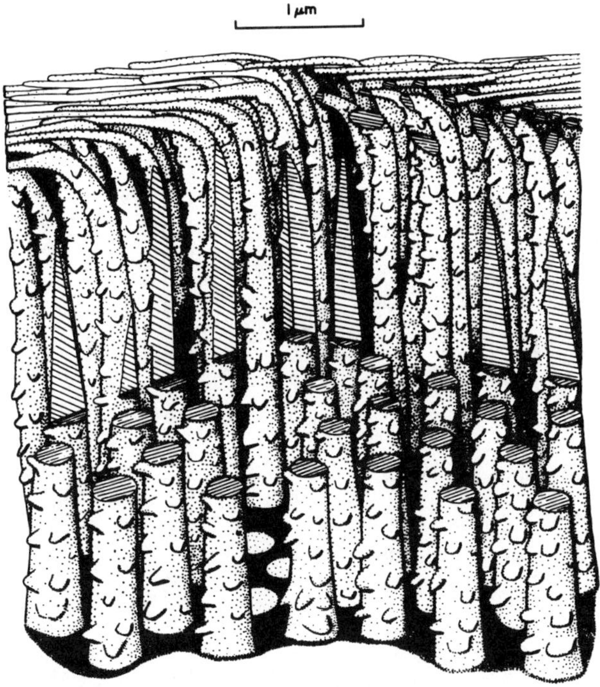

Figure 2.10 A diagram of the plastron hairs of the heteropteran *Aphelocheirus aestivalis*. [From Hinton (1976b). Reprinted with permission of Pergammon Press plc.]

categories in Table 2.3 based on a presumed evolutionary development of hair plastron respiration (Thorpe and Crisp 1949). Riparian forms include many terrestrial or semiaquatic insects for which hydrofuge hairs provide buoyancy and short-term respiration by entrapping air bubbles during temporary, often accidental, immersion. Aerial adult insects that crawl or swim to underwater oviposition sites, may also utilize air stores held by hydrofuge hairs. One might also include marine (*Halobates*) and riverine (*Ventidius*) surface film bugs of the family Gerridae which possess dense hair piles apparently for use during accidental submergence (Cheng 1973b).

Transitional forms include insects dependent on frequent surface visits to satisfy respiratory requirements. Despite well-developed hair piles with from 6

---

trachea). The central spiracular scar is shown as a black dot. The small circles and bud-like projections of the arms are minute spiracular openings. *Middle*: A section through the exocuticle (black) to show the hair-lined channels of the spiracular rosette of *Aphelocheirus* (H = hypodermis P = plastron hairs). *Bottom*: A longitudinal section through the abdominal sternum of the weevil *Eubrychius* (= *Phytobius*) *velatus* showing the overlapping scales bearing plastron hairs. [Top and Middle from Thorpe and Crisp (1947a); bottom from Thorpe and Crisp (1949). With permission of The Company of Biologists Ltd.]

Table 2.3 Some characteristics of beetles and bugs grouped according to the respiratory efficiency of hydrofuge body hairs[a]

Category	Examples	Hair Density	Excess Pressure Withstood	Oxygen Reserve	Surface Visits	Buoyancy
Riparian	*Donacia, Stenoplemus*	Low	(Terrestrial)	Considerable	(Terrestrial)	Positive
Transitional	*Dryops, Hydrophilus*	Medium	<0.5 atm	Considerable	Frequent	Positive
Functionally perfect plastron	*Elmis, Riolus, Macroplea*	High	0.5–2 atm	Slight	Never	Negative
Functionally and structurally perfect plastron	*Aphelocheirus, Eubrychius, Stenelmis, Cylleopus*	Extremely high	>2 atm	None	Never	Negative

[a]Based on the four categories developed by Thorpe and Crisp (1949), but sequenced to indicate increasing evolutionary adaptation.

x 10^4 to 8 x 10^5 hairs cm^{-2}, such species are, therefore, not completely adapted to an aquatic existence. Their positive buoyancy ensures that they do not venture into water too deep, and that they easily rise to the surface to replenish air stores. The air film of transitional forms may function efficiently as an underwater gill, but nitrogen is lost and there is a progressive diminution in the volume of the air store.

Only those insects which can remain submerged indefinitely in well-aerated waters by extracting oxygen at the interface of a thin rigid air film on the body surface truly exhibit plastron respiration. Two evolutionary levels of the hair plastron are recognized (Table 2.3), those that are functionally perfect (in the sense that surface visits are not necessary), and those that are functionally as well as structurally perfect.

The plastron-bearing elmid beetles (except *Stenelmis* and *Cylloepus*) and the chrysomelid *Macroplea* have a functionally perfect plastron as adults. Plastron hair density ranges from 3 x 10^6 to 1.5 x 10^7 cm^{-2}. The elmid beetles in this category exhibit "plastron replacement activities" involving grooming behavior to keep the plastron hairs from clumping and to incorporate bubbles entrained in the water column into the air film. Among plastron-bearing insects only these elmid beetles are able to effect a partial recovery of the plastron when submerged (Thorpe and Crisp 1949). However, their plastrons are wetted by excess pressures of somewhat less than 1 atm, and they must avoid deep water and areas where entrained air bubbles are not available. Most elmid beetles in this category, although normally negatively buoyant, can alter their buoyancy and rise to the surface under adverse environmental conditions.

Aphelocheirus, the weevil *Eubrychius velatus*, and the elmids *Stenelmis crenata* and *Cylloepus barberi* each have a plastron that is nearly functionally and structurally perfect. The plastron is an extremely thin air film of negligible volume and large surface area that is not wetted by excess pressures of 2 atm. Hair density exceeds 10^8 cm^{-2}. The respiratory efficiency of the plastron obviates the need for an oxygen reserve or for special mechanisms of buoyancy control. *Aphelocheirus* ventures into water of considerable depth, but has highly elaborated pressure receptors that apparently prevent incursions into water so deep as to wet the plastron (Thorpe and Crisp 1947c).

In oxygen-depleted waters the plastron works in reverse. That is, oxygen diffuses from the air store of the insect into the ambient medium. The great majority of insects with true plastron respiration inhabit turbulent well-aerated waters, primarily rapid reaches of running waters, but also the wave-swept littoral of large lakes and intertidal habitats. Such habitats always maintain oxygen-saturated waters. *Aphelocheirus,* for example, is unable to survive in low-oxygen waters, but such conditions do not normally occur in their natural environment, rapid stretches of rivers. In contrast, the chrysomelid *Macroplea* and the curculionid *Eubrychius* bear true plastrons, but live in standing waters where oxygen depletions commonly occur. These beetles are, however, remarkably tolerant of low-oxygen conditions and are able to go into oxygen debt yet remain active (Thorpe and Crisp 1949).

Two other examples of plastron respiration are known (Thorpe 1950). The brachypterous form of the female aquatic moth *Acentropus niveus* apparently remains submerged during the adult stage (Berg 1941). The plastron is held by a scale covering on the body. Several hymenopterans parasitize the egg or larval stages of aquatic insects. The females submerge to oviposit on a suitable host. *Prestwichia* females, for example, may remain immersed for periods exceeding 5 days (Thorpe 1950). The waterproofed spiracles protect the tracheal system from flooding, but because of their minute size no special respiratory adaptations are needed. *Agriotypus armatus* exhibits a special type of plastron respiration in pupal and adult stages (Clausen 1950). The females deposit eggs on the pupal and prepupal stages of cased caddisflies. After devouring the caddis pupa (the undulations of which provided respiratory currents while alive), the larval parasite spins a silk cocoon within the caddisfly case. A ribbon of silk about 3 cm long is attached to the anterior end of the cocoon and extends from the case into the water (Fig. 2.11). The larval hymenopteran fills the cocoon and ribbon with gas (Grenier 1970). The ribbon is composed of a double layer of silk and, although not a hair plastron, apparently functions as a unique type of plastron in communication with the air-filled cocoon. Removal of the ribbon results in death of the parasite (Messner 1965).

SPIRACULAR GILLS The aquatic immatures of some dipterans and a few coleopterans have plastron-bearing gills formed by extensions of spiracles or evagination of the body wall adjacent to spiracles. The plastron air film is supported, not by hydrofuge hairs, but by a highly specialized cuticular latticework on the gill surface. The cuticular latticework distinguishes spiracular gills from the ordinary respiratory horns from which they evolved in some dipterans (Hinton 1968). The air held in the plastron is in direct communication with the gas in the tracheal system via functional spiracles, as is true of a hair plastron.

Spiracular gills are primarily possessed by insects, frequently the immobile pupal stage, inhabiting environments that are alternately dry and flooded (Hinton 1953). When immersed, plastron bearing spiracular gills provide the insect with a large air–water interface for respiratory exchange, without the necessity of increasing cuticular permeability. When exposed to air, the insect breathes through functional spiracles that provide only a small surface area, and excessive water loss is avoided as in terrestrial insects. Although spiracular gills are normally associated with aquatic insects, as Hinton points out, the immobile stages (eggs, pupae) of terrestrial species may be covered by water during and following rain. According to Hinton (1968), it "appears that instances of plastron respiration among terrestrial insects will be much more numerous than among aquatic ones."

The cuticular modifications commonly consist of microscopic struts extending from the spiracular gill surface and branching apically to form an open hydrofuge network that retains the air film (Fig. 2.12, top). There are several other structural types (Hinton 1968). For example, the tipulid *Lipsothrix* has a series of perforated grooves in the wall of the spiracular gills (Fig. 2.12, bottom).

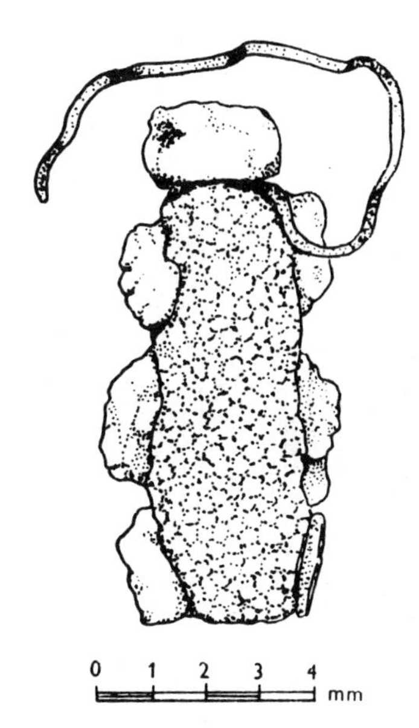

Figure 2.11 The pupal case of the caddisfly *Silo pallipes* inhabited by the parasitic wasp *Agriotypus armatus,* the larvae of which has spun a silk ribbon shown extending from the case. The ribbon functions as a unique type of plastron. [From Fisher (1932). Reprinted with permission of The Zoological Society of London.]

Spiracular gills when present in aquatic dipterans are restricted to the pupal stage, and, except for empidids discussed below, are associated with the first pair of thoracic spiracles. The tanyderid *Eutanyderus* and several genera of tipulids, in which pupation occurs under water, possess spiracular gills. The majority of these occur in running waters (e.g., *Antocha*) or the intertidal zone (e.g., *Idioglochina*). According to Hinton (1968), spiracular gills have a polyphletic origin; the spiracular gills of tipulids alone have been independently evolved from respiratory horns on at least four occasions.

Three families of dipterans are essentially restricted to running waters, being especially characteristic of rapid streams; all possess spiracular gills. Black fly (Simuliidae) pupae have spiracular gills in the form of thoracic respiratory filaments that are usually branched. The Mountain midge (Deuterophlebiidae) pupae possess a pair of normally three-branched spiracular gills. The spiracular gills of net-winged midge (Blephariceridae) pupae are variable in structure, but often consist of respiratory lamellae surrounding the opening of the spiracular atrium.

Two genera of empidids that reside in running waters, *Hemerodromia* and *Chelifera,* have spiracular gills that consist of long filaments projecting from the first thoracic and the first seven abdominal spiracles. Microscopic canals containing the plastron gas run along the gill surface and extend to the spiracular atrium at the gill base.

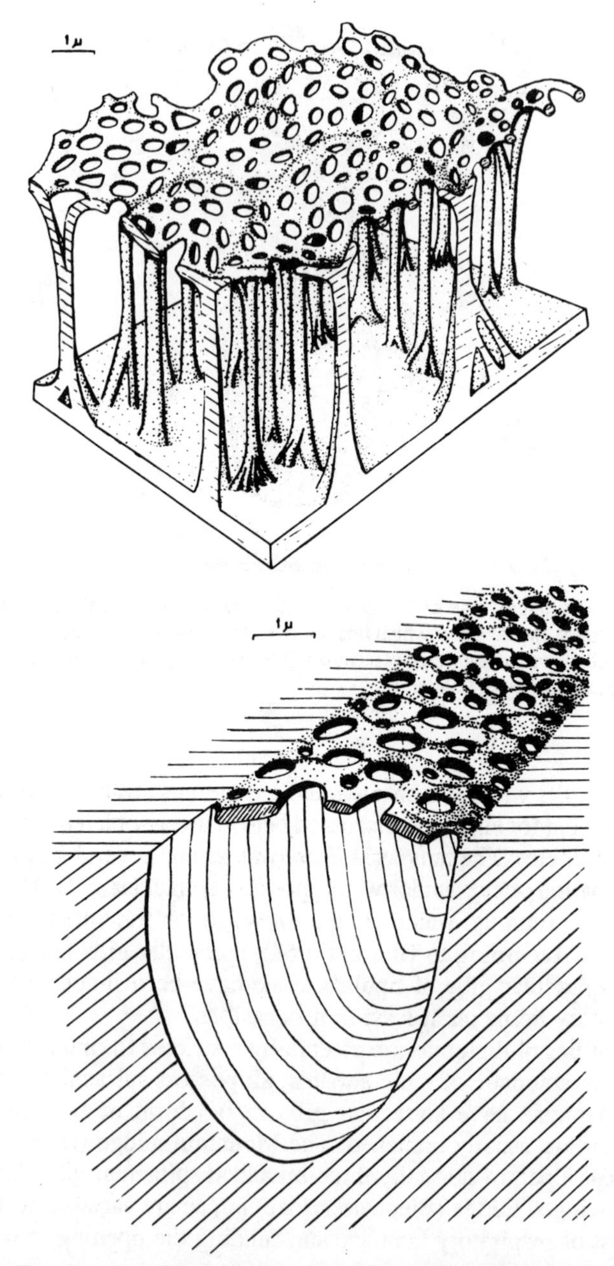

Figure 2.12 *Top*: A portion of the cuticular network forming the plastron on the spiracular gills of the intertidal tipulid *Dicranomyia trifilamentosa*. *Bottom*: A diagram of a plastron canal in the wall of the spiracular gills of the stream-dwelling tipulid *Lipsothrix remota*. [Top from Hinton (1968); bottom from Hinton (1967a),. Reproduced by permission of The Royal Entomological Society.]

Intertidal species from two other families of Diptera, Dolichopodidae and Canaceidae, also possess spiracular gills.

The spiracular gills of the pupae of some chironomid midges (e.g., *Chironomus* and related genera) do not have an associated plastron. However, plastron-bearing spiracular gills occur in the subfamilies Tanypodinae and Podonominae (Hinton 1968).

Spiracular gills have evolved independently in two subfamilies of psephenid beetles (Hinton 1966). Species with spiracular gills pupate in streams or the wave-swept littoral of large lakes, unlike most aquatic beetles which pupate in the terrestrial environment. The gills are enlargements of functional abdominal spiracles 2 to 7.

The pupae of the beetle family Torridincolidae possess spiracular gills on the first two abdominal segments. The long filamentous gills are similar to those of the Empididae and contain plastron gas within superficial longitudinal canals. *Torridincola* is the only insect known to have spiracular gills in both larval and pupal stages.

In addition to Torridincolidae, two other beetle families (Hydroscaphidae, Sphaeriidae) have spiracular gills in larval stages (the pupae are terrestrial). These beetles are the only known insects that have a plastron as larvae (Hinton 1967b).

Spiracular gills are thus remarkable dual organs able to function in respiratory exchange whether the insect is exposed to air or under water. Most aquatic genera that possess spiracular gills occur in habitats liable to rapid changes in water level. Although species with spiracular gills normally shed the pupal cuticle when submerged, all those tested are capable of ecdysis above the water surface (Hinton 1953).

Osmotic Regulation in Aquatic Insects

In this section the osmoregulatory problems and related adaptations exhibited by underwater stages of insects, especially in the context of a secondarily aquatic evolutionary history, are considered. For general principles of osmoregulation, consult Potts and Parry (1964), Prosser (1973), and Rankin and Davenport (1981). Krogh (1939) and Beadle (1943, 1957) deal specifically with osmotic regulation in aquatic animals. The chapter by Stobbart and Shaw (1974) addresses salt and water balance in insects, including freshwater, brackish, and marine species. Komnick (1977) provides an excellent review of chloride cells and chloride epithelia in aquatic insects.

The evolutionary adaptations necessary for the success of insects on land included the development of a high degree of cuticular impermeability. Although posing major respiratory problems, from an osmoregulatory standpoint a relatively impermeable cuticle was generally preadaptive for those insects secondarily invading water. For aquatic insects whose respiratory adaptations are based on the use of atmospheric air (including gas gills, plastron respiration, and extracting air from plants), it was not necessary to reduce cuticular impermeability. For example, larvae of the rat-tailed maggot *Eristalis,* which have a respiratory siphon used to reach atmospheric air, have a highly impermeable body wall and are able to reside in waters with extreme chemical conditions. The chrysomelid beetle *Donacia simplex* is able to inhabit the inhospitable chemical environment of anaerobic muds by tapping the air stores of plant roots.

However, in adapting to life in water other insects increased the permeability of the general body surface and evolved specialized regions or evaginations (gills) to extract dissolved oxygen directly from the water. For these species the freshwater environment posed major osmotic problems. The remainder of this chapter applies to aquatic stages of insects with a relatively high degree of cuticular permeability. There is no clear relationship between internal osmotic concentrations of aquatic insects and the salinity of the habitat (Stobbart and Shaw 1974).

Any organism with permeable surfaces residing in freshwater must be an osmoregulator (homoiosmotic); osmoconformism (poikilosmotic) is precluded because the internal fluids would be diluted below the concentration necessary to maintain the biochemical integrity of an insect in an active state. Aquatic insects thus possess an internal osmotic pressure greater than that of the freshwater environment; they are hypertonic to the medium in which they live. Therefore, water has a tendency to enter the insect, and salts tend to passively diffuse from the insect to the surrounding water. In general terms, osmoregulation includes those mechanisms that excrete the excess water, as well as mechanisms for the differential retention and uptake of salts (Fig. 2.13). There is a cutaneous influx of water, ranging in different freshwater species within 1–15% of the body weight per day (Komnick 1977). The production of urine hypotonic to body fluids functions to remove excess water while reducing ion loss. However, since no animal is capable of excreting distilled water, some ions are lost with the urine. In addition, no mechanism has evolved that is completely effective in ion retention. Even if such were the case, additional ions are necessary for growth, repair, and reproduction. Defecation is another ion efflux pathway, and unless the exuviae is eaten, ecdysis represents yet another loss of ions. Active uptake of ions is thus essential for insects inhabiting freshwater.

There are four types of ion absorption sites in aquatic insects, examples of which are shown in Figure 2.14. The first three types may be arranged in order of increasing histological complexity as follows: (1) chloride cells (unicells or cell complexes), (2) chloride epithelia, and (3) anal papillae. The fourth type involves the epithelium of the gut wall in drinking insects. The known occurrence of ion absorption sites in aquatic insects is shown in Table 2.4.

Komnick (1977) provides considerable data on the fine structure of ion absorption sites, as well as details on the taxonomic distribution of the types and subtypes among aquatic insects. Ion absorption structures are known only from aquatic stages of amphibiotic insects, but are present in both immatures and imagines in those species of Hemiptera that inhabit water throughout their entire lives. Within a species the number of chloride cells is inversely proportional to the salinity of the medium (Fig. 2.15), in a manner analogous to the size

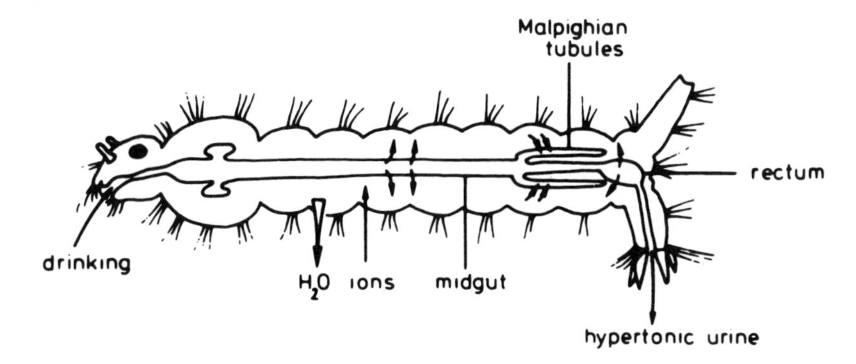

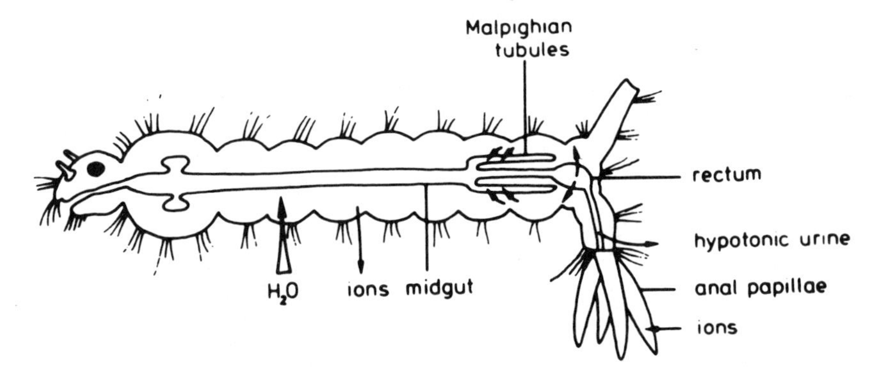

Figure 2.13 *Top*: Hypoosmotic regulation in saltwater insects (mosquito larvae with small anal papillae). *Bottom*: Hyperosmotic regulation in freshwater insects (mosquito larva with large anal papillae). Arrows indicate main pathways for water and ions. [From Komnick (1977).]

Table 2.4 Occurrence of ion absorption sites in aquatic insects

Absorption Site	Occurrence
Chloride cells	Ephemeroptera, Plecoptera, Heteroptera
Chloride epithelia	
Abdominal	Trichoptera: Limnephilidae, Goeridae
Anal	Diptera
	Brachycera: Tabanidae, Stratiomyidae
	Cyclorrhapha: Ephydridae, Muscidae
Rectal	Odonata: Anisoptera, Zygoptera
Anal papillae	Diptera
	Nematocera
	Cyclorrhapha: Syrphidae
	Trichoptera: Glossosomatidae, Philopotamidae
Drinking	
Midgut	Megaloptera: Sialidae
Ileum	Coleoptera: Dytiscidae

Source: Modified from Komnick (1977).

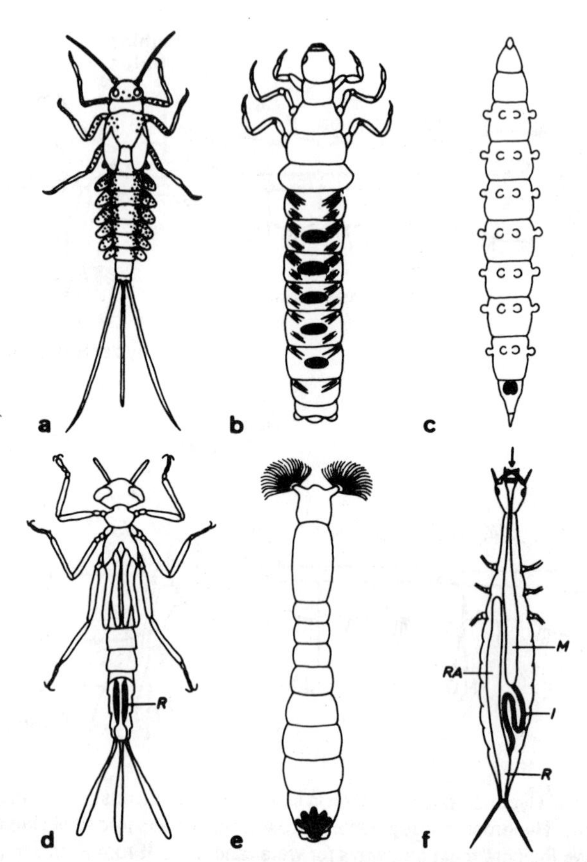

Figure 2.14 Examples of the location of ion absorption sites in freshwater insects: (*a*) chloride cells in Ephemeroptera, Baetidae; (*b*) abdominal chloride epithelia in Trichoptera, Limnephilidae; (*c*) anal chloride epithelia in Diptera, Tabanidae; (*d*) rectal chloride epithelia in conjunction with rectal ventilation in Odonata, Coenagrionidae; (*e*) everted preanal papillae in Diptera, Simuliidae; (*f*) ileum in conjuction with drinking in Coleoptera, Dytiscidae (I, ileum; M, midgut; R, rectum; RA, rectal ampulla). [From Komnick (1977).]

variations of anal papillae (Fig. 2.13). A very few freshwater insects are known to drink the medium as a means to expose ions to transport epithelium lining the gut. According to Komnick (1977), this may represent the most primitive method of osmoregulation in freshwater insects.

Marine and saline inland waters are physiological deserts, the osmotic gradient favoring a passive efflux of water and an influx of ions (Fig. 2.13). Osmotic regulation therefore involves water conservation and the elimination of excess ions. Insects drink the medium to compensate for water loss, and excrete hypertonic urine. The anal papillae of the mosquito *Aedes campestris* are apparently able to absorb ions from a freshwater medium and excrete ions when in saltwater (Phillips and Meredith 1969).

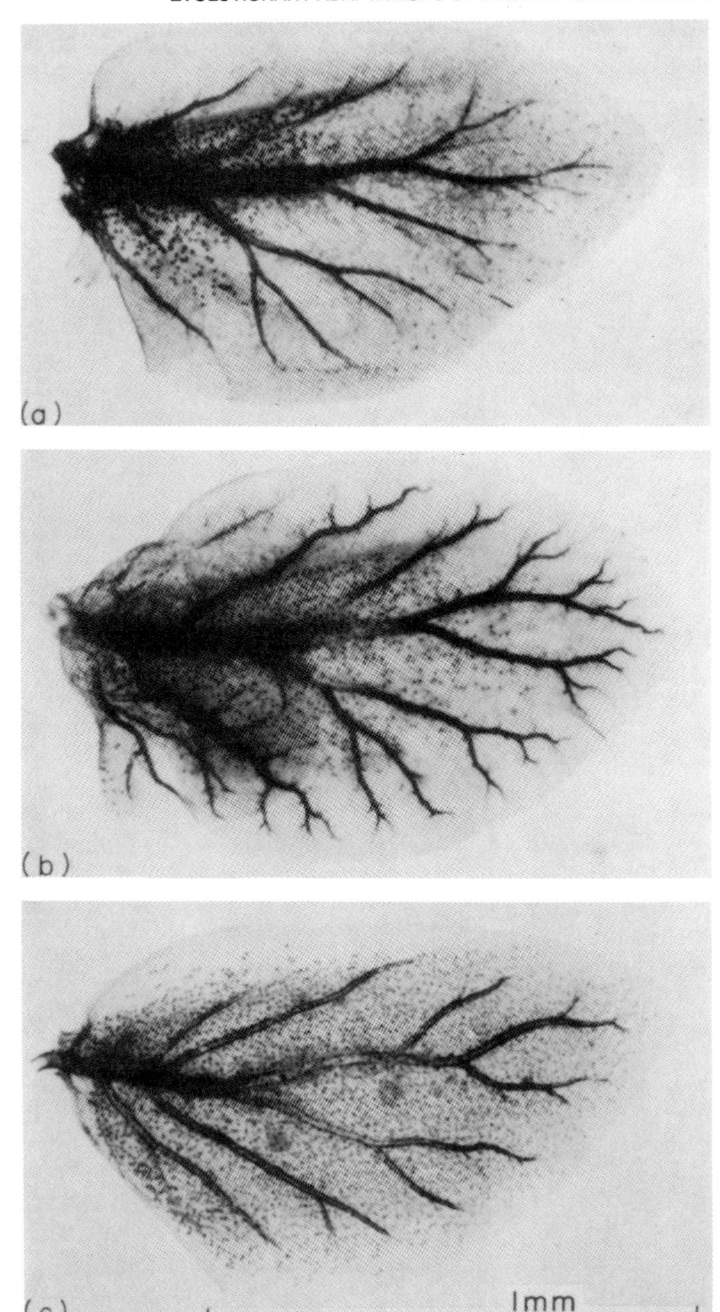

Figure 2.15 Tracheal gills of *Callibaetis coloradensis* maintained at different salinities (*a–c*) for 30 days and subsequently fixed in the osmium tetroxide–silver lactate mixture for chloride precipitation: (*a*) concentrated fresh water (final concentration 130 m*M* NaCl); (*b*) normal freshwater; (*c*) diluted freshwater (1:100). Notice the increase in the number of chloride cells at decreasing salinities. [From Wichard et al. (1973). Reprinted with permission of Pergammon Press plc.]

II

HABITATS AND COMMUNITIES

This section of the book will address the habitats and communities (roughly equivalent to "biotopes" and "biocoenoses" in the European literature) within which aquatic insects reside. Whereas more than 97% of the earth's water is contained in the seas, the vast majority of aquatic insects occur in inland waters which collectively comprise less than 1% of the total (icecaps contribute about 2%). That inland waters play a disproportionately greater role in many aspects of ecology on this planet is well known; no attempt will be made here to reiterate the reasons for their importance.

The ecological classification of aquatic habitats and communities presented below is intended to assist the reader with the terminology and organization detailed more fully in subsequent chapters. Such schemes are, of necessity, oversimplifications of reality in which heterogeneous systems are lumped together for convenience, and where no attempt is made to convey the complex interactions between components. In the tabular classification, designations in parentheses indicate that insects are generally absent from that community.

Marine systems are grouped into three major habitat types. The oceanic habitat, offshore areas of open sea, are nearly devoid of aquatic insects. Insects are nearly as rare in the neritic, or near-shore habitats. Only in intertidal habitats, areas of the shore that are alternately submerged and exposed by tidal fluctuations, do a few aquatic insects become relatively common. The adaptations of some intertidal species are extremely fascinating, as will be seen in Chapter 5.

Estuaries, the ecotones between inland waters and the sea, while also exhibiting a dearth of species, contain a somewhat richer aquatic insect fauna than marine habitats.

System	Habitats	Predominant Communities			
		Plankton	Nekton	Pleuston	Benthos
Marine	Oceanic	(X)	(X)	X	(X)
	Neritic	(X)	(X)	X	X
	Intertidal	—	(X)	—	X
Estuarine	Subtidal	(X)	(X)	X	X
	Intertidal	—	(X)	—	X
Lotic	Crenal	—	(X)	X	X
	Rhithral	—	(X)	X	X
	Potamal	(X)	X	X	X
Lentic	Lacustrine	X	X	X	X
	Palustrine	—	X	X	X
Subterranean	Troglal	—	(X)	—	X
	Stygal	—	—	—	X

The term *lotic* encompasses all inland running waters from tiny brooks to grand rivers. Three major habitat types are considered here. *Crenal* is a term for spring-fed headwaters, whereas *rhithral* includes streams and small rivers. *Potamal* refers to large rivers upstream from the influence of ocean tides. The longitudinal zonation of lotic habitats and resulting distribution patterns of aquatic insects is a subject of considerable interest.

Lentic is an all-encompassing term for inland standing waters. *Lacustrine* refers to lake and pond habitats, whereas *palustrine* includes shallow-water habitats such as marshes and swamps, which form the ecotones between aquatic and terrestrial ecosystems. Aquatic insects reach their greatest diversity in lentic habitats where aquatic vegetation is well developed.

Subterranean waters (caves or troglal and groundwater or stygal) contain few aquatic insects except where groundwater aquifers merge with lentic and especially lotic systems. Hypogean (subsurface, as contrasted with epigean or surface waters) habitats and the inhabitants thereof will be dealt with further in Chapter 4.

A biotic community *(sensu stricto)* comprises all organisms in a given habitat, both producers and consumers, including everything from bacteria to mammals and higher plants. However, current usage (as evinced in ecological literature) is less precise. It is, for example, common to see reference made to the "planktonic community" or even the "crustacean planktonic community." "Community" is used in the latter sense throughout this book.

Plankton refers to those organisms whose normal mode of existence is to remain suspended in water. Both the sea and inland lentic waters contain diverse planktonic communities. Few aquatic insects are, however, truly planktonic.

The nektonic community is also associated with the open water, but its members, in contrast to plankton, are strong swimmers with directional mobility. Nekton includes most fishes, some larger crustaceans, and a very few insects.

The pleustonic community, organisms residing at the air-water interface, while fairly well developed in certain lentic habitats, is poorly represented in the sea. *Halobates,* the marine water strider, is a notable exception. These surface forms normally breathe atmospheric air, but depend on the aqueous medium for other needs (e.g., food, reproduction).

Aquatic insects are best represented in the benthic community, the assemblage of organisms associated with the substratum. The term *benthos* used broadly includes not only species residing on or in bottom materials, but also those living in plant beds or associated with logs or other solid surfaces.

The following three chapters deal in more detail with the habitats and communities of aquatic insects. In Chapter 3 lentic systems are considered, including special habitats such as temporary ponds, bogs, and water-filled tree holes. Chapter 4 deals with lotic waters, including cave streams, intermittent and interrupted streams, and thermal streams. Special consideration is also given to the use by insects of groundwater habitats adjacent to running waters. Marine, brackish, and saline inland waters are all considered in Chapter 5. These chapters describe the faunal composition and adaptations to the general habitat conditions encountered by insect communities in the major aquatic biotopes. The ecological and evolutionary responses of aquatic insects to specific environmental conditions, such as temperature or substrate, will be treated in separate chapters.

3

LENTIC FRESHWATERS

A lake is the landscape's most beautiful and expressive feature. It is the earth's eye; looking into which the beholder measures the depth of his own nature.
—Henry David Thoreau

LAKES

Lakes collectively provide a diverse array of habitats for aquatic insects. Within a given lake, environmental conditions often exhibit distinct spatial gradients, and temporal changes may be pronounced. The composition of aquatic insect communities may change markedly along habitat gradients within a lake and also differs greatly between lake types.

Lake Zonation

The limnetic zone of a lake is the open water area devoid of rooted plants, whereas the littoral zone comprises the shallow areas along the shore characterized by rooted vegetation (Fig. 3.1). The littoral zone may be further divided into three subzones of rooted plants. Emergent hydrophytes, plants rooted in aquatic soil and projecting above the water surface, include sedges (Cyperaceae), grasses (Gramineae), and rushes (Juncaceae). Hydrophytes rooted in the bottom with leaves floating on the water surface occupy the next subzone. Water lilies (e.g., *Nuphar*), the lotus (*Nelumbo*), and several pond weeds (*Potamogeton*) are examples of floating-leaved rooted hydrophytes. Submerged hydrophytes nor-

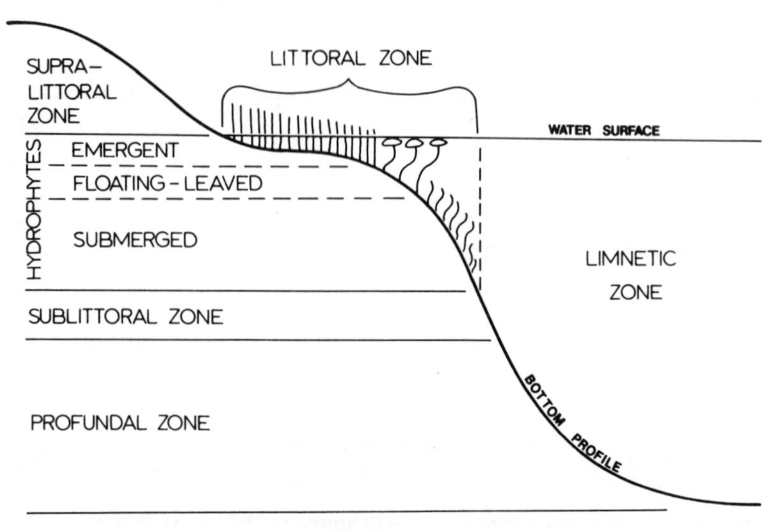

Figure 3.1 A simplified lake zonation scheme.

mally extend farthest from the shore and include rooted plants such as *Elodea, Myriophyllum, Ranunculus,* and many species of *Potomogeton.* Ideally, the three subzones of the rooted hydrophytes form concentric, although overlapping, rings in the littoral zone of a lake.

Although lake zonation schemes are generally based on rooted aquatic angiosperms, free-floating species such as duckweed (e.g., *Lemna*) and water hyacinth (*Eichhornia*) may occur on the water surface, and a few are suspended below the surface (*Ceratophyllum, Utricularia*). In addition, a variety of lower vascular plants (e.g., *Equisetum, Isoetes, Azolla*), and nonvascular aquatic plants, such as mosses and algae, may be present. Not all lakes contain higher vascular plants. Wind-swept lakes may be devoid of higher plants along the shore, although submerged hydrophytes may be present at depths below the major effects of wave action. Rooted plants are also absent from rocky and steep shores and from lakes at high altitudes or latitudes where growing seasons are short and ice scouring along shorelines may be severe. High turbidity from suspended clay particles prevents the establishment of higher plants in some lakes. The composition, abundance, and distribution of aquatic plants markedly influence aquatic insect communities.

The sublittoral zone [the "Littoriprofundal" of Hutchinson (1967)] is the region of transition between the well-illuminated upper strata of the lake and the profundal zone where light is insufficient for photosynthesis. Shade-tolerant plants may be sparsely distributed on the substrate in the sublittoral zone.

Emergent hydrophytes occur in water up to 2 m deep. The maximum water depths reported for floating-leaved plants rarely exceed 4 m. Spence (1982) reviewed data on depth distributions of submerged plants in freshwater lakes located over 130° of latitude. Maximum depths occupied by macrophytes

(Tracheophyta, Charophyta, and Bryophyta) ranged from <1 m to >100 m. Angiosperms occurred to depths of 12 m; the stonewort *Chara* was reported from 65 m; and mosses have been collected from depths of ≤122 m. These are, of course, extreme examples from lakes of exceptional transparency. In the vast majority of lakes, submerged macrophytes rarely colonize depths of >10 m.

Lake Communities

Pleuston

> Surely there must be some truce among these keen competitors of the surface! For here they are all eager for food and not one with scruples against pouncing upon a weaker brother, yet getting on together—a heterogeneous collection from the large Gerrid to the tiny *Microvelia*. How they manage to maintain an armed truce and survive by their alertness only patient observation can discern.
>
> —H. B. Hungerford

The pleustonic community includes organisms associated with the surface film, the air–water interface. Epipleustonic forms, such as water striders, are adapted for support on the upper surface of the water. Hypopleustonic species, such as mosquito larvae and pupae, "hang" from the underside of the surface film. Many hypopleustonic and some epipleustonic species do not continuously reside at the surface film and are thus meropleustonic (partly pleustonic). For example, the whirlygig beetle *Gyrinus* dives below the surface, and mosquito larvae and pupae retreat to greater depths to escape danger. Aquatic insects of several orders make excursions to the water surface to replenish air stores, although such species are not normally considered even temporary members of the pleuston. Many amphibiotic species utilize the surface film for support during ecdysis to the adult; mating may occur on the water surface, and females may rest on the surface film during oviposition. The eggs of some species remain at the air–water interface until eclosion. The terms *epineuston* and *hyponeuston* (with the root *neuston* rather than *pleuston*) refer to the assemblages of microscopic organisms associated with the upper and lower surfaces of the air–water interface. Such organisms, and associated organic compounds, constitute a concentrated food source that is exploited by certain metazoans, including some insects (Rapoport and Sánchez 1963).

The preceding terminology basically follows Hutchinson (1967). See Banse (1975) for additional discussion of terms for organisms associated with surface films.

Pleustonic insects exhibit various adaptations enabling them to reside at the air–water interface (Table 3.1). The small size, springing organ (furcula), and hydrophobic cuticle of collembolans enable them to be supported by and move across water surfaces, and many individuals found in the pleuston are accidentals dislodged from their normal habitats in the soil (Rapoport and Sánchez 1963). Springtails exhibiting adaptations for a freshwater pleustonic existence are apparently restricted to three families (Table 3.1). Known adapta-

Table 3.1 Insect taxa containing species that commonly reside in the pleuston of inland standing-water bodies and some major adaptations

Taxon	Position	Adaptation
Collembola	Epipleuston	Small size
Isotomidae		Hydrophobic cuticle
Poduridae		Lamellate mucrones
Sminthuridae		Paddle-shaped mucrones
		Furcular setation
		Claw modification
Coleoptera	Epipleuston	Hydrophobic dorsum
Gyrinidae		Divided eyes
		Echo location
		Paddle-shaped legs
Hemiptera	Epipleuston	Hydrofuge hairpiles
Gerridae		Retractable preapical claws
Veliidae		Elongate legs and body
Hydrometridae		Echo location
Diptera	Hypopleuston	Hydrofuge hairs surrounding spiracles
Culicidae		Respiratory siphons

tions, all of which increase the area in contact with the surface film, include lamellate or paddle-shaped mucrones, long setae on the furcula, and modifications of the thoracic claws (Waltz and McCafferty 1979).

Adult whirligig beetles (Gyrinidae) possess a hydrophobic dorsum and a wettable ventral surface, and thus rest half in and half out of the water. Each eye is clearly divided into upper and lower halves, permitting vision simultaneously in both the aerial and aquatic media. Each time the jerky swimming movements of gyrinids exceed 23 cm/sec, capillary waves form on the water surface, which, when reflected back to the insect by a solid object, serve as a means of echo location (Tucker 1969). The middle and hind legs of adult gyrinids are paddle-shaped. *Gyrinus*, which can move the hind legs at 50-60 strokes per second, attain speeds exceeding 1 m/sec for short spurts in the surface film. The hind legs of gyrinids function with a high hydromechanical efficiency (Fig. 3.2) and may well be the most effective swimming organ in the animal kingdom (Nachtigall 1974).

The three families of hemipterans adapted for a pleustonic existence (Table 3.1) are collectively called *water striders*. The extremely elongated legs and body of hydrometrids serve to distribute the insect's weight over a large area of the surface film. Water striders are supported by the full length of the tarsi which are covered with hydrofuge hairpiles. The claws, which are wettable, are generally subapical and retractable. On initiating a rowing stroke, the middle tarsi of *Gerris* are abruptly pressed downward and backward, creating a circular

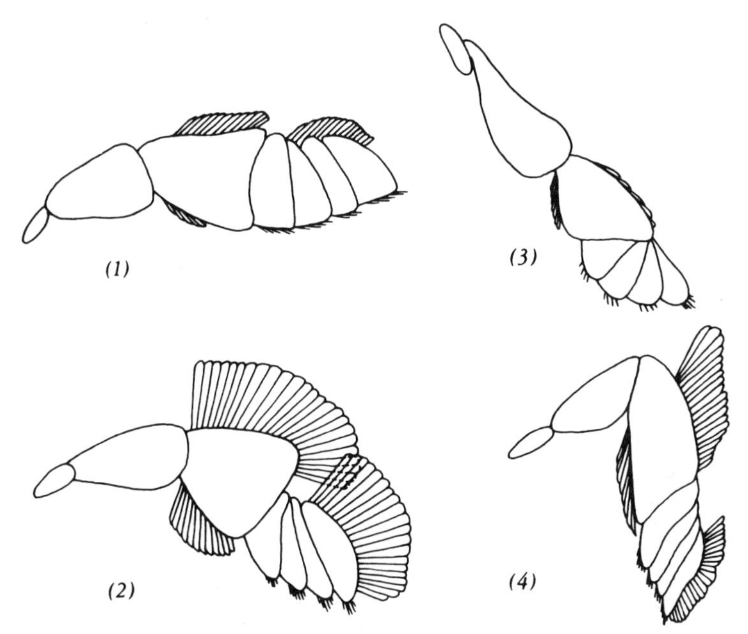

Figure 3.2 The rowing stroke of the hind leg of the beetle, *Gyrinus*. The oarlets fold back during the recovery stroke (3) and (4). [From Life in Moving Fluids: The Physical Biology of Flow by Steven Vogel, © (1981) by Willard Grant Press. Reprinted by permission of Wadsworth, Inc.]

surface wave, the crest of which is used as an abutment to impart forward thrust, much as a runner uses a starting block (Andersen 1976). At least some water striders secrete a substance that lowers the surface tension and propels the bug at high speed for a short distance (Linsenmair and Jander 1963). Sensillae on the femora and trochanters of gerrids detect surface film vibrations caused by potential prey (Murphey 1971). Surface wave signals are also involved in mating behavior (Wilcox 1972).

Among dipterans, only the mosquitoes (Culicidae) are more or less permanent members of the pleuston in lentic waters. The larvae and pupae of most species are part of the hypopleuston, using the underside of the surface film for support. Larval *Anopheles* typically lie horizontally with the spiracles of the eighth abdominal segment in contact with the surface film. Larvae of other genera (except *Mansonia,* which taps air stores of aquatic plants) hang upside down with the elongated terminal respiratory siphon penetrating the surface film. Pupae (except *Mansonia)* hang from the surface film with the cephalothoracic breathing trumpets in contact with the atmosphere. Culicids apparently do not exhibit any particular adaptations for life at the air–water interface that are not also possessed by other insects that come to the water surface to replenish air stores. Their inclusion in the pleuston community is based primarily on the nearly continuous contact with the surface film by the immatures of most species.

There are a few other isolated examples of aquatic insects that are intimately associated with the air-water interface of lentic water bodies. Some species of hydrophilid beetles are capable of "walking" in an inverted position on the underside of the surface film; *Spercheus* and *Amphiops* display special modifications for a hypopleustonic mode of life (Crowson 1981). Lakes Baikal (Siberia), Tanganyika (Africa), and Titicaca (South America) all have flightless species of caddisflies; the legs of the adults are adapted for swimming in the surface film of the open water (Marlier 1962).

Plankton

Although categorized separately, the various planktonic organisms are more appropriately considered with nekton along a continuum of mobility (Hutchinson 1967). *Plankton* and *nekton*, although at opposite ends of the continuum, both refer to organisms that reside in the free water as opposed to those associated with air–water or solid–water interfacies. Passively suspended nonmotile unicellular algae represent one end of this series. At the other extreme are the large fishes that are independent of water movements and are truly nektonic. The remaining free-water organisms of lakes are positioned between the two extremes and are really nektoplanktonic. As commonly used in the limnological literature, however, the term *plankton* is interpreted broadly to include all free-water forms except vertebrates, large crustaceans such as mysid shrimp, and large insects.

Insects are extremely poorly represented as permanent members of the plankton community. The phantom midges (Chaoboridae) are normally regarded as the only planktonic insects. Because they are relatively large and motile, chaoborids could be categorized as nektoplanktonic; because in many lakes they occur in surface waters during the night, but reside in the bottom mud during the day, they could also be classified as meroplanktonic.

Immature chaoborids occur in lakes and other lentic water bodies worldwide (Saether 1972). The nearly transparent larvae have prehensile antennae and prey on zooplankton and mosquitoes. In larval *Chaoborus* the tracheal system is essentially reduced to kidney-shaped air sacs, one pair within the thorax and another in the abdomen (Fig. 3.3). The air sacs have no respiratory role, functioning solely as hydrostatic organs (Saether 1972). Although the composition of gases within the air sacs is in equilibrium with the dissolved gases of the surrounding water, larvae have considerable control over the volume of the hydrostatic organs and, therefore, their own density, buoyancy, and depth in the water column. Volume changes are induced by a poorly understood mechanism involving differential uptake of water by the cuticular walls of the air sacs (see Wigglesworth 1972). *Chaoborus* is thus highly specialized for a planktonic existence and is the only chaoborid commonly occurring in lakes although some species typically reside in smaller lentic habitats (Table 3.2). Other genera of chaoborids with less well-developed hydrostatic organs and more complete tracheal systems than *Chaoborus* are generally restricted to shallower water bodies (Borkent 1981).

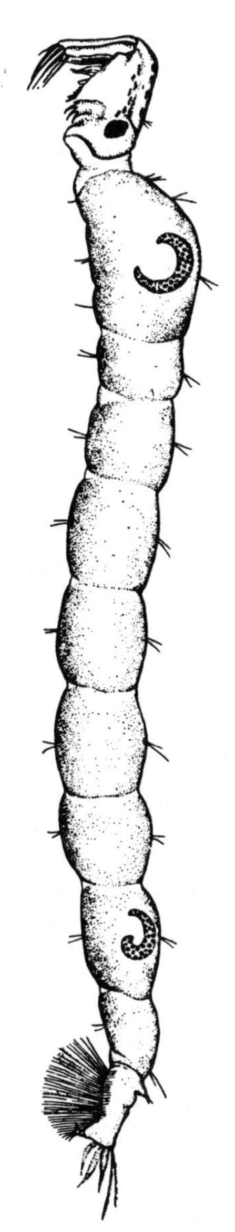

Figure 3.3 *Chaoborus* larva (x12) showing thoracic and abdominal air sacs. [From Johannsen (1934).]

Chaoborus larvae exhibit two general types of diel vertical migration referred to as "full" and "reduced" by Borkent (1981). Lake species typically exhibit full migrations, often involving a planktonic phase in surface strata during the night and a benthic phase by day (e.g., Juday 1921b, Berg 1937, Wood 1956, Woodmansee and Grantham 1961). Populations exhibiting full migrations may be

Table 3.2 Common habitats of larval *Chaoborus* in the Holarctic

	Ponds			Lakes	
		Permanent			
Species[a]	Temporary	Wooded	Exposed	Fish Absent	Fish Present
C. americanus (–)		X	X	X	
C. obscuripes (–)		X	X	X	
C. crystallinus (–)		X	X		
C. flavicans (+)				X	X
C. trivittatus (+)		X		X	X
C. cooki (–)	X				
C. nyblaei (?)	X				
C. albatus (+)				X	X
C. punctipennis (+)				X	X
C. astictopus (+)				X	X
C. pallidus (–)		X			

[a] Types of vertical migration exhibited by larvae are designated by + (full) or - (reduced) in parentheses following species names.

Source: Modified from Borkent (1981).

entirely planktonic in some lakes (Fig. 3.4), the larvae descending to the hypolimnion in the early morning hours, but remaining within the water column (e.g., Teraguchi and Northcote 1966, Fedorenko and Swift 1972). Pond species typically have reduced migrations and remain within the epilimnion when they occur in lakes. The limited migratory behavior of species such as *C. americanus* may indeed explain their absence from lakes containing fishes (Table 3.2), since they have no spatial refuge from fish predation during the day (von Ende 1979). *Chaoborus flavicans* typically coexists with fishes and is the only chaoborid in

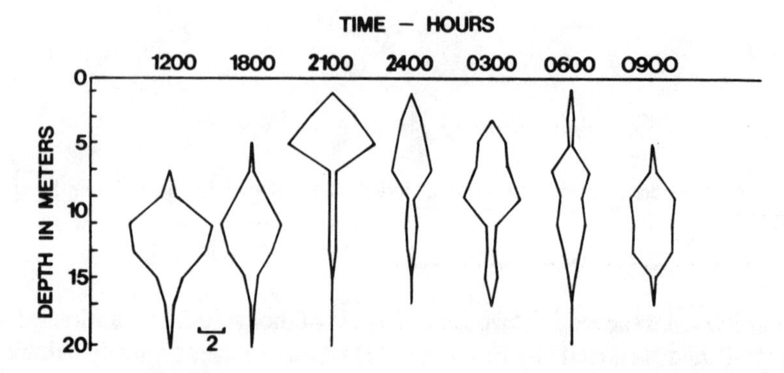

Figure 3.4 Diel vertical migration of fourth instar larvae of *Chaoborus trivittatus* on September 16–17. The scale indicates the number of larvae per 100 liters. [Modified from Fedorenko and Swift (1972).]

the limnetic zone of deep European lakes (Saether 1972). However, *C. flavicans* is the dominant species in a few large and deep fishless lakes in Québec, where it exhibits reduced rather than full migratory behavior (Pope et al. 1973).

Different larval instars may exhibit different migratory behavior. For example, *C. trivittatus* first and second instars are restricted to the surface strata of a lake in British Columbia, whereas third and fourth instars move to deeper water during the day (Fedorenko and Swift 1972). Lewis (1975) found a strikingly regular depth distribution of *Chaoborus* size classes in a tropical lake. During the day, the smallest larvae are near the surface with larger specimens ranked by size below them. Mature larvae migrate upward as much as 80 m at night. Lewis postulates that the size-depth distribution of *Chaoborus* relates to food distribution, as larger food items are generally deeper during the day, acting in concert with the greater vulnerability of larger larvae to visual predation by planktivorous fishes.

Light apparently functions as the entraining agent, the Zeitgeber, in diel migratory behavior of *Chaoborus* larvae. The benthic and planktonic phases may be artificially reversed in the laboratory by providing light during the night and darkness by day (Berg 1937). LaRow (1971) found that third- and fourth-instar larvae were negatively phototactic to all wavelengths of light tested (15), whereas early instars exhibited a slightly positive phototactic response to the blue end of the visible spectrum.

Other variables have been implicated as major regulatory factors determining whether larvae receiving the appropriate light stimulus will exhibit significant migratory behavior. Under high-oxygen conditions (81% saturation) only 1% of larval *C. punctipennis* moved upward, whereas 30% of the larvae exposed to low oxygen (3% saturation) migrated to the surface (LaRow 1970). Temperature and food abundance may modify migratory behavior, and have been used in energetic models to explain seasonal patterns of vertical migration (Giguère and Dill 1980). In laboratory experiments with *C. crystallinus*, Ratte (1979) reported accelerated development of larvae when high or increasing temperatures corresponded to the dark phase of the photoperiod, the normal situation for a vertically migrating population.

Some investigators have reported additional migrations, such as movements of *Chaoborus* larvae to the littoral zone during spring (Wood 1956, Roth 1968). As suggested by Borutsky (1939), such movements may be an adaptation enabling mature larvae to be exposed to warmer water prior to pupation and emergence. Using a sonar system, Franke (1987) also documented a horizontal (adlittoral) component in the diel migration of *C. flavicans*.

Many insects other than chaoborids, however, occasionally have been collected from the open water of lakes. Anyone who has extensively sampled lake plankton has noted the occasional "benthic" organism in the samples. Some collections undoubtedly represent accidentals washed in by rivers or specimens swept from the shore zone during storms. Chironomid pupae, of course, rise to the surface prior to emergence and may be collected in plankton samples. In other cases, however, the occurrences in the free water of species normally considered to be wholly benthic can hardly be considered as accidental. Many of these merobenthic species are poor swimmers, at least partly at the mercy of water movements, and are thus temporary members of the plankton community.

There is a large body of evidence [reviewed by Davies (1976)] that first instar chironomid larvae serve as a passive dispersal stage. First instars of many species are adapted for a planktonic existence by their positive phototaxis and ability to feed in the open water. Planktonic larvae have been reported for a variety of species, the later larval stages of which occur on various substrate types and collectively exhibit a wide range of feeding habits. Davies (1974) recorded 11 species of chironomids from four subfamilies that had planktonic first instars in Loch Leven, Scotland. Several other investigators also stress the morphological and behavioral separation of first instar larvae associated with their planktonic existence (Dorier 1933, Alekseyev 1955, Morduchai-Boltovskoy and Shilova 1955, Kalugina 1959). Dorier suggested that morphological differences may be sufficient to separate immatures into two larval types: the "larvule," first instars residing in the plankton; and "larva," sedentary second, third, and fourth instars. The positive phototaxis of first instars is normally retained for 1–3 days after which larvae become sedentary. Young chironomid larvae exhibit vertical migrations related to diel changes in light intensity. In Rybinska Reservoir, Russia, the number of chironomid larvae in the water column closely tracked the curve of underwater illumination over a diel cycle (Luferov 1966). Abundant species attained densities exceeding 100 first instar larvae per cubic meter of surface water during periods of maximum illumination. Low values in surface waters during the night do not necessarily suggest that first instar larvae settle to the bottom during periods of darkness. Rather, it may be that they are more evenly distributed throughout the water column at night and are concentrated near the surface by day. Diel vertical migrations apparently represent a response to light rather than endogenous rhythms, since midday cloud cover causes a temporary drop in the number of larvae in surface strata.

Although as chironomid larvae mature there appears to be a shift from a highly photophilous phase residing in the plankton to increasingly photophobic benthic inhabitants, late instars may again enter the plankton temporarily, especially at night. In Lac la Ronge, Saskatchewan, late-instar chironomid larvae were collected at night in surface tows taken over various depths of water (Mundie 1959). Vertical migrations in the water column enable larvae to move into well-aerated waters during periods when oxygen is depleted from bottom strata. Whereas the small size and transparency of first instar larvae minimize their vulnerability to visual predators, later instars would be highly visible in surface waters during the day. Inexplicably, however, late instars of some species exhibit the same diel pattern of vertical migration as first instars (Luferov 1966).

Aquatic insects other than chaoborids and chironomids have been less commonly collected from the plankton of lakes. *Ranatra montezuma*, a water scorpion endemic to Montezuma Well, a small spring-fed lake in Arizona, displays nocturnal planktonic behavior (Blinn et al. 1982). As light intensity drops below 100 lux, *R. montezuma* move from plant beds in the littoral zone to the limnetic zone of this fishless lake, where they feed on an endemic planktonic amphipod that concentrates near the water surface during the night. At sunrise the water scorpions return to the marginal plant beds. In a hypersaline crater lake in Turkey, larvae of the beetle *Potamonectes cerisyi* take on a planktonic mode of life in later instars (Dumont 1981). The increased buoyancy of the medium (due to high salinity) enables larvae to exploit the abundant zooplankton in the limnetic zone without expending extra energy for swimming. In another fishless hypersaline lake (Lake Werowrap, Australia), Walker (1973) commonly collected beetle larvae (*Necterosoma penicillatum*) and larvae and pupae of ephydrid flies (*Ephydrella* sp.) in plankton tows. Immature ceratopogonids, mayfly nymphs (*Caenis latipennis, Hexagenia occulata*), and trichopteran larvae (*Triaenodes frontalis*) were collected in surface tows taken during the night in Lac la Ronge (Mundie 1959). Large numbers of the burrowing mayfly *Hexagenia* migrate from shore areas into the open water at night in reservoirs on the Missouri River in central North America (Cowell and Hudson 1967).

Nekton

The word *nekton*, first used by Haeckel (1890), is derived from the Greek word for swimming. Nektonic forms are distinguished from plankton by their directional mobility, and from benthos by their association with the open water. Although many aquatic insects swim, often quite rapidly for short bursts, most are closely tied to the substrate and can hardly be considered nektonic. The term *nektobenthic* is a convenient appellation for such intermediate forms. Only those species especially adapted for sustained swimming in the open water, and that are relatively independent of the substrate for much of their lives, are nektonic in the pure sense. On the basis of this definition, with the exception of isolated cases in other orders, truly nektonic insects are restricted to a relatively few species of hemipterans and coleopterans.

Among the hemipterans, nektonic species are contained within the Notonectidae, Corixidae, and perhaps Belostomatidae, all of which are strong swimmers. A nektonic existence is best developed in two genera of notonectids, *Anisops* and *Buenoa.* These backswimmers use hemoglobin for buoyancy control as already discussed in the last chapter. The attainment of neutral buoyancy has enabled these insects to exploit the limnetic zone of fishless lentic waters where they prey on small arthropods. Other members of this family, such as *Notonecta*, lack hemoglobin and are positively buoyant. They rise to the water surface except when continuously swimming or clinging to underwater vegetation. Gittelman (1974) demonstrated an ecological relationship between leg structure and habitats of three notonectid genera. *Buenoa*, which remains stationary or slowly swims below the water surface and captures free-moving prey, has hind legs designed for rapid pursuit and forelegs and midlegs well developed for prey capture. *Martarega*, which maintains position in the eddies of rivers and eats immobile prey carried to it in the surface film, has long hind legs designed for slow, sustained swimming. The other legs are poorly developed and are easily folded against the body to reduce resistance during swimming. *Notonecta*, which maintains a stationary feeding position on the underside of the surface film and preys on moving prey as well as those caught in the surface film, has legs intermediate in structure between those of *Buenoa* and *Martarega*.

The corixids or water boatmen swim with their elongate, flattened, and hair-fringed hind legs. Because of elaborately developed air stores that function efficiently as a physical gill, corixids are able to remain submerged for long periods. Most species, however, appear restricted to water <1 m deep. A striking exception is the occurrence of *Sigara lineata* at depths >10 m in Lake Erie (Hungerford 1948). Even though corixids are hardly independent of the substrate and are unable to attain neutral buoyancy, their well-developed swimming ability and respiratory mechanisms, coupled in some cases with planktivorous feeding habits, qualifies at least certain species for inclusion in the nektonic community. Many others reside only in the shallow parts of lakes and should be regarded as nektobenthic.

Despite their large size and well-developed swimming ability, it is questionable whether any of the giant water bugs, Belostomatidae, belong to the nekton community. Remarkably little is known of their precise habitat requirements, although Menke (1979) states that most are sedentary hunters, perching on aquatic vegetation while awaiting prey rather than actively pursuing their food. Cullen (1969) found *Belostoma* restricted to vegetation in the shallow margins of tropical lakes, whereas *Lethocerus* occurred in deeper water.

Although most aquatic beetles are intimately associated with the substrate, adults of a few of the larger Dytiscidae (Galewski 1971) and Hydrophilidae are strong swimmers and may be considered part of the nekton. The functional morphology associated with swimming has been examined in detail for a few species of adult beetles and is summarized in the following section.

Swimming Aquatic insects exhibit a variety of mechanisms for moving through the aqueous medium in which they live and many of these involve some type of swimming (Table 3.3). Although swimming is by no means confined to nektonic forms, it is convenient to include the topic as a unit in this section on nekton.

To appreciate the adaptations involved in swimming, it is helpful to briefly consider some simplified hydrodynamic properties [see Nachtigall (1974, 1985) and Vogel (1981) for comprehensive accounts]. The resisting force W of an object moving through a fluid can be represented by the equation

$$W = c_w F \left(\tfrac{1}{2}\right) p \, v^2$$

where c_w is the coefficient of resistance, F is the frontal area of the object exposed to the current, p is the density of the medium, and v is velocity.

For a general understanding of the efficiency of rowing swimmers moving

Table 3.3 Types of underwater swimming in aquatic insects

Swimming Type	Examples
Rowing swimmers	Coleoptera (Dytiscidae, Hydrophilidae, Gyrinidae); Hemiptera (Corixidae, Notonectidae, Belostomatidae)
Curling swimmers	Diptera (Culicidae, Chironomidae, Ceratopogonidae, Dixidae)
Undulating swimmers	Ephemeroptera (Baetidae, Siphlonuridae); Odonata (Zygoptera); Coleoptera (larvae of some species)
Winging swimmers	Hymenoptera (Mymaridae, Trichogrammatidae, Scelionidae, Braconidae)
Jet swimmers	Odonata (Anisoptera)

Source: After Nachtigall (1974).

through water of a given density, it is therefore essential to consider only the frontal areas and resistance coefficients of the body and the propulsion organs (swimming legs). The coefficient of resistance (c_w) is primarily determined by the shape of an object. It is a dimensionless number directly proportional to the degree of resistance to current with approximate mean values as follows (based on Reynold's numbers attained by large aquatic insects during rapid swimming):

	c_w
Minimum resistance (drop-shaped body)	0.07
Acilius sulcatus (Dytiscidae) adult	0.23
Dytiscus marginalis (Dytiscidae) adult	0.31
Hydrophilid adults	0.35
Sphere	0.38
Disc (maximum area perpendicular to current)	1.11
Maximum resistance (parachute-shaped body)	1.35

Swimming by an aquatic insect will be most efficient if (1) the frontal area of the body exposed to the current is small, (2) the swimming legs have a large frontal area during the power stroke and a small frontal area during the recovery stroke, (3) the body shape provides a low c_w, (4) the swimming legs have a high c_w during the power stroke and a low c_w during the recovery stroke, and (5) the main propulsive areas (e.g., hair fans) of the swimming legs are moved distally during the power stroke and proximally during the recovery stroke (Nachtigall 1974). The strongest swimmers among aquatic insects have evolved remarkable adaptations to attain maximum hydrodynamic efficiency, while maintaining stability and maneuverability.

The large dytiscid beetle *Dytiscus marginalis* has been extensively studied from a hydrodynamic perspective (Hughes 1958; Nachtigall and Bilo 1965, 1975). Adults are dorsoventrally flattened and streamlined and are able to attain speeds >100 cm/sec for short bursts. As swimming speed is increased (up to a critical velocity), c_w significantly decreases. This is accomplished by adjustments in body attitude in relation to the direction of movement, and by folding the nonswimming forelegs into depressions in the body, thus increasing the smoothness of the body contour (c_w is less with the forelegs folded back than with them removed). If the lateral body edges (prothoracic and elytral seams) of *D. marginalis* are experimentally rounded off, the coefficient of resistance decreases considerably. The normal edges create eddies that increase c_w. However, the lateral edges improve swimming stability, greatly increase braking efficiency, and enable the beetles to turn rapidly. It is for such reasons that c_w values in nature are higher than those of the least resistant ideal shapes. Deviations in either the horizontal or vertical planes of a rapidly swimming *Dytiscus* are self-stabilizing as a result of the lateral edges that consume the kinetic energy of the oscillation. The body of large dytiscids provides a good comprise: c_w is relatively low, maneuverability is comparatively good, and stability is very good. The whirligig beetle *Gyrinus* which rapidly darts along the surface, in contrast, has high maneuverability, but lower stability and is thus required to expend energy in constantly correcting the body attitude.

The rowing legs of the best swimmers (Fig. 3.5) exhibit several adaptations that greatly increase swimming efficiency. The legs are flattened and are constructed so that forces generated during swimming automatically swivel the broad surfaces perpendicular to the direction of movement during the power stroke and present only the narrow edges to the current during recovery. Thrust is enhanced by elaborate hair fans that spread during the power stroke, but fold back during the recovery stroke. For example, the rowing legs of the water boatman *Corixa punctata* each bear approximately 5000 swimming hairs so arranged that they lie flat along the leg during the recovery stroke (Schenke 1965). During the power stroke a row of long stiff hairs supports finer hairs and

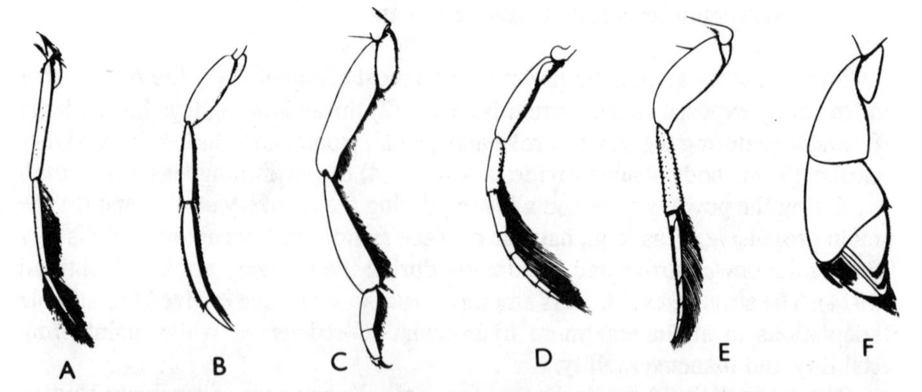

Figure 3.5 Hind legs modified for movement on or in water. Hemiptera: (*A*) Notonectidae; (*B*) Corixidae; (*C*) Belostomatidae. Coleoptera: (*D*) Dytiscidae; (*E*) Hydrophilidae; (*F*) Gyrinidae. [Norris (1970) from CSIRO (ed.)(1970). The Insects of Australia (Carlton: Melbourne University Press).]

are themselves held erect by a row of rotatable spines. In addition to hairs, the whirligig beetle *Gyrinus* has swimming blades on the middle and hind legs (Hatch 1925, Bott 1928). The blades fold back during the recovery stroke, automatically rotate into the same plane during the power stroke, and collectively provide a solid surface perpendicular to the current. The tarsus of *Gyrinus* collapses during recovery and is drawn into a concavity in the tibia; the tibia moves into a concavity in the femur (Nachtigall 1974). The leg of *Gyrinus* therefore not only presents the narrow edge to the current during recovery but also reduces the frontal area and moves the effective area to a more proximal position.

These examples of propulsion mechanisms are all categorized as rowing–swimming, a type that is best developed in certain adult coleopterans and hemipterans. [See Blake (1986) for the biomechanics of rowing.] Less efficient rowing–swimming is also present in the nymphs of water bugs, a few larval beetles, and a very few caddisfly larvae, among others.

The propulsion systems of other aquatic insects are based on somewhat different principles (Table 3.3). The immatures of several families of dipterans are curling swimmers, moving through the water by coiling and uncoiling their bodies. In larval *Chironomus* curling did not develop as a primary means of locomotion, but rather as movements to enhance respiration.

Undulating swimming is a very effective means of locomotion over short distances in some nymphal ephemeropterans (Craig 1990). Vertical undulations of the abdomen provide the major propulsion to move the cerci, which in swimming mayflies have well-developed hair fringes. Locomotion is sometimes augmented by movements of the gill lamellae. Damselfly (Zygoptera) larvae undulate their bodies laterally when swimming, as do a few other aquatic insects, whereas some larval beetles swim in a leech-like fashion by vertical undulation of the body.

The use of wings for swimming is rare and apparently restricted to a few species of aquatic wasps.

Dragonfly (Anisoptera) nymphs possess rectal respiration with a secondary locomotor function (Mill and Pickard 1975). Water is pumped in and out of the rectum, which has become enlarged and is lined with tracheal gills. By rapidly contracting the abdomen, a jet of water is forced through the anus, enabling large nymphs to attain speeds up to 50 cm/sec for short bursts.

Benthos

The term *benthos*, derived from the Greek word for bottom, refers to the fauna (zoobenthos) associated with the solid–water interface. The majority of aquatic insects found in lakes are denizens of the benthos, and along with crustaceans, molluscs, and oligochaetes, comprise the major portion of the benthic community. Hutchinson (1967) distinguishes between species that move through or over the bottom sediment, the herpobenthos, and species associated with solid surfaces, the haptobenthos. While few lentic species are permanently attached to a substrate, many aquatic insects of lakes are haptobenthic in the sense of being intimately associated with plants, logs, rocks, and other solid substrate surfaces, in contrast to the herpobenthos which inhabit soft bottom sediments.

Several orders of insects reach their greatest abundance and diversity (among the aquatic members) in lentic habitats. This is true for the Odonata (dragonflies), Hemiptera (bugs), and Coleoptera (beetles). Some families of Diptera (true flies), an order extremely important in both lentic and lotic environments, are primarily or exclusively found in standing waters. Aquatic Lepidoptera (moths) are best developed in lentic habitats, as are the few aquatic Hymenoptera (wasps), and Neuroptera (spongillaflies). Even among some primarily rheophilic orders, such as Ephemeroptera (mayflies) and Trichoptera (caddisflies), certain groups are extremely important members of benthic communities in standing waters. Insects collectively constitute <10% to >90% of the total benthic fauna. Their composition and relative abundance is dependent on a variety of factors, some of which are integrated along depth profiles.

Depth Distribution Hutchinson (MS) considers the insect fauna of lakes to fall roughly into three depth categories, although not without exceptions. In the first category he includes the hemipterans and coleopterans, the aquatic adults of which have never developed morphological gills. These insects rarely occur in water deeper than a few meters, and most species must periodically surface to replenish air stores. The second group includes all other orders except Diptera. Members of the second category are amphibiotic, with immatures that typically extract dissolved oxygen from the water, often by means of tracheal gills, but are nevertheless restricted to relatively shallow water. With some rare exceptions, only certain dipterans among aquatic insects have successfully colonized the profundal zone of lakes and thus belong to the third category. Aquatic dipterans are also amphibiotic, so it is unlikely that depth poses any insurmountable problems relating to the distance preimagines must traverse to reach the surface prior to the final ecdysis to the aerial adult. It is rather more likely that only in certain dipterans have larvae evolved adaptations necessary to exploit the rather special and sometimes adverse conditions frequently encountered in the profundal zone. Indeed chironomids and chaoborids are virtually the only dipterans to occur at great depths. Chaoboridae is a small family and only a few species in the genus *Chaoborus* occur in the profundal benthos; even they exhibit diel vertical migrations. While the Chironomidae constitute the largest family of freshwater insects, most species are not found in the depths of lakes.

A variety of factors vary as a function of depth in lakes and may influence the depth distribution of aquatic insects. Light is rapidly attenuated with depth and photosynthetic plants are thus restricted to surface strata as already discussed. Temperature exhibits marked variations with depth, and thermal regimes are normally quite different in the littoral and profundal zones of a given lake. In many lakes dissolved oxygen varies significantly with depth; during summer stratification it is not uncommon to have anoxic conditions in the hypolimnion while epilimnetic waters are saturated with oxygen. The organic and inorganic substrate available to insects also exhibits depth profiles, partly related to the restriction of wave action to surface strata. In addition, biotic factors such as predator–prey and competitive interactions vary as a function of depth. These variables and others, alone and in concert, are responsible for the depth distribution patterns of aquatic insects.

The taxonomic richness of benthic insect communities exhibits a general decline with increasing depth (Fig. 3.6). On exposed shores the maximum number of taxa may occur at depths of 1–2 m, below the major scouring action of the waves. Except for a few dipterans, most species of aquatic insects are restricted to shallow water, markedly so in some cases (Fig. 3.7). Smith et al. (1981) found distinct depth zonation of some species within the upper 500 cm of the littoral zone.

Ecological conditions in the littoral zone vary greatly from lake to lake and at different locations within the same lake (Dall et al. 1990). Much of the significant variation is reflected by the substrate type, often determined principally by wave action. At one extreme is the exposed rocky shoreline with coarse

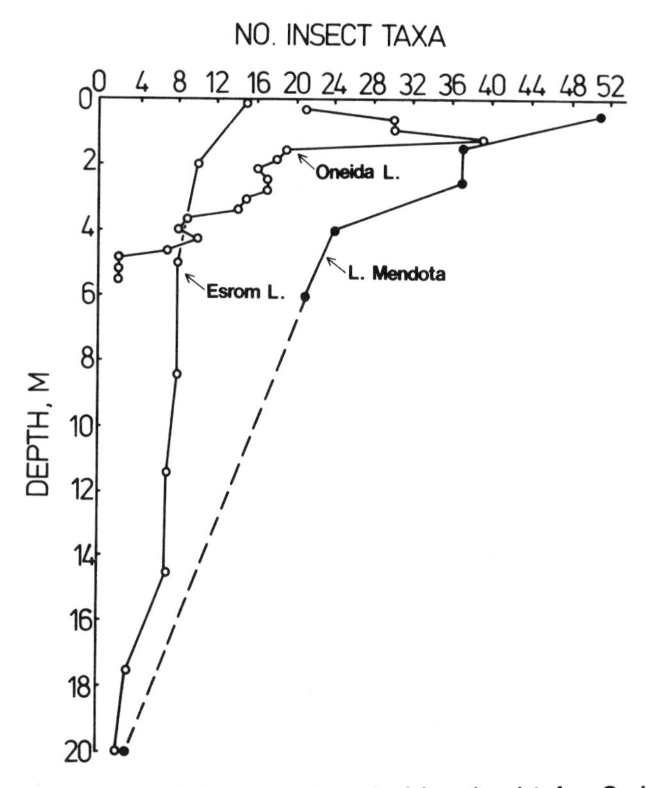

Figure 3.6 Number of benthic insect taxa versus depth based on data from Oneida Lake, New York (Baker 1918), Esrom Lake, Denmark (Berg 1938), and Lake Mendota, Wisconsin (Muttkowski 1918, Juday 1921a).

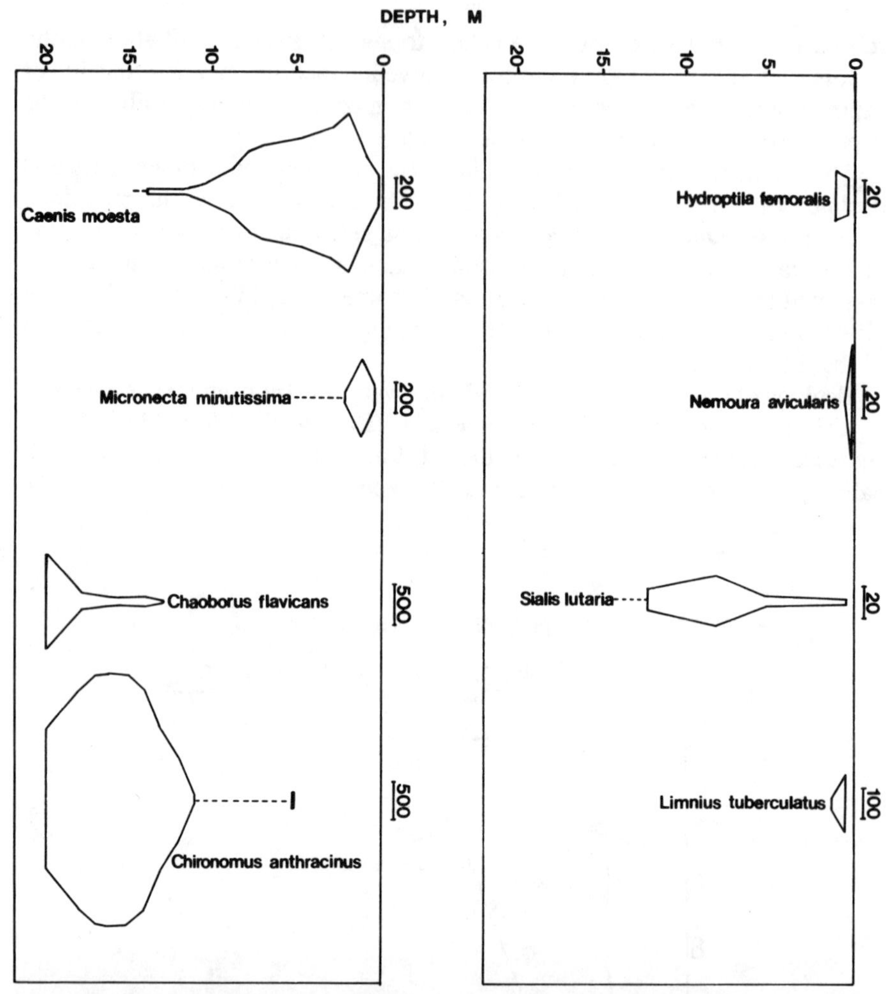

Figure 3.7 Depth distribution of selected benthic insects in Lake Esrom, Denmark. The scales indicate number per square millimeter. [Modified from Berg (1938).]

substrate, relatively silt-free interstices, continual water movement, high dissolved oxygen levels, and plant growth generally limited to attached algae and bryophytes. It has long been recognized that some of the insects and other invertebrates inhabiting wave-swept rocky shores typically reside in rocky streams (Zschokke 1900, Steinmann 1907, Wesenberg-Lund 1908). Ehrenberg (1957) reported that half the species he collected from rocky lake shores in Germany were characteristic of lotic biotopes. The rocky shores of large Scottish lochs are also largely comprised of rheophilic forms (Smith et al. 1981). Stony, wave-swept lake shores are virtually the only lentic habitats occupied by the most rheophilic order of insects, the stoneflies, except at high latitudes or altitudes.

At the other extreme is the sheltered littoral zone exemplified in Figure 3.1, the quiet bays and backwaters well protected from wave action. Such areas are often choked with emergent, floating, and submerged hydrophytes. The well-vegetated littoral zone provides a high habitat diversity for both herpobenthic and haptobenthic species, typically resulting in a rich faunal diversity. Many of these species also occur in ponds. Odonates and aquatic representatives of the hemipterans, coleopterans, and lepidopterans attain their maximum diversity and abundance under such conditions. In addition, some species of trichopterans, ephemeropterans, and dipterans typically inhabit the pond-like conditions of the well-vegetated littoral zone.

The sublittoral zone is the transitional area between the littoral and profundal zones. Under certain conditions the sublittoral may provide ideal conditions for benthic species, such as the mollusc *Dreissena* in Esrom Lake (Berg 1938, Jónasson 1978). In most cases, however, the benthos of the sublittoral is comprised of the lower limits of littoral species, and the upward extensions of species adapted to the special conditions of the profundal zone.

Only a few species of dipterans regularly occur in the profundal zone, especially in eutrophic lakes, where bottom waters become anoxic. Profundal insects are often limited to a few species of chironomids. Chaoborids, if present, are normally represented by a single species of *Chaoborus*. Despite the limited diversity, insects may predominate in the profundal benthos. For example, *Chaoborus flavicans* constituted 96% of the total profundal benthos in a Dutch Lake (Parma 1971); chironomid larvae contributed 76% of the total benthos at 40–45 m in a Canadian lake (Rawson 1930); and insects (*Chironomus anthracinus, Procladius pectinatus, Chaoborus flavicans*) represented 56% of the total benthos by numbers (Jónasson 1978) and 92% by weight (Berg 1938) in the profundal zone of Lake Esrom, Denmark. The relative contribution of chironomids to total profundal benthos tends to decrease, while oligochaetes concomitantly increase, with increasing depth (Thut 1969, Edmonds and Ward 1979).

The benthos of large, well-circulated lakes and less productive lakes without severe oxygen depletion in bottom waters tend to exhibit somewhat different distribution patterns. Insects restricted to rather shallow depths in small productive lakes may penetrate deeper in large lakes where currents carry oxygenated waters to greater depths. Adamstone (1924) found mayflies down to 12 m and caddisflies to 15 m (with a few hydrophilid larvae in dredge samples from 71 and 98 m) in a large Canadian lake. Immature mayflies, stoneflies, and caddisflies were all collected at depths >32 m in Lake Superior (Selgeby 1974). The caddisfly *Phryganea cinerea* occurred down to 100 m. The stonefly *Capnia lacustra* occurs at depths of 80 m in Lake Tahoe, California–Nevada. In Lake Baikal larvae of the endemic chironomid *Sergentia koschowi* inhabit the abyssal zone at depths of >1300 m (Linevich 1971).

Insects tend to be less predominant, although often more diverse, in the profundal benthic communities of lakes without severe oxygen depletion in bottom waters. For example, although at least 14 species of chironomids occurred in the profundal zone of a reservoir where bottom waters exhibited oxygen saturation values >70% during most of the year, they collectively constituted only 2% of the total benthos (Edmonds and Ward 1979). Slack (1965) identified 33 chironomid taxa in the profundal zone of Loch Lomond, a stark contrast to small eutrophic lakes where only two or three species occur. A striking feature of the Laurentian Great Lakes is the paucity of insects in the profundal benthos (Cook and Johnson 1974). Eggleton (1937), for example, found that three noninsect groups (tubificid oligochaetes, sphaeriid clams, and the amphipod *Pontoporeia hoyi*) represented 94% of the total benthos from 24 to 246 m in Lake Michigan. Chironomids were the only insects encountered and apparently represented only a small fraction of the remaining 6%, which was predominantly nematodes.

The benthic communities of several lakes have been subjected to especially thorough and comprehensive examination. Some noteworthy examples of early work include Oneida Lake, New York (Baker 1918), Lake Mendota, Wisconsin (Muttkowski 1918, Juday 1921a), Lake Simcoe, Ontario (Rawson 1930), Lake Esrom, Denmark (Berg 1937, 1938; see Jónasson 1977), and Lake Borrevann, Norway (Ökland 1964). Brinkhurst (1974) provides a lengthy bibliography in his book, *The Benthos of Lakes*. Details from many of these studies will be referred to in later chapters, as topics such as the distribution of insects on different substrates are considered.

Lake Typology Lakes have been classified according to a variety of criteria, including geological origins, bottom types, morphometric features, thermal regimes, chemical conditions, trophic status, circulation patterns, biological components, or various combinations thereof. All the major criteria used in lake classification schemes are meaningful, to at least a limited extent, from the perspective of aquatic insect ecology.

Insects have been used extensively as integrators of environmental conditions in lakes. Most typologies based on insects have attempted to relate the composition of the profundal benthos with the trophic status (e.g., oligotrophic, eutrophic) of lakes (see Brinkhurst 1974), but littoral insects have also been considered (Kennedy 1922, Macan 1955, Savage 1982, Hershey 1985). Subfossil remains (Frey 1964, Walker and Mathewes 1989) as well as contemporary faunas are useful in this regard. Brinkhurst (1974) provides a comprehensive treatment of lake classification based on profundal benthos, and much of the following summary is drawn from his review and synthesis of this topic.

Thienemann (1918) developed the dichotomy of *Tanytarsus* lakes versus *Chironomus* lakes, later referred to as *oligotrophic* and *eutrophic* lake types, based on the predominant chironomid genus in the profundal zone. Oligotrophic (little nourished) lakes tend to be deeper, colder, have fewer nutrients and lower productivity than do eutrophic (well-nourished) lakes. As a result, oligotrophic lakes have abundant oxygen in bottom waters while eutrophic lakes develop oxygen deficits in lower strata during periods of stratification. *Chironomus*, being able to tolerate the periods of oxygen depletion, became the dominant chironomid in eutrophic lakes; *Tanytarsus,* being unable to tolerate low oxygen, predominated in the profundal zone of oligotrophic lakes. Nothing in nature can be quite so simple, however, nor did Thienemann suggest that it was. Considering only temperate lakes, at least one additional major category was needed, the dystrophic (poorly nourished) lake type, the waters of which are heavily stained with humic acids derived from peat. A bog lake represents an extreme and special form of dystrophy. Such lakes are low in nutrients and autochthonous production (production within the lake) is low. Dystrophic lakes have large inputs of organic materials produced outside the lake (allochthonous production) and thus tend to exhibit oxygen depletions in the hypolimnion. The profundal benthos of these three lake types have been described as follows:

Oligotrophic Lakes	Eutrophic Lakes	Dystrophic Lakes
Diverse benthos	Restricted benthos	Very restricted benthos
Tanytarsus	*Chironomus*	*Chironomus* or absent
Chaoborus absent	*Chaoborus* usually present	*Chaoborus* present

Thienemann (1925) eventually developed a rather more elaborate scheme of lake typology, which is presented in abbreviated and modified form as follows:

I. *Tanytarsus* lakes: chironomids of the genus *Tanytarsus* characterize the profundal benthos. *Chaoborus* absent.

II. *Chironomus* lakes: chironomids of the genus *Chironomus* characterize the profundal benthos.
   - A. *Chaoborus* absent
   - B. *Chaoborus* present
     1. *Anthracinus* (= *Bathophilus*) lakes: *Chironomus anthracinus* in the profundal zone
     2. *Plumosus* lakes: *Chironomus plumosus* in the profundal zone.

III. *Chaoborus* lakes without *Chironomus*: chironomid larvae absent from profundal zone.

The subcategories of the above typology are arranged to correspond to a progressive decline of oxygen in deep water. *Tanytarsus* lakes are rich in oxygen year round. *Chironomus* lakes without *Chaoborus* and *Anthracinus* lakes exhibit oxygen depletions in deep water during summer stratification, but have adequate oxygen during winter. The remaining lakes in the series have low oxygen in summer and exhibit progressively declining oxygen levels.

Although *Chaoborus* may reach very high densities in eutrophic and dystrophic lakes, they may also occur in oligotrophic water bodies [see refs. cited in Saether (1972)]. Because of their questionable relationship with a lake's trophic status, and because they are merobenthic, and in some cases truly planktonic, *Chaoborus* has limited value in lake typology schemes based on profundal benthos, as recognized by Lundbeck (1926).

Dealing solely with chironomids, Lundbeck (1936) added categories to the oligotrophic–eutrophic series and arranged lakes according to increasing degrees of eutrophy as follows:

I. Oligotrophic
   A. *Orthocladius*
   B. *Tanytarsus*
II. Mesotrophic
   A. *Stictochironomus*
   B. *Sergentia*
III. Eutrophic
   A. *Anthracinus* (= *Bathophilus*)
   B. *Anthracinus–Plumosus*
   C. *Plumosus*

Lundbeck also added a second dimension (not shown) based on the humus content of the sediment to account for the influence of dystrophy (oligohumic, mesohumic, and polyhumic).

Brundin (1956) developed the following lake classification scheme based on chironomid indicator species:

I. Ultraoligotrophic lakes: Heterotrissocladius subpilosus
II. Oligotrophic lakes: *Tanytarsus lugens*
II/III. Mesotrophic lakes: *Stictochironomus rosenscholdi–Sergentia coracina*
III. Eutrophic lakes:
   A. Moderately eutrophic: *Chironomus anthracinus*
   B. Strongly eutrophic: *Chironomus plumosus*
IV. Dystrophic lakes: *Chironomus tenuistylus*

From studies of the profundal chironomids of North American lakes, Saether (1980) identified seven groups of species along a gradient from ultraoligotrophy to extreme eutrophy (Fig. 3.8).

There have been numerous elaborations, modifications, and criticisms of these typologies developed for the profundal benthos of lakes. Many other schemes for classifying lakes of other regions, including the tropics (Thienemann 1932), have been devised (see Brinkhurst 1974, Saether 1980). Although lake typologies have been subjected to severe criticisms, much of it justified, such schemes have considerably added to our understanding of lakes and their biota. Thienemann never intended lake types to be anything other than very general and idealized abstractions and did not expect a given lake to conform in all respects to any simplistic category.

Most of the lake typologies using insects as indicators of trophic conditions have relied solely or primarily on profundal benthic communities. There are, however, a few notable exceptions. Macan (1938, 1949b) related the habitat occurrence of all known British corixids to stages of lake succession, especially

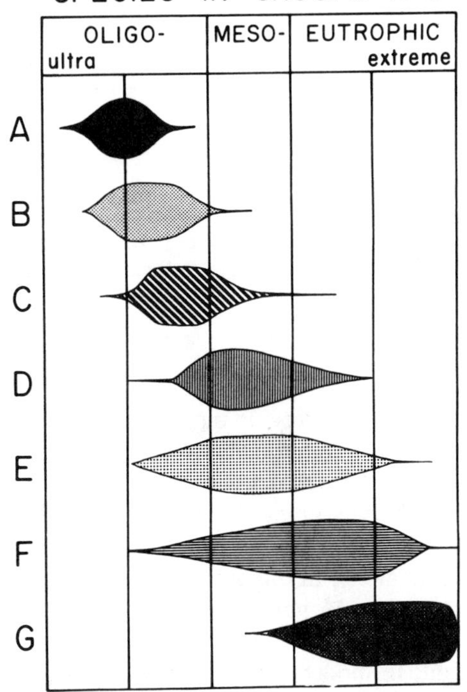

Figure 3.8 Generalized trophic distribution of chironomids in the following groups: (*A*) *Paracladius alpicola, Protanypus hamiltoni*; (*B*) *Paracladopelma galaptera, Heterotrissocladius* sp. D, *Monodiamesa tuberculata*, Gen. near *Trissocladius*; (*C*) *Stictochironomus rosenschoeldi, Tanytarsus*; (*D*) *Chironomus atritibia, Cladopelma* sp., *Heterotrissocladius latilaminus, Monodiamesa* cf. *prolilobata, Phaenopsectra coracina*; (*E*) *Phaenopsectra albescens, Chironomus plumosus* f. *semireductus*; (*F*) *Chironomus decorus, Chironomus plumosus plumosus*; (G) not designated. [From Saether (1980). Reprinted with permission of Pergammon Press plc.]

as indicated by the calcium content of the water, the organic content of the substrate, and the development of aquatic vegetation. For example, species such as *Micronecta poweri,* characteristic of early successional stages (oligotrophy), tend to occur on primarily mineral substrates with few or no higher plants, in lakes low in calcium. He showed that as the percentage of organic matter in the substrate increased, there is a corresponding succession of corixid species sequenced in a predictable series. Considering only the genus *Sigara,* Macan (1955) found species and species combinations that were indicative of a lake's trophic status (Table 3.4). Oligotrophic lakes are characterized by the co-occurrence in exposed littoral zones of *S. dorsalis* and *S. scotti,* both of which are abundant in such habitats. *S. scotti* is either rare (lake No. 8) or absent from

Table 3.4 Occurrence of indicator species of *Sigara* (Corixidae) in exposed littoral zones of lakes with different trophic status

	Oligotrophic					Lakes[a]					Eutrophic
	1	2	3	4	5	6	7	8	9	10	11
S. dorsalis or striata	+	+	+	+	+	+	+	+	+	+	+
S. scotti	+	+	+	+	+			+			
S. falleni							+	+	+	+	+
S. fossarum							+	+	+	+	+

[a] English Lake District: 1, Ennerdale; 2, Crummock; 3, Derwentwater; 4, Bassenthwaite; 5, Coniston; 6, Ullswater; 7, Windermere; 8, Esthwaite; 9, Blelham. Danish Lakes: 10, Esrom; 11, Fure.

Source: After Macan (1955).

lakes of moderate to high productivity. In mesotrophic Ullswater (lake No. 6) *S. dorsalis* occurs alone. In the more productive lakes, *S. dorsalis* (or *S. striata* on the continent), *S. falleni,* and *S. fossarum* co-occur.

In Finland three congeneric species of corixid bugs of the genus *Micronecta* are common in the littoral zone of some lakes (Jansson 1977). In oligotrophic waters *M. poweri* occurs alone; in oligotrophic waters with indications of incipient natural eutrophication, *M. poweri* and *M. minutissima* co-occur, but the former species is clearly dominant; in mesotrophic waters these two species are about equally abundant; in moderately eutrophic waters, *M. minutissima* becomes the dominant with *M. poweri* and/or *M. griseola* usually present; and in strongly eutrophic waters, *M. griseola* is dominant and *M. minutissima* is usually also common. The species associations are also sensitive to changes in trophic state within the same lake over quite short distances caused by localized pollution. In Finland these species are at or near the northern limits of their ranges, and Jansson suggests that such peripheral populations may have stricter ecological requirements than those near the center of their ranges. Jansson generally discounts interspecific competition as a major influence on species associations. He feels that the relation of their distribution to the water's trophic status reflects species specific differences in oxygen and food requirements.

Hutchinson (MS) suggests that the distribution of corixids probably involves a minimum of 10 major niche axes. The ten axes are (1) size of water body, (2) organic content of substrate, (3) calcium, (4) salinity, (5) depth, (6) water movement, (7) temperature, (8) food, (9) oviposition sites, and (10) underwater perching sites.

The distributions of other groups of littoral insects besides corixids have also been found to correspond to trophic conditions or successional stages. Seeger (1971) identified four associations of beetles in the genus *Haliplus* that corresponded to four levels of productivity (oligotrophic, mesotrophic, eutrophic, polytrophic) in lentic habitats in eastern Holstein, Germany. The species formed a series from those of oligotrophic waters that consumed little algae to those of productive waters that depended heavily on algae as a food source. Species with high respiratory rates were associated with well-aerated oligotrophic habitats, while species in more productive waters tended to exhibit lower rates. Kennedy (1922) examined dragonfly communities in lentic water bodies collectively representing a wide range of developmental stages and demonstrated a clear succession of species associations corresponding to the ontogeny of the habitats.

PONDS

The major types of lentic water bodies may be separated as shown in Figure 3.9. Whereas the deepest portions of lakes lie beneath the compensation point, the depth below which there is no net photosynthesis, ponds have rooted vegetation across the entire bottom. Marshes are sufficiently shallow for emergent hydrophytes across the entire surface, leaving a few or no areas of open water during the growing season. In swamps, emergent forbs and grasses are joined by woody plants, such as cypress trees (*Taxodium*), that are adapted to root in water-saturated soils.

Permanent Ponds As stated by Welch (1952), the word "pond" generally connotes "very small, shallow bodies of standing water in which the relatively quiet water and extensive plant occupancy are common characteristics." It is in such habitats that several groups of insects attain their maximum diversity and abundance, especially in the absence of fishes. The dragonflies, bugs, and beetles, along with certain dipterans, caddisflies, mayflies, and aquatic moths encountered in the well-vegetated littoral zone of lakes is in general but an attenuated version of the diverse insect community that may develop in permanent ponds (Table 3.5). Stoneflies are characteristically absent from ponds.

Despite being shallow, ponds may at times exhibit intense stratification of temperature and oxygen. The extent or presence of stratification largely depends on the degree of shelter afforded by the surrounding forest and topographical features. Because of their small volumes and large surface/volume ratios, thermal conditions in ponds are much more responsive to air temperatures than are lakes. Ponds are often highly productive habitats with largely organic substrates

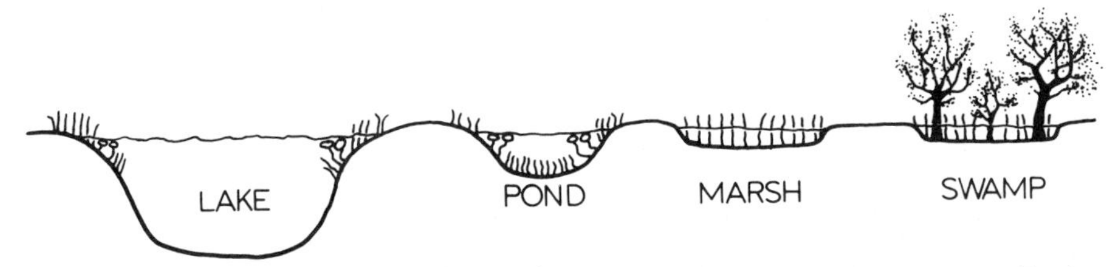

Figure 3.9 Schematic drawing showing the distinguishing features of four major types of lentic water bodies.

Table 3.5 Aquatic insects characteristic of permanent, well-vegetated ponds in the northern Temperate Zone

Taxon	Pleuston	Plankton	Nekton	Benthos
Collembola	X			
Ephemeroptera				
Baetidae				X
Caenidae				X
Leptophlebiidae				X
Siphlonuridae				X
Odonata				
Aeshnidae				X
Libellulidae				X
Coenagrionidae				X
Hemiptera				
Hydrometridae	X			
Veliidae	X			
Gerridae	X			
Belostomatidae			X	X
Nepidae				X
Pleidae				X
Corixidae			X	X
Notonectidae			X	X
Coleoptera				
Dytiscidae			X	X
Hydrophilidae			X	X
Haliplidae				X
Gyrinidae	X			X
Helodidae				X
Chrysomelidae				X
Diptera				
Tabanidae				X
Chironomidae				X
Chaoboridae		X		X
Culicidae	X			
Lepidoptera				
Pyralidae				X
Trichoptera				
Phryganeidae				X
Limnephilidae				X
Leptoceridae				X

so that anoxic conditions rapidly develop at the mud–water interface during periods of stratification.

All the aquatic insect community types of lakes also occur in ponds (Table 3.5). The pleuston community is typically well developed in ponds with luxuriant vegetation. In addition, a relatively large proportion of the benthic and nektonic species (e.g., adult beetles and bugs) must venture to the air–water interface to replenish air stores. Nektonic bugs (notonectids, corixids, belostomatids) and beetles (dytiscids, hydrophilids) are often better represented in ponds than in lakes. The notonectids *Anisops* and *Buenoa*, the only nektonic insects able to fully exploit the midwater zone, typically occur in small fishless water bodies. The benthos includes species that burrow, those that climb among plant beds, and a diverse array of nektobenthic forms. Chaoborids, present in some ponds, are part of the plankton (meroplankton) community. Under some conditions chaoborids may comprise a major portion of the total pond fauna (Young 1974).

Ponds containing fish have less diverse insect fauna than fishless ponds. Kenk (1949) commented on the paucity of beetles from a Michigan (USA) pond containing fishes in contrast to a nearby fishless pond where coleopterans were extremely diverse and abundant. The fishless pond contained 29 species of beetles versus 7 species in the pond containing fish. Insect taxa totaled 84 for the fishless pond and 43 for the pond with fish. Culicids and chaoborids were absent from the pond with fish. The major groups of insects and their relative abundance based on the number of taxa were as follows:

	Fish Absent	Fish Present
Coleoptera	35%	16%
Odonata	13%	9%
Hemiptera	13%	16%
Chironomidae	17%	33%
Other Diptera	13%	19%
Trichoptera	5%	2%
Ephemeroptera	2%	5%
Lepidoptera	1%	0%
Megaloptera	1%	0%

Palmer (1973) compares the results of her investigation of a hard-water pond at Castor Hanglands in Britain with Hodson's Tarn, a soft-water pond studied by Macan (1963). The mayflies *Leptophlebia vespertina* and *L. marginata* and the corixid *Sigara scotti* are soft-water species present in Hodson's Tarn but absent from the pond at Castor Hanglands, whereas numerous species (e.g., *Notonecta glauca* and several corixids) were found only in the calcium-rich pond. A much more diverse insect fauna was recorded from the hard-water pond; the most striking examples are 32 species of beetles and 23 species of bugs, in contrast to 8 species in each of these orders in Hodson's Tarn. However, insects clearly dominated the fauna in the soft waters of Hodson's Tarn where the two species of *Leptophlebia* and the dragonfly *Pyrrhosoma nymphula* comprised the vast majority of the total numbers of macroinvertebrates. In the hard waters of the pond at Castor Hanglands, the noninsects such as snails, sphaeriid clams, and *Asellus* attained greater abundances than any single insect taxon.

Temporary Ponds In contrast to the relative dearth of studies on permanent ponds, there is a considerable body of data dealing with temporary (astatic) lentic waters (Wiggins et al. 1980, Williams 1987). Yet considering the importance of temporary waters over vast regions of the continents, a great deal of work remains to be done (Williams 1985).

The ability to reside in seasonally astatic ponds, lentic water bodies lacking surface water during part of each annual cycle (Decksbach 1929), is restricted to a relatively few aquatic insects. Not only must the fauna of temporary ponds adapt to the alternating wet and dry phases per se, but they may also be exposed to extremes in temperature and progressive increases in salinity during the wet phase (Hartland-Rowe 1966, Daborn 1976, Williams 1983). Insects of ephemeral ponds possess special abilities, each of which apparently evolved as adaptations to withstand drought in species restricted to permanent waters. Only when several of these adaptations are coincident in a single species are insects able to establish reproducing populations in temporary ponds.

It is necessary to distinguish between two basic temporal patterns in seasonally astatic lentic waters. Temporary vernal ponds fill with water from melting snow and spring rains, retain surface water only until midsummer, and are dry throughout late summer, autumn, and winter. Temporary autumnal ponds are dry only during late summer and early autumn. Vernal ponds are much harsher habitats for aquatic insects than autumnal ponds, not only due to the extended dry phase (about 8 months vs. 3 months), but also because overwintering of resident species occurs during the dry phase.

The insect taxa listed in Table 3.6 contain species that are permanent residents of vernal ponds. Two general adaptive strategies are involved. Species that oviposit during the wet phase and estivate, usually as eggs or larvae, during the dry phase are called "overwintering spring recruits" by Wiggins et al. (1980). The other adaptive strategy, exhibited by "overwintering summer recruits," involves species that have evolved the ability to oviposit in moist microhabitats of the pond basin during the dry phase. Most of these species overwinter in the egg stage.

Overwintering spring recruits in some taxa hatch when the pool fills in early spring from eggs that have overwintered in the dry basin. Immatures grow rapidly; they must reach sexual maturity and oviposit while the pond still contains water. Larvae of some dipterans and coleopterans, rather than diapausing eggs, serve as the quiescent resistant stage during the dry phase (Table 3.6). Some beetles survive the dry phase as quiescent pupae or even adults. Quiescent stages, whether eggs, immatures, or adults, normally require exposure to low temperatures and/or desiccation before development or activity resumes. The eggs of some species exhibit differential rates of embryogenesis so that some eggs are able to hatch almost immediately should water appear earlier than normal in the pond basin. It is not uncommon for temporary ponds to exhibit a vernal pattern (winter dry) one year and an autumnal pattern (winter wet) the next. Differential rates of embryogenesis allow species to exploit such year-to-year differences yet not risk local extinction should the pool be flooded for only

Table 3.6 Resident aquatic insects (Groups 2 and 3 of Wiggins et al. 1980) of temporary vernal ponds in northeastern North America

Taxon	Overwintering Stage	Oviposition	
		Wet Phase	Dry Phase
Ephemeroptera			
Siphlonuridae:			
Siphlonurus	Egg	X	
Leptophlebiidae:			
Leptophlebia	Egg	X	
Paraleptophlebia	Egg	X	
Odonata			
Lestidae: *Lestes*	Egg		X
Libellulidae: *Sympetrum*	Egg		X
Coleoptera			
Dytiscidae:			
Agabus	Egg	X	
Hydroporus	Adult	X	
Rhantus	Egg	X	
Haliplidae:			
Haliplus	Egg, larva, pupa, adult	X	
Peltodytes	Larva, adult	X	
Hydrophilidae:			
Anacaena	Adult	X	
Helophorus	Adult	X	
Hydrobius	Adult	X	
Helodidae:			
Cyphon	Larva	X	
Trichoptera			
Polycentropodidae:			
Polycentropus	Egg	X	
Limnephilidae:			
Anabolia	Egg?		X
Limnephilus	Egg[a]		X
Ironoquia	Egg		X
Phryganeidae:			
Ptilostomis	Egg, larva		X
Diptera			
Chironomidae (14 genera)	Larva	X	
Ceratopogonidae (4 genera)	Larva	X	
Culicidae (3 genera)	Egg		X
Chaoboridae (2 genera)	Egg		X
Sciomyzidae (8 genera)	Egg, pupa		X
Stratiomyidae: *Odontomyia*	Larva	X	
Tabanidae: *Tabanus*	Larva	X	

[a] Larvae hatch but remain within gelatinous egg matrix.

Source: Modified from Wiggins et al. (1980).

a short period by a heavy rainstorm. However, species whose food is not present in sufficient quantity until spring would not gain an advantage by hatching in autumn, and exhibit obligatory spring hatching apparently synchronized by temperature and photoperiod.

Overwintering summer recruits typically hatch when water appears in the spring from eggs that have overwintered in the dry pond basin. Larval growth is completed during the wet phase, but sexual maturity and oviposition may be delayed by a larval or ovarian diapause away from the water. Oviposition in the dry basin is thus delayed until autumn so the eggs are not exposed to the severely desiccating conditions of midsummer. Some of the caddisflies adapted to vernal ponds hatch soon after oviposition, but the larvae remain quiescent within the gelatinous egg matrix over the winter (Wiggins 1973) unless water appears sooner (autumnal pattern).

Among Odonata, only a few species of the damselfly *Lestes* and the dragonfly *Sympetrum* are adapted to the harsh conditions of temporary vernal ponds. Figure 3.10 shows the series of adaptive features, some of which are also possessed by species restricted to permanent waters, that enable *Lestes dryas* and *L. unguiculatus* to exploit vernal ponds. A univoltine life cycle (one generation per year) is exhibited by numerous species of *Lestes*, and several species of permanent waters have evolved adaptations enabling them to oviposit in plants above the water surface. Vernal pond species have an egg diapause controlled by thermal and photoperiodic cues and exhibit obligatory spring hatching. *Lestes disjunctus* apparently exhibits all adaptations necessary to colonize vernal ponds successfully, except for the high thermal coefficient for growth that ensures that larval growth rate accelerates with increasing temperature (Corbet 1962). *Lestes disjunctus* cannot therefore fully exploit the rapidly rising temperatures during spring and is unable to complete larval development before termination of the wet phase. The species of *Sympterum* that successfully colonize vernal ponds exhibit similar adaptive strategies.

The aquatic insects considered thus far are permanent residents of temporary ponds. Although capable of dispersal flights, they are able to complete all life-cycle stages in ephemeral waters. In contrast, the "non-wintering spring migrants" of Wiggins et al. (1980) can maintain populations in vernal ponds only by overwintering in permanent water bodies. Aerial adults of spring migrants enter vernal pools after water appears in the spring. Oviposition, eclosion, growth, and the emergence of adults must be completed before the ponds dries. Adult Ephemeroptera (*Callibaetis*) and Diptera (*Chaoborus*, tanypodine chironomids) emerging from vernal ponds disperse to permanent waters to oviposit. Adults resulting from this new generation emerge from the permanent water body the next spring and may disperse to vernal ponds for oviposition. There are thus two generations each year, one in the vernal pond, the other in a permanent water body. Most aquatic families of Hemiptera, and at least Dytiscidae, Gyrinidae, and Hydrophilidae among Coleoptera, contain spring migrant species. The beetles and bugs in this category may complete a second generation in permanent waters as do mayflies and dipterans, but this is not

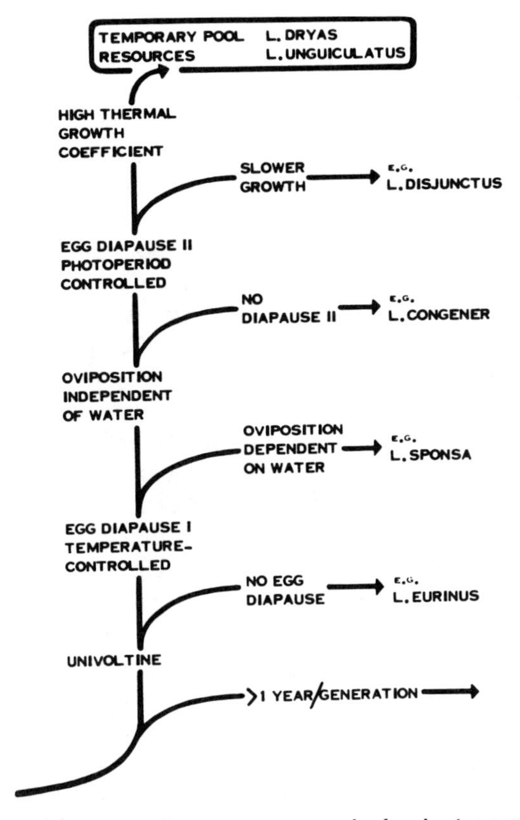

Figure 3.10 Adaptive features leading to occurrence of a few *Lestes* species in temporary ponds. [From Wiggins et al. (1980).]

necessary because adults, being aquatic, can overwinter in the permanent water body. In the tropics adult dragonflies may aestivate in the forest during the dry season and oviposit in ephemeral ponds formed at the beginning of the rainy season (Morton 1977). Even in the Temperate Zone, gerrid bugs characteristically overwinter as adults in terrestrial vegetation along the edges of permanent waters, an ability which preadapts this group to the spring migrant strategy for exploiting temporary vernal ponds. Gerrid species associated with lakes and running waters are often micropterous or apterous and therefore flightless (Brinkhurst 1959, Vepsäläinen 1974). Not surprisingly, flightless morphs do not occur among univoltine gerrids that are spring migrants in temporary ponds, although it is not known whether the macropterous condition is a result of genetic factors or represents an environmental response to temporary waters, or both.

Despite the harsh conditions, several advantages are conferred on species able to tolerate conditions in temporary ponds. A major factor is the absence of fish predation, and generally low invertebrate predation during the early portion of the wet phase. Some insects and other invertebrates of temporary waters are unable to maintain populations in association with fishes. Many of the spring migrants are predators that colonize temporary ponds only after their prey species have attained maximum densities. Detritus provides an especially rich food source in ephemeral waters. Detritus decomposes more rapidly under the alternating dry and wet phases and has a higher protein content when inundated in the spring than if it had been continuously submerged (Bärlocher et al. 1978). The dry phase also enhances nutrient regeneration. Small water bodies reach temperatures favorable for growth earlier than do larger permanent waters. Species with a high thermal coefficient for growth are able to exploit the rapidly increasing spring temperatures in such habitats. Competition is reduced compared to permanent waters.

Little is known, however, regarding the mechanisms used by insects to detect suitable temporary waters. In addition to the "nonwintering spring migrants," macropterous adults of other species of aquatic coleopterans and hemipterans may temporarily enter vernal ponds. Adult beetles tend to enter lentic waters at random but select particular habitats for oviposition (Galewski 1971, Fernando and Galbraith 1973), whereas at least some flying insects are able to visually select water bodies by their size and color (Popham 1943, Fernando 1959). Mosquitoes select oviposition sites from chemical cues (Kalpage and Brust 1973), and the habitats preferred by different species of corixids are closely related to the organic content of the substrate (Macan 1938).

Detritivorous species such as limnephilid caddisflies and some mosquitoes are among the first animals to resume activity when water appears in the spring. Detritivores of vernal ponds generally become active if flooding occurs early (autumnal pattern). This is a decided advantage since detritus will always be present, and a facultative response to early flooding enables immatures to get a head start on growth and development should the pond dry earlier than normal in the spring. Predators, in contrast, tend to exhibit obligatory spring hatching to ensure that adequate prey densities are encountered during the period of maximum growth. Spring migrants, which are largely predators, tend to be the last animals to occur in vernal ponds. Corixids, which tend to be more detritivorous than other spring migrant hemipterans, are generally the first to arrive.

Seasonally astatic waters may contain water for a week or less during an annual cycle or may be dry for less than a month each year. *Vernal* and *autumnal* ponds refer only to two commonly occurring patterns in the Temperate Zone. Autumnal ponds have a less restricted fauna than vernal ponds, and perennially astatic waters that dry up only every several years contain yet additional species. Conversely, pools with a wet phase shorter than in vernal ponds exhibit further reductions in their insect fauna. The duration of the dry phase is not the only factor determining the harshness of temporary waters for aquatic insects. In arid regions temporary ponds tend to be highly saline (Crawford 1981). Atmospheric conditions prevailing during the dry phase also determine the harshness of the habitat.

The chironomid *Polypedilum vanderplanki* resides in African rock pools and larvae are exposed to extremely high temperatures and severe desiccation in sun-baked mud during the dry phase (Hinton 1960). Larvae survive such adversity in a cryptobiotic state, a reversible cessation of metabolism engendering remarkable resistance to environmental extremes. Larvae are able to withstand nearly complete dehydration for extended periods and tolerate temperatures from -270 to +102°C in the cryptobiotic state. Desiccated larvae remain viable for years and can be alternately dried and activated in water numerous times, as in fact occurs in their natural habitats (Adams 1984). Cantrell and McLachlan (1982) found that small rock pools in tropical Africa each typically contained only one of three species of dipterans. *Chironomus imicola*, a species unable to enter the cryptobiotic state, occurs alone in pools with a wet phase of several weeks' duration. Pools lasting less than one week contained either *P. vanderplanki* or *Dasyhelea thompsoni*. The latter species is a ceratopogonid, the larvae of which estivate in watertight capsules during dry phases. It appears that the nature of the substrate determines which of these two species occurs in a given pool. *P. vanderplanki* is competitively excluded from pools of longer duration, which are inhabited by *C. imicola*, whereas the eggs of *D. thompsoni* apparently require more frequent water level fluctuations for successful hatching than the larger pools provide.

Estival Ponds Estival ponds are permanent in the sense of containing water throughout the year, but temporary in a physiological sense because no liquid water is available during the winter (Welch 1952). Estival ponds occur at high latitudes and altitudes where winter temperatures are sufficiently low to freeze shallow ponds from the surface down to and into the bottom substrate (Daborn and Clifford 1974). Temperatures of the ice and frozen substrate may drop well below 0°C for several months each year.

Daborn (1974) reported insects from four orders in an estival pond in western Canada. Chironomids clearly predominated, reaching densities exceeding 25,000 larvae m^{-2}. Fourteen genera of coleopterans were identified, but nine of these were only represented by adults. The following caddisflies occurred in low numbers: limnephilid larvae, *Triaenodes grisea* (Leptoceridae), *Agrypnia pagetana* (Phryganeidae), and *Agraylea multipunctata* (Hydropilidae). The odonate fauna exhibited an interesting transition from the first to the second open seasons, apparently attributable to especially low ice temperatures (-8°C) resulting in reduced survival of nymphs (Daborn 1971). During the first summer, three species of Coenagrionidae were the only odonates present in the pond. Coenagrionids occurred at much lower densities the second summer when large numbers of Lestidae appeared (*Lestes dryas, L. forcipatus, L. disjunctus*). *Lestes* are common members of temporary ponds and are adapted for rapid nymphal growth (see Fig. 3.10). Species replacement was also noted among zooplankton and oligochaetes. Daborn (1974) suggests that the faunal composition of a given summer is "determined as a fortuitous assemblage selected from the successful immigrants or residents of the previous summer by environmental conditions during the ensuing winter and spring."

Nelder and Pennak (1955) and Schmitz (1959) investigated estival ponds located above 3500-m elevation in the Colorado Rocky Mountains, where there is open water for only 3–4 months each year. The insect fauna comprised largely chironomids. Other dipterans included *Aedes* mosquitoes, tipulids, and stratiomyids. Caddisflies (*Limnephilus, Phryganea*), bugs (*Notonecta, Sigara, Arctocorisa*), and beetles (*Agabus, Hygrotus, Ilybius*) also occurred in low numbers.

Chironomid larvae also dominated the bottom fauna of arctic tundra ponds at Point Barrow, Alaska (71°N latitude) that are frozen down to permafrost for about 9 months each year (Butler 1982). Rigler (1978) comments on the absence of metabolic adjustment to low temperatures among aquatic animals of the arctic. Rather than thermal compensation allowing organisms to grow rapidly and reach reproductive age quickly at low temperatures, the adaptive strategy is to survive the harsh conditions long enough to reproduce. Indeed, Butler reports a 7-year life cycle for two species of chironomids in estival ponds in Alaska, where larvae spend 9 months of each year in frozen sediments. Only six species of insects besides dipterans reportedly inhabit the tundra ponds at Point Barrow, the stonefly *Nemoura arctica*, two dytiscid beetles (*Agabus* sp., *Hydroporus* sp.), and three caddisflies (*Limnephilus* sp., *Micrasema scissum, Asynarchus* sp.). Not only were chironomids extremely abundant; they were surprisingly diverse (Butler et al. 1980). The four chironomid subfamilies with the number of species in parentheses are as follows: Podonominae (1), Tanypodinae (4), Chironominae (12), Orthocladiinae (21+). The absence of mosquitoes from the Point Barrow ponds contrasts with their abundance in ponds farther inland. There are no Ephemeroptera, Odonata, Hemiptera, or Megaloptera. In the soft-bottom habitat of the pond's center, chironomids constituted 75–95% of the macrobenthic biomass; in the peripheral macrophytes, beetles, caddisflies, stoneflies, and the snail *Physa* accounted for much of the biomass (Fig. 3.11).

Little is known of the specific mechanisms of freeze tolerance in aquatic insects. Freeze-tolerant terrestrial species survive extensive extra- cellular ice formation within their tissues by synthesizing ice-nucleating agents (Lee 1989). The agents, consisting of special proteins or lipoproteins, prevent lethal intracellular freezing by serving as catalysts for the extracellular nucleation of ice.

OTHER LENTIC HABITATS

The remainder of this chapter will be devoted to the special lentic habitats and their communities. Topics include additional lake types (mountain, polar, tropical, ancient, and bog lakes), palustrine habitats, and phytotelmata.

Lakes at High Latitude In the high arctic, water temperature and duration of ice cover are the overriding factors structuring benthic communities. In contrast to tundra ponds, lakes at high latitudes maintain liquid water beneath the ice during winter. Nonetheless, because of the low temperatures, a large portion of a lake bottom may be frozen for long periods. For example, 20% of the bottom

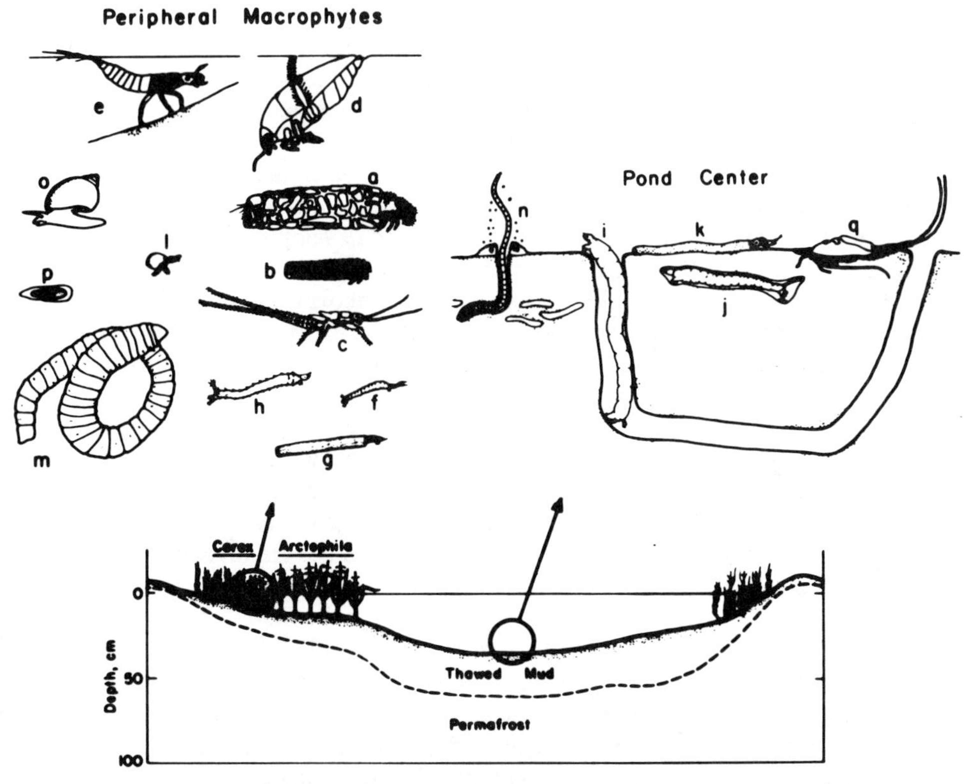

Figure 3.11 Habitats and principal taxa of benthic macroinvertebrates in a thaw pond in the Arctic tundra: (a) *Limnephilus* sp., (b) *Micrasema scissum*, (c) *Nemoura arctica*, (d) *Agabus* sp. (adult), (e) *Hydroporus* sp. (larva), (f) *Corynoneura* sp., (g) *Paratanytarsus penicillatus*, (h) *Trichotanypus alaskensis*, (i) *Chironomus pilicornis*, (j) *Procladius vesus*, (k) *Tanytarsus inaequalis*, (l) *Libertia* sp., (m) *Propappus* sp., (n) *Tubifex* sp., (o) *Physa* sp., (p) *Turbellaria*, (q) *Lepidurus arctica*. [From Butler et al. (1980) with permission of Van Nostrand Reinhold.]

of Char Lake, a polar lake 26 m deep, is frozen for up to 8 months each year (Rigler 1978). Chironomid larvae that overwintered in the frozen littoral sediment exhibited an 84% survival rate (Andrews and Rigler 1985). Littoral insects of lakes may migrate to deeper waters during the period of ice cover, a strategy unavailable to species of tundra ponds.

In general, the benthic community contributes a greater proportion of an arctic lake's total production and biomass than in lakes at temperate latitudes where zooplankton is relatively more important. In arctic lakes chironomids account for a large portion of the energy flow through the benthos (Hershey 1985). Chironomids contribute a progressively greater proportion of the diversity and abundance of lake benthos from the subarctic to the high arctic. In the extreme environments of the most northern lakes, chironomids are virtually the only aquatic insects encountered. Arctic chironomids have several special features, such as highly synchronous seasonal emergence and peak emergence during the warmest portion of the diel cycle, enabling them to survive and prosper under the extreme climatic conditions (Oliver 1968). The chironomids of Char Lake exhibited no indication of metabolic compensation to low temperatures (Welch 1976). However, larvae have the ability to remain active down to 0°C, and most littoral species migrate to deeper water as ice forms. Growth of the species examined by Welch was continuous throughout the year with no indication of diapause or other forms of suspended development during winter. The chironomids of Char Lake required 2–3 years to complete development.

Two species of chironomids and several collembolans comprise the entire insect fauna of the continent of Antarctica, with the exception of ectoparasites on birds and mammals (Downes 1964).

Mountain Lakes Lakes at high altitudes share many common features with arctic lakes, including low water temperatures, extended periods of ice cover, and high winds. Most mountain and arctic lakes are young, having been formed by glaciers several thousand years ago. Few investigators have dealt with the insect fauna of high mountain lakes, although such areas are favored locations for casual collecting.

In a study of 11 lakes (3161–3350-m elevation) in the Sierra Nevada Mountains of California, insects represented an average of 27% of total littoral macroinvertebrates (Taylor and Erman 1980). Chironomids were the only insects reported, but this undoubtedly resulted from sampling with a small core (2.8-cm diameter) which restricted sampling to relatively fine substrates.

Brittain (1974) reported a good correlation between the length of the ice-free period and the number of ephemeropteran species in mountain lakes in Norway. The mean number of species of mayflies was as follows: high alpine lakes, 0; midalpine lakes, 0.2; low alpine lakes, 0.7; subalpine lakes, 3.3; and boreal lakes, 4.8 species. The few mayflies at the highest elevations are primarily summer-growing species such as *Baetis macani* and *Siphlonurus lacustris*. Brittain found a fourfold increase in the number of species and a 100-fold increase in nymphal abundance from low alpine to subalpine lakes.

The heptageniid mayfly *Ororotsia hutchinsoni* was described from a female collected on the ice of a cirque lake in Indian Tibet at an altitude of 5297 m (Traver 1939). In mid-July the lake was still ice-covered except for a narrow zone along the shore (Hutchinson 1937a).

The plecopterans have perhaps been more intensively studied than any other order of insects that occur in high mountain lakes. Although few stoneflies occur in lentic biotopes, those that do are best represented in mountain lakes. For example, Brinck (1949) recorded 18 species of Plecoptera from mountain lakes in Sweden, but only three species from lowland lakes. Most species are typically lotic forms also able to colonize the wave-swept shores of cold mountain lakes.

Seventeen species of stoneflies were collected from 58 of 162 lakes located at elevations of 985–2343 m in the Canadian Rockies (Donald and Anderson 1980). Collecting was restricted to areas well away from inlet or outlet streams.

All lakes containing stoneflies had rocky shorelines. Eight species were collected from a single lake at 1671-m elevation. *Utacapnia trava* was the most frequently encountered species, occurring in 25 lakes. Based on their own collections and literature data, Donald and Anderson identified three species (*Arcynopteryx compacta, Bolshecapnia spenceri, Isogenoides zionensis*) that are restricted to lakes or to streams near lakes.

Stoneflies, in contrast to mayflies, are present in high alpine lakes in Norway, but are poorly represented in lowland lakes (Brittain 1974, Lillehammer 1974). *Capnia atra* and *Diura bicaudata* are often the predominant species in alpine lakes, and in northern Norway these two species tend to dominate the fauna of both lakes and streams. Brinck (1949) lists these two species plus *Arcynopteryx compacta* as the only stoneflies in Swedish lakes above 1000-m elevation.

The aquatic insects of Øvre Heimdalsvatn, a subalpine lake (1090-m elevation) in the Jotunheimen Mountains of Norway, have been extensively studied for a number of years (Aagaard 1978; Brittain 1978a, b; Lillehammer 1978a, b). The exposed zone (shore to 1–2 m) is subjected to waves and ice action. No macrophytes are present in this zone, the substrate of which is typically stony. The macrophytic zone (1–2 to 5 m) is below the major effects of wave action and has finer sediment. The remainder of the lake (5–13 m) is called the nonmacrophytic zone. Øvre Heimdalsvatn is ice-covered for 8 months each year.

Amphipods and five insect taxa (Ephemeroptera, Trichoptera, Plecoptera, Coleoptera, Chironomidae) accounted for an average of 92% of the total numbers of benthic macroinvertebrates in the exposed zone (Fig. 3.12). Tipulids (2.7% of benthos) were the only other insects to occur in any abundance. Ten species of stoneflies occurred in Øvre Heimdalsvatn, and all of these occupied the exposed zone during the summer. The two most abundant species exhibited different winter strategies. *Diura bicaudata*, a species highly adapted to low temperatures, remains in the narrow zone of free water under the ice (0–1.0°C) near the shore (Fig. 3.13). *Nemoura avicularis* migrates to deeper water (3-4°C) during winter, which subjects nymphs to much higher trout predation than do *D. bicaudata*. Mayflies were the most abundant macroinvertebrates in shallow waters. Seven species were collected. *Baetis macani* and *Siphlonurus lacustris* were restricted to water <1 m deep. These are summer species that overwinter as eggs or small nymphs and complete all or nearly all of their growth during the brief ice-free period. *Leptophlebia vespertina* and *L. marginata* are winter species that complete a significant portion of their growth as overwintering nymphs. *Leptophlebia* nymphs are found at various depths down to 7 m, but are more abundant in deeper water during the winter. All species of mayflies migrate to shallow water prior to emergence. The caddisfly fauna of the exposed zone comprised in large part species normally found in running waters. Fourteen species from seven families were collected. The two most abundant species, *Polycentropus flavomaculatus* and *Limnephilus nigriceps*, dominated on stable and unstable stony bottoms, respectively. The leptocerid *Mystacides azureus* was the predominant caddisfly on soft substrate in deep water. Eight coleopter-

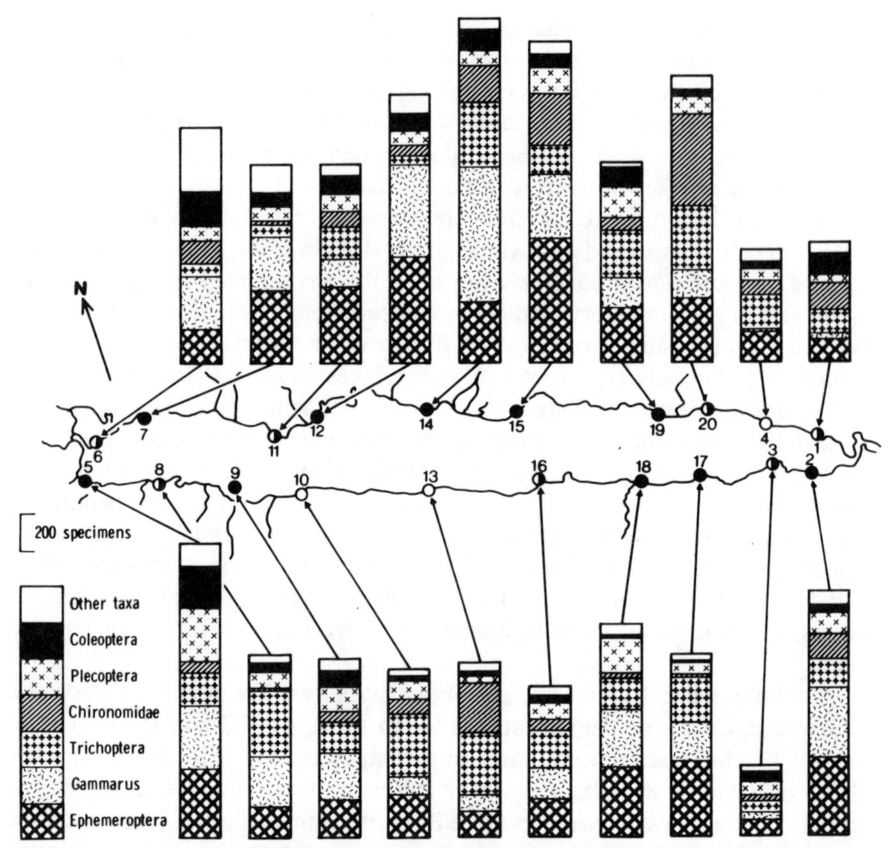

Figure 3.12 Total numbers of the different taxa (adjusted totals) taken during the survey of the exposed zone of Øvre Heimdalsvatn, a subalpine lake in Norway. Stations with high levels of benthic detritus are indicated with solid circles; those with low levels by open circles, and those with intermediate levels, by partly shaded circles. [From Brittain and Lillehammer (1978).]

ans were collected from Øvre Heimdalsvatn and collectively constituted 6% of the benthos of the exposed zone. The families Dytiscidae, Hydrophilidae, and Haliplidae were represented. Larval coleopterans were widely distributed and occurred at all depths, whereas adults were collected only from shallow water. At least 75 species of chironomids inhabit Øvre Heimdalsvatn. The oligotrophic nature and high altitude of the lake is reflected by the species composition and abundance of the common species, many of which are northern or oligothermal forms. Chironomids and amphipods dominate the benthos in deep water.

Tropical Lakes Recognizing the virtual absence of data on tropical waters and their inhabitants, in 1928 two of the world's leading limnologists, August Thienemann of Germany and Franz Ruttner of Austria, set off on an expedition to Indonesia. They sampled lakes, springs, and lotic waters on the islands of

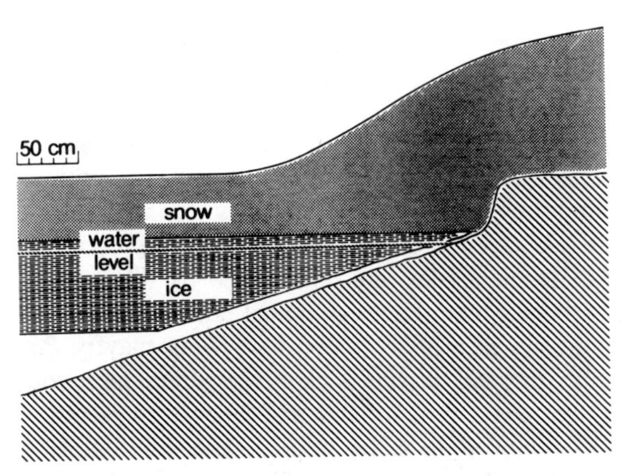

Figure 3.13 Typical profile of the exposed zone of Øvre Heimdalsvatn, Norway, during March showing the narrow zone of free water between the lake bottom and ice cover. [From "The influence of temperature on nymphal growth rates in mountain stoneflies (Plecoptera)" by J. E. Brittain, Ecology, 1983, 64, 440–446. Copyright © by The Ecological Society of America. Reprinted by permission.]

Java, Sumatra, and Bali. Collections of plants and animals were sent to specialists who eventually described 1100 species new to science (Rodhe 1974). The results of the expedition are contained in 11 supplement volumes of the Archiv für Hydrobiologie published between 1931 and 1958. Other expeditions to tropical lakes include the Belgian Hydrobiological Mission to Lakes Kivu, Edward, and Albert in Africa (Verbeke 1957). Verbeke reviews the earlier work on aquatic insects of African lakes. More recent publications are now available for other equatorial regions, a good example is the "Aquatic Biota of Tropical South America," which contains considerable data on aquatic insects (Hurlbert et al. 1981). The development of faunal lists and taxonomic keys will undoubtedly remain major tasks for aquatic entomologists working in tropical regions for some time.

From an ecological context, the aquatic insect communities of tropical lakes are not markedly different from those of lowland temperate lakes. For example, the insect fauna in deep waters of productive lakes in the tropics often comprised nearly exclusively *Chaoborus* and chironomids (frequently *Chironomus*) as in eutrophic temperate lakes (Deevey 1955, Macdonald 1956). Massive swarms of *Chaoborus* and chironomids occur commonly along the African Great Lakes and are described by Beadle (1974, p. 201) as "what appears to be gigantic clouds of smoke, sometimes of more than fifty metres in height and occasionally more than a kilometre in length, rise from the surface of the lake." *Chaoborus edulis* is boiled and made into cakes eaten by the people living along the lake shore.

The insect community of the rocky wave-swept shores of Lake Nyasa (Malawi) in tropical Africa consists of rheophilic forms such as stoneflies, heptageniid mayflies, hydropsychid caddisflies, and psephenid beetles (Fryer

(1959a). These same groups, although usually different genera, typically inhabit rocky shores of temperate lakes and play ecological roles similar to their tropical counterparts in such habitats.

Hydrophytes of the littoral zone provide suitable habitat conditions for a variety of aquatic insects. The major groups include chironomids, mayflies, caddisflies, dragonflies, bugs, and beetles, as is true in comparable habitats in many temperate lakes. Some of the same genera are present, but typically the littoral insects of tropical lakes are represented by different genera, in some cases different families, than their ecological equivalents in temperate waters. Only rarely, as in the case of the wood burrowing mayfly *Povilla adusta* of African lakes, do tropical insects lack a temperate lake counterpart. Free-floating plants, such as the fern *Salvinia* and the water hyacinth *Eichhornia*, are especially well developed in some tropical and subtropical waters and are colonized by a variety of aquatic insects. Pleustonic insects are generally represented by the same families (e.g., Poduridae, Gerridae, Veliidae, Culicidae) as in temperate waters.

Aquatic insects may complete a generation in 2 months or less in tropical lakes (e.g., Macdonald 1956). The emergence of many species exhibit a lunar periodicity (Verbeke 1957), a phenomenon of considerable interest, explored further in Chapter 9.

Bog Lakes Bog lakes most commonly occur in glaciated regions of the northern Temperate Zone, where a particular combination of climatic conditions, vegetation, soil, and topography favors the formation of peat deposits (Welch 1952). Although the pH of bog lakes varies, values are typically acidic, sometimes extremely so. The waters of bogs (called "moors" in most European literature) are brown because of the abundant humic acid colloids, which tend to form a flocculent false bottom overlying the peat deposits. Bog waters are characteristically low in calcium. A few aquatic angiosperms, such as water lilies and pond weed, are often present, and mats of *Sphagnum* moss typically occur on the shore.

Bog lakes tend to have a restricted fauna and low productivity. Herpobenthos may be almost totally absent from deep-water areas that have a flocculent false bottom and extended periods of oxygen depletion. Small, strongly acidic bog lakes without inlets or outlets have an especially restricted fauna with entire groups (e.g., Ephemeroptera) absent. A limited number of species of Hemiptera, Coleoptera, Trichoptera, Odonata, and Diptera are typically present (Griffiths 1973). Some of these are euryokous species, but others are characteristic of bog lakes. *Chaoborus* larvae may be abundant and entirely planktonic in bog lakes lacking fishes.

In a small, strongly acidic bog lake in England, with a bottom composed entirely of peat without an overlying false bottom, the herpobenthos contained oligochaetes and a highly restricted insect fauna (McLachlan and McLachlan 1975). Only 17 species of insects were collected, including 11 chironomids, 2 trichopterans, 1 ephemeropteran, 1 megalopteran, and 2 odonates. The megalopteran *Sialis lutaria* and three chironomid species collectively comprised 96% of the benthic biomass.

Extant Ancient Lakes With only a handful of exceptions, extant lakes are young, having had their inception during the Pleistocene. Lake Superior, for example, is only about 12,000 years old. There are no extant ancient (pre-Pleistocene) lakes in North America. Lakes typically have a life span of a few thousand to a few hundred thousand years, and are thus short-lived from an evolutionary perspective.

There are, however, a very few large, deep lakes that have persisted long enough to be considered truly ancient. Lake Baikal in Siberia, the world's deepest (1620 m) lake and with a larger volume of water than any freshwater lake, originated some 60–75 million years ago (Kozhov 1963). Lakes Tanganyika and Nyasa (Malawi) in the Rift Valley of Africa, Lake Ohrid in Yugoslavia, and Lake Lanao in the Philippines have persisted for a shorter time than Lake Baikal, but nonetheless exhibit remarkable speciation in certain faunal components. Endemism in ancient lakes is best developed among taxa with low vagility, such as malacostracan crustaceans, mollusks, and turbellarians, or in groups with extreme genetic plasticity, such as cichlid fishes (Brooks 1950).

The insects of ancient lakes have received less attention than many other groups. Presumably the presence of an aerial adult stage has to some extent countered the tendency for speciation. In Lake Baikal there are two species of capniid stoneflies in the endemic genus *Baikaloperla*, the adults of which are wingless (Zapekina-Dulkeyt and Zhiltsova 1973). Fourteen endemic species of limnephilid caddisflies in several endemic genera are found in Lake Baikal (Martynov 1924, Bebutova 1941). The adults have reduced wings and are flightless. In some species the metathoracic wings have been lost entirely. Larvae require 2–3 years to complete development and are distinguished by attaining a relatively large size in most species. Some species transform to the adult at the water surface and swim shoreward. Mating occurs in the surface film in some species, but others engage in nonaerial swarms on rocks and vegetation. The adults at times form layers 10 cm thick on rocks along the shore. Kozhov (1963, p. 122) describes how at the height of emergence "the mountain taiga on the north-eastern coast of Baikal becomes transformed, as if donning a new attire," because of the masses of caddisflies. Although poorly studies, 11 endemic species of chironomids have been described from Lake Baikal. Most aquatic insects are restricted to the shallow gulfs and bays of Baikal and do not occur in the lake proper. According to Kozhov, most of these are common representatives of the Siberian fauna.

The insects of other ancient lakes have been subjected to even less investigation than in Lake Baikal. An endemic limnephilid caddisfly, *Limnocoetis tanganicae*, is known from Lake Tanganyika. The adult is flightless and is adapted for swimming in the surface film much like a gyrinid beetle (Marlier 1962). With the exception of the mayfly *Caenis macrura*, the aquatic insects of

Lake Ohrid, Yugoslavia, are restricted to the littoral zone (Stanković 1960). Lake Ohrid is distinctive in that chironomids are absent from the profundal zone. However, no endemic insects have been reported from this ancient Balkan lake. Only the upper 200 m or so of Lakes Tanganyika and Nyasa (Malawi) undergo circulation; the perennially anaerobic bottom waters lack insects and other macroinvertebrates.

Palustrine Waters Freshwater swamps and marshes are among the most stagnant of lentic waters. Aquatic plants impede wind-induced circulation and dense stands of floating and emergent hydrophytes shade the water surface, so that oxygen deficits are common in these shallow, highly organic aquatic habitats. Because of the frequently anoxic conditions in bottom mud, the herpobenthos tends to be poorly developed or even absent (Cantrell 1979). Pleustonic forms that inhabit the air–water interface and nektobenthic species that periodically replenish air stores at the water surface are well represented. The absence of a limnetic zone precludes the development of a true plankton community in palustrine waters.

Haptobenthic forms associated with aquatic plants are especially well developed in swamps and marshes. Some species of chrysomelid and curculionid beetles, aquatic moths, and dipterans in several families supply their respiratory needs by tapping the underwater air stores of aquatic plants, an adaptation enabling them to inhabit low-oxygen waters where there are few aquatic predators or competitors. These and other aquatic insects may reside in burrows within plant tissues and use plants for food and as a source of case-building materials. Plant surfaces are typically colonized by periphytic algae that serve as a major food for grazing insects. Light attenuation by dense stands of emergent plants reduces the periphytic growths on submerged surfaces with corresponding reductions in the insect fauna (Cantrell 1979).

Free-floating aquatic plants such as *Salvinia* and *Eichhornia* may be well developed in palustrine waters especially in tropical and subtropical regions. Rzóska (1974) described an exceptionally diverse and abundant aquatic insect fauna associated with the bushy roots of *Pistia*, a floating plant of Nile swamps. Beetles, aquatic bugs, and mayflies were especially abundant. The beetles were represented by 44 species, 19 in one genus *(Hydrovatus)*.

The aquatic insect fauna of palustrine habitats has been investigated in north temperate regions (e.g., Judd 1949, Mason and Bryant 1974, Henson 1988), in the Great Dismal swamp in the southeastern United States (Matta 1979), and in the tropics (e.g., Cantrell 1979, Mizuno et al. 1982).

Dipterans, coleopterans, hemipterans, and odonates collectively provide the majority of species and individuals in most palustrine habitats. Swamps and marshes are preferred breeding sites for many mosquitoes. For example, 92 species of mosquitoes breed in the swamps of Uganda, Africa, and 26 of these are restricted to palustrine habitats (Goma 1960). Other orders, such as Trichoptera, may also be well represented. Palustrine habitats may provide optimal conditions for minor groups such as aquatic moths.

Phytotelmata The term *phytotelmata* (plant pools), derived by Varga (1928), refers to the aquatic microhabitats formed on or within plants (Maguire 1971). In the Temperate Zone the primary examples of phytotelmata are water-filled tree holes and pitcher plants (Sarraceniaceae), but in tropical regions a diverse array of such habitats exists (Thienemann 1934, 1954; Seifert 1982; Frank and Lounibos 1983). Rainwater collects in the leaf axils of a variety of tropical plants; perhaps the best example is the bromeliads (Laessle 1961), but also in bract axils and flower parts. By feeding on fallen coconuts, terrestrial animals may form hollow husks that fill with water and provide habitats for culicids, psychodids, and ceratopogonids. A relatively diverse fauna, consisting of several families of dipterans and odonate larvae among insects, resides in water-filled cavities of broken bamboo stems that may hold several liters of water. Thienemann (1954) listed the chironomid and ceratopogonid species found in the various types of phytotelmata.

The aquatic insects regularly reported from phytotelmata include dipterans (Calliphoridae, Ceratopogonidae, Chironomidae, Culicidae, Phoridae, Psychodidae, Sarcophagidae, Syrphidae), a few species of odonates, and helodid coleopterans. Some of these are adapted to the special environmental conditions of phytotelmata and are restricted to such habitats.

In Europe, Kitching (1971) lists the following species of aquatic insects as those "specifically associated with water-filled tree holes and seldom, if ever, found elsewhere" [the dendrolimnetobionten of Rohnert (1950)]:

- Diptera
    - Culicidae
        - *Aedes geniculatus*
        - *Anopheles plumbeus*
        - *Orthopodomyia pulchripalpis*
    - Chironomidae
        - *Metriocnemus martinii*
    - Ceratopogonidae
        - *Dasyhelea lignicola*
        - *D. dufouri*
    - Syrphidae
        - *Myriatropa florea*
- Coleoptera
    - Helodidae
        - *Prionocyphon serricornis*

Water-filled tree holes are highly organic habitats bounded by wood and containing leaves and other detritus. The inhabitants must be able to withstand low-oxygen conditions, temperature extremes, and dry periods. In the British tree hole communities examined by Kitching, the aquatic larvae feed primarily on plant detritus and animal remains or the microbes associated with decaying organic matter; aquatic predators are notably absent. Kitching and Callaghan (1982) recorded somewhat more complex food webs in Australian tree hole communities. In addition to sarcophagic species (helodid beetles and culicids among insects), several aquatic predators belonging to the dendrolimnetobiontic component were present. Two of these aquatic predators were insects: *Anatopynia pennipes*, a tanypodine chironomid, and the ceratopogonid *Culicoides angularis*.

In contrast to tree holes, which derive water from rainfall, pitcher plants secrete a liquid containing proteolytic enzymes. Insects attracted by nectar that fall into the reservoir of liquid contained in the modified leaves are digested and amino acids and other breakdown products are absorbed by the plant (Plummer and Kethley 1964). However, the leaf fluid of these insectivorous plants is inhabited by a small community of aquatic organisms, some of them insects, that feed on the insect remains yet are not themselves digested. The digestive fluid of *Sarracenia purpurea*, a widely distributed North American pitcher plant, is typically inhabited by three dipterous larvae, the sarcophagid *Blaesoxipha fletcheri*, the culicid *Wyeomyia smithii*, and the chironomid *Metriocnemus knabi* (Fish and Hall 1978). *Blaesoxipha fletcheri* larvae reside at the air–water interface and feed on recently captured insects floating in the surface film. *Wyeomyia smithii*, unlike many mosquito larvae, swim freely, returning to the surface film only for gas exchange with the atmosphere. The submerged larvae feed on suspended particular matter from decomposed victims trapped by the pitcher plant. The chironomid *M. knabi* feeds on organic matter that accumulates on the bottom of the leaf chamber.

The mosquito *W. smithii* has been the subject of numerous investigations. It is restricted in its preimaginal stages to the leaf fluid of *S. purpurea*. Aquatic stages are highly resistant to desiccation (Bick and Penn 1947) and extremely high summer temperatures (Bradshaw 1980), and overwintering larvae can tolerate freezing. In New Brunswick, Canada, where larvae of *W. smithii* and *M. knabi* remain frozen for 4 months each year, overwintering mortality is <5% (Patterson 1971).

Floral bracts of *Heliconia* are produced about every 7 days, so that large inflorescences, such as shown in Figure 3.14, represent different successional stages of aquatic insect communities (Seifert 1982). The uppermost bract is the youngest and each bract is about one week older than the one above. Three species of mosquitoes succeed one another as the bracts age, although this is confounded somewhat because some larvae move from one bract to another by wriggling up the central rachis when the rachis is wet. One species of chrysomelid beetle larva is replaced by another species in older bracts. Two species of syrphid larvae inhabit the water-filled bracts, one of which occurs only in older bracts. There is a progressive increase in detritus with increasing bract age.

Phytotelmata are aquatic microcosms amenable to addressing ecological questions relating to food webs (Kitching 1987), mutualistic relationships

Figure 3.14 Diagrammatic representation of the insects living in the water-filled floral bracts of *Heliconia aurea* from Rancho Grande, Venezuela. [From Seifert (1982). Quarterly Review of Biology, The University of Chicago Press. Copyright 1982 by The Stony Brook Foundation, Inc.]

(Bradshaw and Creelman 1984), population dynamics (Istock et al. 1975), competitive interactions (Seifert and Seifert 1976), and island biogeography (Maguire 1971, Seifert 1975), among others. Aquatic entomologists have only begun to exploit the information potentially available from these special habitats.

4

LOTIC FRESHWATERS

> . . . rivers and the inhabitants of the watery element were made for wise men to contemplate, and fools to pass by without consideration, . . . for you may note, that the waters are Nature's storehouse, in which she locks up her wonders.
>
> —Izaak Walton

There is a prevalent view that cool running waters served as the ancestral habitat for many groups of aquatic insects (Hynes 1970a, b; Wootton 1972; Wiggins and Wichard 1989). It appears that entire orders and other major taxa, including Plecoptera, Ephemeroptera, Trichoptera, Chironomidae, and even Odonata evolved in cool running waters and that species from these groups now inhabiting the lower reaches of rivers and lentic water bodies are later derivatives of ancestral lines originally adapted to conditions in cool lotic reaches (Ward and Stanford 1982). The contemporary habitat occurrences of these orders range from Plecoptera, which remain largely confined to the ancestral habitat (Hynes 1976), to Odonata that are best represented in lentic waters (Corbet 1962) (Fig. 4.1).

Numerous families of aquatic insects are restricted to running waters (including wave-swept lake shores) and many others attain their maximum development in lotic habitats (Table 4.1). There is an interesting relationship between the presumed ancestral habitat and the aquatic representation of each order: those mentioned above as having evolved in cool running waters are entirely aquatic orders; whereas those such as Hemiptera and Coleoptera, the aquatic members of which likely developed in lentic waters, are primarily terrestrial orders. This dichotomy also roughly separates those orders that extract dissolved

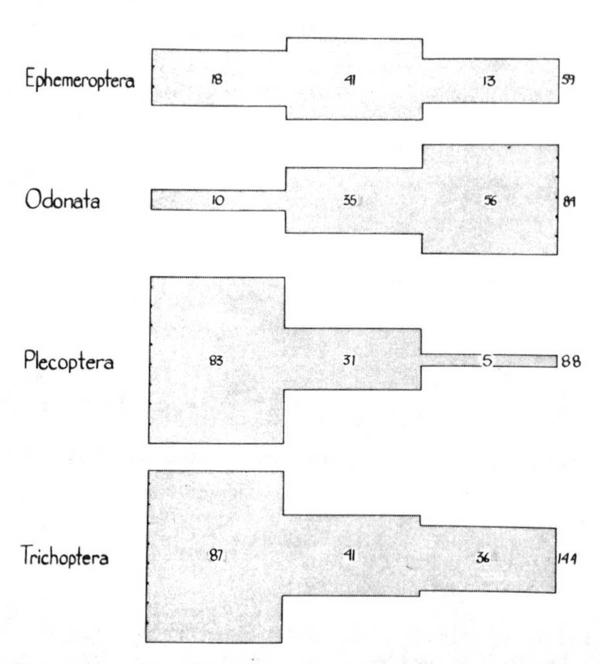

Figure 4.1 Habitats of Nearctic genera in four orders of aquatic insects. Total number of Nearctic genera per order are shown on the right of the figure. Number of genera per major habitat is shown within each block; the sum of these numbers exceeds the number on the right because a few genera are represented in more than one major habitat. [From "Some relationships between systematics and tropic ecology in Nearctic aquatic insects with special reference to Trichoptera" by G. B. Wiggins and R. J. Mackay, Ecology, 1978, 59, 1211–1220. Copyright © 1978 by The Ecological Society of America. Reprinted by permission.]

oxygen from the water from those that rely primarily on atmospheric air to meet their respiratory needs. Those coleopterans and hemipterans best adapted to running waters have evolved highly specialized mechanisms for utilizing the dissolved oxygen contained in water, as described in Chapter 2.

The caddisflies provide an especially good example of an order with an early evolutionary history restricted to cool lotic waters that later exhibited an adaptive radiation to a wide variety of aquatic habitats (Ross 1956, Wiggins and Wichard 1989). Wiggins (1977) points out that all extant North American caddisfly families occur in running waters, whereas only 38% of these are also represented in standing waters.

Cool running waters are indeed the logical place for previously terrestrial organisms to evolve respiratory mechanisms for extracting dissolved oxygen from the water. Cool water contains more oxygen at saturation than does warm water; running water prevents the formation of localized areas of low oxygen

Table 4.1 The percentage occurrence of aquatic insect families in running waters[a]

Order	Occurrence of Families in Running Waters		
		Restricted to Lotic[b]	
	Represented %	%	Example
Ephemeroptera	100	71	Heptageniidae
Odonata	100	18	Protoneuridae
Hemiptera[c]	83	0	—
Plecoptera	100	100	All families[d]
Coleoptera	86	36	Psephenidae
Diptera	86	41	Blephariceridae
Lepidoptera (Pyralidae)	100	0	—
Megaloptera	100	0	—
Neuroptera (Sisyridae)	100	0	—
Trichoptera	100	67	Hydropsychidae

[a]Based on North America from data in Merritt and Cummins (1984), excluding marine representatives.
[b]With rare exceptions found only in running waters, including wave-swept lake shores.
[c]Excluding surface film bugs.
[d]A few stoneflies colonize still waters at high latitudes or high altitudes.

and facilitates uptake by respiratory surfaces. In addition, as Hynes (1970a) points out, river systems tend to be geologically long-lived. They may change dramatically in character concomitantly with long-term climatic changes, but they rarely disappear. Most lakes are ephemeral by comparison and tend to be evolutionary traps for organisms without highly developed dispersal abilities. Aquatic insects as a group are poor fliers. Strong fliers most commonly occur among aquatic insects best represented in lentic waters, which, with the exception of Odonata, tend to be members of largely terrestrial orders.

In the remainder of this chapter attention is devoted to insect communities and the typology of running waters, including special lotic habitat types and adaptations to current. The zonation concept for classifying running waters, prevalent in Europe, is compared to the river continuum concept developed in North America. The reader is referred to Shadin (1956), Hynes (1970a), Macan (1974b), Whitton (1975), Lock and Williams (1981), Barnes and Minshall (1983), Fontaine and Bartell (1983), Davies and Walker (1986), and Minshall (1988) for broad accounts of running-water ecology, and to Hynes (1970b) for a review dealing specifically with the ecology of stream insects.

LOTIC COMMUNITIES

All aquatic insect communities present in lakes also occur, to a greater or lesser degree, in running waters. Water current provides conditions for the development of two additional communities, the hyporheos that reside deep within the substrate, and the madicolous fauna adapted to living in thin films of moving waters.

Pleuston

Although the current velocities of high-gradient streams generally preclude the development of pleuston, this community may be well developed in some lotic situations. Well-vegetated stream margins and quiet backwaters provide lentic-like conditions and the pleuston community includes some of the same spring-tails, water striders, whirligig beetles, and immature mosquitoes encountered in lakes and ponds.

Quiet pools of small, low-gradient streams with overhanging vegetation provide especially suitable habitats for epipleustonic insects such as water striders and whirligig beetles that prey on terrestrial arthropods trapped in the surface film. The widely distributed North American water strider *Gerris remigis,* a typical stream inhabitant, forms aggregations at locations where current patterns concentrate their drifting prey (Riley 1921). *Gerris najas* occurs in similar lotic habitats in Europe. The broad-shouldered water strider *Rhagovelia* is especially adapted for life on the surface of rapidly moving water (Fig. 4.2).

Water surfaces concentrate lipoproteins and other organic molecules and are colonized by microorganisms (neuston). In addition, a variety of allochthonous particles such as flowers, fruit, pollen, and terrestrial insects are retained by the surface film (Rapoport and Sánchez 1963, Fittkau 1976). The current further concentrates these organic films behind obstructions such as those formed by

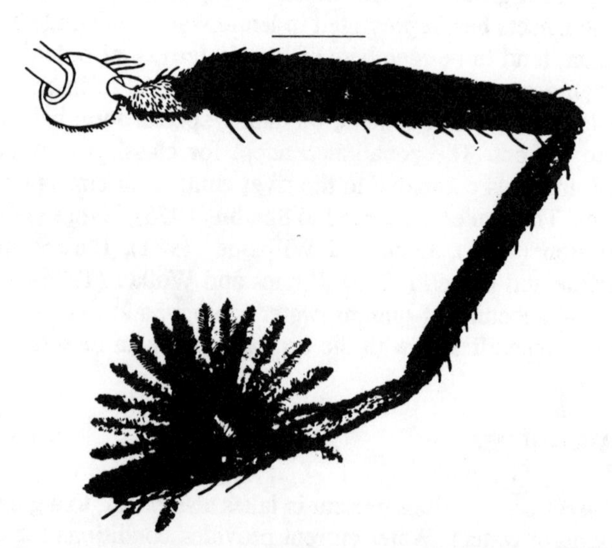

Figure 4.2 The midleg of *Rhagovelia* showing swimming plume. [From Coker et al. (1936).]

wood debris. Fittkau (1976) named the special biotopes formed by these accumulations *kinal* and the communities (neustonic and pleustonic) associated with them *kinon*. Kinal habitats are especially well developed in tropical rivers and may be inhabited by a distinctive aquatic insect fauna (Fig. 4.3), several species of which are essentially restricted to these biotopes (kinobionts).

Plankton

Many normally benthic stream insects are passively transported downstream in the water column. For example, Coutant (1982) reported a positive phototaxis

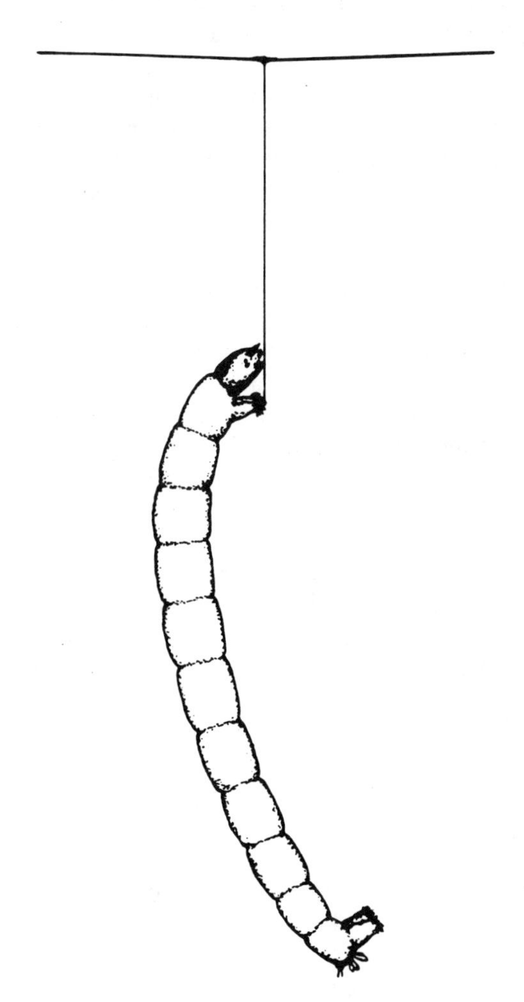

Figure 4.3 A chironomid larva (*Polypedilum* sp.) that hangs from a silk thread in the Potamokinon of the Amazon River. [From Fittkau (1976). Reprinted by permission of Kluwer Academic Publishers.]

in first instar *Hydropsyche cockerelli,* a caddisfly whose later instar larvae are negatively phototactic. He suggested that since adults oviposit in shoreline areas during high water, the positive phototactic response of newly hatched larvae aids dispersal and ensures that larvae are not stranded in shoreline areas as the water level drops. However, such behavior constitutes the phenomenon called "drift," an important functional attribute of lotic ecosystems, that involves benthic not planktonic forms.

Truly planktonic organisms, including *Chaoborus,* occur in running waters, but there is no evidence of a "potamoplankton" distinctive from the plankton of lakes. The plankton found in running waters originate in lentic habitats, although reproducing populations of lentic plankters can attain high densities in rivers (e.g., Eddy 1931; Enǎceanu 1964; Monakov 1964, 1969; Rzóska 1976). Backwaters may supply large numbers of plankton to streams during high water (Clifford 1972), but lakes may provide large, although variable, amounts to downstream reaches virtually year round (Cushing 1964). Where abundant in streams, plankton constitutes an important food for filter-feeding insects that typically occur in high densities (Müller 1956, Illies 1956). The rapid elimination of lentic plankton in lake outlets is partly attributable to the feeding of aquatic insects. There is a characteristic assemblage of insects in lake outlets that rapidly changes with distance downstream concomitantly with the elimination of plankters. Characteristic groups include black flies (Simuliidae), hydropsychid and polycentropodid caddisflies, and in Africa the filter-feeding mayfly *Tricorythus.* Noninsect filter-feeders such as sponges and other insects that do not feed directly on plankton may also be abundant. Most of these species exhibit increased populations in lake outlets, but also occur in other lotic habitats. However, some species of simuliids are apparently restricted to the special conditions in streams immediately below lakes (see Hynes 1970a).

The composition and abundance of insects may dramatically change over very short distances in lake outlets. Illies (1956) found a striking succession of species of filter-feeding insects over distances of only tens of meters below a lake in Swedish Lapland. The aquatic insect fauna rapidly decreased in abundance downstream; from 15 to 215 m below the lake, numbers and biomass decreased 78% and 58%, respectively, and the numerical contribution of filter feeders to total insects declined from 95% to 17%. Even small ponds may cause major alterations in the composition and abundance of the downstream benthic community (Ward and Dufford 1979).

Nekton

Nektonic, indeed even nektobenthic, insects are notably absent from torrential streams. Only vertebrates, primarily fishes, are able to swim upstream for any distance against strong currents. Among insects only a few streamlined baetid mayflies are able to swim against strong currents and then only for short bursts. The adaptations of insects to swift waters generally involve avoiding the current rather than swimming against it. This is well exemplified by elmid beetles, a highly successful family in torrential streams, the legs of which lack swimming hairs but possess large claws for clinging to the substrate. As described in Chapter 2, the elmids have evolved plastron respiration and need not move to the surface to replenish air stores, which would increase the chances of being swept downstream by the current. Swimming among the insect inhabitants of running waters is primarily restricted to those residing in pools, quiet margins, backwaters, or low-gradient reaches.

Madicoles

Thin sheets of water flowing over rock faces, the madicolous or hygropetric habitat, includes splash zones marginal to streams as well as isolated lotic microhabitats where, for example, water from a seep trickles down a cliff face. Vaillant (1956), in an excellent and comprehensive account of these special communities, defined madicoles as organisms residing in films of running water <2 mm thick, to distinguish them from the normal stream organisms (fluicoles) that inhabit deeper water. Madicoles may be intimately associated with the mosses that grow in such areas or may inhabit bare rock surfaces (Fig. 4.4). Diel and seasonal migrations may be undertaken by the fauna to avoid adverse conditions. For example, species living on exposed rock faces move to mosses and crevices during periods of intense isolation and attendant high temperatures.

The madicolous fauna is dominated by insects. Insects accounted for over 80% of the animal species at 74 sites in France, Corsica, and North Africa (Vaillant 1956). Members of all orders of aquatic insects may be found in madicolous situations. Those characteristic of this biotope are designated as members of the "hygropetrische zone" in *Limnofauna Europaea* (Illies 1978). In most cases the species especially adapted for and essentially restricted to the madicolous habitat (the eumadicoles) make up a small portion of the total fauna, although at a site in the French Alps eumadicoles accounted for about two-thirds of the species (Vaillant 1961). Vaillant identified 83 species of eumadicolous insects from the sites in France, Corsica, and North Africa:

Coleoptera	Psephenidae	1 species
Trichoptera	Beraeidae	6 species
	Psychomyiidae	4 species
	Hydroptilidae	6 species
Diptera	Tipulidae	9 species
	Ceratopogonidae	7 species
	Chironomidae	2 species
	Thaumaleidae	10 species
	Psychodidae	24 species
	Stratiomyiidae	13 species
	Dolichopodidae	1 species

Congeneric species are especially common among eumadicolous dipterans.

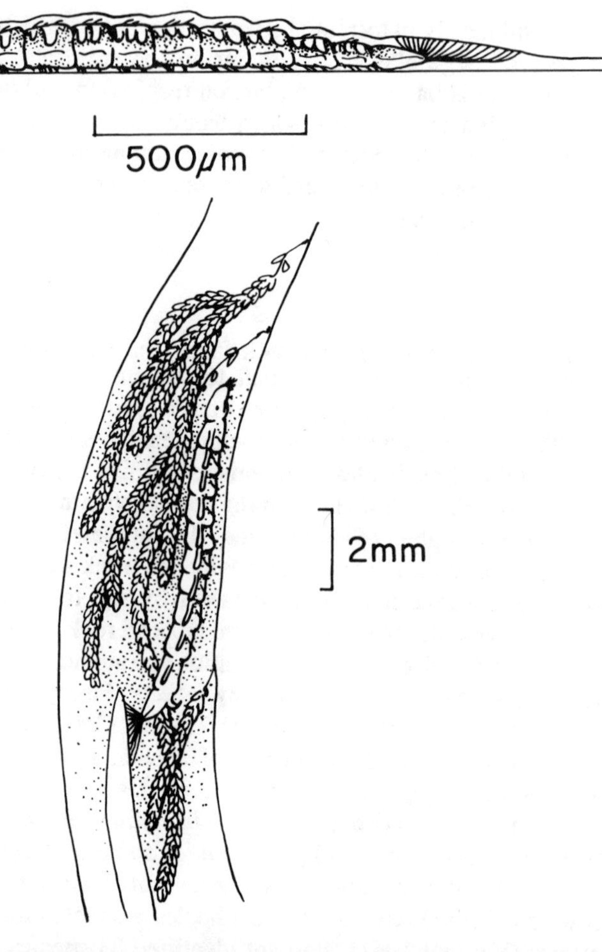

Figure 4.4 Madicolous Diptera. Top: *Pericoma pulchra* (Psychodidae) larva beneath a thin film of water on a rock face. Bottom: *Hermione torrentium* (Stratiomyidae) larva in its natural habitat among moss on a vertical rock face. [Modified from Vaillant (1956).]

The psychodid genus *Pericoma* contributes 20 species to the preceding list; the stratiomyiid *Hermione* contributes 13; and the thaumaleid *Thaumalea* contributes 10. Eumadicolous dipterans breathe atmospheric air in contrast to most rheophilic fly larvae. The problems of surfacing for air in rapid streams would hardly apply to organisms, the bodies of which are frequently thicker than the depth of water they inhabit.

Normal stream samples occasionally contain madicolous insects, such as dixid midge larvae, that typically reside in marginal splash zones. Some madicoles, such as the hydraenid beetle *Ochthebius exsculpus* (Beier and Pomeisl 1959), spend part of their lives in the stream proper and other phases in the marginal hygropetric habitat.

Spangler, who states that isolated madicolous habitats yield rare and new taxa, has described new genera and species of beetles from such biotopes in the tropics (e.g., Spangler 1972). Danecker (1961) presented a detailed analysis of habitat conditions, biology, and ecology of the hygropetric caddisflies *Tinodes* and *Stactobia*.

Hyporheos

The hyporheic biotope, so named by Orghidan (1959), is the subbenthic habitat of interstitial spaces between substrate particles in the stream bed. This biotope is best developed in gravel bed streams with relatively silt-free interstices (Dole 1983, Pennak and Ward 1986, Stanford and Ward 1988). The hyporheic fauna or hyporheos [see Williams and Hynes (1974) for a discussion of terminology] consists of (1) permanent residents that do not normally occur in the surface benthos (crustaceans, mites, nematodes, and others) and (2) species that spend part of their aquatic lives in the hyporheic zone and the remainder in the surface benthos (many stream insects). The early instars of many species of benthic stream insects occur deep within the stream bed if suitable interstices are present (Williams 1984). There are, however, among aquatic insects certain species that are part of the hyporheos for the vast majority of their lives. In Yugoslavian rivers studied by Meštrov and Tavčar (1972), 34 chironomid taxa were identified, 32% of which were collected only in the hyporheic zone. Hyporheic insects must of course move to the surface for the final ecdysis to an aerial adult. Some species the terrestrial adults of which are collected in large numbers rarely occur in samples of surface benthos, then only as mature larvae, and apparently remain within the hyporheic zone until immediately prior to emergence. Of the 42 species of stoneflies in the Flathead River, Montana, 8 spend their entire nymphal existence within the alluvial floodplain aquifer, returning to the river channel only to emerge, mate and oviposit (Stanford and Ward 1988). These hyporheic species, which undertake extensive migrations through the subterranean interstices, have been collected in large numbers in wells up to 2 km from the river and to depths of 10 m below the floodplain surface. At greater distances from the river the stoneflies are replaced by true groundwater forms (stygobionts), such as blind crustaceans, that do not have a terrestrial adult stage. Virtually nothing is known regarding the subterranean food webs of such expansive aquifer systems. The food base probably involves uptake, utilization, and transformation of carbon compounds contained in groundwater by the heterotrophic biofilm coating mineral particles.

The extent of the hyporheic zone (Fig. 4.5), which might best be defined as the area colonized by amphibiotic stream insects, varies according to substrate porosity and dissolved oxygen levels. The most extensive hyporheic zones are associated with alluvial deposits dominated by coarse sand and gravel derived from crystalline rock. Streams flowing over bedrock or compacted clay (hardpan) lack a hyporheic zone, but aquatic insects typically extend 30 cm or more into the coarse substrate of streams (Angelier 1962, Schwoerbel 1967, Coleman

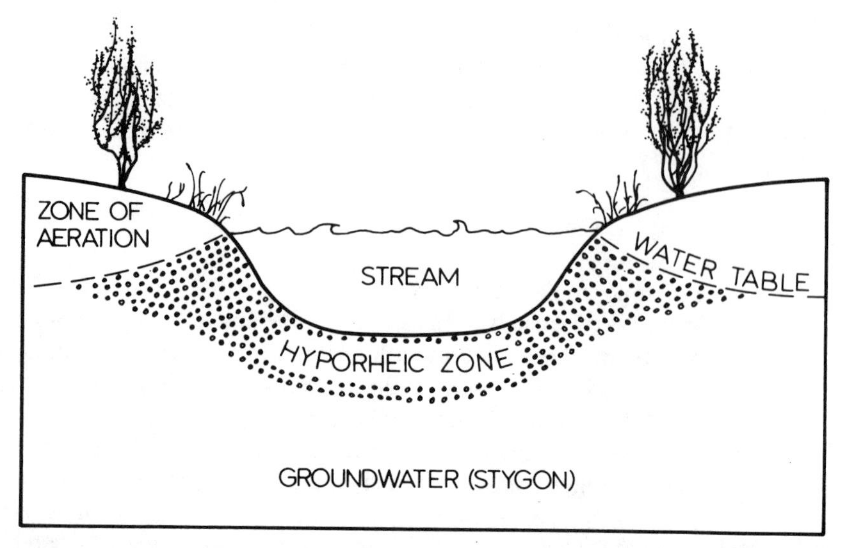

Figure 4.5 A diagrammatic cross section of a gravel bed stream showing the possible extent of the hyporheic zone.

and Hynes 1970, Radford and Hartland-Rowe 1971, Bishop 1973b, Poole and Stewart 1976, Danielopol 1976, Bou 1979, Bretschko 1981). Under optimal conditions, such as described above for the Flathead River, the hyporheic zone has a vertical dimension scaled in meters and a lateral extent on a scale of kilometers.

Hyporheos are exposed to quite different environmental conditions than the surface benthos. Current and light rapidly decline with depth and the temperature range is less than in surface waters. In a Canadian river Williams and Hynes (1974) reported a linear decrease of oxygen with depth with only 5% saturation values at 30 cm below the substrate surface, whereas Schwoerbel (1961) and Husmann (1971) found saturations of ≥75% at 30 cm in German rivers. In an alpine brook in Austria with high substrate porosity, there was no depletion of oxygen in interstitial waters down to at least 60 cm (Bretschko 1981).

Curves of insect abundance as a function of depth in the substrate indicate that a large portion of the organisms of some streams is missed with traditional benthos sampling methods that collect organisms from only the top several centimeters of substrate. Some groups may attain maximum abundance at some depth below the substrate surface (Fig. 4.6). In a small tropical stream Bishop (1973b) found that ≤50% of the benthos occupied the upper 10 cm of substrate. Pugsley and Hynes (1983), using a modified freeze-core technique, found that ~70% of the fauna occurred in the upper 10 cm of substrate. These results contrasted with previous estimates at the same locations in which a greater proportion of the animals were thought to occur at greater depths. However, Bretschko (1985) presents convincing evidence that hyporheic animals are repelled by the advancing freeze front and move vertically upward. When an

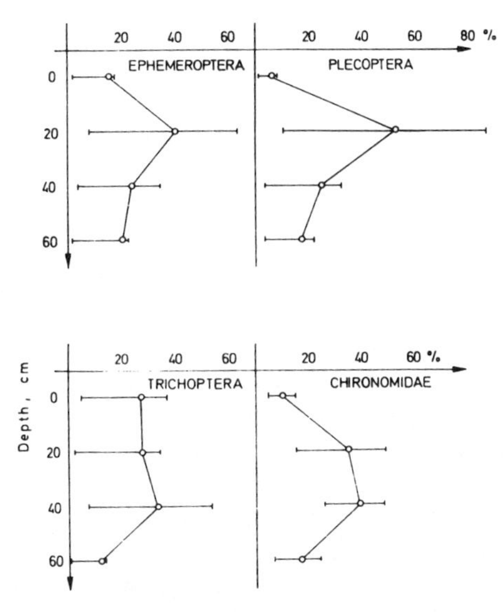

Figure 4.6 Relative vertical distribution of major insect taxa in the substrate of an Austrian mountain stream during April. Horizontal bars show 95% confidence intervals, calculated after a log (X + 1) transformation. [Modified from Bretschko (1981).]

electric field was used to paralyze the fauna *in situ* prior to freeze-coring, the abundance maximum was deeper than when the electric field was not applied.

The hyporheic zone serves as a refuge for benthic insects (Schwoerbel 1967). It offers shelter from floods, drought, and extreme temperatures, and provides suitable and predictable conditions for immotile stages such as eggs, pupae, and diapausing nymphs and larvae. It offers early instars some protection from large predators. The hyporheos forms a faunal reservoir capable of recolonizing the surface benthos should the latter be depleted by adverse conditions (Williams and Hynes 1976b). Although some stream insects such as black flies (Simuliidae) do not normally occur below the substrate surface [see Courtney (1986) for an exception], many species are, at some stage in their life cycles or at some time during the year, more abundant in deep than in superficial strata. The hyporheic zone plays a major ecological role in streams with coarse substrates, a role that remains to be fully understood.

Surface Benthos

The vast majority of stream insects are members of the benthos that, as has been demonstrated, may extend some distance downward and laterally into the hyporheic zone. The separation of benthos into two components, the surface benthos and the hyporheos, although artificial, is necessary for purposes of discussion since most benthic studies have sampled only the top few centimeters of substrate.

Because of the turbulence created by the current, thermal and chemical stratification are not nearly as pronounced in running waters as in lentic habitats. In rapid streams the mixing is complete, although slowly flowing riverine reaches may exhibit temperature stratification and oxygen depletion in lower strata. Environmental conditions in the slow-flowing lower reaches of rivers are similar to conditions in lentic waters and the benthos also has a lentic character. With a few interesting exceptions to be considered later, the insects inhabiting slow-flowing river habitats are facultatively riverine lake and pond species.

The distinctive and highly adapted benthos of high-gradient rocky streams is remarkably similar worldwide and is dominated by insects except in limestone streams where crustaceans and other noninsect groups become abundant (Hynes 1970b). Table 4.2 compares high-gradient rocky streams in New Zealand, Scotland, North America, and Middle Asia with a tropical forest stream, based on the composition of aquatic insects. Aquatic insects comprised the major portion of the surface benthos in all cases. Five orders (Ephemeroptera, Plecoptera, Trichoptera, Diptera, Coleoptera) made up virtually the entire entomofauna, except in the tropical forest stream. In the temperate streams the coleopteran fauna was essentially restricted to Dryopoidea (riffle beetles), of which only members of the family Elmidae occurred at the two high mountain locations in Table 4.2. In high-gradient reaches of the tropical stream, dryopoids (Elmidae, Psephenidae, Dryopidae, Ptilodactylidae) and helodids comprised the majority of the beetle fauna, although members of a few other families were also occasionally collected. Odonates accounted for most members of the "other insects" category in the tropical stream. Hemipterans, corydalids, and aquatic moth larvae occurred in low numbers.

In an upland rainforest stream in tropical Queensland, Australia, insects were highly diverse and dominated the benthos as much as in high latitude streams (Pearson et al. 1986). In a 50-m stretch of riffle 245 species of insects were identified and insects constituted 98% of the macroinvertebrate fauna. Although tropical, the Queensland stream has a seasonal climate and does not exhibit the environmental constancy associated with the wet equatorial tropics.

All aquatic insects of rapid streams (excluding the hygropetric community) utilize dissolved oxygen; species that breathe atmospheric air and require periodic excursions to the surface are typically absent. Nondryopoid coleopterans, hemipterans, and odonates tend to be absent or poorly represented in the benthic communities of high-gradient rocky streams (but to a less marked degree in the tropics). These groups, especially nektobenthic forms or species associated with higher aquatic plants, typically reside in lentic habitats and in stream reaches with low gradient and fine substrate.

Many species of aquatic insects are confined to running waters (rheostenic) because of inherent current requirements associated with their respiratory physiology or feeding mechanisms. Taxa with an evolutionary history in high-gradient streams have had little selection pressure for developing adaptations to cope with low-oxygen conditions, and are, therefore, restricted to well-aerated waters. Some rheostenic species are so dependent on current that they succumb when placed in still water, even if the water is saturated with oxygen. Elmid

Table 4.2 Relative numerical contribution (%) of insects to total surface benthos of selected high-gradient streams

Stream	Percentage Composition[a]							Study
	Total Insects	Eph	Ple	Tri	Dip	Col	Oth	
New Zealand trout stream (Horokiwi Stream)	>93	14	<1	51	15	13	+	Allen 1951 (all quantitative samples)
Scottish trout stream (River Endrick)	97	20	27	19	28	3	+	Maitland 1966 (Station 9)
Rocky Mountain trout stream (N. St. Vrain)	97	61	5	22	7	2	0	Ward 1975 (rubble)
Tien Shan mountain stream (Issyk River)	96	61	6	20	8	1	0	Brodsky 1980 (Point 53)
Tropical forest stream (Sungai Gombak)	>83	13	5	5	32	25	>3	Bishop 1973a (Station I-r)

[a]Eph = Ephemeroptera; Ple = Plecoptera; Tri = Trichoptera; Dip = Diptera; Col = Coleoptera; Oth = Other insects; + = rare occurrences.

beetles, frequently the only coleopterans present, at least in large numbers, in torrential streams, extract dissolved oxygen from the water with a permanent physical gill, the plastron. Although obviating the need to periodically renew air stores at the surface, plastrons work in reverse in oxygen-depleted waters (i.e., oxygen diffuses from the plastron into the water), and elmid beetles are restricted to turbulent well-aerated habitats.

It was suggested by Ruttner (1926) that running waters are physiologically richer than lentic habitats. The unidirectional flow of water provides the energy that would otherwise have to be expended to irrigate respiratory surfaces. The current brings in food and oxygen and carries away waste products. Many lotic insects have adopted a semisessile or sedentary mode of life and exhibit a variety of adaptations for extracting food particles from the water column.

Adaptations to Current In this section attention will be directed primarily toward morphological adaptations to current. Most of these involve adaptations for avoiding current rather than resisting it and act in concert with behavioral mechanisms (Table 4.3). Adaptations solely involved with procuring the food carried by the current will not be considered here.

Of major importance is the fact that aquatic insects in torrential streams are not continually exposed to high current velocities. Low-current microhabitats in rapid streams occur under rocks, in substrate interstices, in the "dead-water" areas on the downstream faces of rocks. Beginning with Steinmann (1907), dorsoventral flattening of the bodies of aquatic insects was regarded as an adaptation to residing in current-exposed habitats of torrential streams. This logical idea was strengthened when Ambühl (1959) introduced the concept that a thin "boundary layer" of low-velocity water occurs on substrate surfaces even in very rapid reaches. Therefore, if insects were flattened or small in size, they could inhabit the upper surfaces of rocks in fast water and not be exposed to appreciable current. Rarely in nature are things quite so simple, and Ambühl's concept of the boundary layer has proved to be a gross oversimplification.

A dramatic change in our understanding of microcurrent patterns and the influence of hydraulic forces on stream insects began in the early 1980s (e.g., Smith and Dartnall 1980; Statzner 1981b; Statzner and Holm 1982, 1989; Chance and Craig 1986; Davis 1986; McShaffrey and McCafferty 1987; Statzner et al. 1988; Davis and Barmuta 1989; Wetmore et al. 1990). On the basis of this body of work, a description of the microcurrent patterns to which lotic insects are exposed is presented in Chapter 8.

Exposure to rapid current is now known to be endured by aquatic insects at a considerable cost of energy. Because rich food resources occur on the tops of rocks, many lotic species feed in current-exposed positions, but only for restricted periods. For much of the time these same species reside in sheltered locations under rocks or in crevices. Stream insects in fact have not evolved optimal morphologies for resisting the forces of the current. This is because evolution of body shape in running waters is influenced by at least five physical factors: diffusion through boundary layers, corrasion, lift, friction, and pressure

Table 4.3 The traditional view of morphological and behavioral adaptations of aquatic insects to current (see text)

Adaptations	Suggested Functions	Lotic Examples
	Morphological	
Flattening	Current avoidance (reside in boundary layer or crevices)	Psephenidae (Coleoptera), Heptageniidae (Ephemeroptera)
Small size	Current avoidance (reside in boundary layer or crevices)	Elmidae (Coleoptera)
Streamlining	Fusiform shape and smooth contours offer least resistance to current	Baetidae (Ephemeroptera)
Suckers	Firm attachment to smooth surfaces	Blephariceridae (Diptera)
Friction pads	Increased body contact with substrate	*Drunella doddsi* (Ephemeroptera), *Rhithrogena* (Ephemeroptera), Psephenidae (Coleoptera)
Claws and hooks	Reduced chance of dislodgement	Elmidae (Coleoptera), Rh*yacophila* (Trichoptera), *Diamesa* (Diptera), Deuterophlebiidae (Diptera)
Silk and sticky secretions	Attach to object in rapid current	Psychomyiidae (Trichoptera), *Petrophila* (Lepidoptera), Simuliidae (Diptera), eggs of many Plecoptera
	Behavioral	
Ballast	Reduced buoyancy	*Silo* (Trichoptera)
Positive rheotaxis	Reduces accidental dislodgement and counters downstream displacement	*Leptophlebia cupida*, (Ephemeroptera)
Negative phototaxis	Current avoidance by cryptic behavior	Heptageniidae (Ephemeroptera)
Current preferenda	Select suitable microcurrent habitats	Many, perhaps most, stream insects

drag. "A simultaneous effective morphological adaptation to all five physical factors in physically impossible . . ." (Statzner and Holm 1989). In addition, because drag forces vary as a function of body size (Statzner 1988), dramatically different morphologies would be required to minimize drag in young and fully grown individuals of the same species.

Dorsoventral flattening of the body is a characteristic feature of several groups of stream insects, being especially apparent among heptageniid mayflies (or their flattened leptophlebiid counterparts in the Southern Hemisphere) and psephenid beetles (Table 4.3). In addition to the traditional view of flattening as a mechanism of current avoidance while remaining in high-velocity microhabitats, flattening also enables insects to move into crevices under rocks where many reside during daylight hours. Predation pressure is reduced by avoidance (cryptic behavior) and by the difficulties of detecting and dislodging a prey species flattened against a solid surface. Dorsoventral flattening is also exhibited by some lentic species and those residing in depositional areas of streams, presumably to enable them to remain on the surface of soft substrates.

Statzner and Holm (1982) analyzed the microcurrent isovels surrounding an aquatic insect (Fig. 4.7) with a laser technique. Based on their results, the boundary layer is thinner than previously thought and the isovels, rather than forming smooth curves over the animal's body as described by Ambühl, exhibit more complicated patterns. They also found evidence that insects may exert active control over their immediate boundary layer by adjustments in body attitude, a conclusion supported by Smith and Dartnall (1980).

Small size has been suggested as an adaptation to current since it enables animals to live in the boundary layer or seek refuge in substrate interstices. The most successful beetles in rapid streams, the elmids (Fig. 4.8), are indeed smaller than most aquatic coleopterans. As with flattening, small size may not have evolved as an adaptation directly related to current, but in this case relating to the plastron respiration of adult elmid beetles.

Streamlining, a fusiform body shape and smooth contours, is highly evolved in a few lentic beetles and bugs as an adaptation for swimming, as described in Chapter 3. Among typical lotic insects, however, streamlining is well developed only in a few baetid mayflies (Fig. 4.9), which are virtually the only insects able to swim for even short distances against strong current.

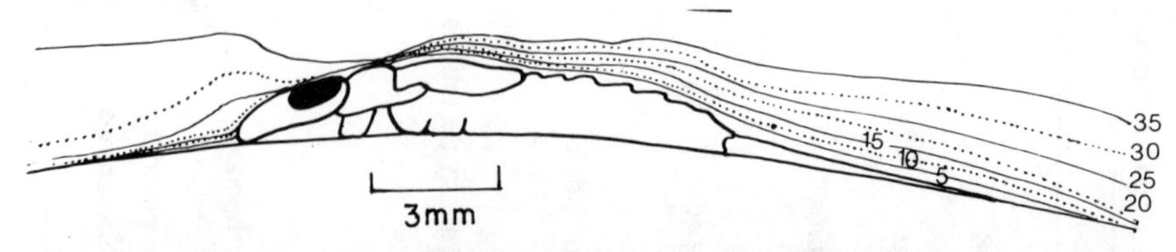

Figure 4.7 Schematic (legs omitted) drawing of the nymph of *Ecdyonurus* cf. *venosus* showing the isovels around the median line of the animal's body. Mean velocity in profile = 38.6 cm/sec; mean velocity within first 10 mm above the substrate = 36.0 cm/sec; water depth above the highest point of the substratum = 2.8 cm. [Modified from Statzner and Holm (1982).]

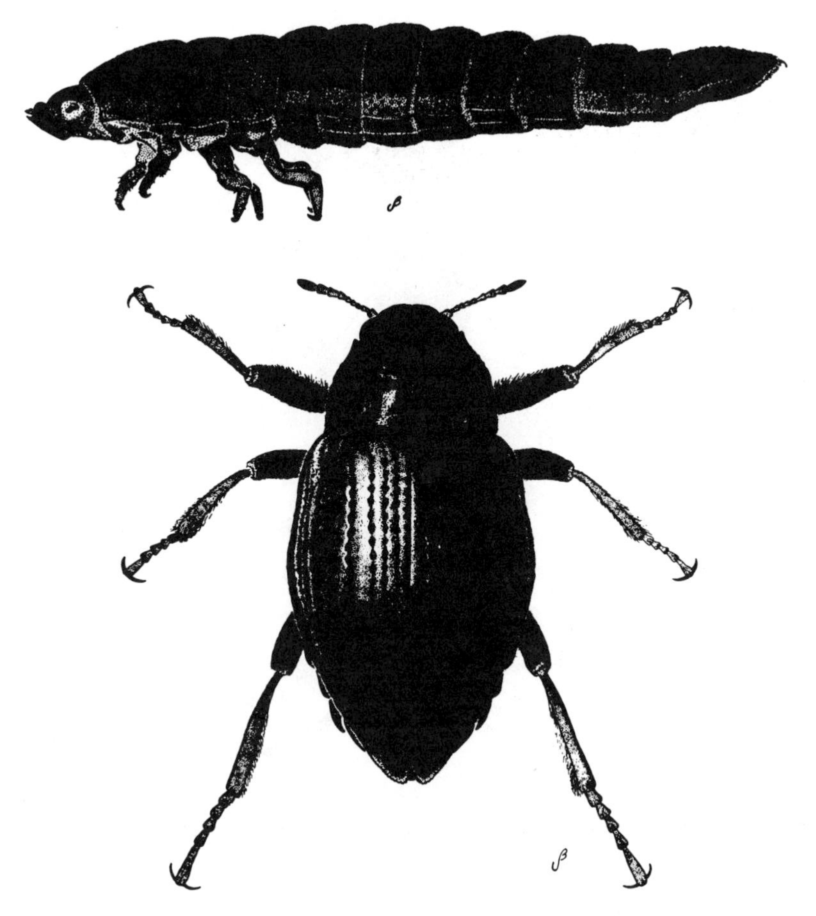

Figure 4.8 Larva and adult *Heterolimnius corpulentus,* an elmid beetle of torrential streams. Adults are <3 mm long.

Among typical stream insects only net-winged midges (Blephariceridae) possess true hydraulic suckers. Larvae reside on the tops of rocks in mountain streams, the habitat to which they are restricted. They maintain position by attaching six ventral suckers (Fig. 4.10) to smooth rock surfaces. In experimental channels larvae were able to move upstream against a current of 240 cm/sec (Dorier and Vaillant 1954). Larvae of the madicolous psychodid *Maruina* have eight ventral suckers enabling them to maintain position on vertical rock faces in hygropetric habitats.

Several insects of rapid streams possess structures that, while not as efficient as hydraulic suckers, increase frictional resistance by increasing body contact with the substrate. The gills of several heptageniid mayflies (Fig. 4.11), and their flattened leptophlebiid counterparts in the Southern Hemisphere, are enlarged and positioned to increase marginal contact with the substrate and to deflect the

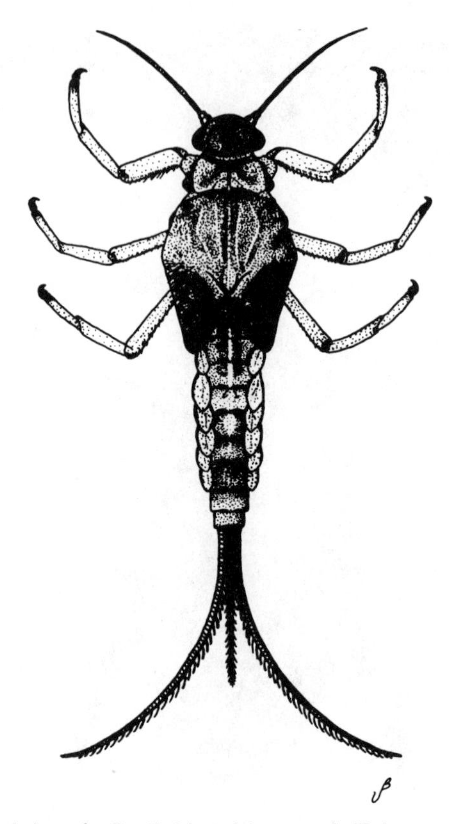

Figure 4.9 Dorsal view of a *Baetis tricaudatus* nymph (Ephemeroptera: Baetidae).

current away from the underside of the body. Many of these species also have expanded and flattened labrums, which may be fringed with hairs, that can be appressed to the substrate enabling the nymphs to feed on periphyton on the tops of rocks in rapid water. Mayflies in other families with expanded labrums include *Drunella doddsi* (Ephemerellidae) (Fig. 4.12) of North America and *Dicercomyzon* (Tricorythidae) from Africa. Both of these mayflies also possess dense hair fringes on the ventral abdominal surface that serve as friction pads. The legs of *Dicercomyzon* are highly expanded and flattened (Ventner 1961). Flattening, variously developed in many stream insects, not only enables insects to reside in the area of reduced current but also tends to increase frictional resistance with the substrate. Psephenid beetle larvae possess peripheral spines that seal the edges against the substrate, and nymphs of the Asian dragonfly *Ictinus* have rows of spines on the ventral surface of flattened and expanded abdominal sterna (Hora 1930). A few zygopterans residing in rapid streams have enlarged caudal lamellae with setigerous margins that serve as adhesive organs (Corbet 1962).

Claws and hooks are employed by many stream insects to maintain position

Figure 4.10 Ventral view of a *Agathon elegantula* larva (Diptera: Blephariceridae). Note the six suckers.

in strong currents. Adult elmid beetles, the most important coleopterans in torrential streams, have large, stout tarsal claws (Fig. 4.8). Well-developed claws occur on the posterior larval prolegs of the free-living (caseless) trichopteran *Rhyacophila* and the megalopteran *Corydalus* (Fig. 4.13). Lotic chironomid larvae may have circlets of large hooks on the anterior and posterior prolegs. The seven pairs of lateral prolegs of mountain midges (Deuterophlebiidae) possess terminal circlets of hooks used by the larvae to cling to the tops of rocks in mountain streams (Fig. 4.14).

Silk and sticky secretions serve a variety of functions among aquatic insects, some of which are current adaptations. Larvae of the stream-dwelling pyralid moth *Petrophila* spin silken canopies over the depressions in rock surfaces where they reside, and some chironomids and psychomyiid caddisflies live in silken tubes attached to the substrate. The trichopterans with fixed larval cases use silk to attach the cases to the substrate; those with portable cases use silk to anchor themselves when they molt, and some species of *Brachycentrus* and

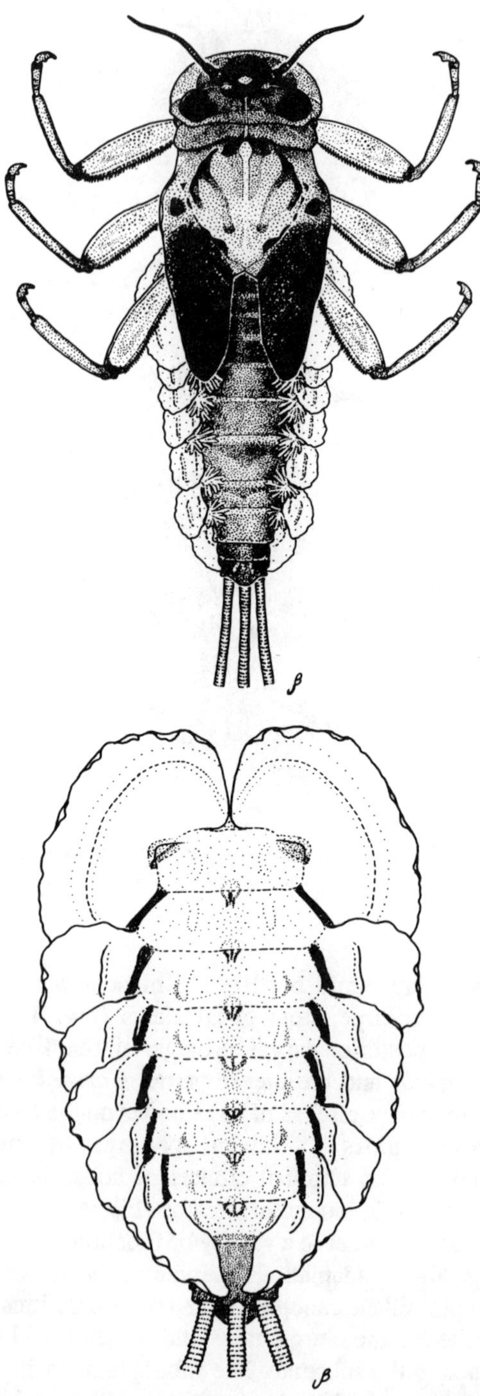

Figure 4.11 Dorsal view of a *Rhithrogena robusta* nymph (Ephemeroptera: Heptageniidae) and a ventral view of the abdomen (enlarged) showing friction disc formed by the gills.

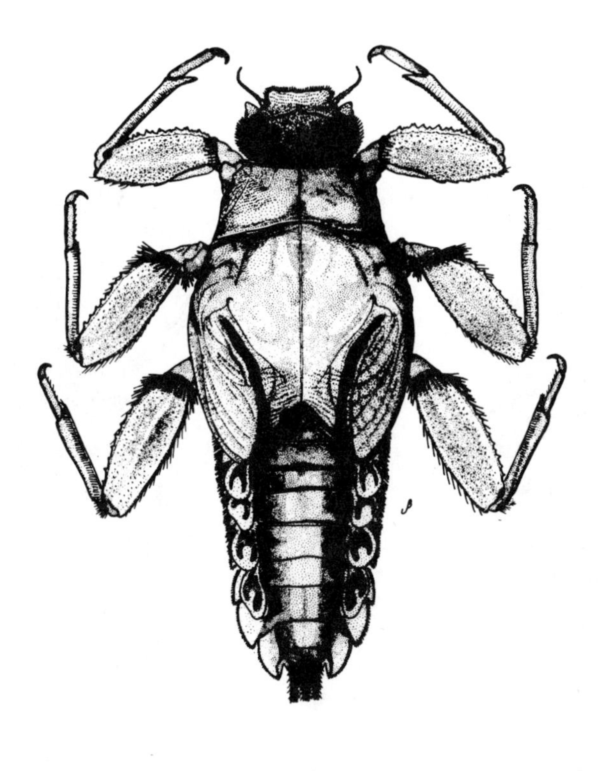

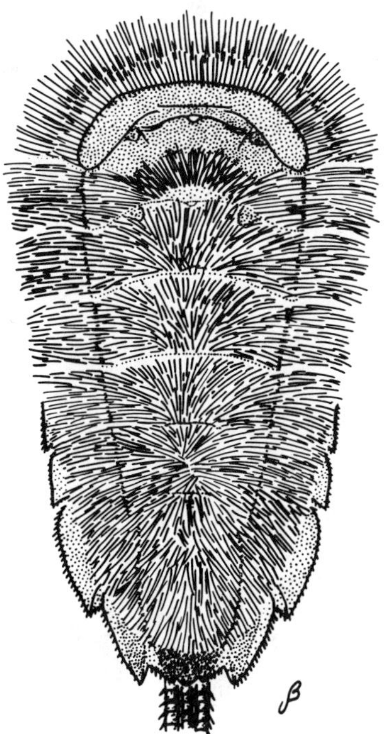

Figure 4.12 Dorsal view of a *Drunella doddsi* nymph (Ephemeroptera: Ephemerellidae) and a ventral view of the abdomen (enlarged) showing the elaborate hair pad.

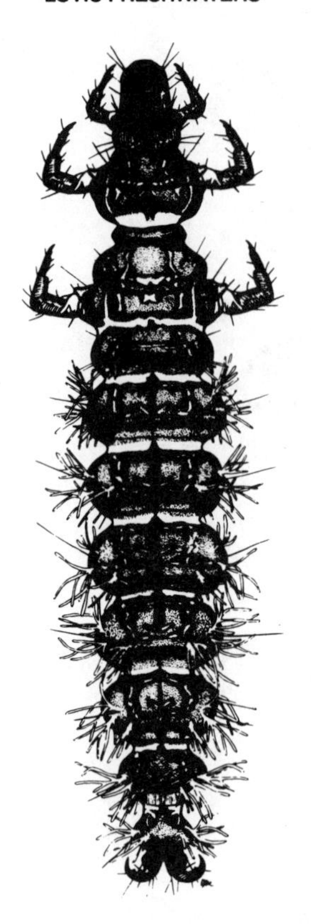

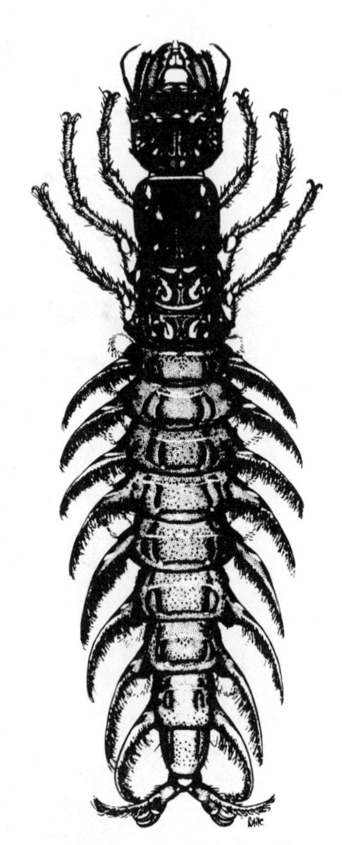

Figure 4.13 *Rhyacophila acropedes* (Trichoptera: Rhyacophilidae) and *Corydalus cornutus* (Megaloptera: Corydalidae). Note the well-developed claws on the posterior prolegs of these larvae.

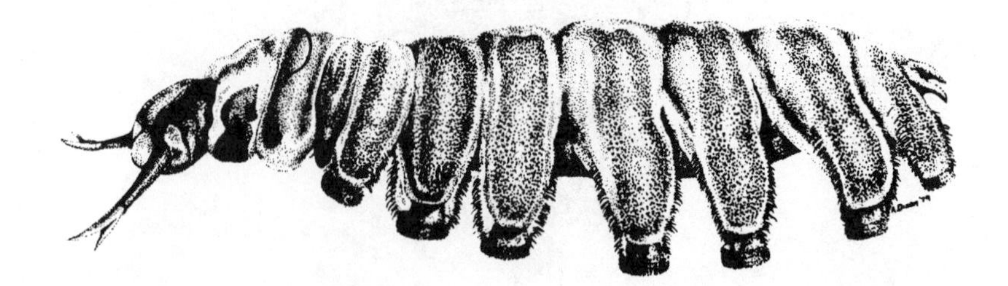

Figure 4.14 *Deuterophlebia coloradensis* larva (Diptera: Deuterophlebiidae) showing the seven pairs of lateral prolegs, each of which terminates in a circlet of hooks.

Oligoplectrum temporarily attach their cases on the tops of rocks while feeding on suspended particles carried by the current. All trichopterans construct pupal cases, which in lotic habitats are affixed to the substrate with silk. Larval simuliids (Fig. 4.15) spin a mat of silk on the substrate surface in which they engage a posterior circlet of hooks. When changing position, the larvae arch their bodies until the head contacts the substrate whereupon a new silk mat is spun by the salivary glands. The anterior proleg located near the head engages its circlet of hooks in the silk mat while the posterior end is released and reattached in the new position. Larvae also spin a safety line of silk to which they remain attached if disengaged from the mat spun on the substrate (Wotton 1986). Pupation occurs in a silk pupal case affixed to the substrate (Fig. 4.15). Various sticky substances are used by stream insects that attach their eggs to underwater surfaces during oviposition. More or less elaborate structures for attachment are associated with eggs laid on the water surface. Some are merely demersal and sticky and lodge in crevices in the substrate, whereas the eggs of some species contain coats of jelly or special adhesive structures projecting from the egg.

Most adult aquatic coleopterans of lentic habitats carry an underwater air store and are positively bouyant. Elmid beetles in contrast exhibit plastron respiration that not only obviates the need to surface, but because the plastron does not serve as an air store, allows them to be negatively buoyant.

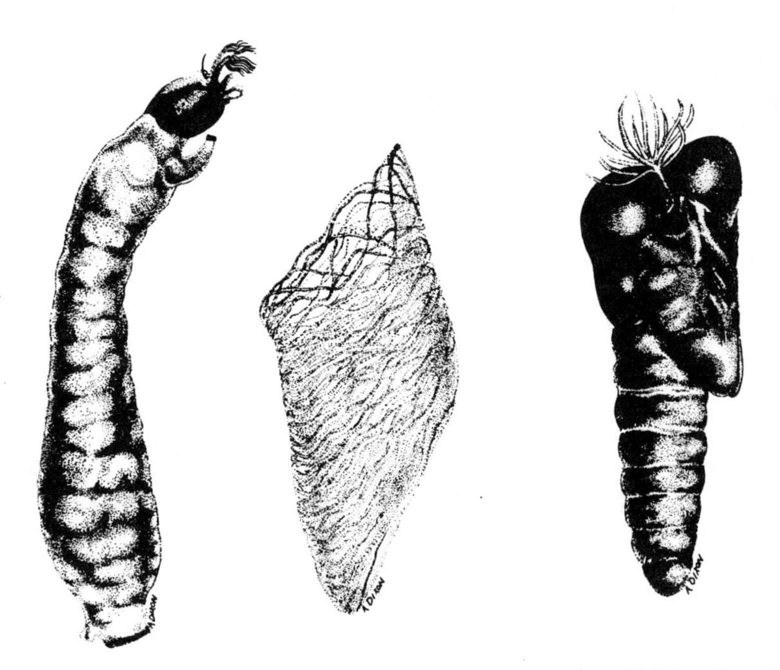

Figure 4.15 *Simulium arcticum* (Diptera: Simuliidae) larva (left), pupa (right), and pupal case (center).

Ballast involves mechanisms to increase density, thus reducing the likelihood of being swept downstream. Stream-dwelling caddisflies tend to construct heavier larval cases than lentic species, and some limnephilids (e.g., *Silo;* see Fig. 2.11) typically attach several larger ballast stones to the outside of their cases. The tendency for lotic limnephilid caddis larvae to construct heavier cases than their lentic counterparts (Webster and Webster 1943, Otto 1982) may merely reflect the availability of case-building materials or may represent different predator avoidance strategies (Otto and Svensson 1980, Otto 1982). Indeed, Statzner and Holm (1989) question whether the ballast stones added to *Silo* cases significantly reduce total lift forces.

Many stream insects are positively rheotactic and tend to orient toward the current. This tendency maintains a body orientation least liable to accidental dislodgement. Corkum and Clifford (1981) suggest that the subapical setae on the claws of *Baetis* inhabiting fast-water areas are mechanoreceptors that detect subtle changes in flow and assist the nymphs in orienting into the current. In addition, a propensity to move against the current counters downstream displacement. An active upstream migration by nymphs of the mayfly *Leptophlebia cupida* was documented in a Canadian stream (Hayden and Clifford 1974).

Negative phototaxis among stream organisms results in current avoidance through cryptic behavior. Many stream insects reside under rocks, appearing on the upper surfaces where current is greatest only during the night. There are, however, few data to indicate whether such behavior is a direct response to light.

Many, perhaps most, stream organisms exhibit current preferenda. A given species is exposed to only a portion of the total range of current velocities in a stream reach. Species-specific current preferenda have been demonstrated for a variety of stream insects (Grenier 1949, Phillipson 1956, Scott 1958, Ambühl 1959). Some members of the surface benthos are poorly adapted to resist current and are restricted to slow-water areas near shore or seek the current refugia within plant beds. Some plant-dwelling stream insects have large curved spines that presumably reduce the chances of their being swept from the preferred microhabitat (Steinmann 1907).

TYPOLOGY OF RUNNING WATERS

Running waters encompass a diverse array of habitats for aquatic insects, varying dramatically in size, current velocity, substrate, and many other factors. "Lotic" and "running waters" are roughly comparable all-inclusive terms for aquatic habitats with unidirectional flow. The term "stream" is somewhat ambiguous. Used broadly, it may apply to virtually any lotic system, as when referring to stream order discussed below. In a narrower context, "stream" applies to lotic reaches intermediate in size between brooks and rivers, the latter terms also being somewhat vague. More precise definitions of lotic systems are sometimes needed and will be introduced in the material to follow.

Stream Network Analysis Stream order is a method of roughly classifying running water segments by size and position within a drainage network (Fig. 4.16). Devised by Horton (1945) and modified by Strahler (1957), this stream ordering technique defines first-order segments as those without tributaries. The confluence of two streams of the same order produces the next highest order. The joining of two first-order streams produces a second-order segment, two second-order streams form a third-order segment, and so on. The headwaters or upper reaches of a drainage network may consist of first- through third-order segments; middle reaches are the fourth- to sixth-order segments; seventh and higher orders are generally riverine in character. The world's largest rivers are eleventh- or twelfth-order at their mouths. Stream ordering provides an objective though imperfect system of classifying running waters (Leopold et al. 1964). One problem is that stream order remains constant when streams of a lower order enter the main channel, irrespective of the number of such smaller tributaries.

Link magnitude (Fig. 4.16), although not as widely used as stream order, may provide a more meaningful comparison of stream segments, especially those in different catchments (Moeller et al. 1979). The link magnitude of a given reach is the sum of all first-order streams above that point in the drainage network (Shreve 1966). Link magnitude analysis thus provides a more sensitive method of classifying streams.

Riffles and Pools The formation of a riffle-pool sequence results from hydrodynamic adjustments of heterogeneous substrate materials to the potential energy of flowing water (Yang 1971). Riffles are shallower, have higher gradients, coarser substrate, and higher current velocities than pools. In natural streams

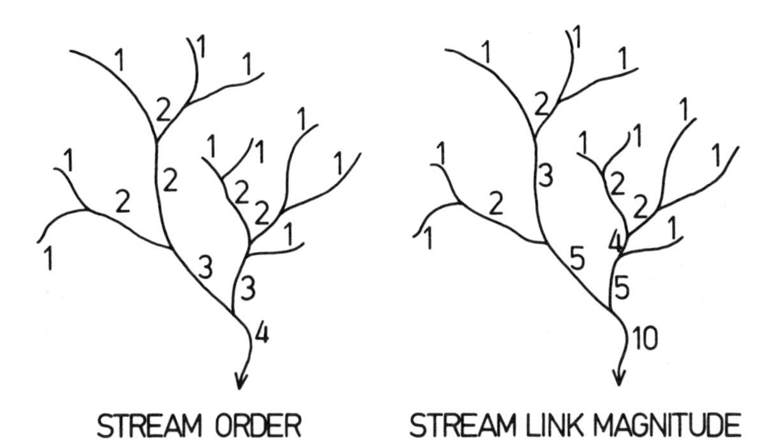

Figure 4.16 Two methods of classifying running water segments by size and position within a drainage network. (See text.)

riffles and pools alternate, with adjacent riffles generally spaced five to seven stream widths apart (Leopold et al. 1964). Riffles and pools provide quite different habitat conditions for aquatic insects and other organisms, although these differences are reduced during high discharge.

In high-gradient reaches of undisturbed forest streams, organic debris dams may create a stairstep profile of alternating riffles and pools. Fallen trees or tree branches lodged in the stream provide a structural framework that traps smaller pieces of organic matter in the pool that eventually forms on the upstream side of the obstruction. This exaggerated riffle-pool sequence created by debris dams is most prevalent in heavily forested headwater streams. In the Hubbard Brook Experimental Forest of New Hampshire (USA), debris dams contained nearly 75% of the coarse benthic detritus in first-order streams, 58% in second-order streams, and 20% in third-order segments (Bilby and Likens 1980). Debris dams reduce downstream losses of organic matter, provide food resources, and increase habitat complexity and living space for aquatic insects.

Longitudinal Patterns

The flow of the river is ceaseless and its water is never the same.

—Kamo No Chomei

From Hojoki (*An Accountof My Hut*) A.D.1212

Longitudinal changes in ecological conditions from the headwaters to the mouth of river systems are profound, especially over large elevation gradients, as are the attendant changes in the structural and functional attributes of lotic insect communities. Table 4.4 lists some studies specifically dealing with downstream changes in the entire aquatic insect fauna. Studies dealing with downstream changes in selected components of the entomofauna include Dodds and Hisaw (1925), Ide (1935), Botosaneanu (1959), Berthélemy (1966), Knight and Gaufin (1966a), Décamps (1967, 1968), Kamler (1967), Mecom (1972), Schoonbee (1973), Lillehammer (1974), Allan (1975), Statzner (1975), Wise (1976), Kownacki and Zosidze (1980), Ward and Berner (1980), and Ward (1981, 1982) among others. The following paragraphs summarize the longitudinal distributions of insects in a lowland river, a hill stream, and a mountain stream.

A LOWLAND RIVER Berg et al. (1948) conducted a comprehensive study of the Susaa, a 90-km-long lowland river in Denmark. Quantitative data on benthos were collected at five main locations along the longitudinal profile (Table 4.5). Some of the between-site variations in the composition of benthic insect communities related more to differences in current and substrate than to the relative position of the site along the length of the river. The uppermost (tributary brook) and lowermost (Susaa at Naaby) locations were faunistically similar in several respects, yet differed from the intervening sites. Elmid beetles and black flies (*Simulium*), for example, were abundant only at these locations which had higher current velocities than the middle sites and were the only ones where large stones occurred. However, whereas the tributary brook had a predominantly stony

Table 4.4 Selected longitudinal studies of river systems with an emphasis on benthic fauna that included all orders of insects

Lotic System	Reference
Lowland Rivers	
River Susaa, Denmark	Berg et al. 1948
Illinois River, USA	Richardson 1921
River Smohain, Belgium	Marlier 1951
Volga River, USSR	Behning 1928
Hill Streams	
Horokiwi Stream, New Zealand	Allen 1951
River Endrick, Scotland	Maitland 1966
River Duddon, England	Minshall and Kuehne 1969
Sungai Gombak, Malaysia	Bishop 1973a
Arima River, Trinidad	Hynes 1971
Eerste River, South Africa	King 1981
River Rheidol, Wales	Jones 1949
River Fulda, Germany	Illies 1953a
Mountain Streams (<1000 m gradient)[a]	
Provo River, USA	Gaufin 1959
Parma Stream, Italy	Bonazzi and Ghetti 1977
N. Boulder Creek, USA	Elgmork and Saether 1970
River Kaunnai, Japan	Okazawa 1975
Furans Stream, France	Bournaud et al. 1980
Vaal and Wilge R., South Africa	Chutter 1970
Mountain Streams (>1000 m gradient)[a]	
River Huallaga, Peru	Illies 1964
St. Vrain River, USA	Ward 1986
Great Berg River, South Africa	Harrison 1964
Streams of Tatra, Caucasus, and Balkan Mts.	Kownacka and Kownacki 1972
Issyk and Akbura River, Central Asia	Brodsky 1980
Streams of Mt. Elgon, East Africa	Williams and Hynes 1971
Tugela River, South Africa	Oliff 1960
Streams of Sri Lanka	Starmühlner 1984
Small German Streams	Braukmann 1987
Ethiopian mountain streams	Harrison and Hynes 1988

[a]Separation of mountain streams into two categories (<1000 m and >1000 m) is based on the elevation gradient encompassed by the study sites.

substrate, at Naaby the stones occurred with gravel, sand, and mud. The caddisfly *Hydropsyche augustipennis* and the mayfly *Baetis* occurred only at the tributary location, with the exception of a few individuals of each at Naaby. At Naaby the mud-dwelling mayfly *Caenis* replaced *Baetis* and *H. augustipennis* was replaced by polycentropodid caddisflies. It is perhaps not surprising that habitat characteristics such as current and substrate to some extent superceded position along the longitudinal profile in determining the composition of the benthic fauna in this lowland river.

Table 4.5 Common insects, listed in order of numerical abundance, at five locations on a lowland river, the Susaa (Denmark)

Tributary Brook	Forest Stream	Meanders	Gunderslevholm	Naaby
1. *Hydropsyche augustipennis*	1. Chironomidae	1. Chironomidae	1. Chironomidae	1. Chironomidae
2. Chironomidae	2. *Caenis*	2. Ceratopogonidae	2. *Micronecta borealis*	2. *Caenis*
3. *Simulium*	3. Ceratopogonidae	3. *Caenis*	3. Ceratopogonidae	3. Eruciform Trichoptera
4. Eruciform Trichoptera	4. Polycentropodidae	4. *Sialis*	4. Eruciform Trichoptera	4. Polycentropodidae
5. Elmidae	5. Eruciform Trichoptera		5. *Caenis*	5. Elmidae
6. *Baetis*	6. *Sialis*			6. *Simulium*
	7. *Sisyra*			

Source: From Berg et al. (1948).

A HILL STREAM The River Endrick, Scotland, rises from the high moorlands at slightly less than 500-m elevation and flows 49 km to Loch Lomond a few meters above sea level (Maitland 1966). Maitland established 12 sampling stations along the longitudinal profile of the River Endrick, which changes from a small brook at its source to a slow-flowing, medium-sized river in the lower reaches. The upper reaches have a steep gradient, are low in dissolved salts, and mosses are the only macrophytes present. The middle reaches have a moderate gradient and medium levels of dissolved salts. Mosses are the predominant macrophytes, but aquatic angiosperms are also present. The lower reaches have a low gradient, relatively high dissolved salt content, soft substrate, and a variety of aquatic angiosperms are present.

Most orders of insects were collected along the entire course of the River Endrick, although odonates, aquatic lepidopterans, and hemipterans (if pleustonic forms are excluded) occurred only in lower reaches where plecopterans were rarely taken. Megaloptera was represented by two species. *Sialis lutaria* was restricted to the extreme lower reaches, whereas *S. fuluginosa* was restricted to the source area. Only a few aquatic insects (e.g., the elmid beetle *Limnius tuberculatus*) occurred at all sampling stations, although several other species traversed nearly the entire length of the river system (Table 4.6). Some species were restricted to a limited segment of the longitudinal profile. Maitland concluded that the benthic fauna of the River Endrick does not exhibit distinct species associations but that the longitudinal pattern is one of transition.

A MOUNTAIN STREAM St. Vrain Creek, a Rocky Mountain trout stream, drops nearly 2000 m from its glacier-fed source in alpine tundra to the plains, over a distance of only 54 km (Ward 1986). Sampling stations were established at 11 locations along the longitudinal profile. Benthic sampling was restricted to rubble riffles to minimize differences otherwise attributable to current or substrate. Mosses and liverworts were the only aquatic macrophytes at all mountain stream stations. Aquatic angiosperms were restricted to the plains location where filamentous chlorphytes were also abundant.

The alpine tundra segment of St. Vrain Creek contained only a few species of insects from four orders (Ephemeroptera, Plecoptera, Trichoptera, Diptera). The mayflies (*Baetis bicaudatus, Ameletus* sp., *Cinygmula* sp.) and stoneflies (*Megarcys signata, Zapada oregonensis*) of the tundra stream were among the most widespread species along the altitudinal gradient (Fig. 4.17). The caddisflies, in contrast, contained a true headwater component consisting exclusively of limnephilids dominated by *Imania* (= *Allomyia*) *tripunctata*.

The general distribution pattern exhibited by ephemeropterans at mountain stream sites was the addition of species in the downstream direction without loss of those present at higher elevations. There was, however, a major faunal discontinuity between the lowest foothills site and the plains location, with very little overlap in the mayfly faunas of these adjacent sampling stations. No stoneflies were restricted to the headwaters or the plains, but some species occurred only in middle reaches. *Pteronarcys californica* exhibited the most

Table 4.6 Distribution patterns of some of the significant species[a] of insects[b] from the River Endrick, Scotland

Species with Restricted Distributions	Widely Distributed Species
Upper Reaches	
Ameletus inopinatus (E)	*Limnius tuberculatus* (C)
Leuctra nigra (P)	*Latelmis volkmari* (C)
Plectrocnemia conspersa (T)	*Helmis maugei* (C)
Middle Reaches	*Esolus parallelopipedus* (C)
Simulium variegatum (D)	*Amphinemura sulcicollis* (P)
Chloroperla tripunctata (P)	*Caenis rivulorum* (E)
Tinodes waeneri (T)	*Baetis rhodani* (E)
Lower Reaches	*Ephemerella ignita* (E)
Notonecta glauca (H)	
Corixa falleni (H)	
Corixa dorsalis (H)	
Ischnura elegans (O)	
Haliplus wehnckei (C)	
Agraylea multipunctata (T)	

[a]Reasonably abundant and identifiable to species in aquatic stages.
[b]*Key*: E = Ephemeroptera, P = Plecoptera, T = Trichoptera, C = Coleoptera, D = Diptera, H = Hemiptera, O = Odonata.
Source: From Maitland (1966).

restricted distribution, being confined to the lower foothills stream segment. Stoneflies were poorly represented at the plains site; only *Isoperla quinquepunctata* occurred regularly at this location. The trichopteran fauna con-

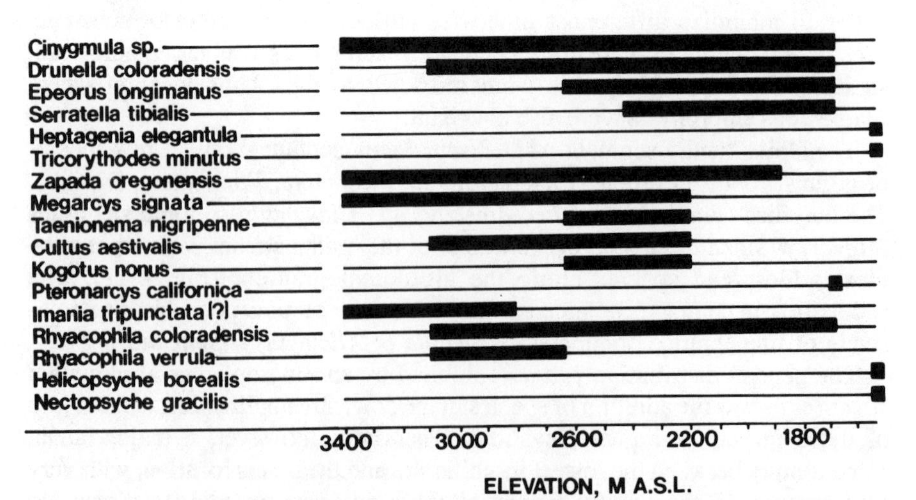

Figure 4.17 Altitudinal distributions of selected species of Ephemeroptera, Plecoptera, and Trichoptera from tundra to plains stations along the longitudinal profile of St. Vrain Creek, Colorado. [Modified from Ward (1986).]

tained species restricted to upper, middle, and lower reaches. Seven species of *Rhyacophila* occurred sympatrically on riffles at Sites 4 (3109 m) and 5 (2816 m). As was true of mayflies, a somewhat distinctive caddisfly community occurred at the plains location. Helicopsychidae and several species from other families were collected only from this site. Décamps (1968) distinguished six main distribution patterns for the Trichoptera of the Garonne River Basin in the French Pyrénées (Fig 4.18).

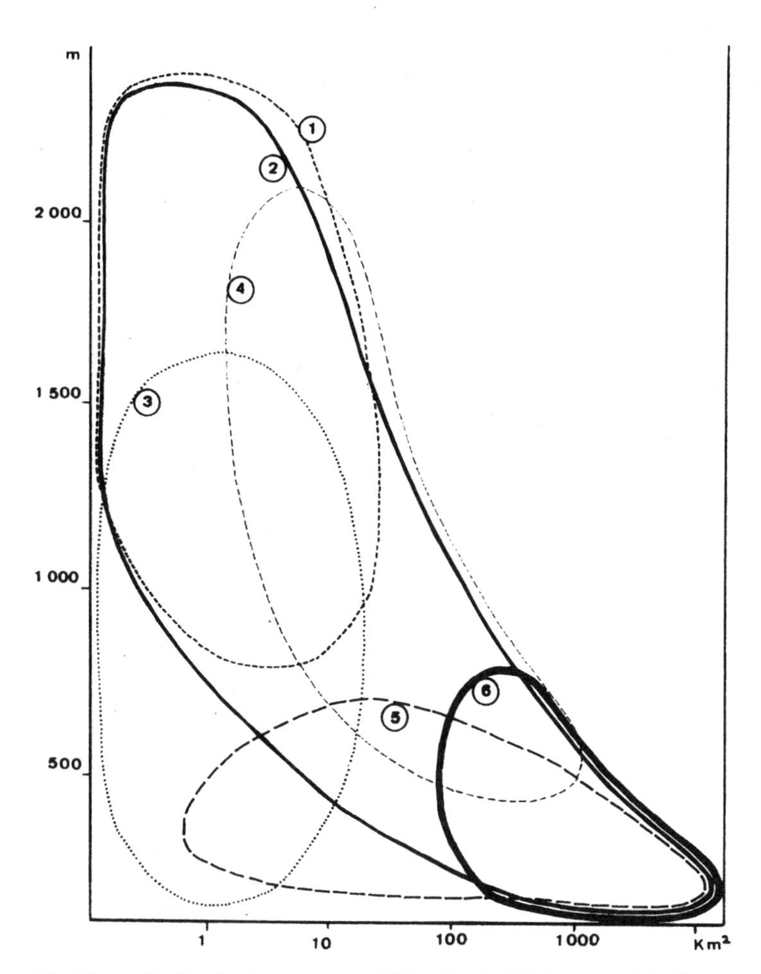

Figure 4.18 The major distribution patterns exhibited by the Trichoptera of the Garonne River Basin, French Pyrénées, according to altitude and stream size (drainage area). Distribution types are indicated as follows: 1, species that are quite frequent >2000 m, less frequent <1000 m; 2, species with a great altitudinal range, extending >1800–2000 m; 3, species in the springs and streams at low and medium altitude; 4, species of mountain rivers; 5, species of rivers at low altitude; 6, species of large rivers and therefore rare or absent >500 m elevation. [From Décamps (1968).]

The Zonation Concept A variety of stream classification and zonation schemes have been proposed to account for observed longitudinal patterns [see reviews by Illies and Botosaneanu (1963), Hynes (1970a), Hawkes (1975), Botosaneanu (1988)]. The longitudinal zonation scheme shown in Figure 4.19 had its inception when Illies (1961a) proposed a worldwide classification system for lotic waters and introduced the terms *rhithral* and *potamal* to divide lotic systems into two major zones, each with upper (epi-), middle (meta-), and lower (hypo-) subdivisions. Illies' earlier studies on the benthos of the Fulda River, Germany (Illies 1953a), demonstrated areas of rather sudden transition in community composition over short distances that corresponded closely to fish zonation patterns (Müller 1951). The faunal discontinuities or nodal points tended to occur at the confluence of segments of the same rank (i.e., where

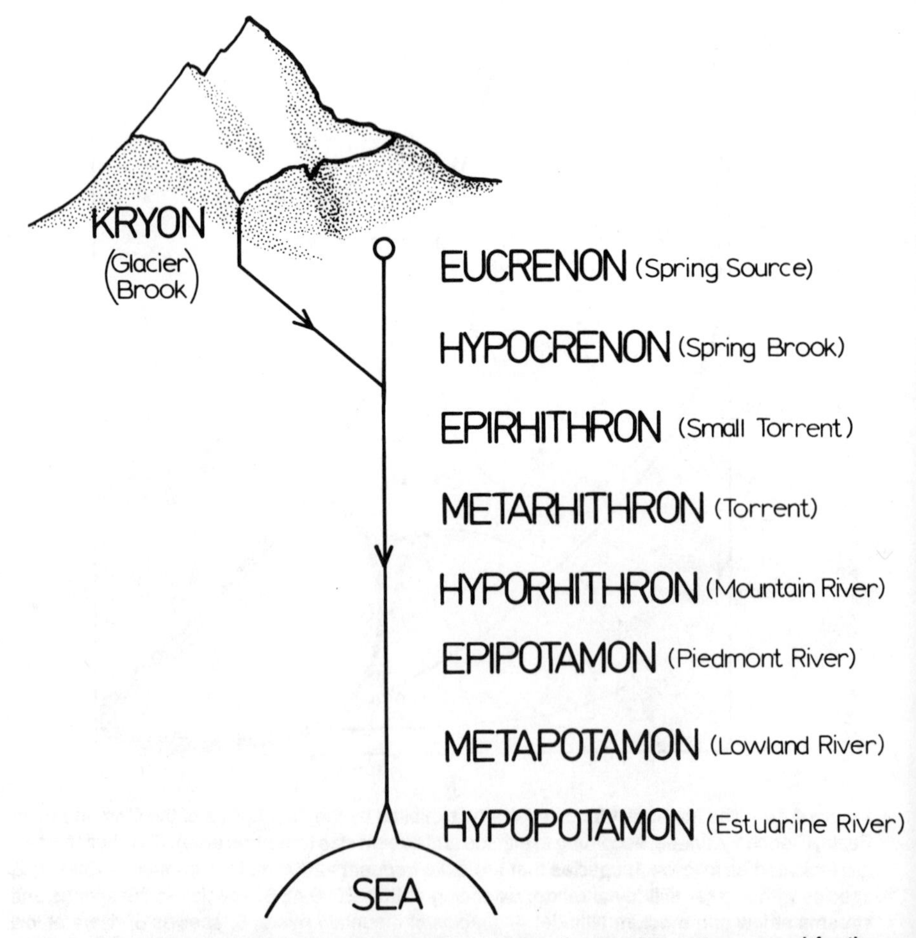

Figure 4.19 Diagrammatic longitudinal zonation scheme for running waters named for the communities occupying the zones proposed by Illies and Botosaneanu (1963) and Steffan (1971).

stream order increased) where there were shifts in factors such as temperature and current [see Thorup (1966) for a criticism of this interpretation]. The most marked faunal changes occurred at the lower end of the salmonid zone; areas upstream from this point were designated rhithral, those below were designated potamal.

Rhithral was defined to include the cooler upper reaches where the annual range of monthly mean water temperatures does not exceed 20°C. Other rhithral characteristics include high gradient, continuously high levels of dissolved oxygen, predominately coarse substrate, rapid current, high water clarity, and the absence of true plankton. The rhithral fauna is typified by morphological, behavioral, and physiological adaptations relating, directly or indirectly, to the cold water and high current velocity.

In potamal segments the annual range of monthly mean water temperatures exceeds 20°C; dissolved oxygen deficits occur at times; the current is slower and the water exhibits greater turbidity, but is less turbulent than in rhithral reaches; and finer substrate materials are more abundant. Reproducing populations of plankton may occur in the potamal. The species occurring in potamal segments frequently do not exhibit any special adaptations for running waters and also inhabit suitable lentic waters.

The following list includes aquatic insect families characteristic of rhithral and potamal zones according to Illies and Botosaneanu (1963) with slight modification:

Rhithral

- Ephemeroptera: Heptageniidae, Ephemerellidae, Leptophlebiidae
- Plecoptera: Capniidae, Leuctridae, Nemouridae, Gripopterygidae
- Trichoptera: Rhyacophilidae, Odontoceridae, Glossosomatidae, Philopotamidae (except *Chimarra*)
- Coleoptera: Elmidae, Psephenidae, Helodidae, Hydraenidae
- Diptera: Blephariceridae, Simuliidae, Psychodidae, Chironomidae (Podonominae)

Potamal

- Ephemeroptera: Siphlonuridae, Potamanthidae, Polymitarcyidae, Caenidae
- Plecoptera: Perlodidae, Perlidae
- Trichoptera: Leptoceridae, Hydroptilidae
- Coleoptera: Dytiscidae, Haliplidae
- Diptera: Culicidae, Tabanidae, Stratiomyidae, Chironomidae (Chironominae)
- Hemiptera: Corixidae, Notonectidae

Because many river systems originate as springs, Illies and Botosaneanu (1963) added the crenal zone above the rhithron. The eucrenal includes the source area, the hypocrenal the springbrook.

GLACIER BROOKS In mountains with maximum elevations exceeding the permanent snowline, yet an additional zone must be added. Streams arising from the meltwater of glaciers and permanent snow fields form a distinct biotope, the kryal, that contains a distinct biocoenosis, the kryon (Steffan 1971). Chironomids of the genus *Diamesa* were the sole animals of the upper zone (metakryal) in the glacier brooks examined by Steffan in northern Scandinavia. The lower zone (hypokryal) is inhabited by *Prosimulium* (Simuliidae) in addition to *Diamesa* in northern Scandinavia, although a zone dominated by simuliid larvae is not universally present (Kownacka and Kownacki 1972). The kryal zone as defined by Steffan extends only a short distance downstream (a few hundred meters). Beginning at about the point where summer water temperatures exceed 2°C, trichopterans, ephemeropterans, and plecopterans appear, and the glacier brook species decline.

Steffan characterizes glacier brooks by their extremely low summer temperatures, strong current, and unstable substrate (it is unlikely, however, that unstable substrate is a universal feature of glacial brooks). In contrast to the very small temperature amplitude of the metakryal, the hypokryal typically exhibits considerable diel thermal fluctuations in summer. Steffan (1971) considers temperature the most distinctive feature of glacier brooks and the factor primarily responsible for downstream faunal changes. It is difficult to find reference to turbidity in the few investigations of glacier brook biota. Not all streams arising from glaciers and snowfields carry loads of glacial flour (clay-sized particles of ground rock), and it appears that most studies of this kind have been restricted to relatively clear kryal habitats. Kownacka and Kownacki (1972, p. 749) state that "the turbidity cannot be a factor responsible for the development of the biocoenosis of the kryon, since e.g. in the Tatra Mts in spite of the lack of glacier streams with turbid water, a typical kryal community develops." They feel, as does Steffan, that low temperature is the essential factor for the development of the kryon. Autochthonous primary production is negligible near the glacier because of the short ice-free period, especially if the substrate is unstable. *Diamesa* larvae feed on wind-blown organic particles entrapped on the surface of the snow and ice and later released on melting (Steffan 1971). However, Kohshima (1984) found *Diamesa* larvae and blue-green algae growing in the tunnel-like meltwater drainage channels of a Himalayan glacier.

Larvae of *Diamesa lindrothi,* the predominant species of the metakryal zone in northern Scandinavia, move to small depressions in rock surfaces as they mature. Final-instar larvae spin mesh canopies over themselves prior to pupation so that pupae are protected from dislodgement by the current and from being crushed if the stone overturns.

In virtually all glacier brooks that have been examined, *Diamesa* is the predominant, usually the only, animal of the metakryal. Such is the case in glacier brooks in the European Alps (Thienemann 1954), Scandinavia (Saether 1968, Steffan 1971), the Tatra, Caucasus, and Balkan Mountains (Kownacka and Kownacki 1972), the Rocky Mountains (Elgmork and Saether 1970), the Himalayas (Kohshima 1984), and even the tropical mountains (Harrison 1965a). Brodsky (1980), however, reports another genus (*Phoenocladius*) in a turbid glacier brook at 3000 m above sea level (a.s.l.) in the Tien Shan Mountains of middle Asia. Figure 4.20 shows the distribution of Diamesinae in glacier brooks of the Otztal Alps of Austria (Kownacka and Kownacki 1975). Some species are restricted to the zone close to the glacier, others traverse a wider range of elevation and may occur in crenal habitats.

SPRINGS AND SPRINGBROOKS Spring sources (eucrenal) and their immediate downstream lotic reaches, springbrooks (hypocrenal), exhibit habitat conditions distinctively different from otherwise similar running waters not directly influenced by groundwater. The distinctiveness of this biotope and its inhabitants was acknowledged by Illies and Botosaneanu (1963), who established a separate category, the crenal, for it in their zonation system. Spring sources are of three general types. Limnocrenes are ponds or lakes formed where groundwater emerges. Helocrenes are marshy areas formed by more diffuse seepage. Rheocrenes emerge as lotic systems (springbrooks).

Rheocrenes are characterized by physicochemical constancy, especially if associated with deep aquifers (Thienemann 1925). Water temperature at the point of egress closely approximates the mean annual air temperature, which, of course, varies as a function of latitude and altitude. In the temperate zone, springs are summer-cool and winter-warm habitats. (Thermal springs will be

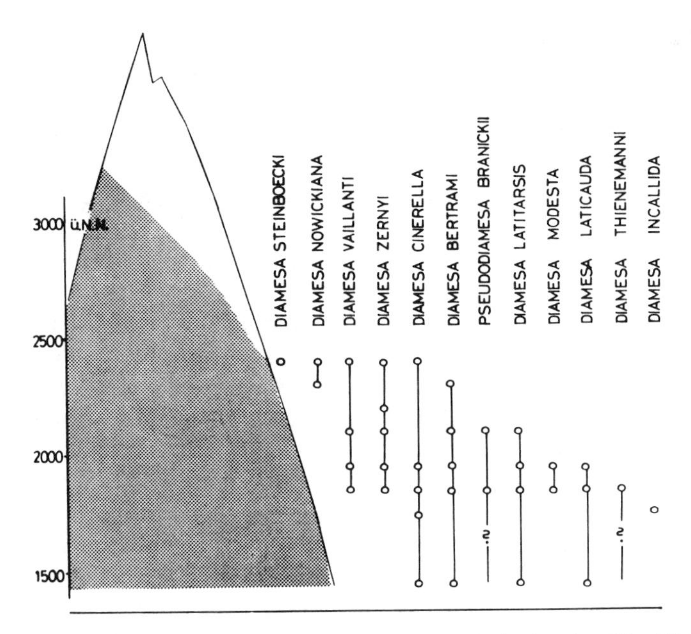

Figure 4.20 Distribution of Diamesinae (Chironomidae) in glacier brooks of the Otztal Alps of Austria. [From Kownacka and Kownacki (1975).]

considered separately.) The constant flow regime and stable substrate is conducive to the establishment of a well-developed hydrophyte community.

A rich fauna may also develop in rheocrenes with well-oxygenated waters, although species diversity is typically depressed (Odum 1957, Minckley 1963, Minshall 1968, Ward and Dufford 1979). Insects are generally superseded in abundance and diversity by other invertebrates such as crustaceans, triclads, and mollusks, especially in limestone springs (Glazier and Gooch 1987). Insects, however, may dominate the fauna of high-elevation springbrooks (Kownacki and Kownacka 1973, Cowie and Winterbourn 1979). Hynes (1970a) delineates four faunal components of springs. The first component, groundwater forms, are primarily restricted to the source area and comprise almost exclusively noninsects. A second component consists of species normally inhabiting wet margins that become more fully aquatic in the uniform conditions of rheocrenes. Among insects in this category are some cranefly larvae typically inhabiting wet soils, and the semiaquatic hydrophilid beetle *Anacaena.* Another category comprises typical crenon forms largely restricted to spring-fed reaches. Insect representatives include the caddisfly *Apatania muliebris,* the stonefly *Leuctra nigra,* and the beetle *Hydroporus ferrugineus.* Those restricted to springs may be relict populations of previously widespread species that found refuge in the summer-cool (or winter-warm) crenon habitats as the climate gradually warmed (or cooled). For example, *A. muliebris,* in Denmark restricted to a few large springs, is apparently an arctic relict unable to tolerate present summer temperatures in other Danish aquatic habitats (Nielsen 1950b). The helicopsychid caddisfly *Rakiura vernale* of Pupu Springs, New Zealand, also appears to be a glacial relict (Michaelis 1977). In addition, species that normally occur in the upper reaches of streams may inhabit and be restricted to springs at low elevations (Michaelis 1977, Ward et al. 1986). The final category includes members of the normal stream fauna that find conditions particularly favorable in springs and springbrooks. The best examples in this group are noninsects, such as amphipods, isopods, and snails. However, the euryokous lotic mayfly *Baetis tricaudatus* attained high density on rubble substrate at the source of a small springbrook in Colorado (USA), where it represented over two-thirds of total invertebrate numbers (Ward and Dufford 1979).

In two spring-fed rivers in Florida (USA), the number of species of insects in each order generally increased then decreased from source to mouth (Fig. 4.21) (Sloan 1956). Sloan attributed the low number of species in source areas (Station 1) to low dissolved oxygen in springheads, and the low diversity at downstream stations to increased salinity resulting from incursions of sea water during high tide.

In the absence of adverse chemical conditions, temperature, flow, and substrate appear to be the primary factors structuring the biotic communities of springs. The constant flow regime allows colonization by certain species unable to maintain populations in streams with highly variable discharge. Because of constant flow conditions the substrate is stable; thus algae, mosses, and higher aquatic plants are well developed during all or most of an annual cycle. The

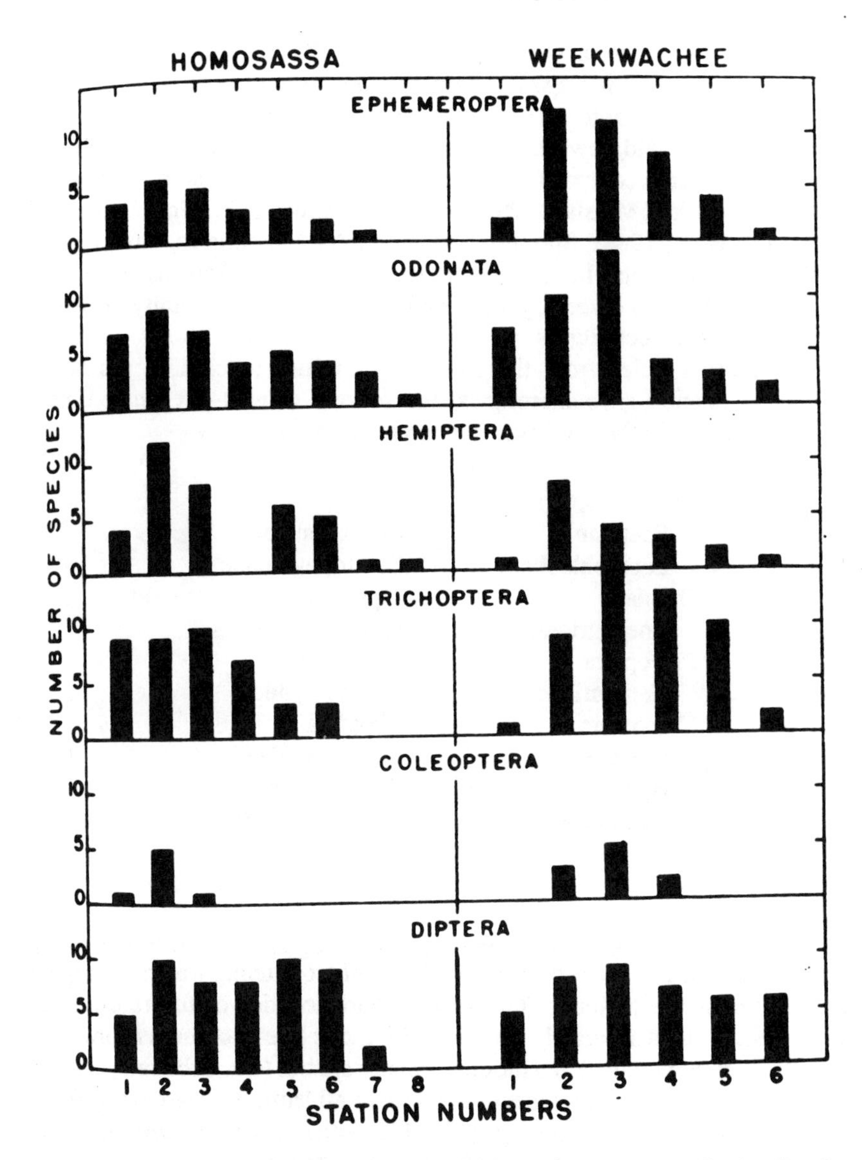

Figure 4.21 The number of species per order of insects from the sources (Station 1) to the sea in two spring-fed rivers in Florida. The Homosassa River is 8 km long; the Weekiwachee River is 23 km long. [From "The distribution of aquatic insects in two Florida springs" by W. C. Sloan, Ecology, 1956, 37, 81–98. Copyright © by The Ecological Society of America. Reprinted by permission.]

plants and inorganic substrate types may provide a mosaic of habitat patches in the source areas of springs (Minckley 1963, Thorup 1966, Michaelis 1977, Thorup and Lindegaard 1977, Ward and Dufford 1979). However, despite spatial heterogeneity and otherwise generally favorable conditions, aquatic insects exhibit depressed species diversity in springs. The constant thermal regime eliminates species requiring a wide range of temperature to complete their life cycles (see Ward and Stanford 1982). The insect fauna of rheocrenes frequently exhibits relatively rapid downstream increases in species diversity and significant changes in faunal composition associated with increased thermal heterogeneity (Minckley 1963, Minshall 1968, Ward and Dufford 1979, Gray et al. 1983, Williams and Hogg 1988). Springbrooks, in contrast to sources, have moderately high environmental heterogeneity, yet are buffered from extremes. An abundant and relatively diverse, yet somewhat distinctive community of insects develops under such conditions.

One or more species from the following families (excluding pleustonic forms) are frequently encountered near the sources of well-oxygenated, nonthermal rheocrenes in the northern Temperate Zone:

Ephemeroptera
- Baetidae
- Leptophlebiidae

Odonata
- Coenagrionidae

Plecoptera
- Nemouridae

Coleoptera
- Dytiscidae
- Hydrophilidae
- Helodidae

Diptera
- Chironomidae
- Ceratopogonidae
- Tipulidae
- Stratiomyiidae
- Empididae
- Psychomyiidae

Trichoptera
- Limnephilidae
- Hydroptilidae
- Glossosomatidae
- Rhyacophilidae
- Sericostomatidae

TROPICAL STREAMS The lotic zonation system developed by Illies (Illies 1961a, Illies and Botosaneanu 1963) was intended to apply to running waters in tropical as well as temperate regions. The thermal characteristics of the rhithral and potamal were defined separately for tropical regions (the maximum of monthly means >25°C in the potamal), but the other characteristics (e.g., substrate, current) of these two major zones were meant to apply to all latitudes. Harrison (1965a) found the zonation system generally applicable, with slight modification, to both temperate and tropical running waters of southern Africa.

Based primarily on the thermal regime, Illies proposed that the transition from rhithron to potamon that occurs near sea level at high latitudes would be ~2000 m a.s.l. in equatorial regions. Illies (1964) described typical rhithron and potamon insect associations for the Huallaga, a tributary of the Amazon that rises high in the Peruvian Andes. However, in areas of low relief, headwater streams may rise at considerably lower elevations. In a hill stream in the West Indies (11°N latitude) Hynes (1971) found a marked faunal discontinuity that he regarded as the transition from rhithron to potamon at only 30 m a.s.l. Upstream from this point the stream exhibited a high gradient, rapid current, and a rocky substrate, in contrast to the lower reach that was deeper and slow flowing with soft substrate. Bishop (1973a) reported a similar lowering of the rhithron in a tropical stream in Malaysia. Harrison and Rankin (1976) proposed the term "pseudorhithron" for tropical montane streams that otherwise conform to the rhithral definition but contain a warm-adapted "pseudorhithric" fauna, in contrast to the cold-adapted "eurhithric" fauna occurring above 2000 m in the tropics. The pseudorhithric fauna, often warm-adapted species from typically eurhithric families, are adapted to high-gradient rocky streams and are intolerant of silt.

The rain forest tributary streams of Central Amazonia typically exhibit three zones before entering the main river (Fittkau 1967). The spring-fed upper course flowing through narrow valleys has a stair-step profile formed by organic debris dams. In the middle course the valley is wider, the gradient is less, and the stream meanders. The lower course, the Igapo (flood wood), or submerged forest zone, is a low-gradient reach that becomes part of the main river during the annual flood.

Aquatic insects are diverse and abundant and comprise the majority of macroinvertebrates in the rainforest tributaries of Central Amazonia. Ephemeroptera, Trichoptera, and Odonata are well represented. The hemipteran fauna is especially well developed. Gerrids and veliids occupy the water surface; submerged forms include *Ranatra* and members of the families Naucoridae, Corixidae, Notonectidae, and Belostomatidae. Among coleopterans, only the families Elmidae and Gyrinidae are abundant. Several species of aquatic lepidopterans are present. Plecopterans are represented by only a few species from the family Perlidae. More than a hundred species of chironomids typically co-occur in these Central Amazonia rainforest streams. Other dipterans include simuliids and ceratopogonids. The special kinal biotope (Fig. 4.3) formed by concentrations of floating flowers, fruits, pollen, and terrestrial insects is inhabited by a variety of aquatic insects. A species of *Polypedilum* (Chironomidae) residing in this habitat hangs from the underside of the water surface by a short thread (Fig. 4.3) that apparently enables the larva to drift to other kinal biotopes.

SUMMARY OF STREAM ZONATION Viewing river systems as a series of zones from source to mouth is a convenient way to classify running waters. Glacier brooks, springbrooks, streams, and rivers tend to have characteristic features and ecologically similar communities on a broad geographic scale. The subdivisions of each zone, while perhaps valid for a given drainage basin, have less general applicability. Even the major zones are not discrete units but are joined by transitional segments with ecotonal properties. The concept of zonation should not be interpreted as a precise system for delineating running-water segments, and it must be recognized that not all lotic waters fit into such a scheme. For example, in areas of recent coastal subsidence (Carpenter 1927) or in geologically youthful regions (Harrison 1965a), the lower reaches of river systems may not exhibit a low gradient or possess soft substrate. A slow-flowing river with soft substrate may be rapidly transformed into a higher-gradient segment with coarse substrate where it traverses an especially resistant geological formation. This "rejuvenation" of the lower reaches creates a running water segment in which rhithral characteristics (e.g., current, substrate), occur concomitantly with potamal characteristics (e.g., temperature regime).

The River Continuum Concept The river continuum concept (RCC) of Vannote et al. (1980) perceives running waters as continuous resource gradients from headwaters to the sea. Downstream changes are considered clinal rather than zonal. Aquatic insects and other stream organisms are hypothesized to be predictably structured along longitudinal resource gradients (Fig. 4.22).

This concept was initially developed for undisturbed lotic ecosystems in the eastern deciduous forest of North America, where land-water interactions play a major role in the ecology of small streams. Headwater streams, according to the RCC, are viewed as heavily canopied, light-limited heterotrophic systems with rocky substrates. Fed largely by groundwater, these headwaters exhibit low-amplitude temperature and flow regimes. The major energy source is leaf litter that enters the stream from the terrestrial system (Fisher and Likens 1973).

As the stream collects water from incoming tributaries and becomes larger, the canopy opens. In middle reaches the water is still relatively clear and shallow, but now the stream bottom receives direct solar insolation. Aquatic macrophytes and attached algae attain maximum development in middle reaches and provide food and shelter for a diverse community of aquatic insects. Temperature and discharge exhibit large temporal fluctuations.

The light-limited heterotrophic conditions in the lower reaches of river systems are engendered by quite different phenomena than those prevailing in headwater reaches. The greater depth, lower water clarity, and unstable substrate of large rivers reduce their suitability for benthic plants, although phytoplankton may contribute to primary productivity. The large volume of water resists short-term temperature changes and the discharge pattern is dampened by the cumulative variations of many tributaries. Temporal and spatial environmental heterogeneity is therefore postulated to be greatest in middle reaches of river systems.

The RCC also predicts downstream shifts in the relative abundance of feeding guilds as food resources change over the longitudinal profile. For example, shredders (species that feed on coarse detritus) are abundant only in the headwaters where their food (leaf litter and associated microbes) is abundant. Collectors, species that feed on fine detritus, are important in all reaches but are virtually the only nonpredaceous invertebrates in the lower reaches (Fig. 4.22). Grazers, species that scrape attached algae from rocks and the surfaces of higher plants, are best developed where autotrophy predominates. Predators are purported to exhibit similar relative abundances along the entire continuum.

Attempts to test the RCC have dealt largely with analyzing changes in organic resources and invertebrate feeding guilds as a function of stream order or link magnitude (e.g., Cummins et al. 1981, Hawkins and Sedell 1981, Bruns et al. 1982, Culp and Davies 1982, Minshall et al. 1982, Ward et al. 1986, Naiman et al. 1987). The generality of the continuum concept has been questioned (Winterbourn et al. 1981, Barmuta and Lake 1982).

Did the authors of the RCC intend that all rivers should conform to the undisturbed eastern deciduous forest river system used as a conceptual model in the original paper? This was clarified in a subsequent paper (Minshall et al. 1983b), in which a "sliding scale" was used to accommodate differences in climate-vegetational-hydrologic settings. The RCC is an evolving concept that will be subject to further refinements and modifications (Minshall et al. 1985, Statzner and Higler 1985). Pristine rivers are highly interactive with the surrounding landscape (Ward 1989a), but not all interactive pathways were incorporated in the original RCC model. The need to include interactions between rivers and their floodplains (Welcomme 1979, Junk et al. 1989, Ward 1989b) and contiguous aquifers (Gibert et al. 1990) in future continuum models has been recognized (Cummins et al. 1984, Minshall 1988, Sedell et al. 1989).

It remains to be seen how broadly the river continuum concept can be applied to lotic systems in different biomes. A basic difficulty relates to the essentially deterministic nature of the RCC model. If the relative importance of deterministic and stochastic processes in structuring lotic communities varies as a function of biome type, one would expect the RCC model to exhibit greater or lesser "fit" with river systems located in different biomes.

OTHER LOTIC HABITATS

The remainder of this chapter will be devoted to special lotic habitats and their communities. Those considered are thermal springs, intermittent streams, and cave streams.

Thermal Streams "Thermal streams" as defined here refer to lotic waters originating from geothermal springs [see Tuxen (1944) for detailed terminology]. Thermal streams occur throughout the world, but are most concentrated in parts of Iceland, New Zealand, Japan, North Africa, Europe, and the western United States. Thermal waters are commonly highly mineralized and even if saturated with oxygen contain lower absolute concentrations than cooler aquatic habitats. Factors other than temperature may, therefore, influence faunal composition and abundance in thermal streams. However, Lamberti and Resh (1983), who devised field experiments to separate the effects of chemistry and temperature, concluded that the thermal component of geothermal waters was largely responsible for the alterations in benthic community structure.

Some bacteria occur right up to the boiling point of water and a few micrometazoans tolerate temperatures around 60°C, although no animal species is known that can complete all phases of its life cycle above 50°C (Brock 1985). The upper limit for insects in an active state is around 50°C (Pennak 1978).

Thermal springs include all waters issuing from the ground at temperatures greater than the mean annual air temperature of a given region. Therefore, the absolute temperature dividing thermal from nonthermal waters varies as a function of altitude and latitude. However, a specially adapted and generally

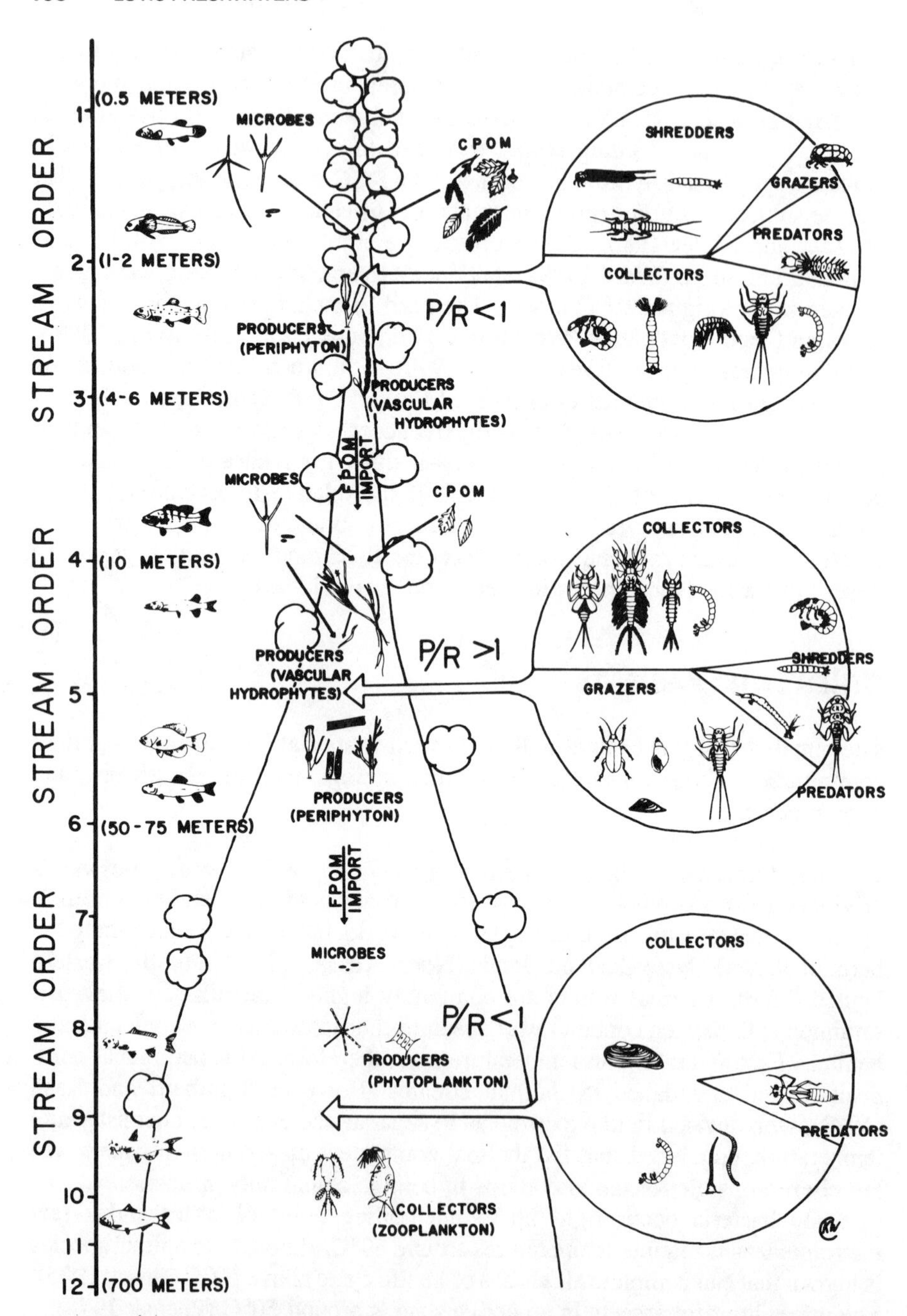

Figure 4.22 Diagrammatic representation of the river continuum shown as a single stem of increasing order. General range of streams widths (in meters) for each order is given, and orders have been roughly grouped into headwaters (orders 1–3), midsized rivers (4–6), and large rivers (7–12), considering the Mississippi River as order 12 at its mouth. The headwaters and large rivers are shown as heterotrophic (P/R, or ratio of gross photosynthesis to community respiration, <1)

similar hot spring animal community occurs in waters >40°C worldwide (Brues 1927, Tuxen 1944). On the basis of their macroinvertebrate inhabitants, Winterbourn (1968) distinguishes between "hot" springs (40°C+) and "warm" springs (>ambient, <40°C). The most important insect groups of the warm and hot springs of New Zealand (Figs. 4.23, 4.24) are generally representative of thermal waters wherever such habitats occur. Any differences between the fauna of hot springs in widely separated areas are usually attributable to zoogeographical factors (e.g., the absence of Stratiomyidae from Iceland precludes their occurrence in hot springs).

Truly aquatic species of insects from only three orders (Diptera, Coleoptera, Odonata) have been reported from thermal streams at temperatures above 40°C (Table 4.7), although a few epipleustonic and semiaquatic hemipterans live in close association with hot springs. Most of the investigators cited in Table 4.7 acknowledge the necessity of measuring habitat temperature as close to the animal as possible. Tuxen (1944), for example, emphasizes microhabitat differences in temperature and confirms Schwabe's (1936) earlier record of 47.7°C for *Scatella thermarum,* but rejects the higher values reported by other investigators. Microhabitat selection enables thermophilic species to inhabit thermal streams with much higher average temperatures than those to which they are exposed. Larvae of the stratiomyid *Hedriodiscus truquii,* for example, reside in a hot spring with a constant temperature of 47°C (well above their lethal limit), yet select marginal microhabitats that do not exceed 27°C (Stockner 1971).

Except for chironomid larvae and the rare occurrence of odonate nymphs at temperatures >40°C, the insects of hot springs breathe atmospheric air, which frees them from dependence on dissolved oxygen and reduces osmoregulatory problems. Some chironomids of thermal springs produce hemoglobin which enhances uptake from low-oxygen waters and have blood gills that function in osmoregulation.

Whereas the insect fauna of hot springs includes species generally restricted to thermal waters, warm springs contain selected members of the Odonata, Hemiptera, Coleoptera, and Diptera, living close to their upper temperature limits, that also inhabit nonthermal waters.

Some thermal streams have source temperatures that are similar to the summer maxima of nonthermal waters. Such a thermal stream is located at a high

---

because of restricted light, a consequence of shading by riparian zone vegetation in the headwaters and attenuation from depth and turbidity in the large rivers. The midsized rivers are depicted as autotrophic, with a P/R >1, through a combination of reduced riparian shading, relatively shallow and clear water. The importance of terrestrial inputs of coarse particulate organic matter (CPOM) decreases and the transport of fine POM increases down the continuum. The ratios of macroinvertebrate functional groups shift from shredder–collector headwaters to collector–grazer (scraper) midsized rivers to collector-dominated large rivers, which are lentic-like, with plankton communities. Fish populations are shown to shift from cold- to warm-water invertebrate feeders. Midsized rivers have piscivorous forms as well, and large rivers have both bottom feeders and planktivorous species. [From Cummins (1979).]

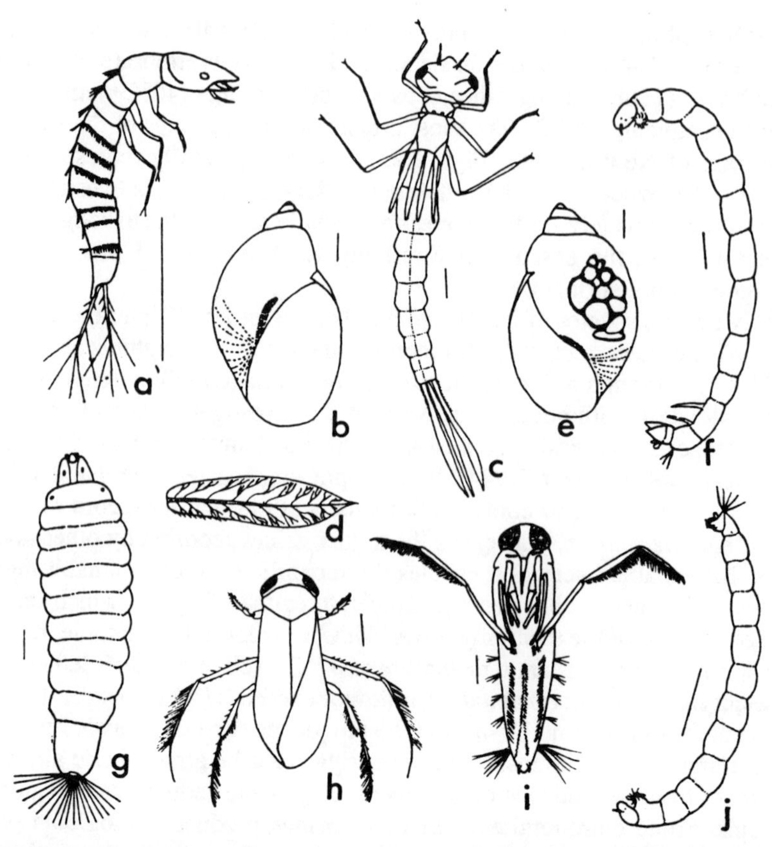

Figure 4.23 Macroinvertebrate inhabitants of warm springs (>ambient, <40°C): (a) *Antiporus* sp.; (*b*) *Simlimnaea tomentosa* (Gastropoda); (*c*) *Ischnura aurora;* (*d*) nymphal gill of *Ischnura;* (*e*) *Physa fontinalis* (Gastropoda); (*f*) *Chironomus zealandicus;* (*g*) *Odontomyia* sp.; (*h*) *Sigara* sp.; (*i*) *Anisops wakefieldi;* (*j*) *Chironomus cylindricus.* Scale lines = 1 mm. [From Winterbourn (1968). With permission from Tuatara 16 (1968).]

elevation (3109 m a.s.l.) in the Colorado Rocky Mountains. The following insects were collected from the source area (25°C) of the stream (Ward, unpubl.):

Helicopsyche borealis (Helicopsychidae: Trichoptera)
Argia prob. vivida (Coenagrionidae: Odonata)
Optioservus divergens (Elmidae: Coleoptera)
Chimarra utahensis (Philopotamidae: Trichoptera)
Euparyphus sp. (Stratiomyidae: Diptera)
Chironomidae (Diptera)

The upward range extension of *Helicopsyche,* a caddisfly normally restricted to low elevations in Colorado (Ward 1981), is noteworthy, as is the substitution of

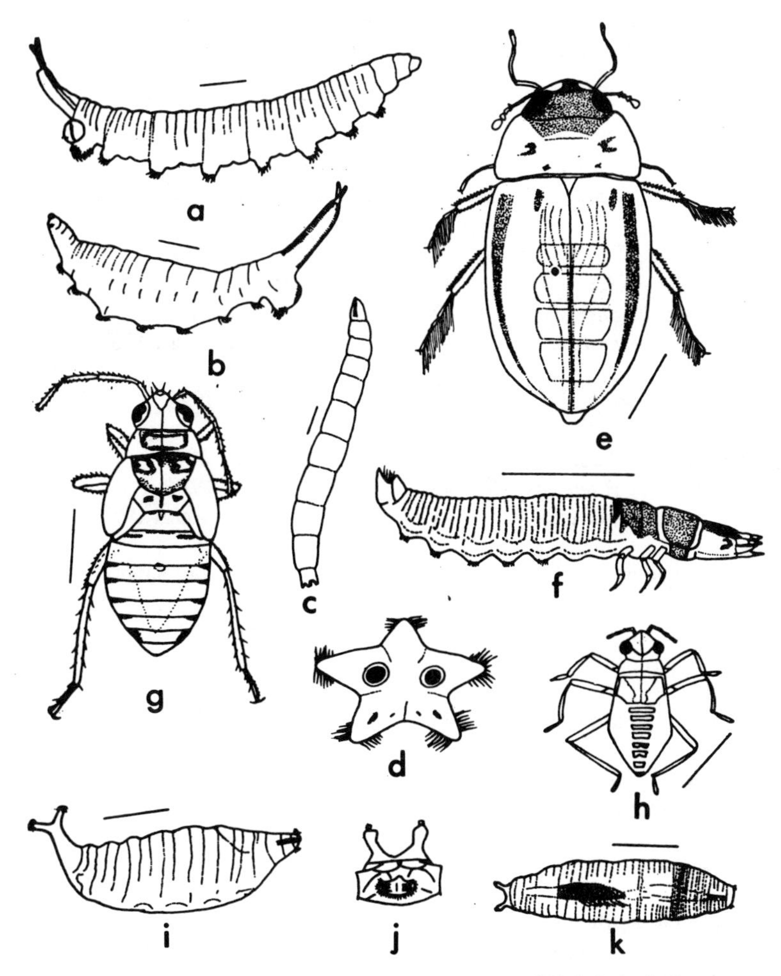

Figure 4.24 Macroinvertebrate inhabitants of hot springs (40°C+), including supraaquatic associates: (a) *Ephydrella* larva; (*b*) *Ephydrella* pupa; (*c*) *Erioptera* sp.; (*d*) spiracular disc of *Erioptera* larva; (*e*) *Enochrus* adult; (*f*) *Enochrus* larva; (*g*) *Saldula* sp.; (*h*) *Microvelia* sp.; (*i*) *Scatella* larva; (*j*) pupal respiratory siphons of *Scatella;* (*k*) *Scatella* pupa. Scale lines = 1 mm. [From Winterbourn (1968). With permission from Tuatara 16 (1968).]

O. divergens for the elmid beetle *(Heterlimnius corpulentus)* normally occurring in high altitude streams of this region.

Mitchell (1974) postulates that the evolution of thermophily did not primarily involve adaptations to higher and higher temperatures by members of the typical rheophilous fauna. Indeed entire orders containing typical inhabitants of cold, well-oxygenated running waters (Plecoptera, Ephemeroptera, Trichoptera) have no representatives in hot springs. Among the insect families represented in hot springs, only chironomids are common in normal running-water habitats, but many species belong to groups residing in madicolous habitats. Because these

Table 4.7 Orders and families of insects, the active aquatic stages of which occur in hot springs at temperatures in excess of 40°, with selected records of maximum habitat temperatures

Taxon	Habitat Temperature (°C)	Source
Diptera		
Ephydridae		
Scatella thermarum	47.7	Schwabe 1936
Scatella nitidifrons	47	Stark et al. 1976
Ephydrella thermarum	47	Winterbourn 1968
Stratiomyidae		
Odontomyia spp.	47	Brues 1928
Chironomidae		
Eucricotopus sylvestris	41	Tuxen 1944
Chironomus near *tentans*	49	Brues 1924
Chironomus cylindricus	41	Winterbourn 1969
Culicidae		
Culex tenagius	42	Njogu and Kinoti 1971
Tabanidae		
Tabanus punctifer	43	Brues 1928
Odonata		
Libellulidae		
Mesothemis simplicicollis	43	Brues 1928
Coleoptera		
Hydrophilidae		
Philydrus hamiltoni	45.5	Brues 1928
Enochrus tritus	45.3	Winterbourn and Brown 1967
Helochares normatus	46	Schwarz 1914
Dytiscidae		
Bidessus thermalis	45	Issel 1910
Coelambus thermarum	44.5	Brues 1932
Hydroscaphidae		
Hydroscapha natans	46	Schwarz 1914

thin films of water may reach temperatures of 40°C during periods of intense insolation, madicolous species are preadapted to the thermal environment of hot springs. Eumadicolous species breathe atmospheric air, in contrast to most insects of rapid streams, which may be regarded as an additional preadaptation to conditions in thermal streams.

Intermittent Streams Intermittent streams, lotic waters exhibiting surface flow for only a portion of each annual cycle, occur throughout the world. In some mesic regions the total length of intermittent stream reaches may exceed the length of permanent running water (Clifford 1966). In arid areas, intermittent streams may constitute the predominant or even sole lotic habitat type (Williams

1987). Some intermittent streams are devoid of surface water during the dry phase, whereas others, although lacking surface flow, retain isolated stagnant pools along the stream course providing suitable habitat for additional insect groups such as mosquitoes (Abell 1959). Some interrupted streams, drainage channels where reaches with surface flow alternate with reaches of subsurface flow (Ward and Holsinger 1981), periodically dry up entirely and are thus intermittent. Despite their wide distribution and their abundance in some regions, temporary streams have received little attention from stream ecologists.

Environmental conditions of intermittent streams tend to be more variable than in permanent lotic habitats. Temperatures may reach high levels in stagnant pools, which are frequently highly organic and exhibit oxygen deficits (Smith and Pearson 1987). When surface water is present, current velocities range from zero during nonflow periods to high values during major periods of runoff.

Clifford (1966) studied an intermittent forest stream in Indiana (USA) in which surface flow was restricted to winter and spring. During the summer surface water was completely absent. Species with the following characteristics were eliminated from the permanent aquatic fauna: (1) those with life cycles longer than one year, (2) those with more than one generation per year, (3) those with the major period of growth during summer, and (4) those with emergence during late summer or autumn. Most of the permanent aquatic species that were present exhibited life-cycle characteristics (univoltine, spring emergence) that preadapted them to the temporary stream habitat. A few species of caddisflies that survived the dry period as estivating prepupae and pupae were the only insects to exhibit special adaptations to the intermittent conditions. Mayflies and stoneflies survived the dry period as resistant eggs or as small nymphs residing deep within the stream bed. Adult aquatic beetles of several species burrowed deep into the substrate during the dry phase. A uniform and apparently stable community of primarily permanent residents characterized the stream. Clifford attributes this to the relatively predictable hydrologic regime and the presence of suitable interstitial spaces in the stream bed. The dense terrestrial vegetation in the catchment reduces silt input to the stream (which would clog interstices), buffers temperature extremes, and maintains high subsurface seepage rates by increasing infiltration. Delucchi (1988) postulates that some of the common adaptations of stream insects—life-history strategies to avoid the summer low-flow period, drought-resistant eggs, well-developed colonization abilities, utilization of the hyporheic zone—"reduce the expected differences in community structure between temporary and permanent stream sites."

In contrast, Harrison (1966) and J. D. Hynes (1975) conclude that oviposition by aerial adults that spent their aquatic stages in permanent waters was the most important recolonization mechanism in intermittent tropical streams. Neither Harrison nor Hynes found evidence that resistant eggs or diapausing larvae were used by insects as mechanisms to survive the dry phase. It is suggested that the high temperatures and associated extreme desiccation during the dry season in the tropics reduce the value of survival strategies based on resistant stages for species capable of flight. However, Statzner (1981a) found that at least one species of hydropsychid caddisfly survived the nonflow period as larvae in a temporary stream in tropical Africa.

Figure 4.25 illustrates eight mechanisms used by temporary stream animals to cope with the summer dry phase, based on studies of intermittent streams in Ontario, Canada (Williams and Hynes 1976a, 1977). Some insects reside in permanent waters during the dry phase and reinvade the temporary stream on resumption of flow in autumn. Others have an aerial adult stage coinciding with the summer dry period. Many species, however, remain within the dry stream channel. Some of these remain active in the few remaining pockets of free water, the isolated pools, and crayfish burrows (if present); others are quiescent in the stream bed or under accumulations of plant litter. Quiescence characteristically occurs in the same life-cycle stage within a major taxonomic group of aquatic insects. For example, the dry phase is passed in the egg stage by ephemeropterans, the nymphal stage by plecopterans, the pupal stage by trichopterans, and the adult stage by coleopterans.

Most aquatic insects of the temporary streams examined by Williams and Hynes were either truly lotic species (the fall-winter stream fauna) or inhabitants

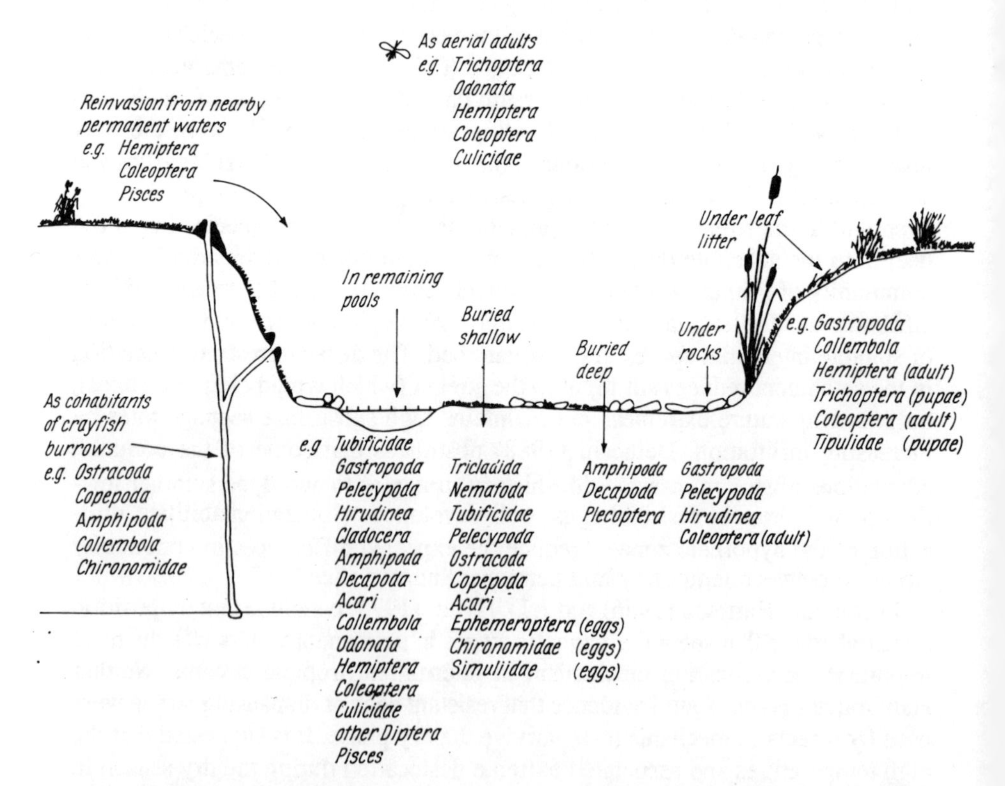

Figure 4.25 Summary of the oversummering methods used by the fauna of a temporary stream. [From Williams and Hynes (1977).]

of the isolated pools (the spring pool fauna). A few species of chironomids and the odonate *Anax junius* are members of both groups since their growth begins when flow resumes in autumn but is completed in the pool phase during spring. Typical members of the fall-winter stream fauna include plecopterans, trichopterans, simuliids, and chironomids, in which all growth and reproduction occur during the period of surface flow. Groups typically associated with the spring pool fauna include ephemeropterans, odonates, hemipterans, coleopterans, culicids, and ephydrids. These spring pool species are of two basic types. One type includes permanent residents, the eggs of which do not hatch (some culicids) or the adults of which do not reproduce (some coleopterans) until flow ceases and the pools become isolated. The other type (several coleopterans and hemipterans) move into the pools as they become isolated, oviposit, and frequently depart. Their eggs undergo but a short incubation period and larval growth is completed before the pools disappear. Additional species, primarily coleopterans and hemipterans, may use the pools as feeding sites, but oviposit in permanent waters.

Williams and Hynes (1977) emphasize, as did Clifford (1966), the relatively stable species composition of the abundant taxa from year to year in a given intermittent stream. Temporary streams also tend to have a distinctive faunal composition with little overlap with adjacent (or even connected) permanent streams. In temporary streams in Spain, for example, *Tyrrhenoleuctra* replaces *Leuctra*, and *Hemimelaena* replaces *Perla* as the predominant stoneflies (Aubert 1963). There is, however, considerable overlap between the fauna of permanent and temporary streams in Australia, where even permanent streams exhibit extreme variations in discharge (Boulton and Lake 1988).

In some regions the timing of the wet and dry phases and their durations are highly unpredictable. In some streams of arid regions the dry phase may be of greater than one year's duration. Aquatic insects of such intermittent streams in Texas (USA), exhibit highly flexible life cycles (Snellen and Stewart 1979a, b, Grant and Stewart 1980). The perlid stonefly *Perlesta placida* produces some eggs that undergo a 5–6-month diapause and others that diapause for 1 year and 5–6 months. Two species of leuctrid stoneflies are apparently able to exhibit several distinctly different growth patterns. A small proportion of the eggs of these species are nondiapausing and hatch shortly after oviposition. Because there is no nymphal diapause (unlike boreal stoneflies), these populations perish except in wet years. The remaining eggs diapause for 2–4 months at which time environmental conditions induce hatching, or undergo an extended diapause, perhaps for as long as several years.

An extreme example of tolerance to desiccation in the larval stage is exhibited by the water penny *Sclerocyphon bicolor*, which inhabits intermittent streams in Australia (Smith and Pearson 1985). Larvae survive up to 4 months out of water and loss of 79% of their body water.

Subterranean Waters Subterranean waters are divided into cave habitats (the troglal) and groundwater biotopes (the stygal or nappe phréatique), based primarily on whether humans have ready access to underground aquifer systems. The interconnectedness of subterranean waters render such anthropocentric separation generally meaningless from an ecological perspective. Indeed, cave systems are much more extensive and widely distributed than generally realized, yet only a minority are open to the surface (Howarth 1983). However, a cave entrance is requisite for colonization by bats and other terrestrial vertebrates, the guano of which provides a major energy resource for cave animals (cavernicoles) that is generally unavailable to the groundwater fauna. In addition, many aquatic insects are amphibiotic and so require an aerial habitat during the adult stage. Even most aquatic beetles, although inhabiting water as adults, depend on aerial dispersal and must pupate above the water level. Subterranean waters have served as important evolutionary pathways and speciation sites for primary aquatic invertebrates such as crustaceans (e.g., Pennak 1968, Husmann 1971, Rouch and Danielopol 1987, Holsinger 1988, Botosaneanu and Holsinger 1991), but apart from the hyporheic zone, aquatic insects only rarely penetrate hypogean (subsurface) waters and their ecology remains virtually uninvestigated.

Insects truly adapted to life in underground waters (stygobionts) are known only among dipterans and coleopterans. In tropical Africa several species of *Anopheles* (Culicidae) occur in caves (Vandel 1965). The larvae reside in pools where they feed on bat guano. The adults remain within the cave and depend on bats for blood meals.

A very few dytiscid, noterid, hydrophilid, and elmid beetles are highly adapted for life in hypogean waters (Spangler 1986). Among dytiscids, two species have been recorded from France (Abeille de Perrin 1904, Guignot 1925), one from Africa (Peschet 1932), two from New Zealand (Ordish 1976), one from Mexico (Franciscola 1979), one from Japan (Uéno 1957), one from Venezuela (Sanfillipo 1958), and one from Texas (USA) (Fig. 4.26) (Young and Longley 1976). All were collected from subterranean waters (caves or wells) and exhibit features distinguishing them from epigean (surface) beetles, such as depigmentation, reduction or absence of compound eyes, loss of ocelli, and reduction of wings. A single female hydrophilid beetle, collected from a cave stream in Ecuador, is the only stygobiontic species known from this family (Spangler 1981a). A stygobiontic noterid occurs in Japan (Uéno 1957). Four stygobiontic elmid beetles include one species from Africa (Jeannel 1950) and three species from Haiti (Spangler 1981b).

There are reports of aquatic insects that complete their entire life cycles underground and are apparently restricted to subterranean waters, but that do not exhibit discernible modifications for a hypogean existence. The caddisfly *Wormaldia subterranea*, the larvae of which inhabit cave streams in Yugoslavia (Radovanovič 1935) is in fact *W. occipitalis occipitalis*, by no means restricted to caves (Botosaneanu 1989). The Trichoptera, Ephemeroptera, and Plecoptera collected from cave streams in Scandinavia all commonly occur in epigean lotic habitats (Hippa et al. 1985).

Some aquatic insects commonly occur at cave entrances (the parietal associa-

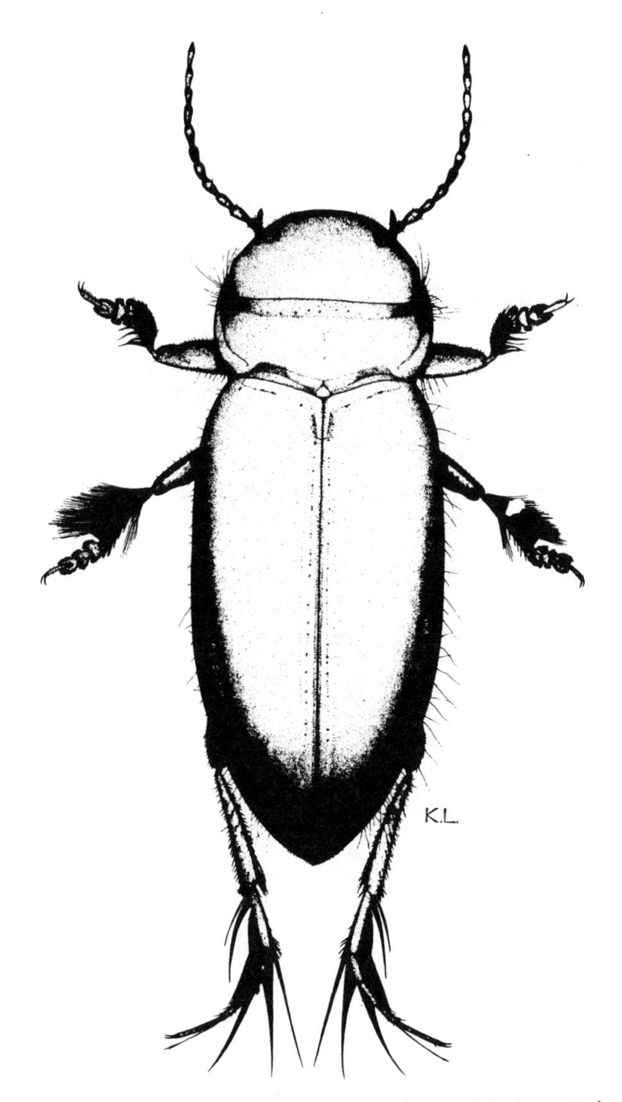

Figure 4.26 Dorsal aspect of *Haideoporus texanus*, a blind and depigmented cavernicolous beetle (Dytiscidae) discovered in an artesian well in Texas. [From Young and Longley (1976). Reprinted by permission of The Entomological Society of America.]

tion), but also inhabit surface waters. Waters (1981) reported that virtually no organisms drifted from the source of a cave stream in Minnesota (USA), but Ward (unpubl.) collected several species of normally epigean stream insects by placing drift nets just inside the entrance of a high elevation cave in the Rocky Mountains. Several species of lotic trichopterans in France typically move into caves immediately on emergence where they spend the summer in imaginal diapause (Bournaud 1971). It appears that the cave environment provides the required low temperature and high humidity.

Additional species occasionally collected in hypogean habitats are "accidentals" that have washed in from surface waters or, in the case of certain beetles, are epigean forms that move into subterranean waters in time of drought (Galewski 1971). Some terrestrial invertebrates of caves are amphibious and temporarily move into water, where they may remain submerged for some time (Vandel 1965). Several cavernicolous collembolans, for example, occasionally move into subterranean pools, and nearly all collembolans are at times found on water surfaces.

Energy flow in the subterranean environment is based on organic material elaborated above ground and imported via groundwater (dissolved and particulate detritus) or by organisms (e.g., bats, oil birds, swiftlets, cave rats). Because photosynthesis cannot occur in total darkness, animals depending on green plants are absent. Among permanently hypogean animals, only detritivory (including coprophagy) or carnivory are possible.

5

MARINE, BRACKISH, AND INLAND SALT WATERS

This chapter deals with the insects of the various aquatic environments having a higher salt content than freshwaters. Marine habitats consist of seawater; brackish habitats contain water of intermediate salinity and are normally ecotones between freshwater and the sea; saline inland waters have intermediate to high salinity, but lack direct connection with the sea. The approximate total dissolved salt content of the major aquatic environments are as follows (in parts per thousand):

	Concentration (ppt)
Freshwaters	0.01–0.5
Brackish waters	0.5–32
Saline inland waters	0.5–250
Seawater (open sea)	34–37

The relative composition of dominant ions also varies between the four major aquatic environments. Sodium and chloride predominate in seawater, which is characterized by its constancy in total salinity and ionic composition. The salinity of the other aquatic environments varies from one water body to the next, and spatially and temporally within the same water body. In fact, the intertidal zone where most marine insects occur is also characterized by highly variable salinity. The term "brackish" refers to mixtures of freshwater and seawater (Williams 1981a). Freshwaters and many saline lakes are described as athalassohaline (Greek: athalassa = not sea) because their ionic composition differs significantly from that of seawater, a fact of major importance in the evolutionary invasion of inland waters.

THE MARINE ENVIRONMENT

Insects have been largely unsuccessful in colonizing marine environments. There are only several hundred species of truly marine insects (Cheng 1976), and most of these are restricted to the intertidal zone (Table 5.1). Insects are unknown from the deep sea and only rarely occur in near-shore (neritic) or offshore (oceanic) waters. Additional species occur in estuarine waters, but these are brackish rather than marine habitats.

Why has the sea, the most extensive habitat on the planet, posed such difficulties for a group that contains more described species than all other animals combined and that possesses such a remarkable repertoire of ecological adaptations? Although a definitive answer is not possible, this question is indeed intellectually tantalizing and has been the subject of considerable speculation (e.g., Walsh 1925, Buxton 1926, Mackerras 1950, Usinger 1957, Hinton 1976a). The following are commonly suggested as factors limiting the invasion of the sea by insects: (1) turbulence, (2) hydrostatic pressure changes, (3) alternating tidal immersion and exposure, (4) depth, (5) salinity, (6) food availability, (7) competition, and (8) predation.

Hinton (1976a) agrees that the physical violence of the sea shore (items 1-3 above) tends to exclude insects, and he attributes their rarity in the sea primarily to the respiratory problems that were "imposed upon them by previous history of adaptation to relatively dry terrestrial environments" (Hinton 1976a, p. 44). Species that require frequent excursions to the surface, a common respiratory adaptation among aquatic insects, are not found in the intertidal zone. Nor are species present that possess respiratory extensions that pierce the surface film while the insect remains submerged. Surface turbulence and the extreme and rapid changes in water level would severely restrict the efficacy of such respiratory mechanisms. None of the intertidal insects have pointed spiracles to tap the air stores of plants, although such adaptations are possessed by species of salt marshes and estuaries; nor do any marine species have pupal cocoons that communicate with the air spaces (aerenchymous tissues) of vascular plants. Empirical evidence from ephydrid distribution patterns supports the idea that insects are excluded from shorelines with violent wave action (Steinly 1986).

Cutaneous respiration, the most common respiratory mechanism among marine insects, involves loss of the waxy layer to increase cuticular permeability. Plastron respiration, which (like cutaneous respiration), frees its bearer from dependence on atmospheric air, is also common among marine insects. Plastron-bearing spiracular gills, an adaptation especially suited for respiration in environments that are alternately dry and flooded, have been independently evolved at least four times among intertidal dipterans (Hinton 1976a) (see Fig. 2.12, above). Spiracular gills are dual organs that form a large surface for respiratory exchange when immersed, yet provide only a small surface area for evaporative water loss when exposed to air (see Chapter 2).

Numerous writers have suggested that salinity and attendant osmoregulatory problems are the major deterrents to colonization of the sea by insects. The marine environment in a sense is a physiological desert since there is a tendency for water to move from the insect to the environment in response to osmotic forces, quite the reverse of the situation in freshwaters (Fig. 2.13).

It does not, however, appear that osmoregulatory problems alone account for the paucity of marine insects, since various species have adapted to hyperhaline inland lakes and to salt marshes exhibiting extreme fluctuations in salinity (Foster and Treherne 1976). Staphylinid beetles of salt marshes select food with low salt content and reduce ingestion rates during periods of high salinity (Bro Larsen 1952). Larvae of the brackish water caddisfly *Limnephilus affinis* drink saltwater equivalent to ≤7% of their body weight each day, in contrast to freshwater species, which consume water equivalent to half their body weight per day (Sutcliffe 1962). Species such as the mosquito *Aedes detritus* and the fly *Ephydra riparia* are extremely efficient osmoregulators that are able to maintain relatively stable internal osmotic concentrations over a wide range of external salinities and can tolerate salt concentrations considerably greater than that of seawater. Brackish water species such as *L. affinis* can maintain internal concentrations lower than that of the medium only at salinities less than that of seawater (Sutcliffe 1961). At higher salinities they become osmoconformers. Their ability to tolerate increased internal osmotic concentrations enables them to tolerate short periods of exposure to higher salinities (Foster and Treherne 1976). Euryhaline species, those capable of hyperosmotic regulation in freshwater and hypoosmotic regulation in saltwater, apparently possess transport epithelia capable of ion absorption in dilute media and ion excretion in waters of high salinity (Komnick 1977).

Some of the adaptations of terrestrial insects relating to water balance (a highly impermeable cuticle, ability to produce hypertonic rectal fluid) preadapt them to the osmotic problems encountered in seawater. It may be no coincidence that the majority of species inhabiting the intertidal zone are apparently from terrestrial rather than freshwater lineages. Retention of a highly impermeable body surface, however, while greatly reducing osmoregulatory problems, precludes certain types of aquatic respiration. As already discussed, physical conditions in much of the potentially habitable areas of the sea appear to select against respiratory mechanisms requiring excursions or extensions to the surface, thus largely eliminating an invasion strategy successfully employed by many freshwater species.

Competition from established marine animals also has been suggested as a major deterrent to insect colonization of marine habitats. Usinger (1957) and others have pointed out that crustaceans and other groups of marine ancestry already had undergone considerable niche diversification in the sea prior to the major period of adaptive radiation of insects. Indeed, primary aquatic animals such as crustaceans would seem to possess many attributes giving them a

Table 5.1 Insect orders and families containing marine representatives, with an estimate of the number of genera (Arabic numerals) in three marine habitats[a]

	Intertidal	Neritic	Oceanic
Collembola			
Onychiuridae	1	0	0
Hypogastruridae	2	0	0
Neanuridae	5	0	0
Isotomidae	5	0	0
Actaletidae	1	0	0
Hemiptera			
Gerridae	0	4	1
Veliidae	2	1	0
Mesoveliidae	1	0	0
Hermatobatidae	1	1	0
Corixidae	0	(1)	(1)
Gelastocoridae	1	0	0
Omaniidae	2	0	0
Saldidae	5	0	0
Dermaptera[b]	1	0	0
Hymenoptera			
Proctotrupidae	1	0	0
Scelionidae	1	0	0
Coleoptera			
Staphylinidae	58	0	0
Carabidae	13	0	0
Hydrophilidae	1	0	0
Hydraenidae	2	0	0
Heteroceridae	1	0	0
Limnichidae	3	0	0
Melyridae	2	0	0
Salpingidae	2	0	0
Tenebrionidae	2	0	0
Rhizophagidae	1	0	0
Curculionidae	2	0	0
Diptera			
Chironomidae	13	(2)	0
Tipulidae	3	0	0
Dolichopodidae	4	0	0
Canaceidae	8	0	0
Dryomyzidae	1	0	0
Trichoptera			
Chathamiidae	2	0	0

Sources: Based primarily on data from Joosse (1976) for Collembola; Gunter and Christmas (1959), Andersen and Polhemus (1976), and Polhemus (1976) for Hemiptera; Moniez (1894) and Masner (1968) for Hymenoptera; Doyen (1976) and Moore and Legner (1976) for Coleoptera; Wirth (1951), Wirth and Stone (1956), Hashimoto (1976), Hinton (1976a), and Burger et al. (1980) for Diptera; and Riek (1976) for Trichoptera.

competitive edge over a group with terrestrial ancestry. For example, even insects able to live in the marine environment must expend large amounts of energy in osmoregulation, which must place them at a competitive disadvantage if competing with groups whose body fluids are essentially isotonic to seawater.

The abundance and diversity of predators has also been postulated as a formidable barrier to the invasion of the sea by insects. Marine invertebrates and fishes exhibit a vast array of predatory tactics. In addition, the intertidal zone is subject to aquatic predators during submersion and terrestrial predators during emersion.

The paucity of angiosperms in the sea may provide yet another explanation for the dearth of marine insects. Many terrestrial insects are phytophagous and depend on one or a few closely related plant species. While proportionately fewer freshwater insects feed directly on living aquatic angiosperms, higher plants provide shelter, current refugia, case-building materials, and surfaces for colonization by epiphytic algae. Buxton (1926) considered it significant that *Pontomyia natans,* then considered the only fully submarine insect, is found in intimate association with one of the few marine angiosperms *(Halophila).*

It is unlikely that the paucity of marine insects is attributable to any single factor. Insects, through various adaptations, have been able to overcome each of the barriers to colonization. One can only conclude, as did Buxton (1926), that it was the combination of impediments to the evolutionary invasion of the sea that proved insurmountable for insects as a group.

Successful Invaders

In this section consideration will be given to the major groups of insects that have successfully colonized the sea and the adaptive strategies employed by successful invaders. Those inhabiting brackish waters, such as estuaries and salt marshes, or inland saline waters are dealt with in later sections of this chapter.

Marine Intertidal Zone

Six, possibly seven, orders of insects have representatives in the intertidal zone (Table 5.1). Dipterans, coleopterans, and collembolans are generally the most abundant groups. Hemipterans are well represented, but there are only a very few hymenopterans and trichopterans. Marine hemipterans and trichopterans were derived from forms associated with freshwater habitats, whereas most members of the remaining groups invaded the sea from land. Most intertidal beetles, for example, are from largely terrestrial families (Staphylinidae,

---

[a]Including the marine intertidal zone, the neritic zone (lagoons, bays, and other nearshore waters), and oceanic (offshore) waters, but not estuaries, brackish marshes, supralittoral habitats, or inland saline waters. However, the various authors cited did not use exactly the same criteria in determining what constitutes a marine occurrence. Records in parentheses are addressed in the text.

[b]Cheng (1976) makes brief reference to a marine dermapteran (*Anisolabis littorea*), known only from New Zealand, that presumably occurs in the intertidal zone.

Carabidae), and the intertidal tipulids probably evolved from the terrestrial rather than freshwater lineages within the Diptera (Hinton 1976a). Some marine chironomids are related to freshwater forms, but the majority are of terrestrial ancestry (Hashimoto 1976, Neumann 1976).

It is, in fact, contentious whether many of the groups in Table 5.1 are even aquatic. The intertidal collembolans are essentially terrestrial animals that have evolved behavioral mechanisms to avoid submersion and wave action during high tide (Joosee 1976). Likewise, the intertidal shorebugs (Gelastocoridae, Omaniidae, Saldidae) either retreat shoreward before the rising tide, hide in entrapped air pockets during high tide, or remain quiescent during submergence (Polhemus 1976). The marine water striders are epipleustonic.

Most marine beetles are members of typically terrestrial families; although a few species of hydrophilids are considered to be marginally marine, they are members of a terrestrial genus (*Cercyon*). Swimming forms are not represented among truly marine beetles, all of which are closely tied to the substrate and exhibit cryptic behavior appropriate to conditions in the intertidal zone. In contrast, freshwater families (e.g., Dytiscidae, Hydrophilidae) are well represented in brackish waters where adaptations for aquatic respiration and swimming are common. With few exceptions, intertidal beetles, although they may tolerate prolonged submersion in seawater, are not in direct contact with water even at high tide. Rather they reside in deep crevices, fissures, or burrows and apparently rely exclusively on entrapped air for respiration while restricting their activities to air pockets during periods of submersion (Doyen 1976). This explains the virtual absence of respiratory adaptations among intertidal beetles.

Intertidal Adaptations In addition to the behavioral mechanisms for avoiding submergence already mentioned, insects exhibit various adaptations to the special conditions of the intertidal environment.

The dryomyzid fly *Oedoparena glauca* is a predator on intertidal barnacles (Burger et al. 1980). Females attach eggs to the opercula of barnacles. Eggs hatch during low tide, and larvae must enter a closed barnacle before the habitat is inundated by the rising tide or be swept away. After consuming their prey, larvae vacate the empty test during a subsequent low tide and search for another barnacle to feed on. The larvae are by then too large to enter the operculum of a closed barnacle. They affix their mouthparts to the operculum of the new prey item where they remain attached until the tide returns and the barnacle opens to begin feeding, enabling the fly larvae to enter. Larvae are protected from predators, turbulence, and desiccation within the barnacle test. Pupae are securely anchored inside an empty test. Emergence is synchronized with the tidal cycle such that adults emerge during a morning low tide.

Marine caddisflies are known only from the Australasian family Chathamiidae, the four species of which occur in the rocky intertidal zone (Riek 1976). One of these, *Philanisus plebeius,* apparently oviposits in starfish, probably through the papular pores (Winterbourn and Anderson 1980). The eggs develop within the coelom of the starfish. On hatching, the larvae escape from their oviposition host, either via papular pores or when the stomach of the starfish is everted through the mouth during feeding. Larvae construct cases from pieces of coralline algae and complete their development in the intertidal zone. Pupal cases are firmly affixed to coralline algae. Oviposition within starfish ensures that embryonic development will take place in a stable, protected microenvironment. The eggs and newly hatched larvae are protected from predators, desiccation, and wave action, and are retained within the intertidal zone rather than being carried into deeper water by tides and currents.

The physical permanence of the sea over geological time (in contrast to most freshwater habitats) reduces the need for dispersal flights, and loss or reduction of wings characterizes many marine insects (collembolans are primitively wingless). Winged species would risk being carried away from preferred habitats or zones by air currents. In addition, the energy that would have been used for development of flight muscles can be channeled into reproductive effort, thus placing flightless forms or species at a competitive advantage.

All truly marine water striders are wingless (Anderson and Polhemus 1976). Intertidal species exhibit behavioral adaptations that further reduce the risk of their being swept out to sea. For example, *Halovelia*, which inhabits the intertidal zone of tropical islands, is active only at ebb tide, remaining in submerged air pockets during high tide and also seeking shelter during low tide, when the rocks are completely exposed. Such behavior, coupled with the loss of flight, accounts for the fact that intertidal species of water striders are often endemic to a particular island or archipelago.

Loss or reduction of wings is also a common morphological modification among marine beetles (Doyen 1976). Some flightless species have abbreviated elytra, whereas in others the elytra are fused along the midline.

Marine intertidal chironomids are flightless or have reduced powers of flight in all but a few primitive groups (Hashimoto 1976). Walking (running) ability is well developed among species that mate on the exposed intertidal habitat. Even those with fully developed wings are distinctly inclined to walk or run rather than fly and tend to mate on the ground. Dislocation of the swarming site to sheltered supralittoral locations is characteristic of those few intertidal species that mate while in flight (Neumann 1976).

Two genera of marine chironomids, *Pontomyia* and *Clunio,* have developed the ability to skate or glide on the surface film. In contrast to the walking genera, which tend to be sexually isomorphic, the gliding genera exhibit striking dimorphism. The wings of the males (Fig. 5.1) are adapted for propelling the insects along the water surface, but are incapable of normal flight. The females lack wings or halteres and the legs are reduced. Females of most species are unable to emerge from the pupal exuviae without assistance from the male (Fig. 5.2). Copulation occurs on the surface film in most gliding species. Oviposition occurs as soon as the pair uncouples. The female dies shortly after mating, but males may copulate with other virgin females. The adaptations evolved by gliding chironomids of the intertidal zone may represent an essential stage in the evolutionary invasion of the open sea.

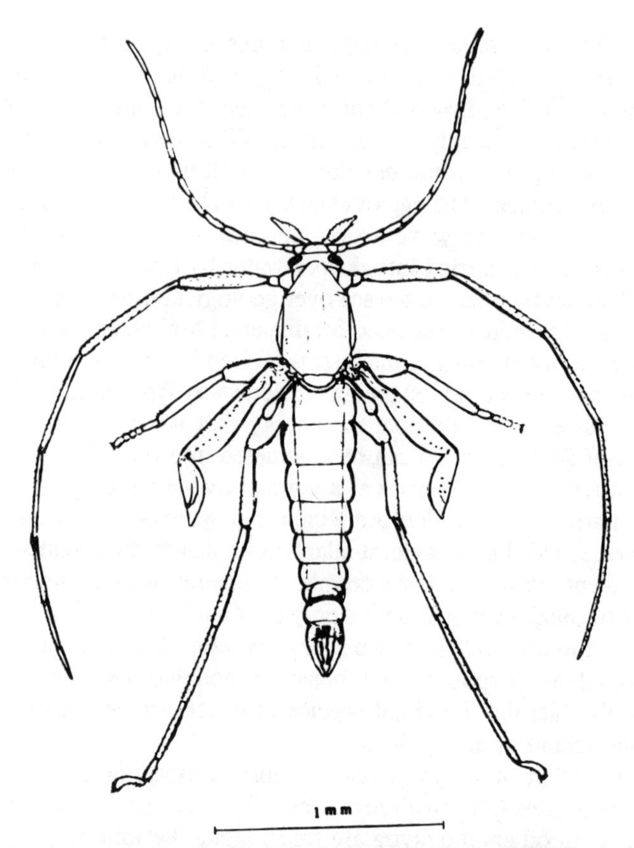

Figure 5.1 Male imago *Pontomyia cottoni*. [From Hashimoto (1976).]

Species with a shortened period of adult life have evolved in all subfamilies of chironomids represented in the intertidal zone. The adult stage of female *Clunio,* for example, has a duration of from 15 minutes to 2 hours, although males may live a bit longer. Abbreviation of adult longevity is necessary if emergence, mating, and oviposition are to be completed during tidal exposure of the larval habitat.

The success of the intertidal species, especially those with short-lived imagos, depends on synchronization of emergence among members of the population as well as synchronization with the tidal cycle (see Caspers 1951). The life cycle of intertidal populations of *Clunio marinus* is precisely synchronized with the semilunar tidal rhythm (Fig. 5.3). During the period of reproduction in the spring, emergence occurs at 2-week intervals (semilunar periodicity) in the afternoon when the *Clunio* habitat at this location is exposed by the tides (diel periodicity). According to Neumann (1976), moonlight is the synchronizing agent for the semilunar rhythm of emergence and the day-night cycle synchronizes the diel emergence pattern. A genetically controlled phase relationship between the circadian rhythm of a given population and the day-

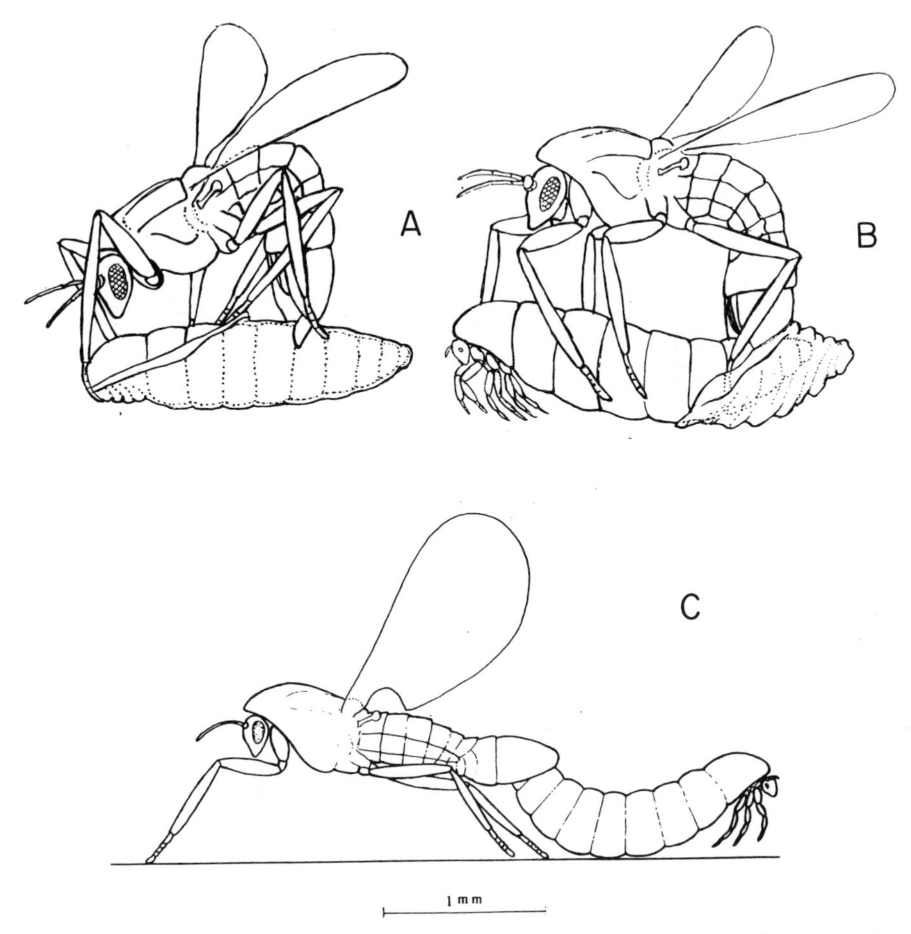

Figure 5.2 Male *Clunio tsushimensis* assists the female with emergence by stripping the pupal skin (*a*, *b*). Copulation then occurs in an end-to-end position (*c*). The entire process takes place at the surface film. [Modified from Hashimoto (1976).]

night cycle has evolved to account for geographic differences in daily tidal patterns (Neumann 1967). On the Pacific coast of Japan, where low tides occur in the morning during summer, but at night during winter, *Clunio tsushimensis* emerges only in the morning in summer and only at night during winter (Hashimoto 1976).

Freshwater Invasions by Intertidal Dipterans The Canaceidae, a small family of dipterans, and the Clunioninae, a subfamily of chironmids, occur in marine intertidal habitats throughout much of the world. These almost exclusively marine groups of insects are, however, also represented in the rapid mountain streams of Hawaii (Wirth 1949, 1951).

Five species of *Telmatogeton* (Clunioninae) occur in mountain streams in

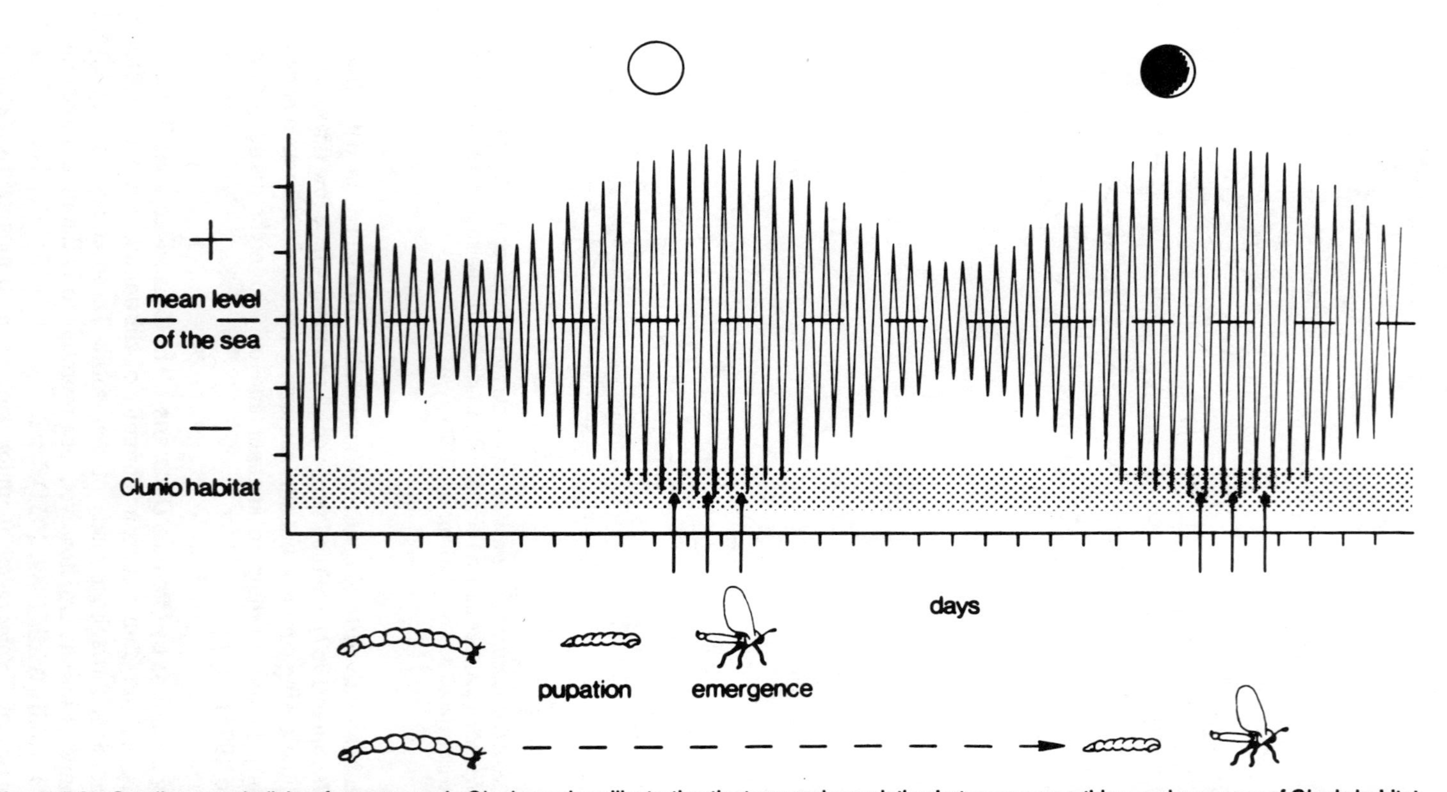

Figure 5.3 Semilunar periodicity of emergence in *Clunio marinus* illustrating the temporal correlation between moon, tides, and exposure of *Clunio* habitat. Stippled strip indicates the level of *Clunio* habitat being exposed during extreme low waters. Arrows indicate emergence. [Modified from Neumann (1975).]

Hawaii. Larvae are apparently restricted to rapid water where they reside in silken tubes. *Telmatogeton torrenticola,* for example, constructs silk galleries on waterfalls and other submerged surfaces where the current is strongest (Terry 1913). These torrential lotic habitats are somewhat analogous to the surf zone of the marine intertidal, where most species of *Telmatogeton* live. The Hawaiian Islands are geologically young and receive high rainfall. Streams are numerous, and many remain torrential all the way to the sea. Because of the isolation of the Hawaiian archipelago, there are few indigenous insects or other aquatic animals in freshwater habitats. As Wirth (1949) points out, because of the paucity of lotic organisms, competitive pressures are relaxed; there are open niches for the evolutionary invasion of freshwaters by marine species that possess some degree of euryhalinity. Indeed, there is evidence that *Telmatogeton* is a transitional genus between freshwater and the sea. *Telmatogeton japonicus,* while inhabiting the intertidal zone, can be reared in freshwater (Tokunaga 1935), and this species and some other Clunioninae in Hawaii prefer coastal habitats where the sea is freshened by stream outlets (Wirth 1949). *Telmatogeton* in the Hawaiian Islands may be an example of a rare phenomenon among insects, the active invasion of freshwater by marine species.

Two species of Canaceidae, both of the genus *Procanace,* are known from streams (Wirth 1951), but very little is known of their larval ecology. One species occurs in Hawaii, the other in Java. As with *Telmatogeton,* there is evidence that *Procanace* invaded freshwaters from the marine environment (Hinton 1967).

Neritic and Oceanic Waters

Only a few chironomids and hemipterans occur in marine environments other than the intertidal zone. For reasons discussed below, it is questionable whether some of the insects commonly referred to in the literature as *neritic* or *oceanic* (those placed within parentheses in Table 5.1) truly belong in those categories.

The corixid bug *Trichocorixa verticalis* has been collected in marine plankton samples from Delaware Bay (USA) at salinities of 24.9-29.4‰ and in the offshore waters of the Gulf of Mexico at salinities of 26.1-32.3‰ (Hutchinson 1931, Gunter and Christmas 1959). It is doubtful that these records are anything other than accidental marine occurrences of species tolerant of highly saline inland waters. As Hutchinson (1967) points out, however, ocean currents may transport such species great distances and enable them to colonize isolated islands. Indeed, *Trichocorixa beebei,* a species endemic to the Galapagos Islands, where it inhabits highly saline lakes (Howmiller 1969), may have differentiated from a species that was carried to that isolated archipelago by ocean currents.

Marine Water Striders Three families of water striders (Hemiptera) have colonized neritic waters (Table 5.2), but only the genus *Halobates* contains truly oceanic species (Herring 1961, Andersen and Polhemus 1976, Cheng 1985). Water striders are epipleustonic insects adapted for locomotion on the surface

Table 5.2 Habitat preferences of marine water striders

Family and Genus	Marine Species[a]	Habitat Preferences			
		Brackish	Intertidal	Neritic	Oceanic
Gerridae					
Asclepios	4	X		X	
Halobates group 1	39	X		X	
Halobates group 2	7				X
Stenobates	1			X	
Rheumatometroides	1	X			
Rheumatobates	6	X		X	
Veliidae					
Trochopus	3	X		X	
Husseyella	3	X			
Xenobates	1		X		
Halovelia	11		X		
Mesoveliidae					
Mesovelia	1	X			
Speovelia	1		X		
Hermatobatidae					
Hermatobates	9		X	X	

[a]Marine environments are defined as those with ≥20‰ salinity. Freshwater species that can tolerate saline waters are not included.

Source: Modified from Andersen and Polhemus (1976) and Cheng (1985).

film of aquatic habitats. The body surface contains hydrofuge hairs at densities of several thousand hairs per square millimeter. The hydrofuge body surface prevents wetting from waves, spray, or rain, thus enabling the insect to maintain an epipleustonic position. Should submergence accidentally occur, the hydrofuge hairs entrap air, which renders the insect highly buoyant and supplies oxygen for underwater respiration. The middle or rowing legs of oceanic species of *Halobates* have an especially well-developed fringe of hairs on the tibia and tarsus (Fig. 5.4) that greatly increases thrust during locomotion (Andersen 1976). The veliid *Trochopus* has a tarsal swimming fan of plumose hairs similar to those of *Rhagovelia* (Fig. 4.2), the lotic genus from which *Trochopus* is thought to have been derived.

The marine water strider genus *Halobates* is comprised of 44 known species (Cheng 1985). Five of these occur on the open sea and are the only truly oceanic insects. *Halobates micans*, the most widely distributed species, is circumtropical. The oceanic *Halobates* are confined to the open ocean; only severe storms carry populations to near-shore waters. Neritic species, many of which have geographically restricted distributions, occur primarily in bays and lagoons, including some brackish waters.

There are numerous accounts of *Halobates* eggs found attached to floating objects (e.g., cork, wood, seaweed) in the open sea. If such objects are required

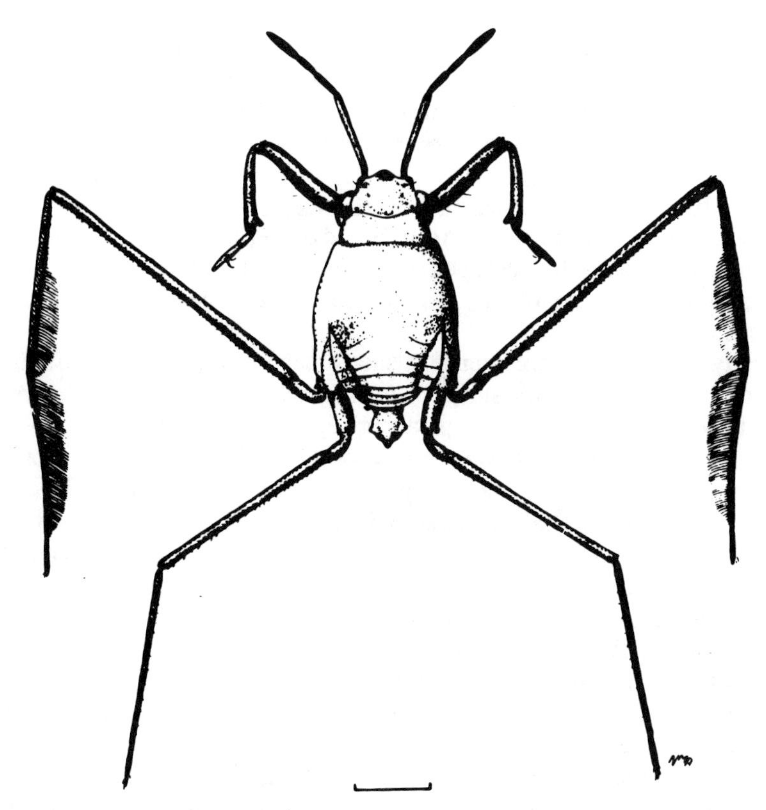

Figure 5.4 Male *Halobates micans.* Scale 1 mm. [From Andersen and Polhemus (1976).]

for the incubation of eggs, suitable oviposition sites may be an important limiting factor for oceanic species.

Halobates are piercing and sucking predators, as is true of all water striders. Oceanic species feed on zooplankton and terrestrial insect flotsam (see Cheng and Birch 1978), whereas near shore species feed primarily on floating terrestrial insects (Cheng 1985).

Based on the few available estimates, the abundance of oceanic *Halobates* is higher than might be expected. For example, Cheng (1974) reported densities of ≤ 0.1 individuals m^{-2} of sea surface. Because *Halobates* tend to avoid towed nets (Cheng 1973a, Cheng and Enright 1973), such density estimates are likely to be underestimates.

The invasion of marine habitats has occurred on at least eight separate occasions during the evolutionary history of water striders (Andersen and Polhemus 1976). Usinger (1957, p. 1178) called attention to the wing polymorphism of freshwater species, in contrast to the permanent aptery of related marine forms: "Thus we have the polymorphic *Microvelia* of fresh waters throughout the world and its wingless counterpart *Halovelia* of tropical oceans, the polymorphic *Metrobates* of inland rivers and the apterous *Halobates* of the ocean and the polymorphic *Rhagovelia* of fresh water streams, with its apterous counterpart, *Trochopus,* of tropical coves and lagoons." There must be strong selection pressures for aptery in marine environments considering that all intertidal, neritic, and oceanic water striders are wingless.

Marine Chironomids It is stated in various sources that two species of chironomids have adapted to the open sea. Based on the known distribution of these species, it appears that by the "open sea" is meant the sublittoral zone. There are no known oceanic chironomids, and only one of the "open sea" species is truly neritic.

Highly synchronized emergence, a short imago life span, and a gliding type of locomotion on the water surface as exhibited by intertidal *Clunio* are considered by Neumann (1976) as preadaptive to colonization of the open sea. *Clunio marinus,* which elsewhere in Europe inhabits the intertidal zone, has invaded a sublittoral habitat far off shore in the Baltic Sea, the world's largest estuary. Larvae of the Baltic population inhabit bottom sediment in water up to 20 m deep where the salinity is normally ≤10‰ (Olander and Palmén 1968). Emergence is highly synchronized, and oviposition occurs within minutes following eclosion, reducing the likelihood of adults being carried by winds to habitats unsuitable for larval development. Unlike intertidal populations of *C. marinus* that have floating egg masses that must be attached to the exposed substrate or be swept away by the incoming tide, the eggs of the Baltic population are demersal and immediately sink to the bottom (Neumann 1976). The life cycle of Baltic populations, which live in cold, deep water, is synchronized by photoperiod, which exerts no influence on intertidal populations of *C. marinus.* The daily timing of emergence of Baltic and most intertidal populations is achieved through circadian rhythmicity.

Pontomyia is the only chironomid other than *Clunio* to have developed a gliding type of locomotion on the water surface. The two genera share several other features (e.g., striking sexual dimorphism, colonization of sublittoral marine habitats), although such characteristics apparently developed independently by convergent evolution (Neumann 1976).

Pontomyia natans was discovered among strands of the aquatic angiosperm *Halophila* in lagoons in Samoa. Buxton (1926) reported that this chironomid was aquatic in all life stages, an erroneous interpretation refuted by Tokunaga (1932), but promulgated in later publications (e.g., Mackerras 1950, Usinger 1957). For example, Mackerras (1950, p. 27) stated that "The early stages and the female belong to the reef fauna, but the males are as truly pelagic as the zooplankton among which they live." The adults of all species of *Pontomyia* are actually aerial (epipleustonic) and remain at the water surface throughout their entire adult lives (Cheng and Hashimoto 1978). The adult life is, however, very short; males live 1-2 hours, females only half that long. The Samoan populations of *P. natans* occur in tidal reef pools. In Japan, there is a subtidal population associated with the algae *Cladophora* and *Sargassum. Pontomyia natans* is the only insect that remains submerged (unlike water striders) in sublittoral marine waters during most of its life, since the Baltic population of *C. marinus* inhabits an estuarine rather than a truly marine environment. Other known species of *Pontomyia,* an exclusively marine genus, are restricted to the intertidal zone.

BRACKISH WATERS

Brackish waters as here defined include aquatic environments of intermediate salinity that have direct connection with the sea (Williams 1981a). Such systems, broadly referred to as *estuarine habitats,* are the transitional zones or ecotones between freshwaters and the sea. Examples include river estuaries, salt marshes, mangrove swamps, and those coastal bays and lagoons diluted by fresh water (Petit and Schachter 1954, Segerstråle 1959, Perkins 1974). Most estuarine habitats are characterized by large temporal fluctuations in environmental conditions such as currents, temperature, salinity, and water level (Day 1951, Dahl 1956, Lauff 1967, Remane and Schlieper 1971). The water-level fluctuations and associated events resulting from tidal action create what Odum (1971) refers to as "pulse stabilization," a phenomenon by which ecosystems are maintained in a highly productive "early successional" stage.

The aquatic estuarine fauna comprises marine, freshwater, and brackish water components (Green 1968). The marine component may be roughly divided into stenohaline marine animals (i.e., those restricted to the estuary mouth where salinity does not fall below around 30‰) and euryhaline marine species (i.e., those that may penetrate some distance into estuaries, the most tolerant of them being occasionally found at salinities as low as 3‰). The brackish water component consists of species that have adapted to waters of intermediate salinity and that do not normally occur in either seawater or freshwater. Euryhaline freshwater species that are able to penetrate some distance into brackish waters constitute the freshwater component. There is also a migratory component of mostly fishes and crustaceans that spend only part of their lives in the estuary.

The aquatic insect fauna of estuarine waters is composed of a few euryhaline marine and true brackish species, but the majority belong to the freshwater component. Remane (Remane and Schlieper 1971) placed the aquatic insect orders into three major groups based upon their overall relationship to salinity (Table 5.3). Ephemeropterans, plecopterans, and megalopterans, comprised nearly exclusively of stenohaline freshwater species, are placed in group 1. There are a few isolated records of species from the group 1 orders in brackish waters. For example, the megalopteran *Sialis lutaria* was collected in a Danish fjord at salinities up to 3‰ (Johansen 1918) and in the Baltic Sea at 5‰ (Segerstråle 1949). Nymphs of the mayfly *Callibaetis floridans* were recorded from water with salinity as high as 8‰ (Berner and Sloan 1954). Although Scott et al. (1952) reported *Cloeon* nymphs at 19.7‰ salinity in a river estuary in South Africa, it is unlikely that any mayfly species can complete its life cycle in water of such high salt concentrations (see Bayly 1972). Some aquatic insects

Table 5.3 Categorization of insect orders based on the salinity occurrences of aquatic members[a]

Group and Order	General Group Characteristics	Salinity Range‰
Group 1 Ephemeroptera, Plecoptera, Megaloptera	Exclusively stenohaline freshwater species, with only isolated exceptions	0–0.5
Group 2 Odonata, Lepidoptera, Trichoptera	Many stenohaline freshwater species; some euryhaline freshwater species; no true brackish species	0–15
Transitional group Hemiptera	Many euryhaline freshwater species; true brackish species rare; some species in hyperhaline waters; few true marine species	0–70
Group 3 Coleoptera, Diptera	Well represented in waters >15‰; common representatives of hyperhaline waters; some true brackish species; some true marine species	0–250

[a]Rare exceptions are discussed in the text.
Source: Primarily from data in Remane (Remane and Schlieper 1971).

collected in estuaries spend their early nymphal life in rivers, although nymphal development is completed in brackish waters. Migrations between fresh and brackish waters have been well documented for a few Group 1 species in the Baltic Sea, as discussed in the next section of this chapter.

Group 2 orders, Odonata, Lepidoptera, and Trichoptera, although consisting largely of stenohaline freshwater species, have freshwater representatives that commonly penetrate brackish waters of 10–15‰ salinity. The damselfly *Ischnura elegans* is a widely distributed species that is commonly collected at salinities around 15‰, and Lindberg (1948) lists 16 species of Odonata, the nymphs of which occur commonly in the salinity range 3-6‰. Larvae of the freshwater moth *Acentropus niveus* are commonly recorded at salinities of ~15‰. *Limnephilus affinis* is the only trichopteran that commonly penetrates brackish waters of moderately high salinity (reference has already been made to the four species of marine intertidal caddisflies). *Limnephilus affinis* is capable of completing its life cycle at salinities of ≤17‰, and larvae tolerate salt concentrations of even higher levels but not pure seawater (Sutcliffe 1961). Several other caddisflies (Lindberg lists nine species), many of them limnephilids, have been reported at salinities of >3‰, and additional species occur in the oligohaline waters of the northern Baltic Sea (Siltala 1906, Gullefors and Müller 1990).

Remane considers the aquatic hemipterans as a transitional order between groups 2 and 3. Hemipterans are commonly represented in moderately brackish waters, and a few species colonize hyperhaline habitats. However, some of these are pleustonic forms (see Table 5.2) rather than being immersed in the aquatic medium, and only rarely have true brackish species (those restricted to waters of intermediate salinity) evolved in this transitional order. Among submerged bugs, the corixids are best represented in saline waters (Fig. 5.5).

The final category, group 3, consists of the orders Coleoptera and Diptera. Many species in this group transgress the 15‰ salinity limit, and a few inhabit highly hyperhaline waters. In addition, there is a significant number of truly brackish species. According to Doyen (1976), at least 20 genera of dytiscids and 15 genera of hydrophilids contain species that characteristically occur in brackish waters. Virtually every family of aquatic beetles contains truly brackish water species. Although some aquatic beetles are highly euryhaline, only a few families of dipterans contain holeuryhaline representatives (apparently valid species with populations in sea water, brackish water, and freshwater). Remmert (1955) lists holeuryhaline species from the families Ceratopogonidae, Culicidae, Chironomidae, Dolichopodidae, Tipulidae, Tabanidae, Stratiomyidae, and

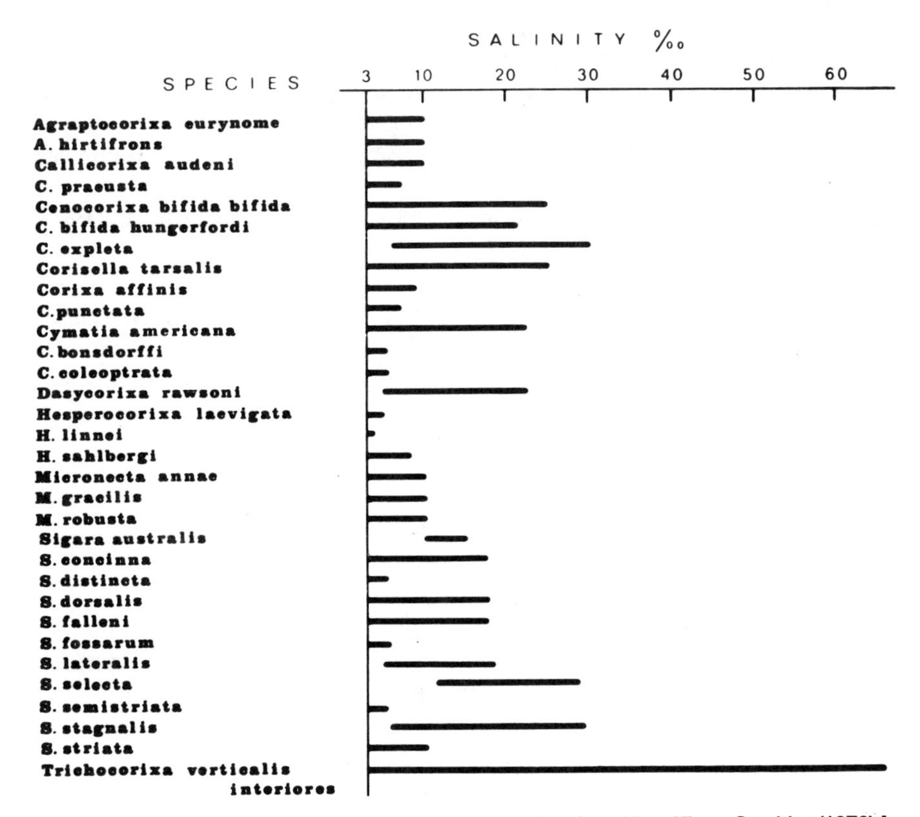

Figure 5.5 The general salinity tolerance of some saline Corixidae. [From Scudder (1976).]

Ephydridae. However, their tolerance of salinities equivalent to or higher than that of seawater does not necessarily indicate that they are able to colonize the marine environment. There are, for example, no truly marine species of Ephydridae, despite the ability of some species to inhabit saturated brine, apparently because larvae are intolerant of the turbulence of the sea (Simpson 1976). Unlike their marine representatives, the majority of which were derived from largely terrestrial groups, the dipterans and coleopterans of brackish waters are of primarily freshwater ancestry.

The Baltic Sea Because tides are insignificant and salinity changes gradually, the Baltic provides an ideal situation for examining the distribution of aquatic organisms along salinity gradients. In addition, a considerable body of information is available on several orders of aquatic insects in the Baltic Sea. The Baltic is connected to the North Sea through straits between southern Sweden and Denmark. It is about 1500 km long [north to south (N-S)] and extends nearly to the Arctic Circle between northern Sweden and Finland. Numerous publications present detailed accounts of the geological history and present hydrography of the Baltic Sea (e.g., Segerstråle 1957, Sauramo 1958, Remane and Schlieper 1971). Surface salinity in the southwestern portion of the Baltic Sea proper is ~10‰; salinity decreases very gradually to 1-2‰ near the mouths of rivers in the northern gulfs. Much of the following data are based on Lindberg's (1948) comprehensive investigations of coleopterans and hemipterans of the Baltic and a book (Müller 1982) that reviews and synthesizes research conducted on the Gulf of Bothnia, the oligohaline (1-6‰) northern portion of the Baltic Sea (61°50′ to 65°50′N latitude).

Lindberg placed the aquatic bugs and beetles that he collected from the Baltic Sea into the following three categories:

1. Halobiontic species—those that occur in the Baltic Sea but are absent from freshwater. It should be noted that Lindberg's use of "halobiont" for species restricted to brackish waters differs from the meaning applied to the term by other workers (e.g., Williams 1981a).

HEMIPTERA

Sigara stagnalis

COLEOPTERA

Ochthebius marinus *Berosus spinosus*
Laccobius decorus *Macroplea mutica*
Enochrus bicolor *Macroplea pubipennis*

2. Pseudohalobiontic species—those that are halobiontic in the northern Baltic, although they also occur in freshwaters in the southern regions.

HEMIPTERA

Gerris thoracicus

COLEOPTERA

Haliplus obliquus *Noterus clavicornis*

H. confinis pallens	*Coelambus parallelogrammus*
H. immaculatus	*Deronectes depressus latescens*
H. apicalis	*Agabus nebulosus*
H. flavicollis	*A. conspersus*
Enochrus melanocephalus	

3. Limnohaline species—those freshwater species that have an affinity for oligohaline brackish waters, thus representing a transitional group between halophilic and limnobiontic species.

HEMIPTERA

Sigara striata
Mesovelia furcata
Gerris odontogaster

COLEOPTERA

Cymatia coleoptrata	*Hygrotus inaequalis*
Haliplus lineolatus	*Ilybius fuliginosus*
Bidessus geminus	*Laccobius minutus*
Coelambus impressopunctatus	*Enochrus testaceus*

Numerous other aquatic beetles and bugs of the Baltic region are essentially restricted to freshwater habitats (limnobionts), although they are occasionally collected in brackish waters. The investigations of Nilsson (1982) on aquatic coleopterans of the Swedish coastal waters of the Gulf of Bothnia differed in some respects with the results of Lindberg's earlier research on Finnish coastal waters. Nilsson comments on the problem of minor differences in color patterns and morphology between freshwater and Baltic populations of some limnohaline species. The aquatic coleopterans of the Baltic Sea are restricted to shallow water near the shore. The only exception is *Macroplea,* a chrysomelid that occurs in water several meters deep. Adult *Macroplea* utilize plastron respiration; larvae tap the air stores of submerged angiosperms.

Table 5.4 shows the percentage of species of the well-studied orders of aquatic insects that occur in northern Sweden that have been recorded from the Gulf of Bothnia. The percentage values for Ephemeroptera and Plecoptera would be lower if migratory species (see below) were excluded.

Lingdell and Müller (1982) have recorded seven species of Ephemeroptera in the Ångerån River estuary in the Gulf of Bothnia (Fig. 5.6). *Heptagenia fuscogrisea, Baetis subalpinus,* and *Cloeon simile* are restricted to the very mouth of the river, which is a freshwater biotope during certain times of the year. Two species of *Leptophlebia (L. marginata, L. vespertina)* extend some distance into the estuary; nymphal development, which began in the river, is completed in the estuary; emerging adults fly upstream and oviposit in freshwater. Only two species of mayflies, *Baetis fuscatus* and *Caenis horaria,* are permanent members of the estuary. The distribution of *B. fuscatus* within the Gulf of Bothnia is determined by its preference for stony coastal areas rather than salinity; *C. horaria* occurs in salinities of ≤4.5‰ in the Ångerån estuary and of ≤6‰ in the

Table 5.4 Percentage of freshwater insect species inhabiting inland waters of northern Sweden that occur in the oligohaline brackish waters (1-6‰ salinity) of the Gulf of Bothnia, northern Baltic Sea

	Total Freshwater Species	Percent Occurring in Gulf of Bothnia
Ephemeroptera	37	19[a]
Plecoptera	29	21[a]
Odonata	44	50
Trichoptera	200	33
Hemiptera	29	59
Coleoptera	213	9
Megaloptera	4	50
Chironomidae	200	10
Total	756	21

[a]Some of the ephemeropterans and plecopterans (and possibly species from other orders) hatch from eggs laid in freshwater streams, drift downstream into the estuary, and complete nymphal development in brackish water (Müller and Mendl 1979).

Source: Modified from Müller (1982).

southern portion of the Gulf of Bothnia. Both of these species (and Baltic Sea species from the orders Megaloptera, Trichoptera, and Coleoptera) also have populations confined to inland freshwater habitats, thus drawing conjecture regarding the extent to which the brackish populations have genetically differentiated from freshwater forms.

The nymphs of six species of stoneflies occur in the oligohaline waters of the River Ängerån estuary and represent the only known records of Plecoptera nymphs developing in brackish water (Mendl and Müller 1982). Only two species, *Nemoura cinerea* and *Leuctra digitata*, tolerate salinity of ≤4‰. As is true of the mayfly genus *Leptophlebia*, the stonefly nymphs hatch from eggs laid in the river, but complete nymphal development and emerge from brackish water (Müller and Mendl 1979).

Salt Marshes Salt marshes are a distinctive type of intertidal ecosystem marginal to river estuaries or sheltered sea coasts that are characterized by an angiosperm community able to tolerate periodic inundation by the tides (Pomeroy and Wiegert 1981). The extent and frequency of inundation depend largely on the height above sea level and the tidal amplitude at a given location. The plants exhibit more or less distinct zonation patterns along inundation and salinity gradients. Cord grass (*Spartina*) is typically the most conspicuous plant of salt marshes worldwide, and many marshes have large expanses dominated by a single species of *Spartina*. Much of the entomofauna of salt marshes is intimately associated with higher plants (Foster and Treherne 1976). *Spartina* detritus and algae form the energy base of most salt marsh ecosystems. Mangrove swamps are the tropical analogues of salt marshes.

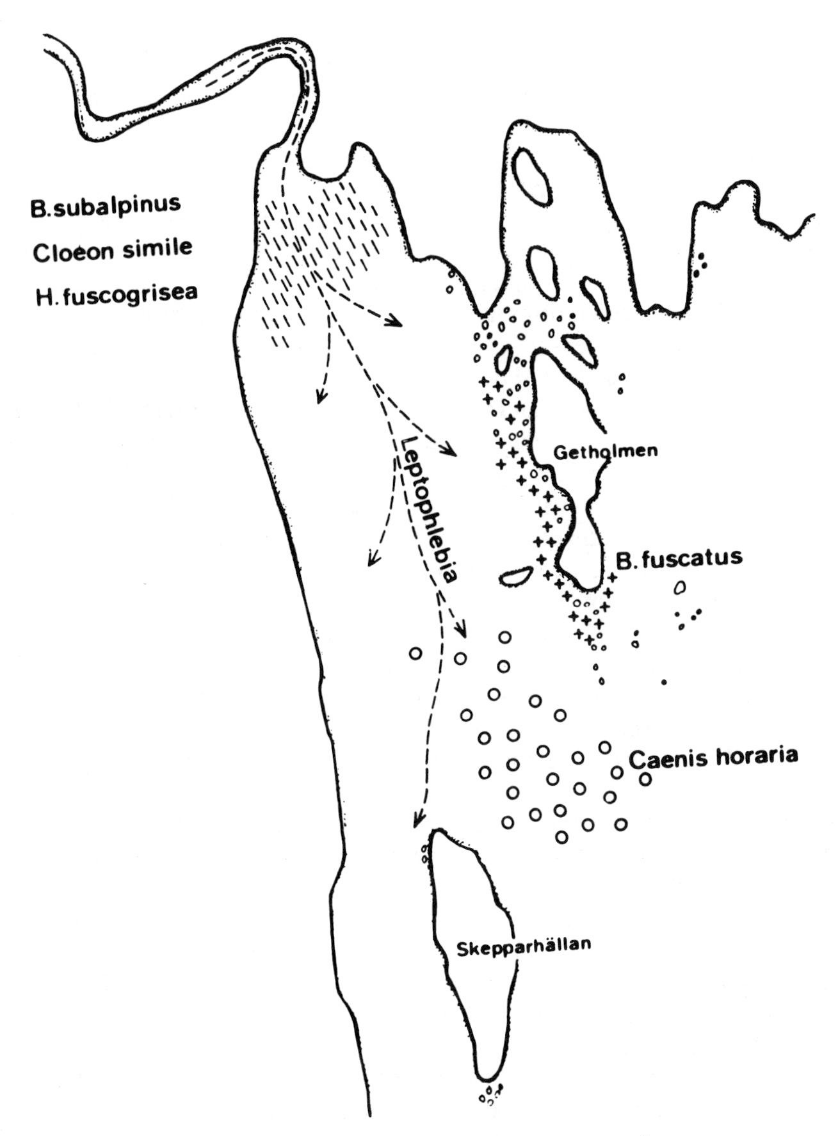

Figure 5.6 Distribution of mayfly nymphs in the estuary of the River Ängerån, Gulf of Bothnia, northern Baltic Sea. [From Lingdell and Müller (1982). Reprinted by permission of Kluwer Academic Publishers.]

The lowest portions of the marsh are typically submerged twice each day, but upper levels are inundated only rarely. Aquatic habitats, especially those at higher levels, may exhibit large temporal fluctuations in salinity. Aquatic insects of salt marshes are necessarily euryhaline; some species reside in habitats that are hyperhaline for extended periods, yet exhibit salt concentrations approaching that of freshwater at other times.

Dipterans, hemipterans, coleopterans, and to a lesser extent odonates and trichopterans, represent the majority of aquatic insects in salt marshes. With rare exceptions (e.g., Campbell and Denno 1978), investigators have not examined the entire aquatic insect community of salt marsh habitats. Data have been largely derived from investigations of economically important groups such as culicids (O'Meara 1976), ceratopogonids (Linley 1976), and tabanids (Axtell 1976). Even for these groups, ecological data on the immature stages may be fragmentary. Axtell (1976), for example, stresses that whereas numerous tabanids are reputed to occur in salt marshes, most records are based on aerial collections of adults with little or no knowledge of the habitats of the immatures. Some otherwise comprehensive ecological studies have sampled only the aerial insects (e.g., Davis and Gray 1966).

Because of their economic and medical importance, there is a relatively large body of information on brackish water mosquitoes. According to O'Meara (1976), nine major genera of mosquitoes *(Aedes, Anopheles, Culex, Deinocerites, Opifex, Aedeomyia, Uranotaenia, Psorophora, Culiseta)* contain species that pass their immature aquatic stages in tideland habitats. Most species exhibit quite specific aquatic habitat preferences, in some cases to a remarkable degree. Certain species of *Deinocerites,* for example, are confined to crabholes during aquatic stages with different species partially segregated on the basis of crabhole size (Peyton et al. 1964). *Aedes dasyorrhus* breeds almost exclusively in the brackish water tree holes of mangrove swamps (Belkin 1962). It is apparent that ovipositing females exhibit a high degree of habitat selection. Ikeshoji and Mulla (1970) demonstrated that at least some mosquitoes produce species-specific chemical attractants that enable ovipositing females to detect water containing immatures of their own species, an indication that habitat conditions are probably suitable for development. In some cases the chemical substance(s) repelled females of other species.

Salt marshes contain several aquatic habitat types. Mud flats and vegetated marsh surfaces are alternately exposed and submerged by the tides. The insects inhabiting such areas are mainly burrowers or species intimately associated with aquatic plants. Most exhibit behavioral adaptations to avoid submersion during periods of inundation. Only some members of the dipterous families Ceratopogonidae, Tabanidae, Ephydridae, Stratiomyidae, and Dolichopodidae that inhabit these areas should be considered even marginally aquatic. The tidal creeks (Fig. 5.7) represent a discrete and more or less permanent aquatic habitat type, but contain few aquatic insects (Campbell and Denno 1978). Pools that form in depressions in the marsh surface (Fig. 5.7) represent the most important habitat type for truly aquatic insects. The larger and deeper pools are generally permanent and provide aquatic habitats for a variety of organisms. Campbell and Denno (1978) examined the aquatic insect communities in several such pools in an eastern North American salt marsh. They found a generally impoverished aquatic insect fauna. Although 20 species were identified, some of the dipterans (tabanids, ephydrids, stratiomyids) were apparently carried into the pools by tidal action, their normal habitat being the mud flats and vegetated marsh

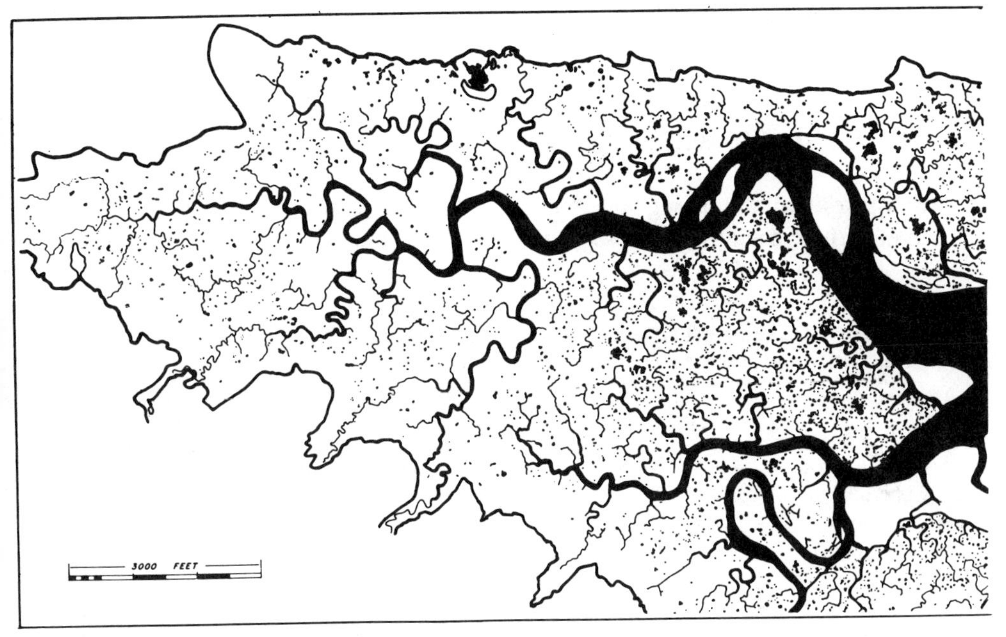

Figure 5.7 Distribution of pond holes (pannes) and tidal creeks in the high marsh at Barnstable, Massachusetts. [From "Development of a New England salt marsh" by A. C. Redfield, Ecological Monographs, 1972, 42, 201–237. Copyright © by The Ecological Society of America. Reprinted by permission.]

surfaces. The corixid *Trichocorixa verticalis* and the chironomid larva *Chironomus* sp. were the predominant insects. Hydrophilid beetles (*Enochrus hamiltoni, Tropisternus quadristriatus*), ceratopogonids (*Culicoides hollensis, C. furens*) and odonate nymphs (*Eurythrodiplax berenice*) were also abundant. The mosquito *Aedes sollicitans,* and species of tabanids, stratiomyids, ephydrids, and dolichopodids were less abundant. Although trichopterans typically occur in salt marsh pools (e.g., Sutcliffe 1962), none were encountered in the pools examined. Campbell and Denno (1978) consider the overall physical harshness of pool habitat (large variations in depth, temperature, and salinity) as largely responsible for the depauperate insect community, but attribute between-pool differences in species diversity to differences in the intensity of fish predation. The fact that corixids, which possess metathoracic repugnatorial glands, were more equitably distributed among the pools would appear to support the role of fish predation in structuring the insect communities of the pools.

Butler and Popham (1958) summarize data on aquatic insects (excluding dipterans, caddisflies, and water striders) collected over several years from pools and dykes adjacent to a river estuary in England. Because the habitats collectively exhibited a range of salinities, the authors were able to assess the salinity tolerances of 62 species based on the maximum salinity at which they occurred

(Table 5.5). The most saline pools, however, exhibited lesser salt concentrations than sea water.

INLAND SALINE WATERS

Salt lakes (saline running waters are not considered here) are largely restricted to endorheic basins (i.e., those without outlets, which typically occur in arid or

Table 5.5 Maximum salinities (% sea water) at which aquatic insects[a] were collected in pools and dykes along the Humber River Estuary, Yorkshire, England

	Number of Species at Maximum Salinity of				
Taxon	<10%	10–20%	20–25%	25–50%	>50%
Ephemeroptera					
Cloeon	1	0	0	0	0
Odonata					
Aeshna	1	0	0	0	0
Libellula	1	0	0	0	0
Sympetrum	0	0	0	1	0
Coenagrion	1	0	0	0	0
Ischnura	0	0	0	0	1
Pyrrhosoma	0	0	0	1	0
Hemiptera					
Corixa, Sigara	6	2	5	1	3
Notonecta	0	0	2	0	0
Plea	0	0	1	0	0
Coleoptera					
Acilius	1	0	0	0	0
Agabus	1	0	0	1	1
Colymbetes	0	0	0	1	0
Dytiscus	0	0	1	0	0
Hydroporus	0	2	2	1	0
Hygrotus	1	0	0	2	0
Ilybius	1	0	1	0	0
Laccophilus	0	1	0	0	0
Rhantus	0	1	0	0	0
Enochrus	0	0	0	0	1
Helophorus	0	0	0	0	1
Laccobius	1	0	0	0	0
Ochthebius	0	2	0	0	1
Haliplus	2	1	2	0	1
Hygrobia	1	0	0	0	0
Gyrinus	0	0	0	1	0

[a]Excluding Diptera, Trichoptera, and Gerroidea (Hemiptera).

Source: Butler and Popham (1958).

semi-arid regions) (Cole 1968, McCarraher 1972). Their elevated salinity results primarily from evaporative concentration processes. Other characteristics of salt lakes include large fluctuations in water level [many are ephemeral; see McLachlan (1979)], physicochemical instability, variable and often low levels of dissolved oxygen, high temperatures, and turbid conditions.

The salinity of natural water bodies ranges from near zero to saturated brine (a saturated solution of sodium chloride has a salinity of ~350‰). Saline lakes, however, rarely approach the saturation values of the most soluble salts (Hutchinson 1957a). Great Salt Lake, Utah (USA), has a total salinity somewhat less than 250‰. The salinity level separating fresh from saline waters has been somewhat arbitrarily placed anywhere from 0.5‰ to 5‰ (Segerstråle 1959, Bayly 1972). Salt lakes are athalassic (nonmarine) water bodies and many, but by no means all, are althalassohaline (differing in ionic composition from seawater). The fauna consists almost exclusively of forms of freshwater derivation (Beadle 1959). The proceedings of a symposium on salt lakes (Williams 1981b) provides an excellent account of the biota of athalassic waters and the physicochemical variables of biological importance.

Most groups of insects that colonize salt marsh pools are also represented in inland saline waters. Most abundant are hemipterans (especially Corixidae), coleopterans, odonates, tripchopterans, and several families of dipterans. Publications containing data on the insects of salt lakes include Hutchinson (1937b), Rawson and Moore (1944), Bayly and Williams (1966), Hammer et al. (1975), Scudder (1976), Timms (1981, 1983), Williams (1981c), Galat et al. (1981), and Halse (1981). Beadle (1943, 1959, 1969), Bayly (1972), and Kokkinn (1986) specifically address the topic of osmoregulation in the context of inland saline lakes.

Rawson and Moore (1944) conducted comprehensive investigations of 50 lakes, most of them saline, that are located in the Canadian Province of Saskatchewan. The salinity maxima for the aquatic insects were reported as follows:

	Salinity Maximum (‰)
Ephemeroptera	20
Trichoptera	20
Odonata	29
Hemiptera	29
Coleoptera	118
Diptera	—
Chaoboridae	8
Ceratopogonidae	14
Chironomidae	29
Ephydridae	118
Dolichopodidae	118

Dipterans and coleopterans exhibited the greatest diversity in the saline lakes, and the members of each of these orders collectively traversed the entire salinity range (0.3-118‰). There was surprisingly little overlap between the beetle faunas of the freshwater and saline lakes. Hydrophilids, haliplids, gyrinids, and especially dytiscids represented the majority of beetles in the saline lakes. The dytiscid *Hygrotus salinarium* and the hydrophilid *Enochrus diffusus* were characteristic of saline waters, although *E. diffusus* occurred in lakes over the entire range of salinity. The dipteran *Chaoborus* was generally restricted to freshwater lakes, although Rawson and Moore collected a few larvae in a lake with salinity of 8‰. In the most saline lake examined (118‰) ephydrids, dolichopodids, and the beetle *Enochrus diffusus* were the only insects present.

Three genera, *Notonecta, Trichocorixa,* and *Arctocorixa,* accounted for the majority of hemipterans collected from the saline lakes. Hemipterans were abundant in lakes of moderate salinity although none were collected from the most saline lake. *Trichocorixa verticalis interiores,* a characteristic subspecies of saline lakes (see Fig. 5.5), occurred in all except the most saline lake in the Saskatchewan series (no lakes were examined with salinities of 29–118‰, however).

Among odonates, zygopterans rarely occurred in saline lakes, whereas anisopterans were collected from all except the most saline lake. *Enallagma clausum,* also reported from saline lakes in western North America (Hutchinson 1937b), exhibited a preference for saline waters. The shorter period of nymphal development exhibited by certain species of odonates when in more saline waters is apparently an adaptation to complete growth and emerge before salinity levels (which increase over the season in some habitats) become intolerable (Bayly and Williams 1966, Cannings et al. 1980).

Mayflies were for the most part restricted to the freshwater lakes of the Saskatchewan series. *Lachlania saskatchewanensis,* however, was described from a lake with a salinity of 8‰, and *Callibaetis* nymphs, common in moderately saline lakes, occurred up to 20‰.

Caddisflies were abundant in the freshwater lakes and common in moderately saline waters. The most common of the 16 species collected from saline lakes were *Molanna flavicornis, Mystacides longicornis, Phryganea cinerea,* and *Helicopsyche borealis.*

A more recent examination of the saline lakes of Saskatchewan (Hammer et. al. 1975) generally substantiated the conclusions of Rawson and Moore (1944) regarding the distribution of insects. Concomitant with the overall reduction in species diversity that accompanies elevated salinities, there is an increase in the relative abundance of hemipterans, coleopterans, and dipterans. Lakes of higher salinity than the most saline lake (118‰) examined by Rawson and Moore did not contain a benthic fauna.

Although Rawson and Moore (1944, p. 187) concluded that "salinity is the chief factor" affecting the abundance of benthic animals, the lakes they studied differed considerably from one another in factors other than salinity. Timms (1981) studied the animal communities of three Australian lakes that are physiographically and physicochemically similar, apart from distinct differences in salt concentrations (Table 5.6). Neither ephemeropterans nor odonates were present in the saline (8‰) lake. In the hypersaline (58‰) lake the insect fauna was limited to dipterans (ceratopogonids and *Tanytarsus barbitaris)* and dytiscid beetles *(Rhantus pulverosus* and *Lancetes lanceolatus).* The absence of corixids and ephydrids from the hypersaline lake contrasts with the occurrence (often predominance) of these groups at comparable salinities on other continents. *Ephydra cinerea* and *E. hians* are, for example, the only insects able to tolerate the high salinity of Great Salt Lake (Stephens 1974). Ephydrids are, however, uncommon in the saline lakes of Australia (Williams 1981c), and corixids are restricted to much lower salinities than on other continents (Knowles and Williams 1973).

Factors other than the direct effects of salinity tolerance are rarely implicated by investigators dealing with the biota of saline lakes to explain the distribution of species along salinity gradients. Williams (1981c, p. 249) concludes that salinity "does appear to be a major determinant of field occurrence," but then elaborates on the possible influence of other limiting factors. A variety of factors other than salinity (e.g., dissolved oxygen, turbidity, temperature, ionic proportions, habitat stability) may influence species distributions directly or by altering salinity tolerance (see, e.g., Herbst and Bromley 1984), which indicates the need for additional comparative field studies of otherwise similar aquatic habitats that exhibit large differences in salt concentrations.

Wide discrepancies between the maximum salinity tolerated in laboratory tests and the salinity at which a species occurs in nature would suggest that factors other than salinity are important determinants of field distribution patterns. Although there is a general concurrence between the salt concentration of the natural habitat and laboratory tolerance levels (Williams 1981c), it is far from universal. The ability of many halophiles and halobionts to tolerate lower

Table 5.6 Insect fauna of three Australian lakes that are similar to each other (morphometrically, physicochemically), but differing in salinity[a]

	Number of Taxa		
	Fresh	Saline	Hypersaline
Ephemeroptera	1	0	0
Odonata	5	0	0
Trichoptera	3	3	0
Lepidoptera	1	1	0
Hemiptera	5	1	0
Coleoptera	12	1	2
Diptera	8	5	2
Insect taxa	35	11	4

[a]Lake Purrumbete is fresh (0.4‰ salinity), L. Bullenmerri is saline (8‰), and L. Gnotuk is hyperhaline (58‰).

Source: Timms (1981).

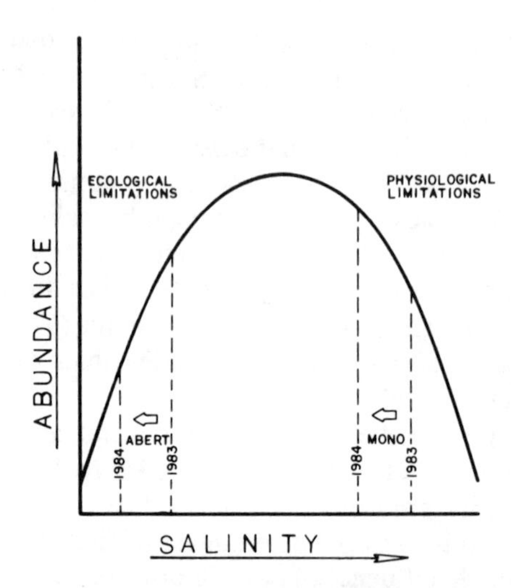

Figure 5.8 Model of hypothetical relationship between salinity and abundance of *Ephydra hians*. Abundance is proposed to be maximized at salinities intermediate between the physiological limitations of high-salinity stress, and the ecological limitations imposed by biotic interactions (e.g., predation and competition) at low salinity. Salinity dilution at Mono Lake was accompanied by increased abundance of E. *hians*, while dilution at Abert Lake was accompanied by a decrease in abundance. [From Herbst (1988). Reprinted by permission of Kluwer Academic Publishers.]

salinities than those to which they are exposed in the field suggests that biotic factors (predation, competition, parasitism) are excluding them from habitats of lower salinity. An interesting example involves the influence of a parasitic mite on the distribution of congeneric species of corixids in saline Canadian lakes (Smith 1977, Scudder 1983). *Cenocorixa bifida* and *C. expleta* occupy different, although overlapping, salinity ranges (Fig. 5.5). The mite occurs over the lower salinity range of its primary host, *C. bifida,* but cannot apparently tolerate the upper portion of the salinity range, where *C. expleta,* the secondary host, predominates. The mite imposes a much higher mortality rate on the secondary host, which, however, has a high salinity refugium from the parasite that is unavailable to the primary host. It thus appears that this tripartite relationship of two closely related species of corixids and a mite that preferentially parasitizes one of them is a coevolved system based on (or initiating) the salinity tolerances of the species involved and enabling them to coexist.

The conceptual model shown in Figure 5.8 was developed to explain the abundance of larval brine flies *(Ephydra hians)* along a salinity gradient (Herbst 1988). Populations of *E. hians* in two salt lakes exhibited opposite responses to dilution of their habitats resulting from high winter precipitation. In Mono Lake, California, salinity dilution from 90 to 80 g/liter total dissolved salts (TDS) resulted in the increased relative abundance of *E. hians,* whereas relative abundance values decreased in Abert Lake, Oregon, where TDS declined from 30 to 20 g/liter during the same period. This led to the hypothesis that abundance is limited by physiological stress at higher salinities and biotic interactions at low salinities due to increases in predators and competitors.

Athalassic waters possess several attributes amenable to ecosystem level studies. Their limited diversity is an advantage, especially given the taxonomic difficulties involved with the immature stages of aquatic insects. Saline lakes in endorheic basins (the majority) are discrete habitats, which, in concert with their simplified trophic pathways, facilitate studies of energy flow and material transfer. In addition, saline lakes provide exceptional opportunities to elucidate biotic interrelationships.

ENVIRONMENTAL CONDITIONS

A variety of interrelated environmental variables determine distribution and abundance patterns of the species that collectively constitute aquatic insect communities. The chapters in Part III deal primarily with abiotic variables, such as temperature and oxygen, that are of major importance to aquatic insects. A description of the natural spatial and temporal ranges exhibited by each of these variables serves as a prelude to discussions of their importance to aquatic insects.

The accompanying figure [modified from Ruttner (1963)] portrays the idealized distribution patterns of three species along an environmental gradient. Each species occupies a distinct portion of the gradient beyond which the biotic

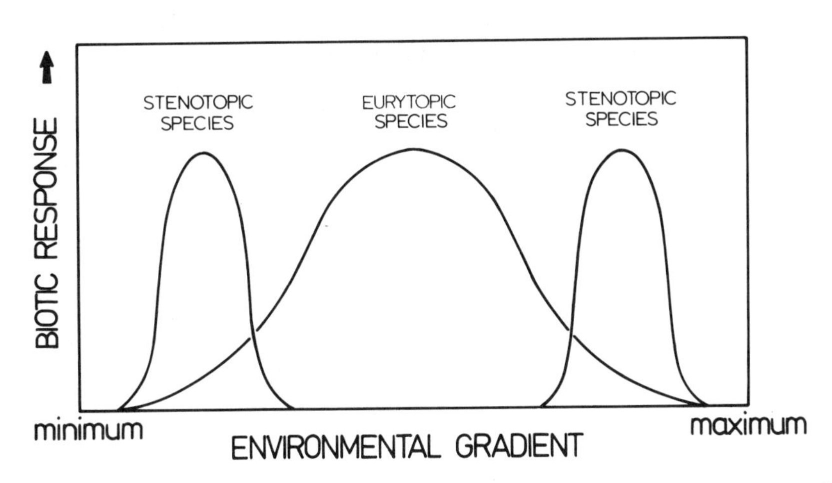

responses (growth, activity, abundance, etc.) are zero. Optimal conditions (maximum biotic response) may or may not occur midway along the portion of the gradient occupied by a species. Eurytopic species are those with broad limits of tolerance; stenotopic species are restricted to a small portion of the environmental gradient. *Eurytopic* and *stenotopic* are general terms used for unspecified environmental factors. More specific terms are used in the treatment of a single factor as in the following examples:

Temperature	stenothermal-eurythermal
Salinity	stenohaline- euryhaline
Food	stenophagic-euryphagic
Oxygen	stenoxybiontic-euryoxybiontic
Altitude	stenozonal-euryzonal
Depth	stenobathic-eurybathic

Additional terms are necessary, however, to distinguish stenotopic species at opposite ends of, for example, a thermal gradient. Species such as the chironomids restricted to glacier brooks where summer temperatures are <6°C are referred to as *cold stenotherms, oligostenotherms,* or *oligotherms* by various authors. Species restricted to a narrow portion of the upper range of temperature are designated *warm stenotherms, polystenotherms*, or *polytherms. Mesotherms* are stenotherms that occupy the middle portion of a thermal gradient. The term *euryoky* is applied to species with relatively broad ranges of tolerance to a variety of environmental factors; euryokous species tend to be widely distributed and to occur in a variety of habitat types. *Stenokous* organisms are those that have narrow tolerance limits for many environmental variables. As Odum (1971) points out, the evolution of narrow tolerance limits can be viewed as a form of specialization by which species increase the efficiency of resource utilization while sacrificing adaptability. According to this perspective, environmental gradients become niche axes [*sensu* Hutchinson (1957b)] along which the various species partition resources. Biotic factors (e.g., competition, predation), as well as tolerance per se, determine the range of environmental conditions occupied by a given species in nature.

6

TEMPERATURE

> With respect to life as a whole, temperature is presumably the most important single environmental entity.
>
> —Otto Kinne (1963, p. 304)

Temperature has been ascribed a primary role in the ecology of aquatic insects (Hynes 1970b, Vannote and Sweeney 1980, Ward and Stanford 1982). The following description of the thermal heterogeneity to which aquatic insects may be exposed precedes consideration of the various ecological implications of temperature.

THE TEMPERATURE REGIME

Aquatic insects respond to the entire temperature regime, which includes absolute levels, seasonal and diel ranges, rate functions, and the timing and duration of thermal events (Fig. 6.1). Only in special aquatic habitats (e.g., spring sources) do constant temperatures prevail; thermal heterogeneity is a characteristic feature of most lotic waters or the surface waters of lentic habitats.

Lotic Temperatures A variety of hydrological, topographical, and meteorological factors are responsible for thermal patterns in lotic systems (Smith 1972, Ward 1985). The headwaters of spring-fed streams are dominated by groundwater temperatures, which are usually within 1°C of the mean annual air temperature of the region. In streams or stream segments not generally influenced by groundwater, there is a close relationship between water and air temperatures;

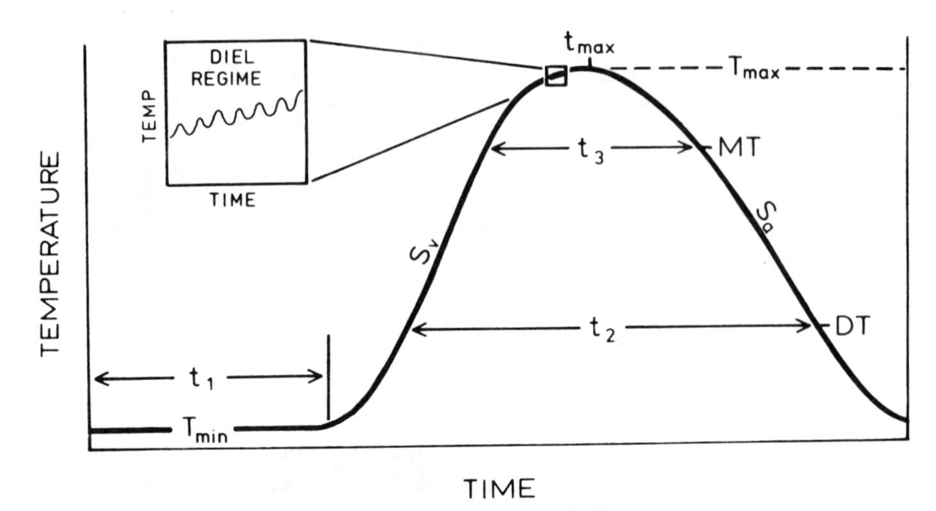

Figure 6.1 An idealized annual temperature pattern illustrating the major components that may influence aquatic insects. Inset shows the diel regime over a 7-day period. The duration (t_1) of T_{min}, the annual minimum temperature, is the period of ice cover at high and midlatitudes. The growing season is represented by t_2, the period of time that temperatures exceed the developmental threshold (DT). In degree-day calculations, DT is developmental zero. For some species a second level, the maturation threshold (MT), must be exceeded (sometimes for a specified time, t_3) to enable them to complete development. Some species require an even higher temperature as an emergence cue. Habitat temperature may exceed an upper level, not indicated in the figure, beyond which growth and development cease, and a lower limit may be required to break diapause. The time of year (t_{max}) at which the maximum temperature (T_{max}) is attained varies between water bodies and between years in the same water body. Rates of vernal temperature increase (S_v) and autumnal decrease (S_a) are additional considerations. [From Ward and Stanford (1982). Reproduced, with permission, from the Annual Review of Entomology, Vol. 27, © by Annual Reviews Inc.]

major discrepancies may occur, however, during periods of ice cover, snowmelt, or spate events. Because larger volumes of water are less responsive to atmospheric vagaries and are traveling at higher velocities, the thermal conditions of source waters are carried farther downstream during high discharge.

The aspect of the drainage basin, streamside vegetation, and channel form influence the relative importance of direct solar radiation on stream temperature. In Swedish Lapland, higher elevation streams (above the birch forest) reach higher summer temperatures than shaded streams at lower elevations (Ulfstrand 1968). On clear summer days water temperatures up to 3.3°C greater than air temperatures were recorded in an Ontario river (Ricker 1934). Substrate type and degree of exposure to direct sunlight play major roles in determining temperature regimes of tropical streams (Geijskes 1942).

Other factors may also influence lotic temperatures. Tributaries may warm or cool the main stream, depending on several variables including season. Precipitation may also have a warming or cooling influence. Thermal conditions in lakes determine the temperature regime of downstream lotic habitats (Humpesch 1979); the annual temperature range of a lake outlet stream in Austria (2–22°C) contrasts greatly with that of a nearby brook (0.5–10°C).

ANNUAL RANGES Natural streams of middle and high latitudes are characterized by temperatures that range from 0 to 25°C, or less, annually (Table 6.1). Lotic waters in regions of continental climates may exhibit wider ranges, whereas the moderating influence of marine climates suppress annual variations. Only in tropical or desert regions do running waters reach or exceed 30°C.

The absolute value and duration of the minimum temperature attained during an annual cycle is of considerable ecological importance. Streams in regions of marine climates may not freeze even at quite high latitudes. However, most temperate streams drop to 0°C during part of the year unless they receive a substantial influx of groundwater. The duration of ice cover exceeds seven months per year in streams at high altitude or latitude.

In addition to surface ice, running waters in cold climates may develop underwater ice (Ashton 1979). Anchor ice (an attached type of underwater ice) forms on clear cold winter nights in rapid sections that lack surface ice, as a result of radiant heat loss from the substrate. Often a diel cycle of formation (night) and release from the substrate (day) is apparent. Thicknesses of ≤60 cm above the stream bed have been reported (Brown et al. 1953). When anchor ice is present, stream waters remain at 0°C irrespective of such factors as air temperature or time of day (Maciolek and Needham 1951).

Table 6.1 Annual temperature ranges and maximum diel amplitudes of selected lotic reaches

Stream and Location	Annual Range (°C)	Diel Amplitude (°C)	Source
Estaragne tributary, France	0–4.2	1.5	Lavandier 1974
River Laxa, Iceland	0–12.4	5.9	Ólafsson 1979
Akbura River, USSR	1–13	5.0	Brodsky 1980
Teichbach, Austria	0–19	10.0	Malicky 1976b
White Clay Creek, Pennsylvania	0–20	5.5	Vannote and Sweeney 1980
Mad River, Ontario	0–22	6.1	Ricker 1934
River Lissuraga, France	10–17.5	6.0	Thibault 1971
Black Brows Beck, England	2.5–16.1	>10	Crisp and LeCren 1970
River N. Tyne, England	2.5–17.5	7.0	Boon and Shires 1976
Spring source, Colorado	8–10	<1.0	Ward and Dufford 1979
Sycamore Creek, Arizona	10–30	10.0	Gray 1981
Silver Springs, Florida	22.2–23.3	<1.0	Odum 1957
Sungai Gombak, Malaysia	22.5–32.5	5.0	Bishop 1973a

TEMPERATURE SUMMATION Aquatic insects may respond to the summation of thermal units (i.e., degree days) as well as absolute temperatures. Annual degree days in aquatic habitats are normally calculated by summing daily mean temperatures above 0°C for one year, without an upper limit. While this is a suitable method for comparing habitats, it may not be appropriate for a given species of aquatic insect. Baskerville and Emin (1969) present a method in which the daily temperature cycle is represented by a trigonometric sine curve in degree-day calculations. Only diel maxima and minima are required (rather than continuous recordings), and both a lower and upper temperature threshold may be utilized in their formulation.

Selected values of annual degree days from specific lotic reaches are shown in Table 6.2. Differences are primarily attributable to differences in latitude and altitude. Stream reaches exhibiting markedly different seasonal temperature patterns may accumulate comparable annual degree days [Fig. 6.2; see also Gray et al. (1983)]. The seasonal distribution of degree days has, however, rarely been considered in temperature analyses. Thermally constant source waters in which degree days are evenly distributed throughout the year contrast with the majority of aquatic habitats. For example, in the Bigoray River, Alberta, where water temperatures are near 0°C for 6 months of the year, 58% of the total degree days occur during July, August, and September (Clifford 1978). In a Michigan headwater stream (Cummins and Klug 1979), the annual degree days were distributed as follows: spring (21%), summer (51%), autumn (22%), and winter (6%). The accumulation of a given number of degree days, an important thermal cue for some aquatic insects, may be attained three to 4 weeks earlier after a warm than a cold winter (Macan 1958).

Table 6.2 Annual degree days of selected lotic habitats

Stream and Location	Degree Days	Source
River Estaragne, France		Lavandier 1974
2350 m a.s.l.	500	
1850 m a.s.l.	1000	
St. Vrain River, Colorado (USA)		Ward 1986
3414 m a.s.l.	450	
1544 m a.s.l.	4220	
Gorge Creek, Alberta (Canada)	1250	Hartland-Rowe 1964
Little Lost Creek, Idaho (USA)		Andrews and Minshall 1979
Headwaters	1510	
Mouth	3160	
Bigoray River, Alberta (Canada)	2190	Clifford 1978
King's Well Beck, England	3267	Crisp and LeCren 1970
Augusta Creek tributary, Michigan (USA)	3324	Cummins and Klug 1979
White Clay Creek, Pennsylvania (USA)	4170	Vannote and Sweeney 1980
River Lissuraga, France	4470	Thibault 1971
Silver Springs, Florida (USA)	8304	Odum 1957

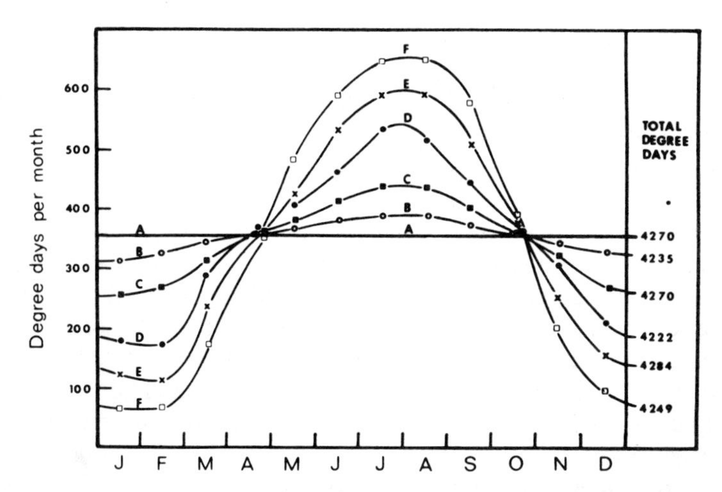

Figure 6.2 Distribution of monthly degree-day accumulations at various recording stations along White Clay Creek, Pennsylvania. Total degree days are the annual sum of monthly records for each station. *A*, outflow of groundwater; *B*, woodland spring seeps; *C*, first-order springbrooks; *D*, second-order streams; *E*, upstream segment of third-order stream; *F*, downstream segment of third-order stream. [From Vannote and Sweeney (1980), American Naturalist, The University of Chicago Press. © 1980 by The University of Chicago.]

DIEL AMPLITUDE Several authors have suggested that the diel temperature amplitude (Table 6.1) is an ecologically significant variable for aquatic insects (Kamler 1965, Décamps 1967, Brodsky 1980, Vannote and Sweeney 1980, Ward and Stanford 1982), although underlying mechanisms remain obscure. Headwaters exhibit little diel variation if they are well shaded and fed largely by groundwater. In such stream systems, the diel range increases downstream until the water volume becomes sufficient to buffer short-term temperature changes (Fig. 6.3). This has caused some confusion in the literature since the length of a river and the segment under consideration determine, in part, whether the temperature range increases or decreases downstream.

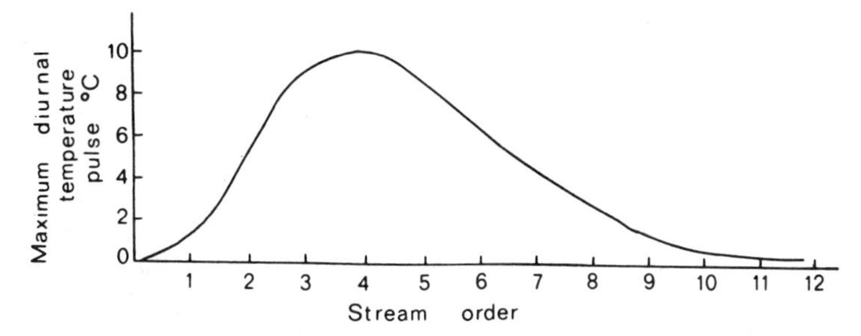

Figure 6.3 Maximum diurnal change in temperature as a function of stream order in temperate North America. [From Vannote and Sweeney (1980), American Naturalist, The University of Chicago Press. © 1980 by The University of Chicago.]

Diel temperature amplitudes are <1°C at spring sources or under ice, but variations of 15°C or greater have been recorded in midlatitude streams (Smith and Lavis 1975). Tropical streams may exhibit diel amplitudes of ≥5°C (Bishop 1973a). Large diel temperature ranges are nearly always associated with small exposed streams with minimal influence from groundwater.

Kamler (1965) regards temperature as the major factor controlling the distribution of aquatic insects in streams and the degree of thermal fluctuation as its most important manifestation. Kamler's coefficient of daily thermal astatism (daily maximum temperature divided by daily minimum temperature) indicates the degree to which waters vary temporally independent of absolute temperatures. Lotic waters in the Carpathian Mountains of Poland exhibited coefficients of daily thermal astatism from 1.09 to 2.66. The longitudinal distribution patterns of hydropsychid caddisfly abundance and production were highly correlated with downstream changes in diel temperature fluctuations in a Georgia (USA) stream system (Ross and Wallace 1982).

The rate of temperature increase is greater than the cooling rate in natural waters. In an English river, the maximum observed rate of increase (1.17°C/hr) occurred in July; the maximum rate of decline (0.75°C/hr) occurred in April (Boon and Shires 1976). Small exposed streams may have variations greater than 3°C/hr, whereas in large rivers, rates of change average less than 0.2°C/hr (Langford 1972). Wurtz (1969) recorded changes of nearly 2°C/hr in a tropical stream. Few studies have considered seasonal rates of rise (spring) or decline (autumn), despite their importance for some species of aquatic insects (Davies and Smith 1958, Macan 1960, Lehmkuhl 1974).

SMALL-SCALE SPATIAL VARIATIONS Although lotic systems are generally regarded as turbulent and well mixed, there may be ecologically significant spatial variations in temperature over short distances. Slow-flowing pools may attain higher temperatures in summer than adjacent rapids [≤4.5°C higher in an English river; see Boon and Shires (1976)]. Differences as high as 13°C between thalweg and bank temperatures were measured in a Czechoslovakian river (Penáz et al. 1968). Greater diel variations also may occur in water near the bank.

Only rarely do running waters exhibit vertical differences in temperature above the substrate. However, insects residing in the hyporheic zone are exposed to different temperature conditions than are the surface benthos (Schwoerbel 1967, Williams and Hynes 1974, Shepherd et al. 1986, White et al. 1987, Irons et al. 1989). Hyporheic waters exhibit reduced annual and diel thermal amplitudes compared to water above the substrate. Temperatures in the hyporheic zone are lower in summer and higher in winter than surface water, and such differences intensify with increasing depth in the substrate. In situations where the source of hyporheic water is infiltration of stream water, diel fluctuations within the gravel are dampened and lag behind stream temperatures (Fig. 6.4*A*), whereas if the source is upwelling groundwater, little or no diel fluctuation will be apparent in the hyporheic zone (Fig. 6.4*B*). The hyporheic zone thus provides a temperature refuge for cold stenothermal insects as well as protection from adverse conditions such as floods, drought, and anchor ice.

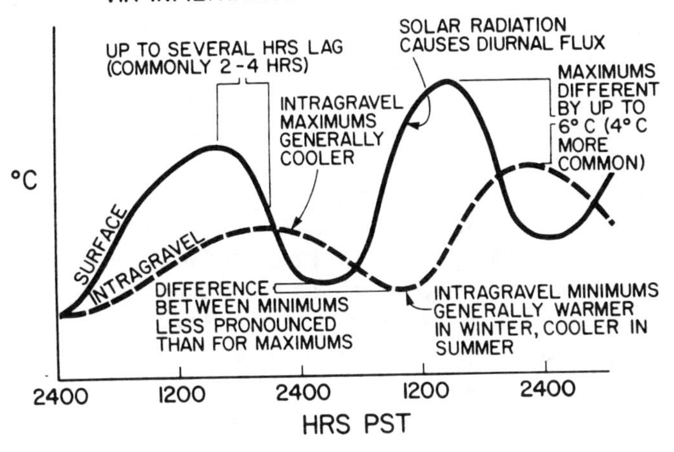

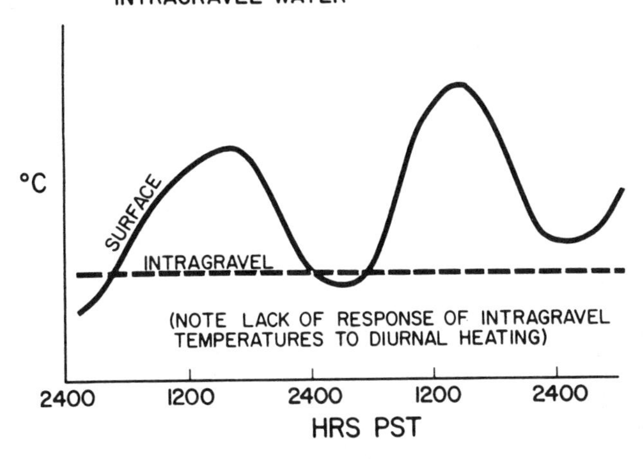

Figure 6.4 General diel temperature patterns for stream (surface) water and hyporheic (intragravel) water during summer under different hydrologic situations. [From Shepherd et al. (1986).]

Lentic Temperatures Hutchinson and Löffler (1956) developed a worldwide thermal classification system for lakes [revised by Lewis (1983)], and Hutchinson (1957a) reviewed the factors (e.g., altitude, latitude, continentality, lake morphometry) that determine the temperature regime of a lentic water body. The greatest source of heat to lakes is direct absorption of solar radiation by the water, although in shallow water the sediments also absorb significant quantities of heat (Dale and Gillespie 1977a, Wetzel 1983).

Temperatures in small water bodies and the littoral zone of lakes are generally similar and are greatly influenced by air temperatures, at least during major portions of the year (Gieysztor 1961). The daily maximum air temperature was a good indicator of the temperature of the exposed littoral of a Norwegian subalpine lake during much of the ice-free season (Fig. 6.5).

The littoral zones of lakes are generally warmer in summer and colder in winter than the open water of deep lakes. In early spring, thermal conditions near the substrate of shallow areas may be much more favorable for aquatic insects than open water lake temperatures would indicate (Dale and Gillespie 1977b).

Lewis (1979) emphasizes the importance of nonseasonal weather changes, such as periods of cloud cover, on thermal conditions of tropical lakes. In equatorial ponds, a rainy period depressed water temperatures 2–3°C and reduced diel amplitudes from 4°C to 1°C (Young 1975).

Not only do lakes influence downstream lotic temperatures, as indicated by the previous section, but incoming rivers may thermally modify portions of even very large lakes. For example, entering rivers are cooler than Lake Baikal, Siberia, in winter and warmer in summer (Kozhov 1963). During the summer large areas near river mouths may be 10°C warmer than the open lake.

ANNUAL RANGES AND PATTERNS The annual ranges of surface temperatures for lentic water bodies are exemplified by the following: 0–24°C for Lake Erie, Ohio (USA) (Britt 1962); 0–20°C for Lake Windermere, England (Macan 1970);

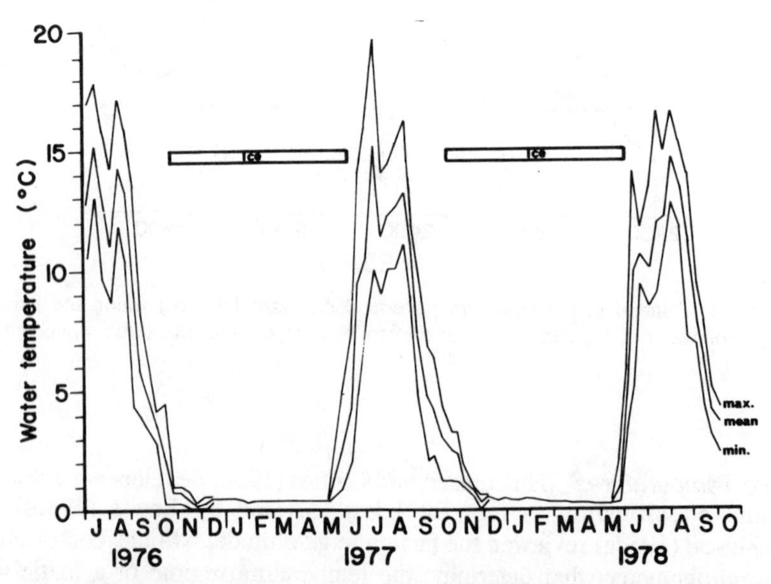

Figure 6.5 Water temperatures in the exposed zone of Øvre Heimdalsvatn, a subalpine lake in Norway. [From "The influence of temperature on nymphal growth rates in mountain stoneflies (Plecoptera)" by J. E. Brittain, Ecology, 1983, 64, 440–446. Copyright © by the Ecological Society of America. Reprinted by permission.]

0-12.5°C for an arctic lake (Alimov and Winberg 1972); 0–15°C for a subalpine lake (Brittain 1983); and 14-28.5°C for Lake Kinneret, Israel (Serruya 1978). Whereas the open waters of Lake Baikal in Siberia vary from 0 to 14°C, the annual range is 0 to 26°C in shallow bays (Kozhov 1963).

The annual range is much less in the depths. For example, there is no seasonal change below 300 m in Lake Baikal (Kozhov 1963). Below 60 m in Lac Léman, Switzerland, temperatures are 4 ± 0.5°C throughout the year (Forel 1895). Annual ranges are further reduced with depth in the sediment (Birge et al. 1928, Butler 1982).

Examples of extremely high temperatures include 41°C in African desert pools (Rzóska 1961); >40°C in lentic habitats in Death Valley, California (USA) (Brown and Feldmeth 1971); and 42°C in the water contained in pitcher plants in Florida (Bradshaw 1980).

Deep temperate lakes in regions of continental climate have bottom waters near 4°C and surface temperatures around 20 to 25°C during summer stratification (Fig. 6.6). In temperate areas with marine climate, deep waters may not drop below 810°C, nor will surface waters normally exceed 18-20°C. Tropical lakes usually exhibit differences between the surface and depths of only 3–4°C with the temperatures of deep waters ~20°C higher than in temperate lakes (Rodhe 1974).

Although shallow water bodies are often well mixed by wind-induced circulation, during periods of calm (or in sheltered locations) they may exhibit extreme thermal stratification. For example, Eriksen (1966) reported a 16°C difference between surface and bottom temperatures in a 35-cm-deep pool in California. In an alpine tundra pond at 3507-m elevation, water was up to 20°C colder at the bottom (1 m) than the surface in summer (Nelder and Pennak 1955). Kushlan (1979) recorded gradients of ≤12°C/m depth in an alligator pond in Florida. Although these are extreme examples, insects in small water bodies

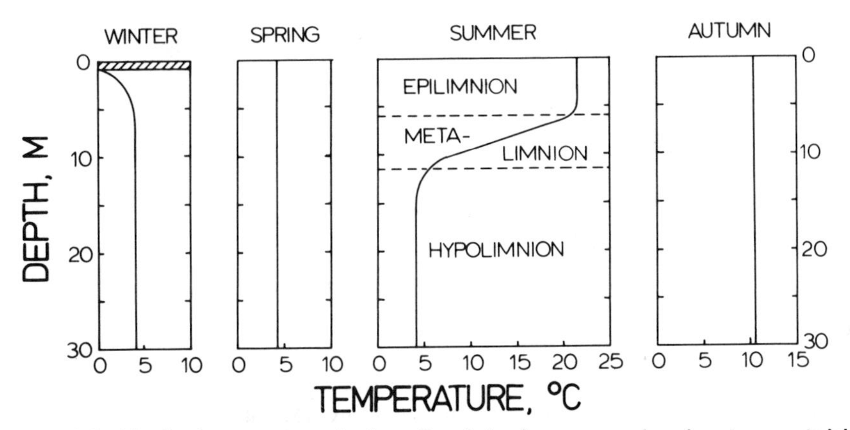

Figure 6.6 Idealized temperature–depth profiles during four seasons in a deep temperate lake [first class dimictic; *sensu* Hutchinson and Löffler (1956)].

may, by moving very short distances, encounter a wide range of thermal conditions.

Temperature patterns differ between lakes, and vary spatially and temporally within the same lake. In Welsh lakes, the seasonal maximum temperature in the littoral zone occurs 2–4 weeks later than in smaller water bodies (Brittain 1976a). From 7 years of continuous temperature records for Hodson's Tarn, Macan and Maudsley (1966) found that dates in spring on which a certain temperature was attained varied by as much as one month between years. Aquatic plants may reduce the diel temperature amplitude, slow the rate of temperature change, and intensify thermal stratification (Martin 1972). Dale and Gillespie (1977b) recorded a 10°C decrease per meter depth in the portion of a lake with dense macrophytes, but only a 0.2°C change per meter where macrophytes were sparse.

Annual degree days in lentic habitats and their seasonal distribution are influenced by spatial factors to a much greater extent than in running waters. Insects inhabiting the profundal zone are exposed to a nearly even distribution of degree days throughout the year, in contrast to the temporal variability in thermal units that occurs in the littoral zones of temperate and high-latitude lakes. Shallow bays may exhibit quite different temperature regimes than the exposed littoral zone of large lakes. Sheltered gulfs of Lake Baikal accumulate 2–3 times more degree days than the open littoral, and exhibit marked differences in the seasonal distribution of heat (Kozhov 1963).

Ice and snow cover may play major roles in determining the thermal regimes of water bodies. Brittain (1974), for example, found a strong positive correlation between the lengths of the ice-free period and the maximum temperatures attained in 34 Norwegian lakes. The number of species of Ephemeroptera inhabiting the lakes was directly correlated with the duration of the ice-free period. Insects are absent from lakes with perennial ice cover, but may constitute the major portion of the macrofauna in arctic ponds which are ice-free for only a few weeks each year (Danks 1971a, Rigler 1978).

Although liquid water does not drop to less than 0°C, aquatic insects of high latitudes or altitudes may be exposed to much lower temperatures. In a shallow lake in Alaska, the annual temperatures ranged from 0 to 12°C in the open water, -16 to 12°C on the sediment surface at 51-cm depth, and 0–9°C at 234 cm in the bottom mud (Brewer 1958). Arctic ponds up to 2 m deep freeze to the substrate, and bottom mud may remain at -20 to -30°C for several months (Danks 1971a). Chironomids, which overwinter as larvae in bottom mud, make up the vast majority of the macrofauna of such habitats.

DIEL AMPLITUDE In lakes, significant diel fluctuations are normally restricted to surface strata and are more intense in the littoral zone than at comparable depths in limnetic regions. In a Polish lake, diel amplitudes of 9°C were recorded in the littoral zone in May, when open waters exhibited daily variations of less than 2°C (Gieysztor 1961). Brittain (1976a) found diel fluctuations of ≤5°C in the littoral of a lake in Wales.

The smaller the volume of a lentic water body, the greater is the diel temperature fluctuation, the higher is the maximum temperature, and the earlier (daily and seasonally) is the maximum temperature attained. Examples of maximum diel amplitudes in small lentic water bodies include 6.5°C in the bottom of arctic ponds (Danks and Oliver 1972a); 18°C (with rates of change up to 4°C/hr) in temporary ponds in Alberta (Hartland-Rowe 1972); 7–20°C in alpine ponds (Pesta 1933); and 4.7°C in a pond in Kenya (Young 1975). Bradshaw (1980) recorded diel fluctuations of ≤26°C in the water contained within pitcher plants. Diel amplitudes over 20°C occur in shallow salt lakes in Australia (Williams 1978). Martin (1972) recorded diel fluctuations of ≤19°C in shallow areas of English ponds, but only 3°C at 70-cm depth.

THERMAL ECOLOGY OF AQUATIC INSECTS

The thermal history of an individual or species of aquatic insect shapes responses operative at the organismic, population, and community levels of organization, which are manifest on both ecological and evolutionary time scales. The remainder of this chapter deals with the influence of temperature on distribution patterns, life-cycle phenomena, behavioral responses, and trophic relationships. Much of this material is based on a review of the thermal ecology of aquatic insects (Ward and Stanford 1982).

That aquatic insect ecology is best understood when placed in an evolutionary perspective is an underlying theme of this book, and this is especially apparent when considering the topic of thermal ecology. As already mentioned, entire orders and other major taxa of aquatic insects evolved in cool habitats; members of these groups now inhabiting warmer water bodies are thought to be later derivatives of cool-adapted ancestral lines. This is perhaps best exemplified by the Trichoptera, based largely on the work of Ross (1956, 1963, 1967; see also Schmid 1955, Wiggins and Wichard 1989). Ross presented convincing evidence that the entire order evolved in cool lotic waters (maximum temperature <20°C) and that extant primitive genera remain largely confined to the ancestral habitat (Fig. 6.7). Warm-adapted caddisflies are specialized offshoots from cool-adapted ancestral lines or are specialized genera within cool-adapted lineages. Based solely on respiratory considerations, it is reasonable to assume that the oxygen-rich nature of cool running waters provided most suitable conditions for the evolutionary invasion of freshwater by trichopterans (and other groups with apneustic larvae that meet respiratory needs by extracting dissolved oxygen from the water). The remarkable adaptive radiation and habitat diversification achieved by larval caddisflies is attributed largely to the possession of silk that enabled tube-case makers to increase respiratory efficiency (Fig. 9.10), allowing exploitation of warm and stagnant waters (Wiggins 1977, Mackay and Wiggins 1979).

Many aquatic insects grow at or near 0°C and possess mechanisms to avoid or withstand high summer temperatures (Pleskot 1958; Kamler 1965; Landa 1968; Ulfstrand 1968; Khoo 1968a, b; Knight et al. 1976; Hynes 1976; Brittain

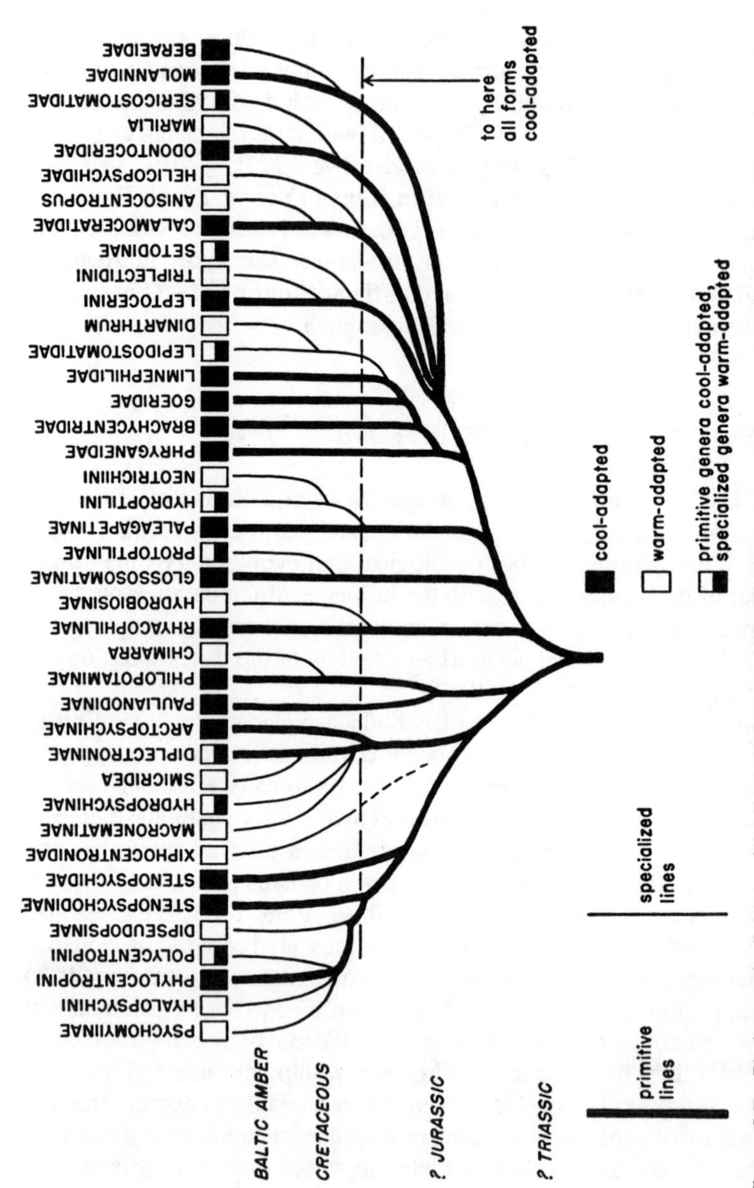

Figure 6.7 Phylogenetic diagram of the Trichoptera showing coincidence of primitive lines with cool-adapted affinities. [From Ross (1956). © 1956 by The Board of Trustees of the University of Illinois. Reprinted by permission of The University of Illinois Press.]

1978a; Bengtsson 1981), providing further evidence for an evolutionary history associated with cool habitats.

Latitudinal Distribution Patterns

The latitudinal distributions of some aquatic insects appear to be determined, in large part, by temperature. For example, the mayfly *Ephoron album,* generally restricted to 40–50°N latitude, has southern limits related to winter chill requirements for the eggs, whereas the necessity of 2.5 months above 18°C for nymphal maturation limits northern range extensions (Britt 1962). Oriental and Palaearctic mosquitoes in Iran are ecologically separated along a "line" below which freezing temperatures regularly occur during winter (Macan 1974a). The distribution of the yellow fever mosquito *Aedes aegypti* roughly corresponds to the equatorial band delimited by the 10°C isotherms of the Northern and Southern Hemispheres (Christophers 1960). Because aquatic insects generally lack well-developed temperature compensation mechanisms (Bullock 1955, Rigler 1978, Vannote and Sweeney 1980), the thermal regime of the habitat is of greater importance than for those animals able to maintain a relatively constant metabolic rate over a wide range of environmental temperatures. According to the thermal equilibrium hypothesis (Vannote and Sweeney 1980), adult body size, metabolic efficiency, fecundity, and abundance will be maximized near the center of a species' latitudinal range where the thermal regime is optimal (Fig. 6.8). Conversely, populations living further north or south where thermal conditions are suboptimal will be smaller and less fecund with a correspondingly reduced competitive position within the community. Decreased body size and fecundity in habitats that are warmer than optimal result from increased maintenance costs, whereas decreased body size and fecundity in cooler than optimal habitats result from reduced assimilation rates, with more energy allocated to adult tissue maturation and less available for larval growth. Body size is significant because of the close relationship between adult female size and egg production (i.e., fecundity) for most aquatic insects (Britt 1962, Thibault 1971, Vannote and Sweeney 1980).

Data from studies of the dominant mayflies at sites on the Colorado River support predictions of the thermal equilibrium hypothesis regarding growth rates, body size, and fecundity (Rader and Ward 1990). Sites where species attained largest body size and fecundity , however, did not always correspond to locations exhibiting the largest population sizes. Apparently other biotic and abiotic factors, in addition to temperature, may interrupt the translation of high fecundity into population density.

Continentality, through its influence on the thermal regime, may modify the latitudinal effects of temperature. Lillehammer (1974) reported a progressive increase in the species diversity of Plecoptera from coastal to inland regions of Norway, in lotic habitats selected to vary only with respect to temperature. Winter minimum temperature, which is higher near the coast, was the major component of the thermal regime that varied along the gradient. Plecoptera may

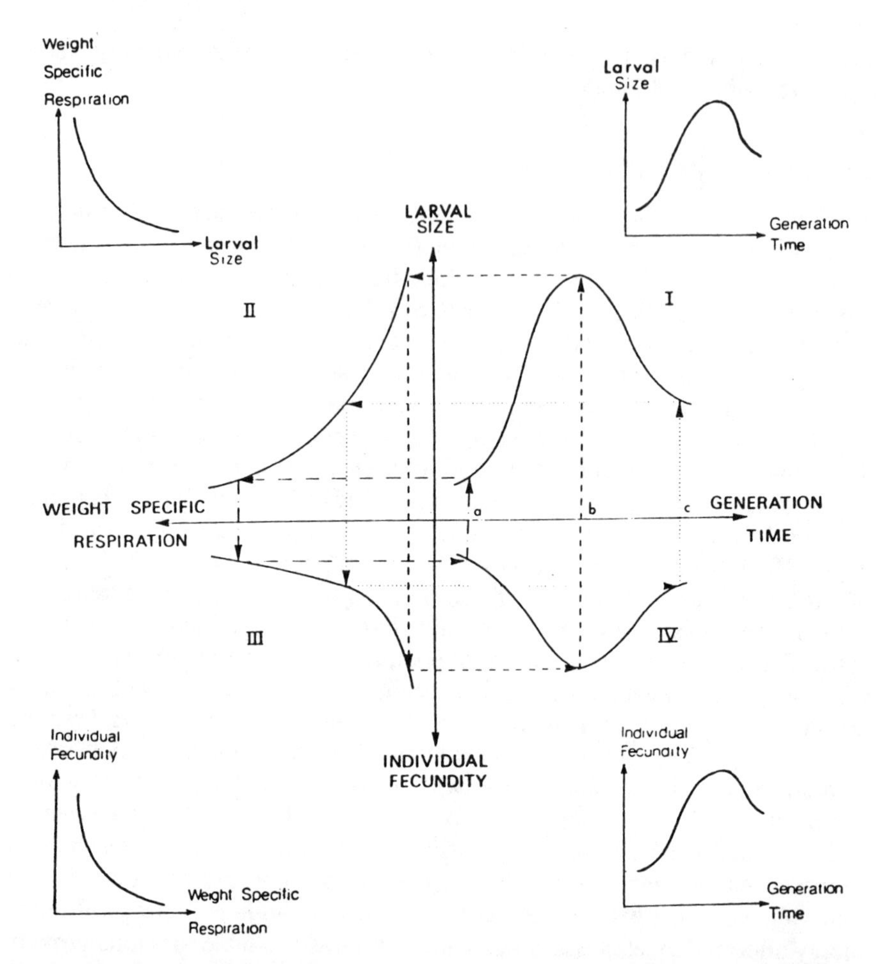

Figure 6.8 A thermal equilibrium model, based on experimental data for White Clay Creek, Pennsylvania, indicating the population interactions between bioenergetic (growth and metabolic rate) and developmental (fecundity and generation time) parameters for insects reared in optimum and nonoptimum thermal regimes. The individual components of the model are described at the periphery of each quadrat (e.g., the inverse relationship between weight-specific respiration and larval size; quadrat II). Pathway *b* (an optimum thermal regime) shows that maximum larval size is associated with an intermediate generation time, low weight-specific respiration, and high adult fecundity. Larval size and adult fecundity are intermediate in cold, nonoptimal regimes (pathway *c*). Generation time is shortest in warm regimes (pathway *a*), but high respiration cost and accelerated development of adult tissues (wing pads, reproductive system) results in small larvae and reduced adult fecundity. [From Vannote and Sweeney (1980), American Naturalist, The University of Chicago Press. © 1980 by The University of Chicago.]

exhibit reduced wing size to body size ratios at the limits of their range, thus reducing the dispersal abilities (gene flow) of marginal populations (Lillehammer 1976).

Major periods of climatic change, including Pleistocene glacial and interglacial periods, played a major role in the evolution and macrodistribution of aquatic insects (Illies 1953b; Ross 1956, 1958; Macan 1962; Harrison 1965b; Nimmo 1971; Danks 1978). During warm periods, mountain waters acted as refuges for cool-adapted forms. Geographical peculiarities of Europe (Thienemann 1950, Illies 1953b) resulted in greater fragmentation of aquatic fauna during the Pleistocene than in North America, where the mountain ranges provide expansive and more or less continuous N-S (north to south) dispersal routes.

Comprehensive analyses of life cycles along latitudinal gradients (see Sweeney 1984) have been made for Ephemeroptera (Clifford et al. 1973), Plecoptera (Brinck 1949), and Odonata (Corbet 1958, 1980; Corbet et al. 1960). A most dramatic example of the relationship between latitude and voltinism has been described for *Ischnura elegans* in Europe (Corbet 1980), which is trivoltine at 44°N, univoltine at 54°N, and semivoltine at 58°N latitude. Some Odonata require 4–6 years to complete one generation near the Arctic Circle.

Altitudinal and Longitudinal Patterns

Temperature is a factor of primary importance in determining the distribution, diversity, and abundance patterns exhibited by aquatic insects over elevation gradients in lentic and lotic waters (Dodds and Hisaw 1925, Brinck 1949, Illies and Botosaneanu 1963, Kamler 1965, Décamps 1967, Brittain 1974, Donald and Anderson 1980, Brodsky 1980, Ward 1986, Stanford and Ward 1983). Temperature may not, however, be a major factor controlling aquatic insect distribution in water courses with small elevation gradients, especially in regions of equable climate (Macan 1974b). Nonetheless, downstream (longitudinal) faunal changes attributable to temperature may occur even without appreciable changes in elevation (Ide 1935, Sprules 1947, Minckley 1963, Minshall 1968, Ward and Dufford 1979).

In considering distribution patterns of aquatic insects along any environmental gradient, it is essential to remember that zonal boundaries are not normally determined by physical factors alone, but "correspond to values of the environmental variables at which the outcome of competition changes" (Hutchinson 1967). Although there are relatively few known examples of competitive displacement constricting the temperature range occupied by aquatic insects (Berthélemy 1966, Hynes 1970a, Gíslason 1981), reductions of a species' fundamental niche space by such mechanisms must be common.

Later emergence at higher elevations has been largely attributed to lower temperatures (Brinck 1949, Pleskot 1951, Lillehammer 1975a, Brittain 1978a, Wise 1980). Exceptions to the altitudinal pattern of emergence in special thermal habitats, such as springs and lake outfalls, support the role of temperature as the major controlling factor (Nebeker 1971c, Lillehammer 1975a). Many Plecoptera exhibit a decrease in wing length with increasing elevation (Brinck 1949, Hynes 1970a), although the effect is also influenced by position within the geographic range of a species (Lillehammer 1976).

The number of generations per year decreases with increasing altitude both inter- and intraspecifically (Corbet 1980). For example, the caddisfly *Rhyacophila evoluta* exhibits a 1-, 2- or 3-year life cycle depending on altitude (Décamps 1967). Nymphal development of the mayfly *Rhithrogena loyolaea* requires 3 years above 2100 m, but is completed in 2 years at lower elevations (Lavandier 1981).

Several authors have reported a serial succession of closely related species downstream along a water course (Pleskot 1951, Zahner 1959, Berthélemy 1966), which apparently reflects niche segregation along the longitudinal thermal gradient. One of the best documented examples involves the spatial segregation of hydropsychid caddisflies in the River Usk, Wales (Hildrew and Edington 1979). Three species, *Diplectrona felix, Hydropsyche instabilis,* and *H. pellucidula,* form a downstream series along a gradient of progressively increasing summer temperatures. *Diplectrona felix,* the species restricted to the immediate headwaters, and *H. pellucidula,* which occupies the river proper, are exposed to relatively small ranges of weekly temperatures, whereas the tributaries inhabited by *H. instabilis* exhibit relatively large weekly fluctuations. Laboratory tests revealed striking differences in the temperature-metabolic-rate responses of these species, suggesting that differential adaptations to thermal conditions provide each species with a zone within which it is competitively superior. These species exhibit no evidence of spatial partitioning by food type or food availability.

Temperature as an Isolating Mechanism

The influence of temperature on aquatic insects has in some cases resulted in spatial or temporal isolation and reduced gene flow, leading to speciation or at least incipient speciation. Lillehammer (1975a) located a population of the stonefly *Leuctra hippopus* that is apparently restricted to a 200-m-long stream connecting two lakes. Because of the differences in the thermal regime, *L.*

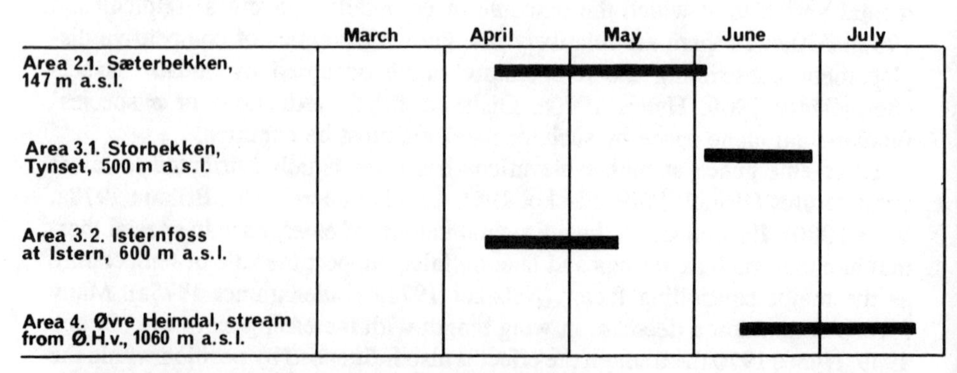

Figure 6.9 The period when adult *Leuctra hippopus* (Plecoptera) were present in localities at different altitudes in Norway. [From Lillehammer (1975a).]

hippopus in the lake outlet (the Isternfoss population in Fig. 6.9) exhibited a marked divergence in emergence period compared with the Storbekken population occurring in a stream of similar elevation. In addition to temporal isolation, the Isternfoss specimens exhibit a higher degree of short wingedness than other populations of *L. hippopus* (Lillehammer 1976).

The very different thermal regime prevents invasion of the open waters of Lake Baikal by the littoral and sor (shallow lagoons) species, since these Siberian faunal elements are unable to complete their life cycles there. Likewise, the endemic Baikalian fauna is restricted to the open lake partly because of thermal requirements (Kozhov 1963).

The isolated nature of the high mountain areas of Europe and their east-west orientation enhanced the geographical isolation of oligothermal aquatic insects during periods of glacial recession in the Pleistocene (Illies 1953b). The highly diverse stonefly fauna of Europe is largely attributable to a high degree of endemism (restriction to a single mountain mass) among many groups (Illies 1969). The tendency of brachyptery in high-altitude populations further enhances the isolation of stoneflies (Lillehammer 1976). The apparently anomalous occurrence of the stonefly *Dinocras cephalotes,* generally of a more southern distribution, at higher elevations in Swedish Lapland (but not in lower reaches) is due to the higher summer temperature in exposed streams above treeline (Ulfstrand 1968). A large population of *Heterlimnius corpulentus,* an elmid beetle characteristic of high mountain streams, occurs in the source of a cold spring at a low elevation site in Colorado (Ward et al. 1986). Conversely, the warm-adapted caddisfly *Helicopsyche,* normally restricted to low-elevation sites (Ward 1986), occurs in a warm spring (25°C) at 3109 m a.s.l. (Ward, unpubl.).

During warm periods mountain waters provide refuges for cool-adapted faunal components (Ross 1956). In the Mesozoic, most of southern Africa was in the southern Temperate Zone (Harrison 1965b). As the continent moved northward toward the equator, the temperate aquatic fauna became restricted to mountain waters. Since no significant Pleistocene glaciation occurred in this region, the ancient faunal elements have survived. The remains of the temperate Mesozoic fauna (cold stenotherms) have "Gondwanaland" affinities with related forms on other southern continents.

Hot springs tend to maintain extremely constant thermal conditions for tens or even hundreds of years, and many exist for thousands of years (Mitchell 1974). Because of the thermal constancy and predictability, organisms of hot springs often occur at or near their temperature optima, a generally nonadaptive strategy in habitats with large fluctuations in temperature because lethal temperatures are frequently only a few degrees above the optimum (Brock 1985). There is a remarkable worldwide similarity of the insect fauna of hot springs (Winterbourn 1969), and tropical and subtropical relicts have been reported from these special thermal habitats. The disjunct population of the subtropical dragonfly *Orthetrum albistylum* in a hot spring on the coast of Lake Baikal is apparently a relict population of a warm-adapted Tertiary fauna (Kozhov 1963).

The disjunct population of *Helicopsyche* in a warm high mountain spring has already been mentioned.

The structure and function of enzymes are generally highly sensitive to variations in temperature on both ecological and evolutionary time scales (Somero 1978). Electrophoretic analyses of *Argia vivida* (Odonata) distributed along a steep thermal gradient suggest quantitative and qualitative enzyme responses to temperature (Schott and Brusven 1980). The adaptation, an example of a temperature compensation mechanism in an aquatic insect, involves an isozyme response [lactate dehydrogenase and leucine aminopeptidase enzyme systems (LDH and LAP)] and greater enzyme production at higher temperatures (glucose-6-phosphate dehydrogenase system).

Climatic changes incurred since the last glacial epoch have isolated relicts of three types in Danish nonthermal springs (Nielsen 1950b). Arctic or glacial relicts, such as the caddisfly *Apatania muliebris*, survived postglacial warming in the summer-cool environments of the Danish springs to which they are now restricted, far south of their major contemporary distribution. *Apatania muliebris* maintains an arctic life cycle, hardly exploitive of the constant thermal conditions, except in a spring where two parthenogenetic subspecies occurred [these endemics have since been exterminated (Nielsen 1976)]. Oceanic relicts extended their ranges northward into Denmark during a time when winters were less severe than at present; such species are restricted to Danish springs because of the winter warm conditions. Continental relicts require higher summer temperatures than now occur in most surface waters. They entered Denmark during the Subboreal Period of more continental climate and are now restricted to shallow seeps that are well exposed to solar radiation where on sunny days water temperatures are 5–9°C above air temperatures. In a later paper, Nielsen (1976) reduced the number of species he originally thought to be relicts. He confirms the glacial relict status of *A. muliebris*, but indicates that many insects of Danish springs are euryokous species not restricted to springs or springbrooks.

Life-Cycle Relationships

Eggs and Fecundity The effects of temperature on fecundity and the responses of eggs to thermal conditions influence distribution patterns of aquatic insects and the competitive position of a species at a given locale. Temperature may influence the egg incubation period, hatching success, duration of hatching, and the induction and termination of diapause. Many of the recent investigations on the effects of temperature on aquatic insect eggs have dealt with Ephemeroptera and Plecoptera (Brittain 1982, 1990; Schmidt 1984; Mutch and Pritchard 1986).

The relationship of temperature to incubation period, within the range of egg viability, generally follows a hyperbolic (Elliott 1978) or power-law function (Humpesch 1980, Waringer and Humpesch 1984). Under constant- temperature laboratory conditions the following data were recorded for eggs of the heptageniid mayfly *Ecdyonurus picteti* (Humpesch 1978):

Temperature °C	Days to Hatch
20	15
15	23
10	42
6	80

Eggs of some species hatch at very low temperatures [2°C for *Rhithrogena loyolaea* (Elliott and Humpesch 1980)], whereas others require higher temperatures to stimulate hatching (10°C for *Ephoron album*; see Fig. 6.10). Upper temperature limits for hatching also vary considerably between species (Brittain 1982). Hatching success may be optimal at a lower temperature than that which produces the most rapid egg development (Benech 1972), although some investigators fail to detect temperature effects on hatching success (Humpesch 1980). Interspecific differences in responses of eggs to temperature may serve as temporal isolation mechanisms for sympatric congeners.

Most laboratory studies of egg development in aquatic insects have been conducted at constant temperatures, despite the inapplicability of such conditions to most aquatic biotopes. Under diel cycles of temperature, eggs may hatch sooner than under constant thermal conditions. In the words of Humpesch (1978, p. 2606) "... it appears that there exist special physiological mechanisms, which accelerate the speed of development under conditions of changing temperatures and which exceed a simple summation of temperature effects." The greater the amplitude of the diel cycle, the fewer degree hours were required for eclosion of corixid eggs (Sweeney and Schnack 1977). However, neither Elliott (1972) nor Humpesch (1982) found any significant differences in the rate of embryonic development at constant versus fluctuating temperatures in the spe-

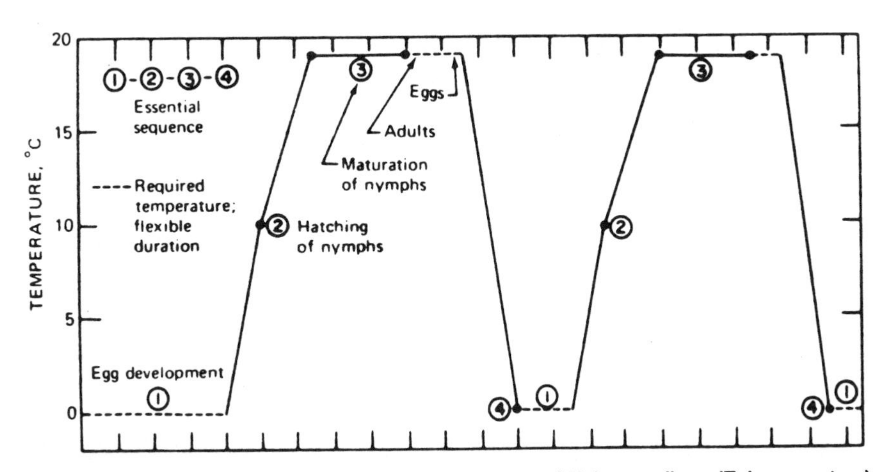

Figure 6.10 Graphic model of the thermal requirements of *Ephoron album* (Ephemeroptera) spanning two generations. [From Lehmkuhl (1974).]

cies they examined. Howe (1967), in reference to terrestrial insects, concludes that at lower temperatures egg development will be accelerated by fluctuating temperatures, whereas development will generally be retarded at higher levels. Thermal heterogeneity may be especially important for species with embryonic diapause (Schaller 1968).

Duration of hatching is highly variable within and between species (Hynes 1970a), and may be correlated with temperature (Bohle 1969, Elliott 1978; see also Fig. 6.11). At 3°C the eclosion of *Baetis rhodani* eggs began after 119 days and hatching extended over a 34-day period; at 22°C eclosion began after only 7 days and hatching was completed in a 3-day period (Elliott 1972).

As previously mentioned, fecundity in aquatic insects is directly related to adult female body size, which is influenced by temperature. Body size is greatest, and therefore competitive abilities are enhanced by prior exposure to optimal thermal conditions (see Fig. 6.8). The pitcher plant mosquito, *Wyeomyia smithii,* exhibited a sevenfold greater fecundity when exposed to fluctuating as opposed to constant temperatures (Bradshaw 1980).

Dormancy Temperature may play a major role in the induction and termination of dormancy, which is often an adaptive response to thermal conditions. Quiescent stages have been reported for all life-cycle stages of aquatic insects (Wiggins et al. 1980). The caddisfly *Rhyacophila evoluta,* because it can enter diapause at any stage, has attained a life-cycle flexibility enabling colonization of waters with a wide variety of thermal conditions (Décamps 1967). Although rising temperatures tend to terminate diapause and falling temperatures tend to induce diapause in terrestrial insects (Howe 1967), the opposite may be true of aquatic species (Khoo 1964, Schaller 1968, Bohle 1972, Oberdorfer and Stewart 1977, Colbo 1979).

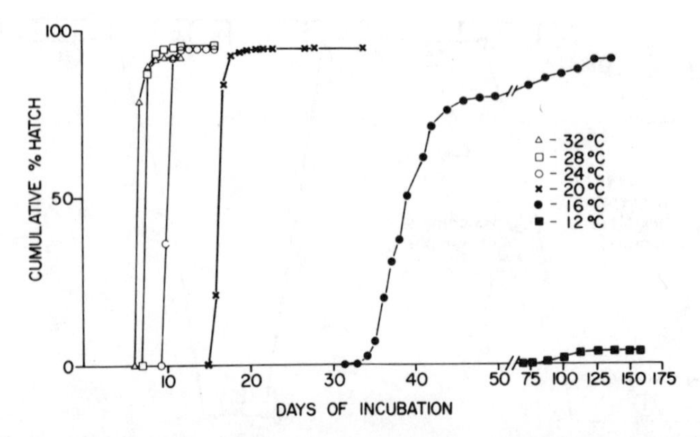

Figure 6.11 Cumulative percentage hatch of *Hexagenia rigida* (Ephemeroptera) eggs versus days of incubation at constant temperatures. [From Friesen et al. 1979). Reprinted from The Canadian Entomologist, Vol. 111 (1979), p. 667.]

Diapause, as a response to avoid warm periods, is especially common among stream insects (Pleskot 1961, Khoo 1968a, Dosdall and Lehmkuhl 1979) and preadapts them to survive drought or colonize temporary waters (Hynes 1970b, Harper and Hynes 1970). Khoo (1968b) found that stream populations of the stonefly *Diura bicaudata* produced only diapausing eggs, whereas lake populations laid mainly nondiapausing eggs. Some species require temperatures at or near freezing to break egg diapause (Fig. 6.10), and eggs of aquatic insects may remain viable at temperatures well below freezing (Bohle 1969). A drop in temperature may (by stimulating diapause development) shorten the time to eclosion following a return to a higher temperature (Schaller 1968). The mayfly *Ephemerella ignita* exhibits an obligatory diapause in Germany, but eggs from populations studied in England and France have direct development (Bohle 1972).

In addition to serving as a resistant stage, diapause may function to synchronize life cycles (Corbet 1958, Lutz 1974b), temporally segregate closely related species (Carlsson et al. 1977), and provide the flexibility needed for survival in unpredictable habitats (Snellen and Stewart 1979b). Increased water temperature downstream from a power plant eliminated the quiescent larval stage of *Hydropsyche pellucidula* and advanced emergence 3–4 months (Fey 1977), but warm temperatures may also retard emergence by delaying the drop in temperature necessary to stimulate hatching (Wise 1980). Photoperiod and temperature may operate independently or in concert to exert control over diapause (Sawchyn and Church 1973, Corbet 1980, Bradshaw 1980).

Table 6.3 Total numbers (denominator) of aquatic insects contained in ice and frozen sediment collected from the River Vindelalven, northern Sweden, and the numbers (numerator) alive after thawing samples at 2°C

	Date of Collection[a]						Σ%
	Jan. 8	Jan. 22	Feb.7	Mar.12	Apr.10	May 5	Survival
Trichoptera							
Agrypnia obsoleta	1/2		1/1	1/1			75
Phryganea bipunctata					0/1		0
Oecetis ochracea				6/6		1/1	100
Molanna albicans					1/1		100
Molanna augustata					3/4	1/1	80
Chironomidae							
Tanypodinae	4/4	1/3			2/3		70
Diamesinae		0/1					0
Orthocladiinae	1/2	1/2	2/2	5/5	0/1	10/10	86
Chironominae	146/147	88/90		4/4	54/57	7/8	97
Σ% Survival	97	94	100	100	90	95	

[a]At the location where samples were taken the river froze solid the last week in November.

Source: Modified from Olsson (1981).

The ability of some aquatic insects to survive being frozen has already been mentioned. Most data of this sort are based on studies of shallow lentic water bodies at high latitudes that freeze solid during the winter (e.g., Danks 1971b; Daborn 1971, 1974; Butler 1982; Solem 1983). However, samples of ice and frozen sediment collected from the hydrolittoral zone of a northern Swedish river revealed numerous chironomids and several trichopteran larvae among insects (Table 6.3). Survival was high for the abundant taxa even after being frozen for ≤5 months, indicating that those collected were not residual specimens from populations that move into deeper water during the winter. Ephemeropterans and plecopterans, abundant in shallow water during the summer, did not occur in frozen samples. Some of these undoubtedly overwinter as eggs, but immatures of other species apparently move to deeper water during winter to avoid freezing. For cold-hardy species, the frozen hydrolittoral zone provides a predator-free winter refugium with lower mortality than the animals would suffer if they migrated to deeper water (Olsson 1985).

Growth and Maturation Temperature plays a major role in regulating seasonal changes in growth rates of aquatic insects and operates, at least to some extent, independently of nutritional factors (Brittain 1976b, 1983; Minshall 1978; Vannote and Sweeney 1980; Sweeney and Vannote 1986; Oemke 1987) or photoperiod (Lutz 1974b, Brittain 1976b). Temperatures optimal for high growth rates may be suboptimal for growth efficiency, emergence success or adult longevity (Heiman and Knight 1975).

As mentioned earlier in this chapter, some species of both lentic and lotic waters remain active and grow at or near 0°C, an ability perhaps best exemplified by certain stoneflies (Fig. 6.12). A subalpine lake population of *Diura bicaudata*, for example, completes two-thirds of its nymphal growth under ice at temperatures <1°C (Lillehammer 1978a).

Some aquatic insects are unable to grow in cold water yet are intolerant of high temperatures. The mayfly *Heptagenia lateralis* is absent from streams with summer maxima >18°C (Macan 1960). Because it cannot grow in cold water, summer-warm streams do not provide sufficient time between the winter minimum and the upper temperature limit of the species for completion of nymphal growth. *Heptagenia lateralis* does, however, occur in Lake Windermere which has a summer maximum of 20°C; the lake warms more slowly than the streams (i.e., there is more time between the lower growth threshold and the upper temperature limit), enabling the mayfly to emerge before temperatures reach 18°C. Likewise, the absence of winter growth and intolerance of temperatures >16°C restrict the black fly *Simulium hirtipes* to locations where temperatures do not rise from the winter minima to 16°C so rapidly that the life cycle cannot be completed (Davies and Smith 1958).

Differential temperature-growth responses have been ascribed a temporal niche segregation role allowing coexistence of species in the same habitat (Ide 1935, Vannote and Sweeney 1980). Coexistence of univoltine mayflies in a subalpine lake was attributed to the portion of nymphal growth (0–96% for different

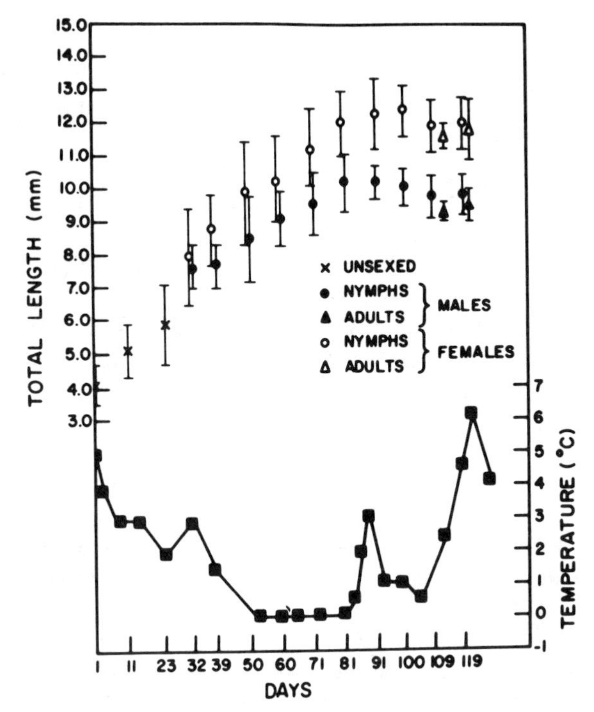

Figure 6.12 Growth of *Taeniopteryx nivalis* (Plecoptera) over a 4-month period from its reappearance in the benthos in November, following a nymphal diapause, to emergence of adults in March. Points represent means; bars indicate standard deviation. Water temperatures ranged from 0 to 6°C during this period. [From Knight et al. (1976).]

species) occurring during ice cover (Brittain 1978a). Other authors attribute coexistence to species specific temperature thresholds for growth (Ide 1935, Spence et al. 1980a) or to temporally segregated periods of maximum growth (Ide 1935, Landa 1968, Vannote and Sweeney 1980). Differential growth at the same temperature by individuals taken from populations inhabiting different thermal biotopes indicates an ecotypic response (Lillehammer 1975b).

Some species exhibit precise temperature thresholds for growth (Lavery and Costa 1976, Thompson 1978), thus supporting the "Entwicklungsnullspunkt" theory (Illies 1952); others respond to thermal summation (Konstantinov 1958, Bar-Zeev 1958, Ross and Merritt 1978, Cudney and Wallace 1980), as exemplified in Figure 6.12. For an aquatic hemipteran, days to first adult as well as degree hours required to reach adulthood both decreased with increasing temperature (Sweeney and Schnack 1977). It is likely that many species respond to both absolute values and thermal summation in different stages of the life cycle (Fig. 6.10).

Although maintenance costs increase with rising temperatures, homeostatic control of nymphal growth may be achieved by reduced weight-specific respiration as body size increases. This counters the tendency of the respiratory rate to increase with rising vernal temperatures (Vannote 1978, Grafius and Anderson 1980). This homeostatic adjustment to temperature may represent an evolutionary response, operating through selection of species or individuals able to adjust growth to predictable thermal heterogeneity.

Some species in constant temperature springs maintain distinct seasonal cycles (Nielsen 1950b, Thorup and Lindegaard 1977), whereas others grow almost continuously throughout the year (Kamler 1964, Thorup and Lindegaard 1977, Grafius and Anderson 1980). *Baetis rhodani* grows at the same rate in an isothermic stream, despite many fewer degree days, as in a stream exhibiting a normal range of thermal heterogeneity (Fahy 1973). *Ephemerella ignita,* however, requires a certain number of degree days to complete development and thus exhibits different growth patterns in the two streams. The degree day requirements from Fahy (1973) are as follows:

	Isothermic Stream	Normal Stream
Ephemerella ignita	1730	1802
Baetis rhodani		
Winter generation	890	1194
Autumn generation	1010	2209

Nymphal development of *Isonychia bicolor* (Ephemeroptera) was positively correlated with the magnitude of diel temperature fluctuations (Sweeney 1978). Although the pitcher plant mosquito *Wyeomyia smithii* develops more slowly under fluctuating than constant temperatures, thermal heterogeneity results in a sevenfold increase in fecundity and a 50% greater capacity for increase. As suggested by Bradshaw (1980, p. 13), "maximum fitness in *W. smithii* is achieved through the action of, and not despite, thermal heterogeneity."

Lutz (1974b) found major photoperiod responses on odonate larval development under constant temperature regimes, but immatures maintained under natural temperatures responded similarly to long, short, and natural day lengths. However, Corbet (1980) emphasizes the important interaction of temperature and photoperiod in regulating growth of temperate zone Odonata.

Temperature-growth adaptations in special habitats include continuous growth at extremely high rates (200–350 degree days per generation) in desert streams (Gray 1981) and the high thermal coefficients of temporary pond insects (Wiggins et al. 1980). A trichopteran that occurs in both England and Iceland is restricted to permanent waters at the latter location because there is insufficient time to complete development during the aqueous phase in temporary habitats at the prevailing lower temperatures (Gíslason 1978).

Voltinism A considerable flexibility in the number of generations per year (voltinism) occurs both intra- and interspecifically in some aquatic insects. Such variability is normally attributed to thermal differences between habitats at different latitudes or altitudes (Pleskot 1961, Gose 1970, Harper 1973, Dodson

1975, Corbet 1980, Lavandier 1981, Trapp and Hendricks 1984), although nutrition may operate in concert with temperature to influence voltinism (Lillehammer 1975b, Lavandier and Pujol 1975, Mackay 1979). As previously mentioned, the caddisfly *Rhyacophila evoluta* has a 1-, 2-, or 3-year life cycle depending on habitat temperature (Décamps 1967). The winter-warm conditions of springbrooks often increase the number of generations (Thorup and Lindegaard 1977, Newell and Minshall 1978). The generation times of the same species may vary within basins of the same lake because of spatial differences in temperature (Flannagan 1979). In a pond in England, the mayfly *Cloeon dipterum* produces an extra generation during especially warm summers (Macan and Maudsley 1966), and the thermal characteristics of lake outfalls (Ulfstrand 1968) or streams below dams (Rader and Ward 1989) may allow some species to complete an extra generation each year. *Wyeomyia smithii* living in pitcher plants in the shade are univoltine, but a portion of the population residing in plants exposed to the sun exhibits bivoltinism (Kingsolver 1979). Given the rapidity of growth coupled with continuous reproduction of certain desert stream insects, there are 25–35 potential generations per year, if conditions would remain suitable (Gray 1981). Stoneflies adapted to intermittent streams in hot climates exhibit extreme flexibility in patterns of voltinism (Snellen and Stewart 1979b), as do the temperate stoneflies in southeastern Australia (Hynes and Hynes 1975), where climatic conditions are likewise highly unpredictable. Many other species, however, exhibit the same number of generations per year, irrespective of thermal conditions, although there is some flexibility within the specified pattern of voltinism.

Emergence The timing and duration of emergence in aquatic insects involves responses to temperature (Harper and Pilon 1970, Danks and Oliver 1972b, Lutz 1974a), often interacting with photoperiod (Nebeker 1971c, Elvang and Madsen 1973, Shepard and Lutz 1976, Sweeney 1984). Light appears to play a more important role under constant thermal conditions (Humpesch 1971). Artificially disrupting the phase relationships between temperature and day length, as has been reported for streams below hydroelectric dams (Ward and Stanford 1982), may disrupt emergence patterns (Khoo 1964, Peters and Peters 1977).

Emergence normally occurs earlier at lower latitudes and elevations (Brinck 1949, Pleskot 1951, Thibault 1971, Nebeker 1971c, Brittain 1975, Wise 1980), and in warmer years (Sprules 1947, Britt 1962, Lillehammer 1975a, LeSage and Harrison 1980); the length of the emergence period may be increased or decreased depending on the species. Emergence also tends to occur earlier, and the emergence period is often extended in winter warm habitats such as springs (Brinck 1949, Tilly 1968, Nebeker 1971c, Schwarz 1973, Thorup and Lindegaard 1977). Artificial elevation of winter temperatures also results in earlier emergence (Lillehammer 1975b, Rupprecht 1975, Brittain 1976a; but see Langford 1975, Armitage 1976); advancements >3 months have been reported (Nebeker 1971b, Fey 1977). Higher than normal discharge during snowmelt may depress water temperatures and retard vernal warming, thus delaying the onset of emergence (Canton and Ward 1981). The ability to directly respond to temperature in this way may represent an essential adaptation for species in mountain streams with a severe period of runoff, the timing, intensity, and duration of which varies from year to year.

Temperature differences also undoubtedly account for the earlier emergence in shallow water bodies and from shallower depths of deeper lentic habitats (Flannagan and Lawler 1972, Danks and Oliver 1972b, Moore 1980).

Precise temperature thresholds for ecdysis have been reported for some aquatic insects (Macan and Maudsley 1966, Danks and Oliver 1972b, Trottier 1973, LeSage and Harrison 1980, McCafferty and Pereira 1984, Peters et al. 1987). The exact timing of emergence and day-to-day variations in emergence intensity have been related to differences in air temperature and irradiance as these influence water temperatures (Macan and Maudsley 1966; Brittain 1976a, 1978a, 1979; LeSage and Harrison 1980). Migration of mature larvae to shorelines where spring and summer temperatures are higher and more responsive to terrestrial conditions has been reported for lentic and lotic species (Macan and Maudsley 1966; Brittain 1976a, 1978a; Stanford and Ward 1979). However, whether rising or falling temperatures stimulate ecdysis depends on the season of emergence and the importance of photoperiod.

Rising spring temperatures serve as the emergence cue for the mayfly *Dolania americana*, although the temperature of the water on the morning of emergence is irrelevant (Peters et al. 1987). Rather, mature nymphs respond to changes in water temperatures 48–24 hours before emergence; dawn water temperatures the previous day best predicted emergence of this species.

In warm streams few species emerge during midsummer (Ide 1935, Macan 1974a). At low elevations in the French Pyrénées, stoneflies exhibit either vernal or autumnal emergence. With increasing elevation, the interval between the two flight periods diminishes until in high mountain streams stoneflies are on the wing throughout the summer (Berthélemy 1966). The earlier emergence of a mayfly from warmer streams and in warmer years is thought to be related to the higher oxygen requirements of emerging nymphs, which cannot be met above a certain temperature (Pleskot 1953).

Differential responses to temperature temporally segregate closely related species within a habitat, resulting in a successional pattern of emergence (Ide 1935, Sprules 1947, Pleskot 1961, Harper and Pilon 1970, Vannote and Sweeney 1980, LeSage and Harrison 1980). Although year-to-year changes in temperature alter the emergence times, the order at which species emerge does not change (Hynes 1976, Brittain 1978a).

The evolution of temporal segregation of sympatric congeners involved development of mechanisms to synchronize emergence patterns. The thermal regime may exert control over synchronization in several ways. Mechanisms generally involve accelerating development of slow individuals and temporally retarding further growth of those that are at a larger size or later stage of development (Bradshaw 1973). Corbet (1957a) devised a model to explain emergence synchrony in summer species of Odonata based on an ascending series of lower temperature thresholds (LTTs) for progressively later instars.

Accordingly, rising vernal temperatures would stimulate growth of early instars (with lower LTTs) before the LTT of larger individuals (which reached a later stage before winter growth cessation) is attained. This theory, which has since received empirical support (Lutz 1968a, b), provides an explanation for synchrony in species that emerge in the spring. A single threshold temperature for emergence, if high enough, would give the same result (Danks and Oliver 1972b). Emergence may be more synchronous during warm years or in warmer habitats (Ide 1935, Britt 1962, Harper and Pilon 1970, Illies and Masteller 1977), presumably because most individuals are able to attain a stage of development immediately responsive to rising vernal temperatures. However, warmer regions may not have winter temperatures low enough to uniformly stimulate diapause development, and so result in less synchronous emergence for certain species (Macan 1974a). This may also partly explain the extended emergence of many aquatic insects in springbrooks.

The closer an individual *Chaoborus americanus* is to emergence, the more its rate of development is retarded by a small drop in temperature (Bradshaw 1973). Such a response synchronizes emergence and reduces the chance of emergence during periods of unsuitable aerial temperatures. Co-occurring morphs with different response times to temperature fluctuations further extend the individual thermal homeostasis and increase the probability of the species surviving severe year-to-year climatic variations.

Vannote and Sweeney (1980) suggest that adult tissue maturation, at the expense of larval growth, is initiated over a wide range of larval sizes when temperature exceeds a critical threshold. This serves to synchronize adult metamorphosis and provides an explanation for the progressive decrease in size of

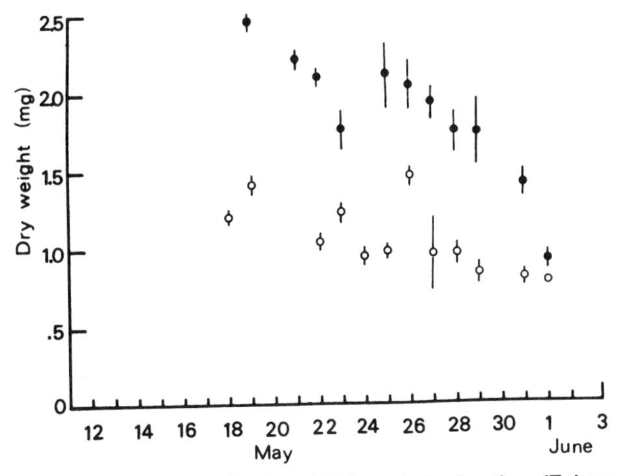

Figure 6.13 Progressive decrease in size of *Ephemerella dorothea* (Ephemeroptera) adults emerging from cool (solid circles) and warm (open circles) tributaries of White Clay Creek, Pennsylvania. [From Vannote and Sweeney (1980), American Naturalist, The University of Chicago Press. © 1980 by The University of Chicago.]

adults over the emergence period (Fig. 6.13). The individual biomass of early-emerging *Ephemera strigata* was about twice that of individuals emerging late in the emergence period (Takemon 1990).

In Odonata there is a tendency for nocturnal emergence in the tropics and daytime emergence in temperate regions; a species near the limits of its range may switch from nocturnal to diurnal emergence depending on day-to-day differences in air temperatures (Corbet 1980). The diel patterns of emergence of some chironomid species closely track diel water temperature fluctuations (Wartinbee 1979). For other species light provides the major diel emergence cue, and water temperature determines the intensity of emergence. In the high arctic, where light is not a reliable measure of air temperature, diel emergence patterns of chironomids are cued to water temperature (Danks and Oliver 1972a). A temperature threshold for emergence coupled with a direct response to diel temperature stimuli restrict emergence to midday during the warmest periods of the year. At lower latitudes, chironomids emerge during midday during spring, but mainly at dusk during summer (Morgan and Waddell 1961, Kurek 1978). Adults thus avoid cold air temperatures during spring and reduce desiccation and predation losses during summer. *Chironomus thummi* emerges in the evening if water is >16°C, near noon if water is <14°C, and divides emergence between noon and dusk at 15°C (Kurek 1978).

Longevity of the adult stage of aquatic insects is determined in part by the water temperatures to which the immatures were exposed (Nebeker 1971a, Macan 1974a) and by air temperatures encountered after emergence (Lyman 1945). At higher water temperatures, there may be a considerable temporal separation in the emergence of males and females (Nebeker 1971b).

In a series of congeneric species, the first to emerge in spring tends to be the largest, with later species being progressively smaller (Ide 1935, Vannote and Sweeney 1980). Summer generations of multivoltine species generally emerge at a much smaller size than do other generations (Grenier 1949, Harker 1952, Macan 1957, Pleskot 1958, Thibault 1971), which does not appear to relate to quality or quantity of food (Vannote and Sweeney 1980). These differences in adult size are believed to be primarily a function of the water temperatures to which the immatures were exposed. Since fecundity is positively correlated with adult size, deviations from optimal thermal conditions relate to the competitive potential of a species at a given location as previously discussed.

Behavioral Relationships

Relatively few data are available concerning the relationships of temperature and behavior in aquatic insects. Much of what is known deals with the aerial adult stage.

Thermoregulatory behavior of adult Odonata involving selection of perch sites, postural adjustments, and alterations in the type of flight employed are important adaptations at higher latitudes but are less developed in tropical species (May 1976, 1978; Corbet 1980). Adult arctic mosquitoes position themselves in the focal points of parabolic flowers where daytime temperatures are

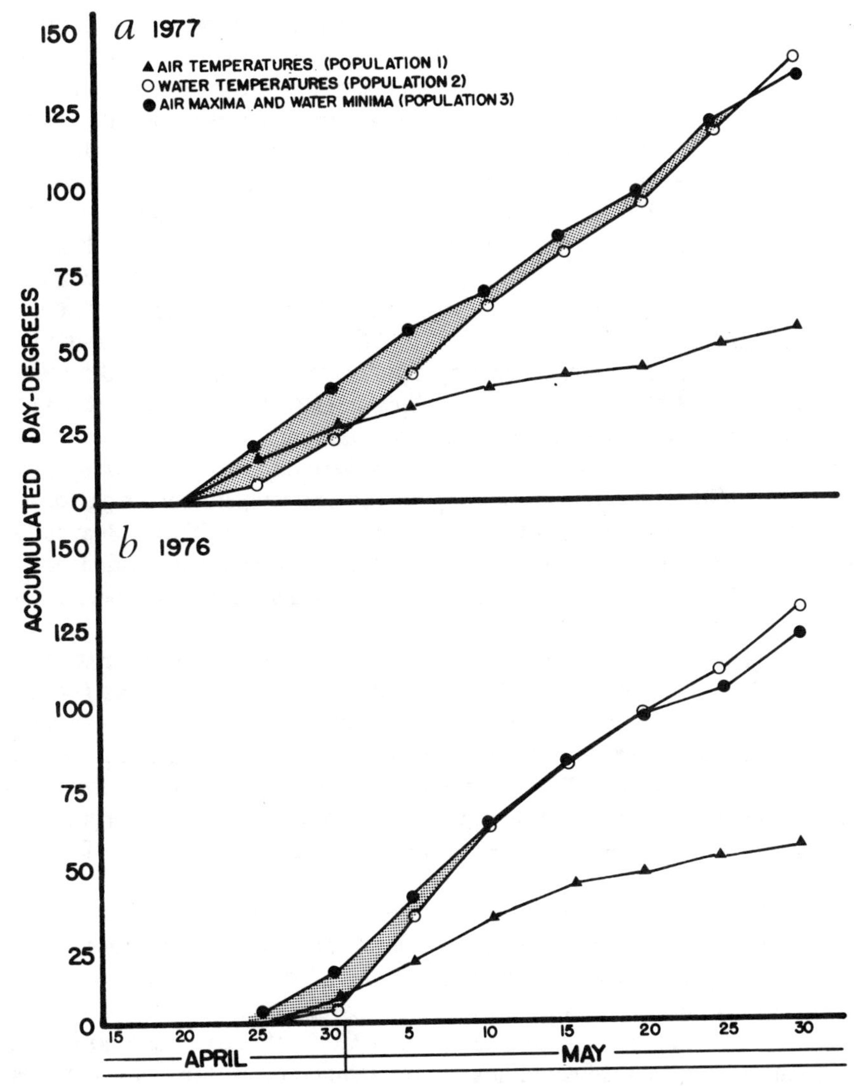

Figure 6.14 Degree days accumulated by three hypothetical populations of water striders (Gerridae) based on prevailing air and water temperatures during two spring periods. Stippled areas indicate differences in accumulated thermal units between submerging gerrids and those experiencing only water temperature. [From Spence et al. (1980b).]

≤6°C higher than ambient (Hocking and Sharplin 1965). Overwintering adult water striders (*Gerris* spp.) exhibit "underwater basking" in the spring (Spence et al. 1980b). By submerging during portions of the diel cycle when water is warmer than air, accumulation of degree days is enhanced (Fig. 6.14), which is thought to increase the rate of gonad development. Adult stoneflies *(Zapada cinctipes)* behaviorally thermoregulate by entering the water during the night when springtime air temperatures in their montane habitat drop to well below water temperatures [mean minimum air -6.2°C, water +8.6°C (Tozer 1979)]. Caged adults allowed to submerge exhibited significantly greater survival than those prevented from doing so. Experimental manipulations demonstrated that the stoneflies were responding to temperature, not light. If preemergent nymphs of *Anax junius* do not encounter air temperature above the threshold for ecdysis, they crawl back in the water to emerge the following day (Trottier 1973).

A variety of other behavioral phenomena exhibited by the aerial adult stages of aquatic insects are temperature-dependent. A 2°C increase in water temperature resulted in emergence of the stonefly *Perla marginata* during winter, a time when air temperatures were below the 10°C threshold required for drumming behavior (Rupprecht 1975). Drumming signals are apparently altered by temperature, as demonstrated by Ziegler and Stewart (1977), who found that female stoneflies repeatedly responded only to male drumming signals recorded within 2°C of the temperature to which the females were exposed at the time of the experiment. Other investigators have reported critical temperature thresholds for flight, swarming, mating, and oviposition (Lutz and Pittman 1970, Solem 1976, Sweeney and Schnack 1977, Zalom et al. 1980). Major modifications in feeding and mating behavior in arctic black flies, compared to temperate species, are attributed to low air temperatures (Downes 1964).

Although temperature has only rarely been implicated as the entraining agent (Zeitgeber) in the diel pattern of behavioral drift of aquatic insects (Waters 1968), high temperatures may increase drift amplitude in some species (Waters 1972, Cowell and Carew 1976), and abrupt temperature changes may induce catastrophic drift (Fey 1977). Upstream migrations of immature insects into tributary waters during spring may (Olsson and Söderström 1978) or may not (Hayden and Clifford 1974) expose aquatic stages to a greater number of degree days.

Thermoperiods with higher temperatures during the dark period or scotophase (the normal situation encountered during diel vertical migration) caused faster development and inhibited induction of dormancy in *Chaoborus crystallinus* compared to constant temperatures or thermoperiods with lower temperatures during scotophase (Ratte 1979). Crane fly larvae in a horizontal temperature gradient selected maximum temperatures (16–18°C) during scotophase and minimum temperatures (12–14°C) during photophase; under conditions of constant light, temperature selection continued as a free-running circadian rhythm of behavioral thermoregulation (Fig. 6.15). Starved mosquito larvae selected a greater range of temperature than did fed individuals (Linley and Evans 1971). Later instars selected progressively higher temperatures, presumably as an adaptation to temporary pool habitats. Vertical migrations of chironomid larvae within the hyporheic zone "follow an optimum temperature for development" (Williams and Hynes 1974).

Other behavioral adaptations or responses include the influence of temperature on attack coefficients of Odonata (Thompson 1978), filtering and net spinning behaviors of Trichoptera (Philipson and Moorhouse 1974, Gallepp

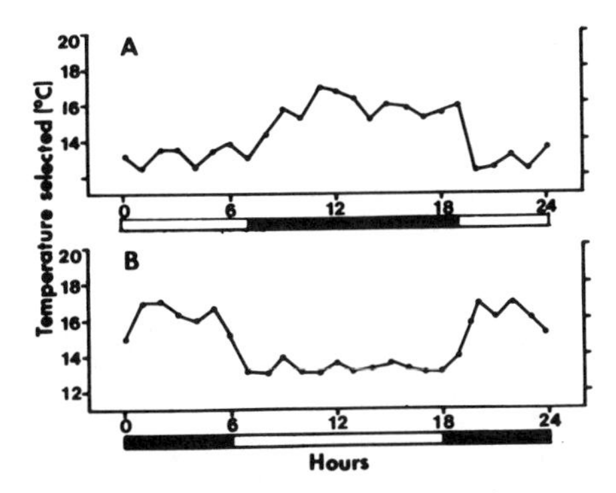

Figure 6.15 Examples of the diel rhythm of preferred temperature selection and behavioral thermoregulation of individual *Tipula* (Diptera) larvae under normal (*A*) and reversed (*B*) 12-hour light: 12-hour dark cycles (LD 12:12). Dark horizontal bars represent the dark portions of the LD cycles. [From Kavaliers (1981). Reproduced from Can. J. Zool., Vol. 59, p. 556, 1981, with permission from the National Research Council of Canada.]

1977, Fey and Schuhmacher 1978, Wallace and Merritt 1980, Cudney and Wallace 1980), and cocoon construction of Chironomidae (Danks 1971b).

Trophic Relationships

Temperature influences trophic dynamics directly through its effects on phenomena such as feeding and assimilation rates, and indirectly by determining the composition, quantity, and quality of food available to aquatic insects. Processing of terrestrial leaf litter, a major energy source for aquatic insects of woodland streams, is greater at higher temperatures (Kaushik and Hynes 1971, Petersen and Cummins 1974, Iversen 1975, Anderson and Sedell 1979, Short and Ward 1980b); although as Cummins (1979) points out, processing per unit of temperature is higher in the cooler headwaters where the community best adapted to utilize this energy source evolved. Rapid processing of leaf detritus was documented for a mountain headwater stream, despite water temperatures at or near 0°C (Short et al. 1980).

Growth of some aquatic insects is regulated by the interaction of temperature and food quality (Anderson and Cummins 1979). Seasonal changes in food quality, partly a function of temperature, may determine the effect of a given temperature on growth (Ward and Cummins 1979, Grafius and Anderson 1980, Johannsson 1980).

Temperature thresholds for feeding have been determined for aquatic insects (Gose 1970, Trottier 1973, Cudney and Wallace 1980, Nagell 1980), and feeding rates, assimilation efficiency, and egestion are often a function of thermal conditions (Nebeker 1971a, Heiman and Knight 1975, Winterbourn and Davis 1976, Brown and Fitzpatrick 1978, Zimmerman and Wissing 1978, Pandian et al. 1979, Wallace and Merritt 1980, Short and Ward 1981b).

7

SUBSTRATE

The majority of the aquatic entomofauna exhibit a primarily benthic mode of existence; that is, they are intimately associated with the substrate during at least a portion of their lives. It is, therefore, not surprising that substrate type is a major determinant of the distribution and abundance of aquatic insects (Minshall 1984). The substrate provides habitat space, food (directly, or a surface where food concentrates), and protection (e.g., from current, predators). Substrate variables of ecological importance include physical structure, organic content, stability, and heterogeneity. This chapter deals with the classification, composition, and distribution of substrate materials in lotic and lentic waters, and the influence of substrate parameters on aquatic insects.

SUBSTRATE CLASSIFICATION

A wide variety of inorganic and organic materials of autochthonous and allochthonous origins comprise the substrate of aquatic systems. The composition and distribution of substrate materials are a function of the parent bedrock of the region, terrestrial vegetation characteristics, topographical and meteorological features, chemical and biological factors, and limnological phenomena including the action of waves and current. The substrate of aquatic habitats is often a composite of various materials and particle sizes arranged in a mosaic fashion, although a single uniform substrate type such as bedrock may predominate in some situations. Solid substrate surfaces (rocks, logs, living plants, etc.) and soft-bottom sediments are inhabited by haptobenthic and herpobenthic insects, respectively.

An elaborate terminology for bottom deposits (Lundqvist 1927) evolved in association with the development of lake typology schemes. Many of these terms are no longer in common usage, but a few of them continue to appear in the literature. *Gyttja* refers to a soft, hydrous mud of circumneutral pH made up of various finely-divided inorganic and organic components, especially the remains of sedimented plankton. Gyttja is coprogenic, meaning that its characteristics are partly a result of the material having been passed through the digestive tracts of bottom organisms. Indeed, fecal pellets make up a major portion of gyttja. *Bioturbation*, the reworking and mixing of bottom materials by benthic organisms (Krantzberg 1985), is an important process in the formation of gyttja (Robbins 1982). Gyttja occurs in both oligotrophic and eutrophic lakes, although deposition rates and the organic content are higher in eutrophic lake sediments (Cole 1983). *Dy* (a characteristic sediment in dystrophic lakes) is a brown gelatinous material comprised of gyttja plus abundant humic acid colloids. Dy has a greater organic content than gytta and a higher carbon to nitrogen ratio (Hansen 1959). Whereas gyttja is a characteristic bottom deposit of large lakes dominated by phytoplankton productivity, dy is typically associated with small lakes (especially bog lakes) having large inputs of acidic humic matter from littoral production or allochthonous sources (Wetzel 1983). *Sapropel*, a black, highly organic sediment, forms when bottom deposits are subjected to anaerobic conditions for extended periods. Sapropel contains reduced compounds such as H_2S and CH_4 and is not normally colonized by macrobenthos.

Other sediment types include marl, a precipitate of calcite that forms crusty deposits in hard-water lakes and streams; silicic acid, derived from diatom frustules and sponge spicules; and ochreous mud, red sediments with a high content of limonitic iron.

Mineral Grain Size The Wentworth Scale has been widely used to partition the inorganic fraction of the substrate into particle sizes (Table 7.1). The size classes of the Wentworth Scale form a geometric progression. The phi (Ø) scale developed by Krumbein (1936) converts the Wentworth size classes into arithmetic intervals (phi is the negative $\log_2$ of the particle diameter in millimeters). Cummins (1962) recommends median phi (MdØ), the value on the phi scale above and below which 50% of the sample (by weight) is distributed, as a convenient quantitative description of a sediment sample.

Detrital Size Classes The particulate organic matter (detritus) of aquatic habitats is separated into two broad categories, CPOM (>1 mm) and FPOM (<1 mm), based on particle size (Table 7.2). Detritus serves as food and physical habitat for a variety of aquatic insects. RPOM consists of large woody debris such as logs and branches that have residence times in aquatic systems of decades or even centuries. LVOM consists of terrestrial leaves that enter the water and, especially in lotic systems, tend to form accumulations or leaf packs. LFOM consists of leaf, twig, and bark fragments; conifer needles; fruits; buds; and flowers. LPOM, the smallest category of coarse particulate organic matter,

Table 7.1 A classification of inorganic substrate materials based on particle size

Wentworth Classification [a]		
Name	Size (mm)	Phi Scale [b]
Boulder	>256	-8
Cobble	64–256	-6, -7
Large pebble	32–64	-5
Small pebble	16–32	-4
Coarse gravel	8–16	-3
Medium gravel	4–8	-2
Fine gravel	2–4	-1
Very coarse sand	1–2	0
Coarse sand	0.5–1	1
Medium sand	0.25–0.5	2
Fine sand	0.125–0.25	3
Very fine sand	0.0625–0.125	4
Silt	0.0039–0.0625	5,6,7,8
Clay	<0.0039	9

[a]*Based on Cummins' (1962) modification of the Wentworth Scale (Wentworth 1922).*
[b]Negative $\log_2$ of the particle size (Krumbein 1936).

consists of plant and animal detritus and whole feces or fecal fragments partitioned into subcategories according to particle size. Microorganisms are an integral part of detritus in all size categories; UPOM includes those microorganisms not attached to detrital particles.

The size categories of mineral and detrital particles presented in Tables 7.1 and 7.2 are not entirely arbitrary. Aquatic insects and other organisms selectively colonize and feed on substrate materials based partly on particle size composition.

Table 7.2 Size categories of nonliving organic matter

Detritus Categories and Subcategories	Acronym	Approximate Size Ranges
Coarse particulate organic matter	(CPOM)	(>1 mm)
Large resistant particulate organic matter	RPOM	>64 mm
Whole-leaf organic matter	LVOM	>16 to <64 mm
Leaf fragment organic matter	LFOM	>4 to <16 mm
Large particulate organic matter	LPOM	>1 to <4 mm
Fine particulate organic matter	(FPOM)	(>0.5 μm to <1 mm)
Medium particulate organic matter	MPOM	>250 μm to <1 mm
Small particulate organic matter	SPOM	>75 to <250 μm
Ultrafine particulate organic matter	UPOM	>0.5 to <75 μm
Dissolved organic matter	(DOM)	(<0.5 μm)

Source: Modified from Cummins (1974).

SUBSTRATE DISTRIBUTION PATTERNS

Lentic Waters The substrate of lakes results from the dynamic interactions of wind and wave action, morphometric parameters, and entering rivers (Fig. 7.1). The reader is referred to Sly (1982) for a detailed account of sediment-freshwater interactions in lentic waters.

The littoral substrate is greatly influenced by the degree of exposure to wind. Shores exposed to wave action tend to be rocky with few or no higher aquatic plants. Because wave action rapidly diminishes with increasing depth, the littoral substrate of exposed shores may grade from bedrock or boulders to clay over relatively short distances. For example, on exposed points of Oneida Lake, New York, boulders and stones extend from the shore to depths of <1m, are replaced by gravel (to depths of 1.5 m), which is followed by sand (to 1.8 m), sandy clay (to 2.4 m), and clay (Baker 1918).

Under certain conditions sand or gravel constitutes the predominant substrate particles of exposed shores. In extremely hard-water lakes, marl may cover large areas of the bottom in the littoral zone. Extensive portions of the lake bottom near the mouths of streams and rivers may be modified by fluvial deposition and erosion, providing further substrate heterogeneity in the littoral zone. Fine materials of autochthonous and allochthonous derivation accumulate in the littoral zone along sheltered shores and in quiet bays and backwaters. It is in such places that aquatic macrophytes attain maximum diversity and abundance (Spence 1982), providing additional habitat complexity and substrate surfaces for aquatic insects.

In the profundal zone, the original materials of the lake basin are typically covered by a layer of "ooze" (Schlamm). Some of the bottom deposits have been carried in by streams and atmospheric deposition or were eroded from the shoreline. A portion of the profundal ooze is derived from autochthonous production, primarily by planktonic and littoral organisms. Herpobenthic insects, especially chironomid larvae, play a significant role in determining the physical (McCall and Tevesz 1982) and chemical characteristics (Fisher 1982) of the medium they inhabit. Although the sediment of deep waters is generally more uniform than the substrate of the littoral zone, habitat conditions for zoobenthos are by no means homogeneous (Milbrink et al. 1974, Reynoldson and Hamilton 1982). For example, a gradient in the composition and organic content of deep-water sediments from the inlet to the outlet of a Colorado reservoir significantly influenced density and distribution patterns of profundal benthos (Edmonds and Ward 1979).

Woody debris (RPOM) is generally less prevalent as a substrate for aquatic insects in lakes than in streams. This primarily relates to the depositional nature of lakes and their lower degree of interaction with the terrestrial environment. Only in swamps and small lentic water bodies (beaver ponds, forest pools), or in lakes to which man has introduced woody debris (inundation of trees by impoundment, use of waterways to transport logs), does wood provide a major substrate surface for aquatic insects of lentic habitats.

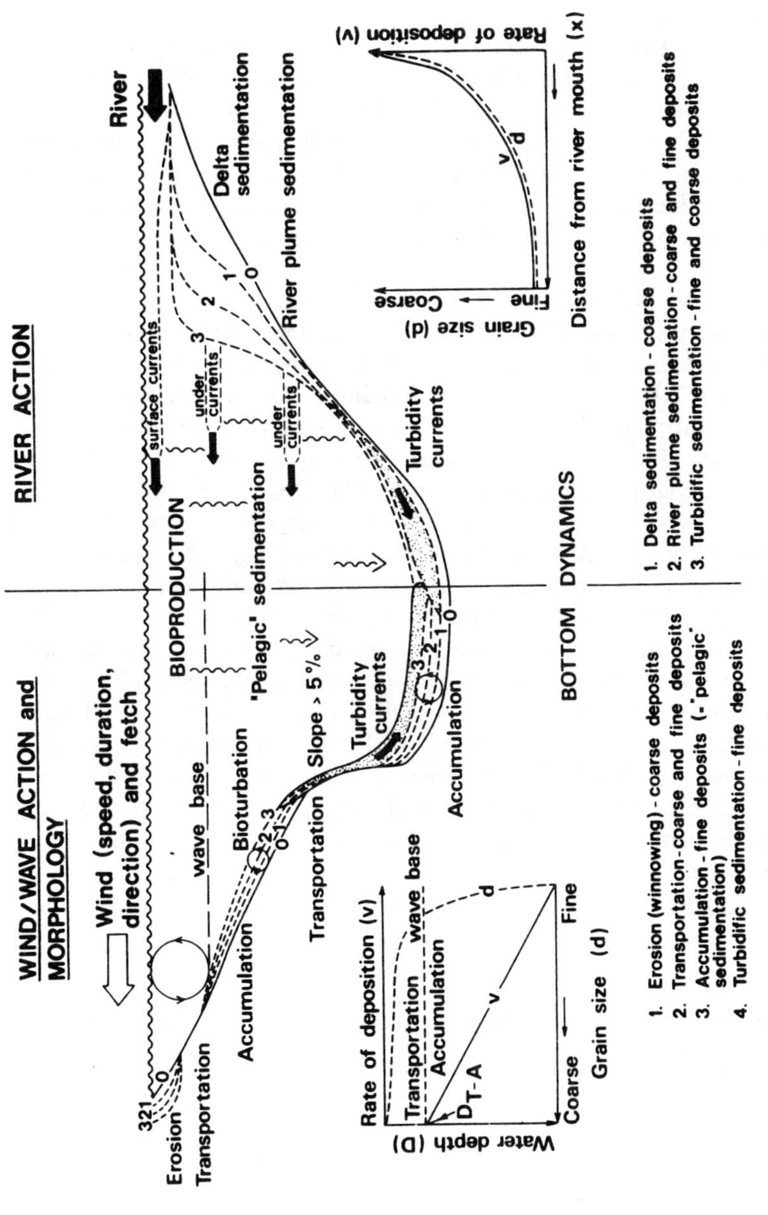

Figure 7.1 Schematic illustration of major sedimentological and bottom dynamic processes in lakes. [From Håkanson (1982). Reprinted by permission of Kluwer Academic Publishers.]

Lotic Waters The substrate of running waters is largely structured by physical processes related to the unidirectional flow of water interacting with basin geology and allochthonous organic debris (Leopold et al. 1964, Keller and Swanson 1979, Bilby and Likens 1980, Rust 1982). Benthic detritus of terrestrial origin plays a greater role in lotic than in lentic waters, and is especially important to the ecology of insects in small woodland streams. It is convenient to consider the mineral and organic substrates of running waters separately, although most sediments are mixtures of both components.

MINERAL SUBSTRATE The substrate composition of running waters, including spatial and temporal variations in sediment characteristics, bears a close relationship to current velocity specifically and to the flow regime in a general sense (Dudgeon 1982a). Table 7.3 indicates the approximate current velocities required to displace mineral particles of a given diameter. Displaced particles may be entrained by the moving water to become part of the suspended load or may merely roll along the bottom as bed load. As expected, higher current velocities are required to move larger substrate particles, although compact clay is entrained less readily than sand despite the smaller diameter of individual clay particles. In general, however, the mineral substrate in areas with current velocities <20 cm/sec tend to be largely clay and silt; at 20–40 cm/sec fine sand predominates; at 40–60 cm/sec coarse sand and fine gravel predominate; at 60–120 cm/sec bed materials range from gravel to small cobble; and at current velocities >120 cm/sec cobbles and boulders predominate (Einsele 1960). Larger substrate materials (e.g., cobbles) may protect underlying fine particles from entrainment (a phenomenon called *armoring*), and dead water zones such as occur in the lee of boulders or other large objects tend to accumulate sand and gravel even in high-gradient reaches. Marl deposition may play a major role in structuring the substrate of hard-water streams (e.g., Minckley 1963).

There is normally a decrease in the mean size of substrate materials from the

Table 7.3 Mean minimum current velocities required to displace substrate particles of different size classes

Particle Size Class	Current Velocity (cm/sec)
Clay	30
Compacted clay	60
Fine sand	20
Coarse sand	30–50
Fine gravel	60
Medium gravel	60–80
Coarse gravel	100–140
Angular stones	170

Source: Modified from Schmitz (1961).

headwaters to the mouth of river systems. Boulder and cobble typically predominate in high gradient upper reaches, whereas sand and gravel are normally the major substrate materials of lower reaches. Particles finer than sand are not ordinarily a major component of the substrate in the main channels of rivers, although silts and clays occur in quiet backwaters or other areas of reduced current, and may form a temporary film on the river bed during low flow periods (Hynes 1970a). In general, increases in mean substrate particle size are associated with increases in physical habitat complexity and bed stability, important factors in determining the distribution and abundance of aquatic insects.

The substrate of running waters thus provides a dynamic mosaic of habitat types for benthic insects partly because of spatial differences in current. For example, pools tend to be depositional habitats, at least during low-flow periods, which tend to accumulate fine bed materials, whereas riffles are erosional habitats from which fine particles are exported [the erosional-depositional dichotomy of Moon (1939)]. Within a given riffle or pool there is a good deal of substrate heterogeneity because of microhabitat differences in current (Dudgeon 1982a). Sedimentary organic matter adds further structure and heterogeneity.

ORGANIC DETRITUS The organic detritus of lotic ecosystems is composed primarily of plant litter derived either from *in situ* production by aquatic plants or allochthonous inputs from the riparian vegetation (Fig. 7.2). The emphasis on plant detritus of terrestrial origin (e.g., Hynes 1963, Egglishaw 1964, Minshall 1967, Fisher and Likens 1973, Petersen and Cummins 1974, Malmqvist et al. 1978, Anderson and Sedell 1979, Short et al. 1980, Dudgeon 1982b, King et al. 1987a) partly reflects a concentration of research effort on small woodland streams where from 60% to nearly 100% of the energy input is from allochthonous sources (Minshall 1978).

Figure 7.3 shows seasonal changes in the standing crop of benthic detritus for a canopied third-order Rocky Mountain stream (Short and Ward 1981). The autumnal and vernal peaks in benthic detritus were associated with leaf abscission and ice out, respectively. The flushing effect of snowmelt runoff largely eliminated sedimentary detritus by June. VPOM (= UPOM in Table 7.2) was the most abundant size fraction except in autumn, when large amounts of leaf litter entered the stream (Fig. 7.4). Headwater streams in deciduous forests, while containing larger standing crops of benthic detritus, exhibit seasonal trends that are similar to those of the Rocky Mountain stream (Boling et al. 1975). In coniferous forest streams examined by Naiman and Sedell (1979), wood debris constituted over 90% of total detritus.

According to the river continuum concept (Fig. 4.22), fine organic particles are an important component of the detritus pool in all stream reaches, but increase in relative abundance in the downstream direction. In large rivers FPOM is the primary organic resource. In heavily canopied headwaters, coarse detritus (primarily leaf litter) attains maximum relative abundance, whereas aquatic plants make major contributions to the detrital pool of the middle reaches. In the headwaters of woodland streams there is a net production of

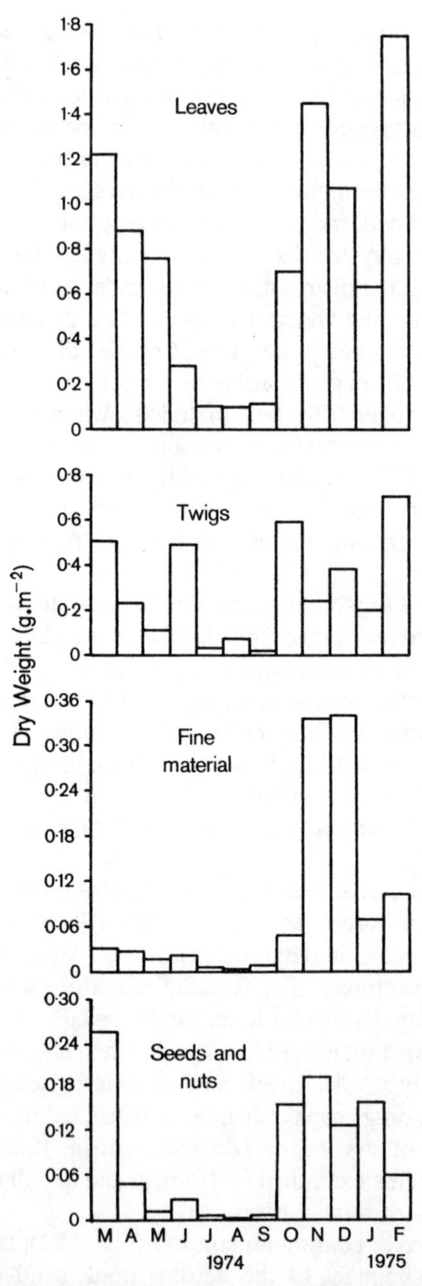

Figure 7.2 Dry weights of leaves, twigs, seeds, nuts, and fine materials collected in litter-fall traps beside a beech forest stream in New Zealand. (From Figure 4 in Winterbourn, M. J. 1976: Fluxes of litter falling into a small beech forest stream. New Zealand Journal of Marine and Freshwater Research 10: 399–416.)

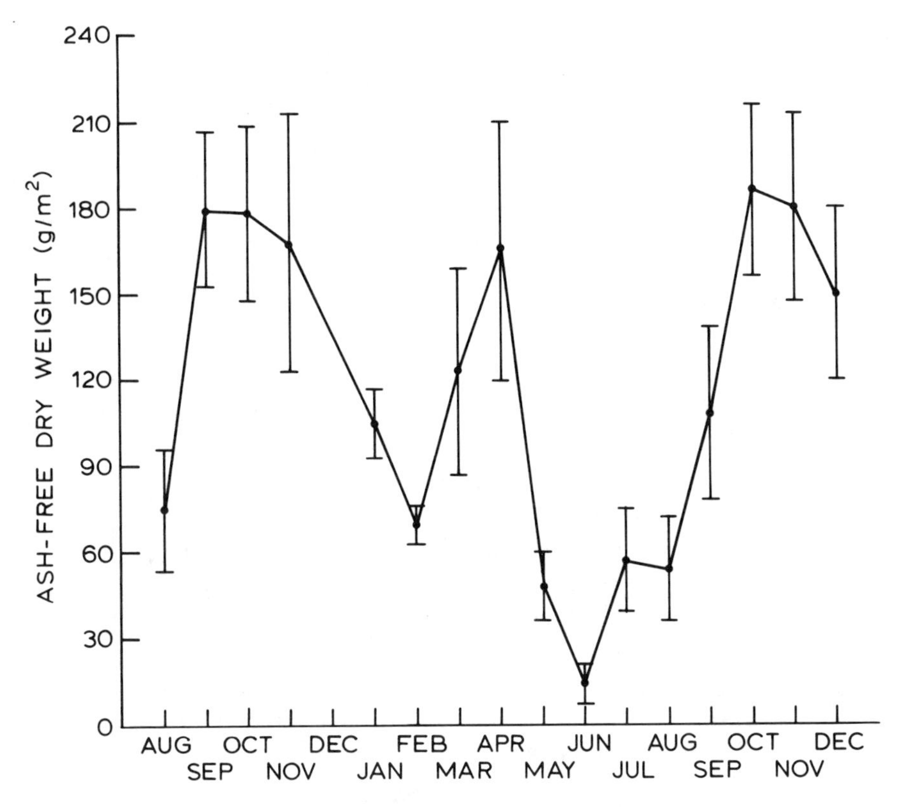

Figure 7.3 Seasonal changes in the standing crop of benthic detritus in a Rocky Mountain headwater stream ($\bar{x} \pm 1$ s.d.). [Modified from Short and Ward (1981).]

FPOM, partly a result of the processing of CPOM by aquatic insects; whereas in the lower reaches there is a net consumption of FPOM (Iversen et al. 1982).

INSECT-SUBSTRATE RELATIONSHIPS

Few aquatic insects are restricted to a specific substrate, but many, probably most, benthic species exhibit distinct preferences for one or another general bottom type (Table 7.4). The close association of a particular species with a given substrate may reflect current preferenda, requirements for shelter, respiratory needs, or food habits, rather than directly indicating an affinity for a specific bottom type. Some investigators of bottom fauna-substrate relationships have examined a variety of bottom types, whereas others deal primarily or solely with a single category of substrate (Table 7.5).

The following overview of the characteristic insect associations on the general substrate categories precedes a comparative analysis of the diversity, abundance, and composition of the aquatic entomofauna as a function of substrate type.

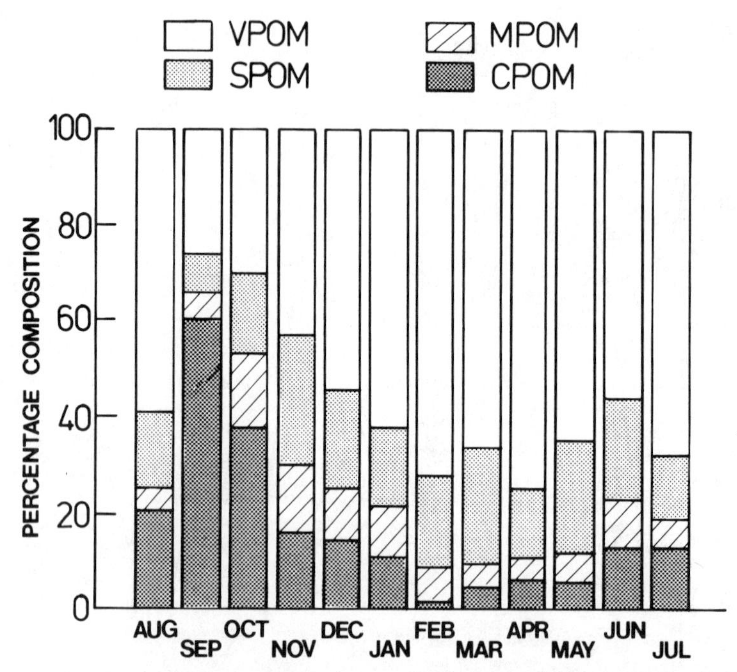

Figure 7.4 Relative contributions of size classes of benthic organic matter in a Rocky Mountain headwater stream. Size classes correspond to acronyms in Table 7.2, except VPOM (= UPOM). [Modified from Short and Ward (1981).]

Lithophilous Fauna

Lithophilous insects are those species characteristically occurring in the stony littoral of lakes or in rocky streams and are thus associated with the large size classes of mineral particles. Environmental conditions normally associated with the habitats in which such substrates occur include ample water movement, high levels of dissolved oxygen, relatively silt-free interstices, and the absence or paucity of higher aquatic plants. Mosses and algae may, however, be well

Table 7.4 Some general categories of benthic insects based on substrate preference

Substrate Type	Faunal Category
Hydrophytes	Phytophilous
Wood	Xylophilous
Stones	Lithophilous
Gravel	Psephophilous
Sand	Psammophilous
Mud	Pelophilous

developed on rock surfaces, and some species of insects are associated with these plants rather than the mineral substrate per se. Others colonize the plant detritus that accumulates between the stones.

The Stony Littoral Zone The rocky shores of Lake Mendota, Wisconsin, are inhabited by a distinctive insect fauna (Muttkowski 1918). *Siphlonurus alternatus* and three heptageniid mayflies are characteristic residents. Coleopterans include the water penny, *Psephenus lecontei,* and a "parnid" larva (probably *Dryops lithophilus).* The rocky shore also serves as optimal habitat for the trichopterans *Leptocerus ancylus* and *L. dilutus.* Additional caddis larvae in the family Hydroptilidae are associated with the *Cladophora* filaments attached to the rocks, and the spongillafly *Sisyra umbrata* is associated with the sponges that inhabit the rocky shore. Additional species occur in this biotope but do not find optimal conditions on the rocky shore.

Muttkowski points out that some of the lithophilous species commonly occur on rocky substrate at depths as great as 6 m, suggesting that in these cases it is substrate per se and not current that is the major controlling factor. Submerged rock outcrops ≥3 m below the lake surface and ≤0.8 km from the nearest shore of Lake Mendota contain elements of the lithophilous fauna. The presence of *Psephenus lecontei* is especially interesting. How do the beetle larvae locate and reach such isolated habitats, and how do they move to terrestrial pupation sites?

The fauna of the stony littoral of Lake Windermere, the largest lake in the English Lake District, has been extensively examined (Moon 1934, 1936; Macan and Maudsley 1968, Macan 1970). Macan and Maudsley identified 48 species of insects (excluding Diptera) in stony shore samples, 14 species of which (Table 7.6) they considered "true inhabitants of the stony substratum." The remaining species collected from the stony littoral were thought to be unable to maintain populations in this biotope without continual recolonization from streams, finer substrates in adjacent deeper water, reed beds, or other nearby habitats.

Baker (1918), in a detailed study of the bottom fauna of Oneida Lake, New York, designated the coleopteran *Psephenus lecontei,* the trichopteran *Helicopsyche borealis,* and the mayfly *Heptagenia* as the insects characteristic of rocky bottoms. Baker reported considerable overlap between the fauna of stone and gravel bottoms.

The littoral rock fauna of Tasmanian lakes exhibited a shift from insect dominance to dominance by noninsect invertebrates with increasing trophic status (Leonard and Timms 1974), a phenomenon also reported for English lakes (Macan and Maudsley 1969). In an oligotrophic Tasmanian lake, 9 of the 14 species encountered were insects that collectively constituted 73% of the total benthos of the stony shore. In a mesotrophic lake, only 8 of 20 species were insects and none were especially abundant. In some instances the species of the Tasmanian lakes are congeners of Northern Hemisphere insects (*Helicopsyche*, some chironomids). In other cases, northern families are represented by different genera of lithophilic insects (e.g., the Tasmanian psephenid beetle *Sclerocyphon*

Table 7.5 Some studies with an emphasis on aquatic insect-substrate relationships

	Habitat		Substrate			
Reference	Lentic	Lotic	Mineral	Hydrophytes	Detritus	Wood
Anderson et al. 1978		X				X
Baker 1918	X		X	X		
Barton and Hynes 1978	X		X			
Barton and Lock 1979		X	X			
Behning 1928		X	X	X		
Berg et al. 1948		X	X	X		
Carpenter 1927		X	X	X		
Chutter 1970		X	X	X		
Egglishaw 1964, 1969		X	X	X	X	
Gregg and Rose 1985		X		X		
Greze 1953		X	X			
Harrod 1964		X		X		
King et al. 1987b		X			X	
Krecker 1939	X			X		
Krecker and Lancaster 1933	X		X	X		
Kuflikowski 1970, 1974	X			X		
Lindegaard et al. 1975		X		X		
Macan 1965	X			X		
Macan and Maudsley 1968	X		X			
Mackay and Kalff 1969		X	X		X	
Maitland 1966		X	X	X		
Malmqvist et al. 1978		X	X		X	
McGaha 1952	X	X		X		
McLachlan 1970	X					X
Moon 1934, 1936	X		X	X		
Müller-Liebenau 1956	X			X		

Muttkowski 1918	X		X	X		
Percival and Whitehead 1929		X	X	X		
Petr 1970b	X					X
Reice 1980		X	X		X	
Russev 1974		X	X			
Short et al. 1980		X			X	
Smith et al. 1981	X		X			
Soszka 1975a, b	X			X		
Sprules 1947		X	X			
Sublette 1957	X		X	X		
Ulfstrand 1967		X	X		X	
Ward 1975		X	X			

Table 7.6 Lithophilous insects (excluding Diptera) of the stony littoral of Lake Windermere, English Lake District (see text)

Taxa	Numbers Collected[a]
Plecoptera	
Nemoura avicularis	129
Diura bicaudata	35
Chloroperla torrentium	28
Leuctra fusca	4
Capnia bifrons	32
Ephemeroptera	
Ecdyonurus dispar	691
Heptagenia lateralis	290
Centroptilum luteolum	108
Trichoptera	
Polycentropus flavomaculatus	369
Agapetus fuscipes	11,696
Cyrnus trimaculatus	10
Tinodes waeneri	176
Coleoptera	
Stenelmis canaliculata	+
Platambus maculatus	2

[a]Based on totals from timed collecting effort. The + sign indicates that the species, although present, was not taken at the sampling stations.

Source: Macan and Maudsley (1968).

and the Tasmanian elmid beetle *Simsonia*). The plecopterans in the stony littoral of the Tasmanian lakes are in families that are ecological equivalents of Northern Hemisphere stoneflies. The mayfly family Leptophlebiidae has expanded to fill the ecological role of the Heptageniidae, a family absent from much of the Southern Hemisphere. The lithophilous insect fauna is, therefore, ecologically, if not taxonomically, similar to the assemblage of species residing in comparable habitats in northern waters.

Such is also the case for the rocky shores of Lake Nyasa, a tropical lake in Africa (Fryer 1959a). The insect fauna is dominated by the stonefly *Neoperla spio,* two species of the heptageniid mayfly *Afronurus,* hydropsychid caddisflies, and the psephenid beetle *Eubrianax.*

In a report on the macrobenthos of the wave-swept Canadian shores of the St. Lawrence Great Lakes, Barton and Hynes (1978) stress the rheophilic nature of the insect fauna in such habitats. They examined exposed shores where the substrate in different areas ranged from sand to bedrock. Along the substrate gradient, which represents a progressive increase in exposure to wave action, insects from typically lentic groups were replaced by lotic forms, such as the following:

Ephemeroptera
  Heptageniidae
Plecoptera
  Capniidae
  Chloroperlidae
  Perlodidae
Trichoptera
  Hydropsychidae
  Polycentropodidae
Coleoptera
  Elmidae
  Psephenidae

Several of the species in these groups had previously been recorded only from lotic habitats.

Granitic bedrock comprises large portions of the shores of Lake Superior. The nets and retreats of hydropsychid caddisflies (*Hydropsyche recurvata* and *Cheumatopsyche* sp.) occurred wherever there were cracks or crevices in the bedrock surface. Larvae of the tipulid *Antocha* were also commonly associated with such crevices. The fauna of the smooth bedrock surfaces was dominated by the chironomid taxa Tanytarsini and especially Orthocladiinae. The lotic caddisfly *Lepidostoma* was also abundant. The mayfly *Baetis,* one of the few insects able to rapidly move about on the upper surfaces of boulders in rapid streams, also occurred on smooth bedrock areas where it moved about actively even in heavy surf.

Barton and Hynes reported a direct relationship between the diversity and relative abundance of macroinvertebrates, and substrate stability. Krecker and Lancaster (1933), who studied the benthos on a variety of substrate types along the western shore of Lake Erie, found that faunal diversity was highest on shores of flat cobble, a substrate that combines the attributes of stability and shelter.

Rocky Streams The insect fauna of rocky streams is distinctive and remarkably similar nearly everywhere such habitats occur (Hynes 1970b). As emphasized in the preceding section, many of the insects inhabiting exposed lake shores are typical lotic species able to colonize lentic situations where wave action maintains a rocky substrate.

Not all insects residing in rocky streams are lithophilous, however. Some species are found exclusively or primarily in depositional habitats, while others are associated with mosses or accumulations of leaf litter. Nielsen (1950a) distinguishes between animals living on bare stone surfaces in streams, which he refers to as the "torrential fauna," and other rheophilous forms that occupy lotic microhabitats such as tufts of moss.

A wide variety of lotic insects characteristically occurs on clean rock surfaces. Some of these are free-ranging forms. Larvae of net-winged midges (Blephariceridae), mountain midges (Deuterophlebiidae), and glossosomatid caddisflies slowly move over the surfaces of cobbles and boulders. Psephenid beetle larvae typically occur on the undersides of rocks. More active free-ranging lithophilous forms include plecopterans from many families, heptageniid mayflies and their flattened leptophlebiid counterparts of the Southern Hemisphere, rhyacophilid caddisflies, certain elmid beetles, and the mayfly *Baetis.* Black flies (Simuliidae) occur on the upper surfaces of rocks; the larvae, although capable of locomotion, are chiefly sedentary, and the pupal cases are permanently affixed to the rock. Larvae of the brachycentrid caddisflies *Brachycentrus* and *Oligoplectrum* fasten their cases to the tops of rocks with silk, thereby maintaining their position and freeing the legs for feeding. Mayfly nymphs of the genus *Isonychia* perch on the tops of rocks and hold their hair-fringed forelegs in the current to filter food particles from the water.

Several stream insects reside in fixed shelters attached to rocks. Some species of orthoclad chironomids construct tubes on the surfaces of stones. Rock-dwelling pyralid moth larvae and pupae live under silken canopies in depressions in rock surfaces. Caddisfly larvae from several families inhabit shelters affixed to rocks. Many species of hydropsychids reside in a shelter associated with their feeding net. Philopotamid larvae construct their finger-shaped nets on the undersides of large rocks. Polycentropodids and psychomyiids also affix their tubular retreats to rock surfaces. Caddisflies from several families attach their pupal cases to rocks.

On the basis of her studies of Welsh Cardiganshire streams, Carpenter (1927) identified the following insects as characteristic of the lithophilous association:

Ephemeroptera —*Ecdyonurus lateralis, Baetis tenax, Baetis vernus*

Plecoptera —*Isogenus sp., Nephelopteryx sp., Leuctra inermis, Dictyopterys mortoni, Isopteryx torrentium*

Trichoptera—*Rhyacophila obliterata, Agapetus fuscipes, Silo nigricornis, Apatania fimbriata, Hydropsyche angustipennis, Mesophylax impunctatus*

Diptera —*Simulium latipes, Dixa maculata* (madicolous)

Hora (1928, 1930) describes dragonfly nymphs *(Ictinus, Zygonyx)* from the torrential streams of India that are limpet-like in form and live on rocks in rapid water. The majority of rocky stream Odonata, however, are members of primitive families of the Zygoptera (damselflies) (Corbet 1962). The dorsoventral flattening of the nymphs includes the three caudal lamellae that are appressed against the substrate surface. As with many insects of rapid streams, flattening among zygopterans often serves as a current avoidance mechanism since most of these species reside in crevices or the undersides of rocks.

The mayflies *Baetis, Ecdyonurus* and *Rhithrogena,* the glossosomatid caddisfly *Agapetus* and chironomids were the chief benthic organisms associated with loose stones in several Yorkshire streams (Percival and Whitehead 1929). On the upper surfaces of stones (without moss) in Scottish streams, Egglishaw (1969) found primarily chironomids, the microcaddisfly *Hydroptila,* the mayfly *Baetis,* the empidid fly *Clinocera,* and elmid beetles. On the bare upper surfaces of boulders in a Rocky Mountain stream, the most common insects were baetid mayflies, the caddisflies *Brachycentrus* and *Hydropsyche,* chironomids, *Simulium,* and *Deuterophlebia* (Ward 1975).

The microdistribution of benthic insects in stony streams partly reflects the mosaic distribution of plant detritus. Egglishaw (1964) found a large variation in the amount of detritus contained within benthic samples collected from different areas of a single riffle. There was a close correlation between the numbers of individuals in a given sample and the amount of detritus collected with the sample for some species of insects, whereas the distribution of other species was unrelated to detritus. Those that increased with increasing plant detritus were primarily detritivores. One species of carnivorous stonefly *(Isoperla grammatica)* exhibited a spatial pattern similar to that of the common detritivores, whereas the distribution of another *(Chloroperla torrentium)* was unrelated to the distribution of plant detritus. Only predators whose preferred prey species tend to concentrate in patches of detritus would be expected to exhibit a similar pattern. The abundance of *Simulium,* a filter-feeder, and *Hydroptila,* an algal-feeder, were also unrelated to the distribution of plant detritus. Based on data derived by varying the amount of detritus in colonization trays placed in the riffle, Egglishaw concluded that the insects were responding primarily to detritus and not the associated physical habitat conditions of depth and stone size. By substituting shredded rubber for detritus, he determined that plant detritus was not merely serving as shelter for the insects.

Other experimental manipulations of stream substrate suggest that accumulations of plant litter are treated as a single resource separate from the inorganic substrate. Reice (1980) found that some stream insects preferentially colonized terrestrial leaf litter (leaf packs) irrespective of the grain size of the underlying mineral substrate (gravel, pebble, or cobble). Rabeni and Minshall (1977) present evidence suggesting that much of the differential colonization of mineral substrate by lotic insects may actually relate to the detritus storage capacity rather than mineral particle size per se. This conclusion is supported by manipulative field experiments showing that when the source and quantity of sedimentary detritus are standardized, the abundance of detritivores was not affected by differences in mineral substrate composition over the range large gravel to small pebble (Culp et al. 1983).

Psephophilous Fauna

The insects associated with gravel substrate exhibit considerable overlap with the lithophilous species of pebble and cobble, and it is contentious whether they warrant a separate faunal category as designated in Table 7.4. Baker (1918) emphasized the faunal similarity between cobble and gravel substrates, and the differences between the insect fauna of gravel and stony substrates in the Scottish lochs studied by Smith et al. (1981) were largely obscured by variations in other factors, such as trophic status. Barton and Hynes (1978) lump the faunal data from gravel and cobble, at least in part because these substrate types are commonly intermingled rather than forming distinct habitat patches. Nonetheless, gravel substrates provide optimal conditions for some aquatic insects.

Gravel deposits of extensive area and considerable depth occur in certain lotic environments. It is in such reaches that hyporheic habitats are best developed. As already discussed, the hyporheic zone, a primary habitat for the psephophilous fauna of streams, is colonized by many lotic insects at least during their early instars, and provides a refuge from drought, floods, and other adverse surface conditions. While the surficial layers of gravel are relatively unstable, deeper layers are disturbed only by severe floods; if the interstices are relatively free of silt, gravel may provide suitable habitat conditions for aquatic insects at depths of several meters below the surface of the stream bed and for some distance laterally beneath the stream banks (Stanford and Ward 1988). Plecopterans with somewhat vermiform nymphal morphology (Leuctridae, Capniidae, Chlorperlidae), elmid beetle larvae, and certain chironomids and tipulids are commonly encountered deep within stream gravels. The New Zealand mayfly *Coloburiscus humeralis* is restricted to gravel bed streams (Wisely 1962). Larval *Pycnopsyche lepida* (Trichoptera) move from silt to gravel as they grow, presumably as a mechanism to partition habitat space with the congener *P. guttifera*, a species that remains in silty areas throughout larval life (Cummins 1964). Although young larvae of *Pycnopsyche scabripennis* respond only to organic substrates, mature larvae ready to enter prepupal aestivation select only 4–8-mm gravel in which to burrow (Mackay 1977).

Some burrowing mayflies characteristically occur in gravel substrate. Nymphs of *Ephemera danica* burrow more readily in gravel than finer substrates (Percival and Whitehead 1926); whereas early instar nymphs occur in both sand and gravel, larger nymphs prefer gravel (Tolkamp and Both 1978). Nymphs of the Nearctic species *E. simulans* most frequently selected gravel [−1 phi (Ø) units] in laboratory tests (Eriksen 1964). Although able to burrow in a variety of substrates, the relatively low interstitial oxygen levels in fine sediments limit this species to relatively coarse particles in field situations (Eriksen 1968). In contrast, *Hexagenia limbata*, an efficient burrower in fine sediment, is unable to burrow in gravel. *Hexagenia limbata* is tolerant of low oxygen and has large gills that are capable of creating a strong current through the U-shaped burrow.

Most insects collected from gravel in a Colorado mountain stream also occurred on cobble, the predominant substrate (Ward 1975). However, the chloroperlid stonefly *Triznaka signata* was absent or rare on all substrates except gravel, and *Attenella margarita*, a mayfly abundant in gravel, was not collected from cobble. Sprules (1947) frequently collected *Serratella deficiens*, another ephemerellid mayfly, from gravel in an Ontario stream, but did not encounter it on cobble or any other substrate.

An odonate (*Argia moesta?*) and an ephemeropteran (*Stenonema femoratum tripunctatum*) were characteristic inhabitants of gravel in Lake Texoma, Oklahoma-Texas, being found on no other substrate type (Sublette 1957). Both species typically occur in gravel bed streams.

Psammophilous Fauna

The psammophilous fauna considered here consists of aquatic insects that burrow in sand. They should not be confused with the psammon community (*sensu* Pennak 1968) comprised of micrometazoans that live in the interstitial spaces between sand grains.

Bare sand, because of its instability and low organic content, is generally characterized by an extremely poor macrofauna, a fact noted long ago by Wesenberg-Lund (1908). Although sand was the most extensive littoral substrate type in Lake Texoma, samples from shore areas of nearly pure sand were commonly devoid of macrobenthic organisms (Sublette 1957). The only insect collected was the sand-cased caddisfly *Oecetis inconspicua.* However, relatively diverse and abundant fauna occurred where detritus and silt were mixed with sand. Of 59 benthic taxa collected, 25 taxa, chiefly insects, were found only in such areas. The following insects were restricted to sand:

Ephemeroptera
Brachycercus lacustris, Callibaetis montanus?, *Siphlonurus* sp.
Odonata
Pantala hymenea, Enallagma civile?, *Erpetogomphus* sp., *Gomphus plagiatus*
Coleoptera
Dineutus sp., *Haliplus* sp., *Tropisternus* sp.
Megaloptera
Sialis sp.
Chironomidae
Calopsectra neoflavellus, C. bausei?, *Stenochironomus macateei*
Other Diptera
Stratiomyia sp., *Tabanus* sp., *Chrysops* sp.

Chironomids are by far the most abundant insects inhabiting areas of sandy shore in the St. Lawrence Great Lakes (Barton and Hynes 1978). Psammophilous chironomids included species of *Polypedilum*, *Paracladopelma*, *Saetheria*, *Robackia*, and *Monodiamesa*.

Specialized psammophilous chironomids dominated the benthos of a large primarily sand-bottom river, the Athabasca, in Canada (Barton and Lock 1979, Barton 1980). In the coarse sand (>0.3 mm) of the main channel, *Rheosmittia* sp., a member of the Orthocladiinae, constituted virtually the entire macrofauna. Larvae of this species are extremely small (final instars are <1.9 mm in length), and pass through the mesh commonly used in benthic studies. In fine and medium sand the fauna was dominated by Chironominae, including species of *Polypedilum*, *Cryptochironomus*, *Robackia*, and *Paracladopelma*, all of which are predaceous forms morphologically adapted for burrowing in shifting sand (Shadin 1956).

Shifting sand provides severe habitat conditions suitable for few benthic organisms. Many of the species or species groups of specialized psammophilous chironomids in the Athabasca River occur in sand bed rivers in Europe and Asia, as well as North America (Saether 1977). It is likely that *Rheosmittia* sp. has been previously overlooked because of its small size and will prove to be widespread in suitable habitats. In fact, subsequent investigations of another Canadian river demonstrated that *Rheosmittia* sp. and *Robackia demeijerei* constituted 80.6% of the biomass and 92.8% of the total macroinvertebrate density in the shifting sand dunes of the main channel (Soluk 1985). The predaceous mayfly *Pseudiron centralis* forages on the chironomids (Soluk and Clifford 1985).

Most sand-dwelling dragonfly nymphs only partially bury themselves in the substrate. They are characterized by short robust legs, pale coloration, and elongated abdomens that are concave ventrally and strongly convex dorsally (Corbet 1962). Certain gomphid nymphs, however, reside well below the substrate surface in sand-bottom streams. Their abdomens are cylindrical, facilitating rather than retarding sinking in sand, and the tenth abdominal segment is greatly elongated to form a respiratory siphon that projects to the substrate surface (Fig. 7.5). Hora (1928) found dragonfly nymphs with elongated respiratory siphons buried deeply in the sand deposited a short distance downstream

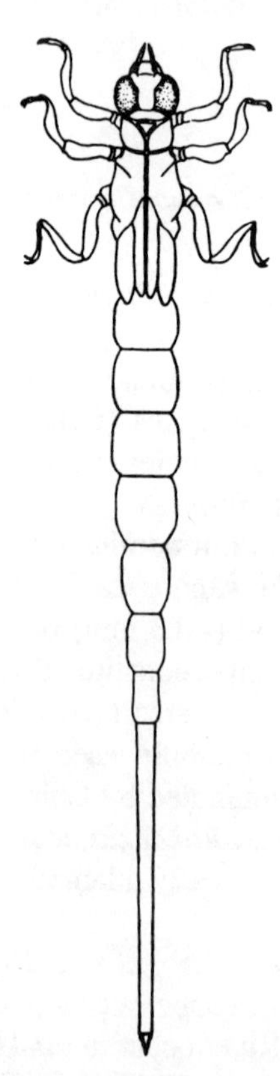

Figure 7.5 *Lestinogomphus africanus* (drawn from a nymphal exuviae), a deep-burrowing dragonfly with a long respiratory siphon. [From Corbet (1962).]

from the plunge pools below waterfalls in India. Similar deep-dwelling psammophilous dragonflies have since been reported from various parts of the world (Corbet 1962). Nymphs buried deep within the substrate would be less readily available as food for bottom-feeding fishes, which may at least partially explain the evolutionary significance of such habits.

Psammophilous trichopterans include *Molanna* and *Macronema*. The sand grain cases of *Molanna* have prominent lateral flanges and an anterior hood, making the larvae nearly invisible from above as they move slowly over the sandy substrate. *Molanna* typically occurs on sandy lake shores and in sand-bottomed rivers. They burrow into the substrate prior to pupation. Species of the hydropsychid genus *Macronema* typically inhabit large rivers (Wiggins 1977). Larvae of a species that occurs in sandy woodland streams of the Amazon River Basin construct a chimney-like intake structure of silk and sand grains that projects well above the substrate surface (Sattler 1963). This ensures that sand grains and debris rolling along the stream bed will not enter the feeding net.

A remarkable psammophilous mayfly fauna occurred in a portion of the Green River, a major tributary of the Upper Colorado River, that has since been inundated by Flaming Gorge Reservoir (Ward et al. 1986). Edmunds and Musser (1960, p. 122) described the preimpoundment mayfly fauna of the Green River as "one of the most unusual and interesting ones known to occur in any part of the world." The carnivorous siphlonurid *Analetris* lived in association with nymphs of the sand-dwelling dragonfly *Ophiogomphus intricatus* and fed on psammophilous chironomids. *Pseudiron* and *Raptoheptagenia*, unusual carnivorous heptageniids, were also present. *Ametropus albrighti* nymphs also burrowed in the sand bottom. Several other unusual riverine mayflies also occurred in this portion of the Green River, but were not associated with sand-bottom habitats.

Other specialized psammophilous ephemeropterans include *Behningia* of Eurasia and *Dolania* of southeastern North America. Nymphs of both genera are highly specialized for burrowing in riverine sand and have dense patches of hairs that apparently function to keep respiratory surfaces free of fine particles (Hynes 1970a). Some species of the unusual mayfly *Baetisca* also burrow in riverine sand. The greatly enlarged mesonotal shield of *B. rogersi* has prolonged lateral spines that apparently act as stabilizers for the nymphs when they are on the substrate surface as well as when burrowing in shifting sand (Berner 1940, Pescador and Peters 1974).

Perkins (1976) studied the microdistribution of adult semiaquatic psammophilous beetles in the sand banks along the edge of the San Gabriel River, California. A sampling system was devised whereby the position of each individual could be graphically presented (Fig. 7.6), based on horizontal distance from the waterline (*H*) and vertical distance above the water table (*V*). Each sample cell represents a section of stream bank 2.5 cm high (less for cells at the water's edge), 2.5 cm wide, and 33 cm long parallel to the waterline. The results of one set of samples from an area of shoreline consisting of homogeneous sand grains is shown in Figure 7.6. The hydrophilid *Laccobius* (desig-

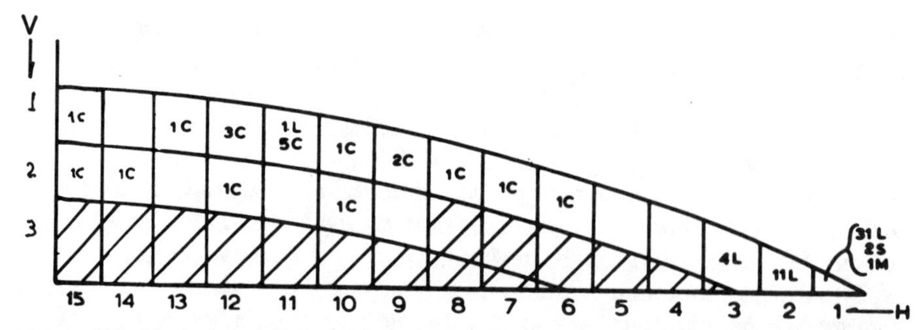

Figure 7.6 Vertical and horizontal distribution of psammophilous beetles in a section of shoreline consisting of homogeneous sand grains along the San Gabriel River, California. C = *Chaetarthria;* L = *Laccobius;* M = *Limnebius;* S = *Ochthebius.* See text for further details. [From Perkins (1976).]

nated "L") was concentrated in a zone at the immediate waterline, where it was accompanied by the hydraenids *Ochthebius* and *Limnebius*. The hydrophilid *Chaetarthria* exhibited a decided preference for the damp shore zone beginning about 13 cm from the water's edge. The genera *Laccobius*, *Ochthebius*, and *Chaetarthria* were each represented by at least four species in the study area. Perkins concluded that the microdistribution patterns of these psammophilous beetles are probably controlled largely by physical factors, such as bank slope, size of interstitial spaces, and substrate moisture content, rather than biotic interactions.

Pelophilous Fauna

Pelophilous insects are mud-dwellers, herpobenthic forms associated with sediments in which silt- and clay-sized particles predominate. Because fine organic particles have hydrological properties similar to those of silt (Inman 1949), food quantity is rarely a limiting factor in silt-bottom habitats. Oxygen deficits commonly occur in water in contact with highly organic sediments; even when the overlying water is well aerated, oxygen levels exhibit a rapid decline with depth in the sediment.

The herpobenthic entomofauna associated with the littoral mud of lakes is generally similar to that inhabiting soft lotic sediments. Chironomids, and chaoborids in some lentic situations, typically outnumber all other benthic insects in mud-bottom habitats devoid of higher plants. These two families of dipterans are virtually the only insects present in the profundal ooze of lakes, and they form the basis for lake typology schemes, as discussed in Chapter 3.

A few mayflies are adapted for residing in silty habitats. *Caenis* spp. and *Timpanoga hecuba,* although not closely related, both have dorsally positioned gills and a pair of sclerotized opercular gills that cover the remaining gills and protect them from silt deposition. The nymphs of *T. hecuba* are highly flattened, apparently as an adaptation for living on soft substrates. *Caenis* is unusual in that the gills beat out of phase with the other member of that gill pair (Eastham 1934). This results in a current that pulses back and forth across the dorsum of the nymph, thereby reducing the tendency of the respiratory current to suspend silt particles. Some mayflies that construct U-shaped burrows inhabit mud bottoms. For example, *Hexagenia* spp. are well adapted for burrowing in mud (Lyman 1956), and are more tolerant of the low-oxygen conditions that prevail in fine sediments than species such as *Ephemera simulans* that inhabit coarser substrates (Eriksen 1964).

The megalopteran *Sialis* is commonly reported from soft-bottom habitats of lentic and lotic waters. *Sialis* larvae and *Caenis* nymphs may penetrate muddy bottoms to some depth (Fig. 3.7), unlike most nondipterous forms, which are restricted to quite shallow water.

A very few species of stoneflies, the nymphs of which are distinctly pubescent, occur in silty habitats. *Leuctra nigra*, for example, occurs only in silty streams (Hynes 1941).

Many species of Odonata burrow in fine sediments. Dragonfly nymphs typically lie buried in silt with only the eyes and respiratory aperture above the sediment surface as they wait in ambush for prey while concealed from their own predators. In contrast to sand dwellers, pelophilous odonate nymphs are dark-colored and have longer legs (Corbet 1962). The abdomen is much broader and often contains long, curved lateral spines, all of which reduce the tendency of the nymphs to sink passively into fine sediments. Mud-dwellers do not possess respiratory siphons, but the terminal tergites are formed such that the rectal aperture faces upward rather than backward.

A very few odonate nymphs appear able to alter not only their color but the shape of the abdomen based on the sediment of the habitat (Corbet 1957b). Such is the case for *Brachythemis leucosticta* that occur on mud or sand substrates in habitats ranging from stagnant pools to rapid rivers. Nymphs residing on sand exhibit pale coloration and have elongated abdomens with small spines that are directed posteriorly. In contrast, those inhabiting mud are darker, have dorsoventrally flattened abdomens with large spines that are directed laterally, and rectal apertures that are in a more dorsal position. Nymphs on mixed substrates exhibit intermediate characteristics. If these morphotypes encompass a single species, as appears to be the case, one can envision numerous experimental manipulations to investigate the ecological implications and underlying mechanisms of such remarkable plasticity.

Xylophilous Fauna

Xylophilous aquatic insects typically occur on or within submerged wood. Many species collected from wood also occur on other substrates and may or may not show a preference for submerged wood (Marlier 1954, Nilsen and Larimore 1973, Anderson et al. 1978, Dudley and Anderson 1982, Winterbourn 1982, Harmon et al. 1986). Wood provides food, living space, concealment, oviposition and attachment substrate, refuge from predators, protection from adverse abiotic conditions, and emergence sites for aquatic insects (Fig. 7.7).

In a survey of invertebrates associated with wood debris in aquatic habitats of the United States, 56 taxa were identified as "closely associated" with wood, and an additional 129 taxa were designated "facultative associates" (Dudley and Anderson 1982). Most of the taxa in both of these categories are insects. Wood debris is most abundant in headwater streams of forested regions, and insects restricted to submerged wood habitats are most prevalent in such situations. The greatest diversity of xylophilous insects was found in mesic forest streams on the western slope of the Cascade Range. No aquatic insects of the lentic waters surveyed by Dudley and Anderson were restricted to wood debris, although several species used submerged wood as a preferred substrate. Xylophilous insects were identified from five orders.

Among the ephemeropterans only *Cinygma*, a heptageniid, is designated as "closely associated" with wood by Dudley and Anderson. Chessman (1986) recorded substantial amounts of wood fragments in the guts of some leptophlebiid mayflies from Australian streams.

Several Nearctic plecopterans (e.g., *Yoraperla, Pteronarcys*) ingest wood and many species use wood as a substrate, but all associations are considered facultative (Dudley and Anderson 1982). At least three Southern Hemisphere stonefly families have xylophagous representatives (Anderson 1982, Chessman 1986).

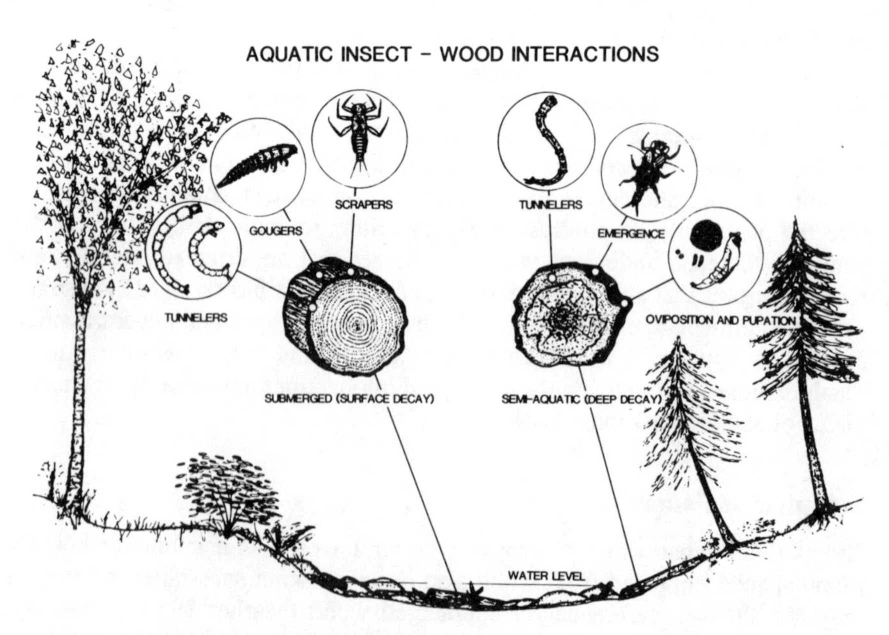

Figure 7.7 A model of aquatic insect–wood interactions in a small stream. [From Anderson et al. (1984).]

Trichopterans from several families are xylophilous according to Dudley and Anderson. Numerous species that feed on leaf litter are known to ingest wood also. Those that use wood to construct cases include species from the families Brachycentridae, Limnephilidae, and Lepidostomatidae. Hydropsychids and other filter-feeders may attach their nets to submerged wood. In soft-bottomed riverine habitats submerged wood may provide the only suitable substrate for net-spinning caddisflies or any other haptobenthic insects (Shadin 1956, Benke et al. 1984). Several species of *Lepidostoma* pupate in wood crevices. *Heteroplectron californicum,* a xylophagus North American caddisfly, hollows out small pieces of wood to form portable cases (Anderson et al. 1978), as does *Triplectides obsoleta* in New Zealand (Anderson 1982).

Among xylophilous coleopterans, the elmid *Lara avara* was most commonly encountered by Dudley and Anderson. Adults frequently occur on wood just above the waterline (Anderson et al. 1978). Larvae gouge superficial channels in wood or reside in crevices or under loose bark. Larvae from beetle families considered wholly terrestrial were found by Dudley and Anderson boring in submerged wood.

Dipterans associated with wood include aquatic and semiaquatic forms as well as a few species from families regarded as wholly terrestrial. Many dipteran larvae tunnel in wood. The chironomid *Brillia* is xylophagus and is one of the first insects to bore into new wood when it first enters a stream. Dipterans such as the tipulid *Lipsothrix* that occur in partially submerged logs do not encounter oxygen gradients such as occur from the surface to the center of fully submerged logs. Truly aquatic dipterans that tunnel in wood may exhibit respiratory adaptations. *Stenochironomus* possesses hemoglobin; *Axymyia* and syrphids have extensible respiratory siphons.

In first- and second-order streams of Oregon's Cascade Range, about 25% of the stream bed is composed of wood and another 25% is wood-created habitat (Anderson and Sedell 1979). The feeding activities of wood-inhabiting insects contribute to the decomposition of wood in aquatic habitats. Species that feed by shredding coarse organic particles frequently ingest wood. Insects feeding on the aufwuchs associated with wood substrate abrade the superficial layers of wood by their scraping and rasping activities and ingest some wood tissue in the process. Newly exposed wood surfaces are rapidly colonized by the microorganisms that are primarily responsible for decomposition. Fine particles of wood are utilized by collectors. The boring activities of dipterans and other tunneling insects contribute to the breakdown of wood directly, and by exposing new wood surfaces and transporting inocula, further enhance microbial decomposition.

The inundation of woodlands by impounding river valleys to form reservoirs provides a large substrate for haptobenthic insects in the form of submerged trees. McLachlan (1970) calculated that submerged trees provided at least twice as much surface area as the lake bottom in Lake Kariba, a large reservoir in central Africa.

In temperate zone reservoirs chironomids are the major animal colonizers of submerged trees (Luferov 1963, Cowell and Hudson 1967, Claflin 1968). Cowell and Hudson (1967) recorded densities of chironomids on submerged trees that were 11 times greater than on the bottom substrate.

In tropical waters, some burrowing mayflies in the family Polymitarcidae bore into soft wood (Illies 1968). *Asthenopus* nymphs, for example, construct U-shaped dwelling tubes in wood in Amazon waters. *Povilla adusta* of tropical Africa is the best known of the wood-boring mayflies. In natural lakes *Povilla* nymphs occur mainly in the dead stems and rhizomes of *Papyrus*, an emergent reed of the littoral zone (Petr 1973). In artificial lakes with drowned trees, *Povilla* may become extremely abundant. Nymphs burrow into the bark of the submerged trees, although wood is not used as a food source. The nymphs remain in their burrows during the day and feed on plankton by filtering the stream of water created by gill movements (Hartland-Rowe 1953). Burrowing behavior apparently evolved as a predator-avoidance mechanism. At night nymphs leave their burrows and feed on the aufwuchs associated with wood surfaces (Petr 1970b).

Povilla adusta nymphs are only capable of constructing burrows in tree species with soft wood. The filling of Lake Kariba inundated hardwood trees that *Povilla* was unable to penetrate. The bark surfaces of these trees was colonized primarily by chironomids. However, the dead trees were heavily attacked by terrestrial wood-boring beetles when the lake level dropped during the drawdown phase (McLachlan 1974). Subsequent inundation of the dead, but standing trees resulted in a quadrupling of faunal biomass as species not previously present invaded the new habitat created by the terrestrial wood borers (Fig. 7.8). In addition to *Povilla* and the trichopteran *Amphipsyche senegalensis*, new chironomid species also appeared in the resubmerged trees.

Phytophilous Fauna

Phytophilous insects, as used here, include those species associated with living aquatic macrophytes (Harper 1986). The association may be intimate and obligatory as in the case of leaf miners, or casual and facultative as when vegetation serves merely as the preferred substrate. Different species of aquatic insects depend on macrophytes for food, shelter, and habitat space. In addition, living plants provide a substrate for epiphytic algae, a primary food source for many phytophilous species.

Moss Fauna Submerged mosses typically attain maximum development in small stony streams, and most investigations of the insects inhabiting moss have been conducted in such habitats (e.g., Thienemann 1912, Carpenter 1927, Hubault 1927, Percival and Whitehead 1929, Frost 1942, Hynes 1961b, Egglishaw 1969, Glime and Clemons 1972, Maurer and Brusven 1983, Suren 1988). Many aquatic insects find suitable microhabitat conditions within tufts of moss. Such species, while they also occur on other substrates, reach maximum densities in moss (Brusven et al. 1990). However, only a relatively few species of aquatic

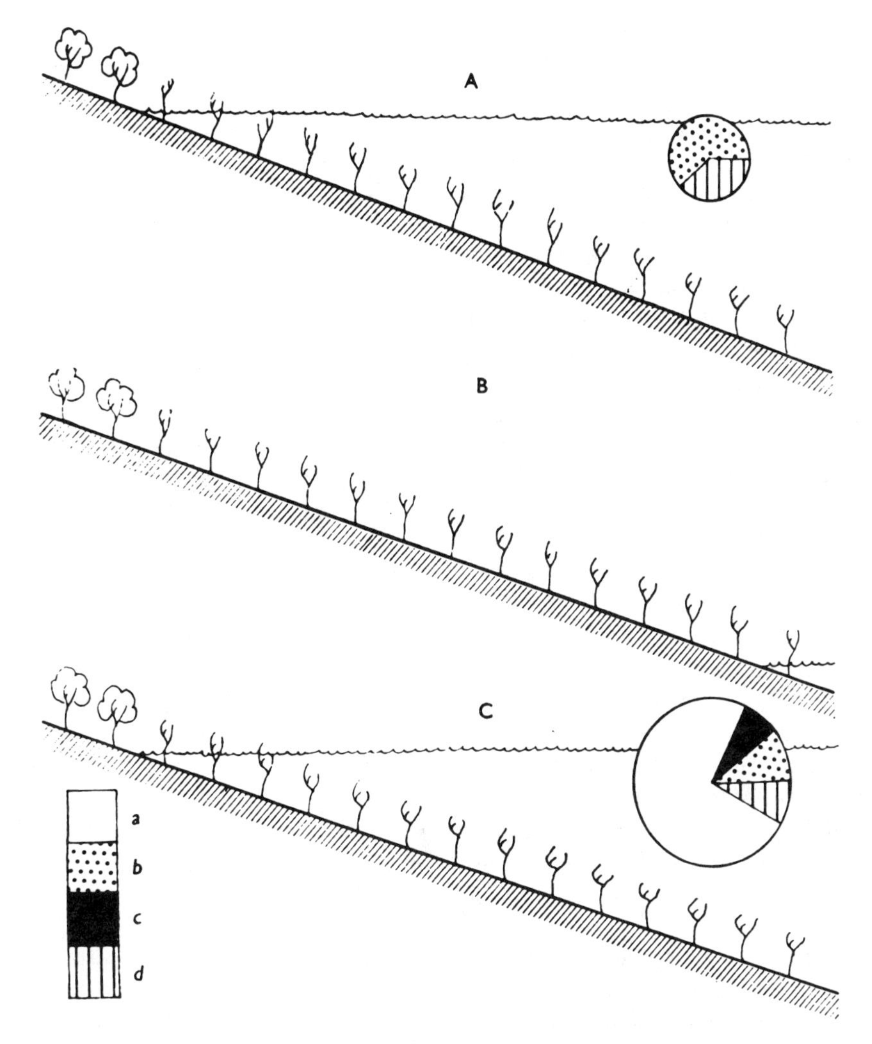

Figure 7.8 Xylophilous invertebrates associated with submerged woodland during the filling and postfilling phases of a tropical impoundment: (*A*) end of filling phase; (*B*) forest exposed during drawdown; (*C*) forest reflooded. Area of circles indicates the relative biomass of total aquatic invertebrates (based on dry weight per square meter of tree surface). Relative contributions of major components are symbolized as follows: (*a*) Ephemeroptera (*Povilla adusta*); (*b*) Chironomidae; (*c*) Trichoptera; (*d*) Oligochaeta. [From McLachlan (1974), Biological Reviews. Reprinted by permission of Cambridge University Press.]

insects exhibit a high selectivity for mosses, and it is questionable whether any species is restricted to submerged moss to the exclusion of other substrates. Some common members of the stony stream fauna, such as the flattened heptageniid mayflies, are rarely collected from mosses and may be eliminated from areas where moss covers much of the available substrate.

Chironomids are frequently the most abundant and diverse faunal component in moss. Percival and Whitehead (1929) found that larvae of the *Orthocladius* group resided in gelatinous cases attached to the leaves and stalks of the moss. Other dipterans frequently associated with moss include psychodids (especially *Pericoma*), tipulids, and empidids. Simuliid larvae may occur in large numbers on moss, using it as an attachment site for filter feeding (Niesiolowski 1980).

Among stoneflies, only nemourids are consistently reported as being abundant in moss. *Amphinemura sulcicollis,* for example, was the commonest species of macroinvertebrate collected from moss in streams in Scotland (Egglishaw 1969) and Wales (Hynes 1961b). Some stoneflies and other aquatic insects that are normally considered part of the lithophilous association commonly occur in moss as early instars, tending to move to the surrounding stones as they become larger (Thienemann 1912, Carpenter 1927, Egglishaw 1969).

The mayflies *Baetis* and *Ephemerella* are commonly reported from moss and may, in some cases, be more abundant in moss tufts than on the stony stream bed. The common European ephemerellid, *E. ignita,* is a moss-dwelling species (Hynes 1970a).

Caddisflies are frequently reported from submerged moss, but only rarely are they more abundant there than on the surrounding substrate. Hydropsychids sometimes construct their nets and larval retreats within mosses and may occur at high densities under certain conditions. Microcaddisflies (Hydroptilidae) are frequently encountered in moss. In an English stream *Hydroptila* was common in moss, although larvae were more abundant on *Cladophora* covered stones (Percival and Whitehead 1929). In the same stream, another hydroptilid, *Ithytrichia lamellaris,* is a typical moss-dweller, only rarely occurring on any other substrate. Yet another caddisfly, the brachycentrid *Crunoecia irrorata,* attained maximum abundance on moss. Most of the trichopterans found in moss leave this habitat as mature larvae and pupate under stones (Percival and Whitehead 1929).

Aquatic beetles from various families have been reported from moss. A few species of elmid beetles are most abundant on submerged mosses and are decidely members of the moss fauna. *Limnius tuberculatus* was one of the commonest species of invertebrates on moss in a stream in Scotland (Egglishaw 1969).

Although few members of the moss fauna exhibit special morphological adaptations, some possess backwardly directed projections, the presumed function of which is to reduce the likelihood of being swept from their preferred microhabitat. The South American moss-dwelling stonefly nymph illustrated in Figure 7.9 has especially long dorsal spines. *Ephemerella ignita,* a common moss-dwelling mayfly, also possesses dorsal abdominal spines (Hynes 1970a). Among the moss-inhabiting insects in torrential streams of India, Hora (1930) described backwardly directed projections for species of stoneflies, mayflies, and dipterans in the families Psychodidae and Tipulidae. Larval *Phalacrocera* and *Triogma,* tipulids that reside in aquatic or semiaquatic mosses, have long fleshy projections on the thorax and abdomen.

Figure 7.9 A moss-dwelling stonefly (Gripopterygidae) from Chile showing an extreme development of dorsal spines. [From Illies (1961b).]

Mosses provide physical habitat, attachment sites, and a concentrated food source in the form of attached epiphytes and accumulated detritus. It is contentious whether the living mosses themselves serve as an important food source for aquatic insects. Glime and Clemons (1972) found that the community structure and overall species diversity of insects were similar among natural moss (*Fontinalis* spp.) and artificial mosses constructed from string and plastic strips. The higher densities of insects on *Fontinalis* is attributed to its more complex habitat structure and greater surface area compared to string or plastic. Glime and Clemons conclude that "moss may in fact be only a physical surface for the insects...."

A somewhat different set of habitat conditions is found in the carpet of emergent mosses that occur in the source areas of certain springs where the current is slow, and temperature and flow are relatively constant (Lindegaard et al. 1975, Ward and Dufford 1979). In the Danish spring examined by Lindegaard et al., the insects were associated with the vertical zones within the moss carpet (Fig. 7.10). The dry zone is inhabited by terrestrial species including collembolans and coleopterans. Most of the species in the madicolous zone, the narrow stratum wetted by capillary water, are listed as members of the "faune madicole" by Vaillant (1956). This ecotone between the terrestrial and aquatic environments is inhabited by trichopterans *(Crunoecia irrorata, Beraea maurus)*, beetles *(Helodes minuta)*, and dipterans from several families (Psychodidae, Dixidae, Chironomidae, Ceratopogonidae, and Thaumaleidae). The chironomid *Chaetocladius laminatus* occurred in this zone at densities exceeding 1000 m^{-2}. The fauna of the water-covered moss zone was the richest both in species and individuals. Species exceeding 100 m^{-2} include an ephemeropteran (*Baetis rhodani*), two plecopterans (*Protonemura hrabei, Nemurella picteti*), and dipterans from several families (Tipulidae, Psychodidae, Chironomidae, Ceratopogonidae, Simuliidae). Most species of this zone are rheophilous, although they tend to avoid rapid current and seek refuge in vegetation. *Nemurella* and *Pericoma* possess long bristles thought to reduce the chances of their being dislodged from vegetation by the current. A few species were associated with the accumulation of detritus on the tops of the stones to

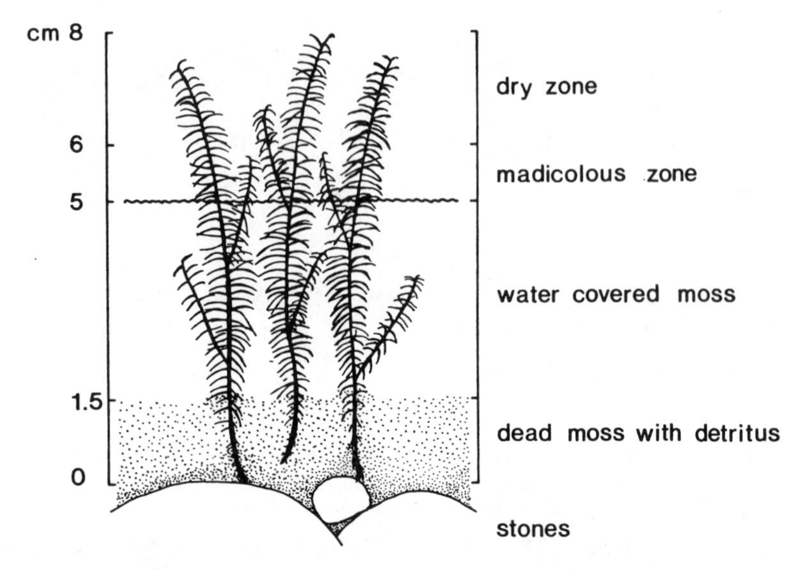

Figure 7.10 Vertical zonation of the moss carpet in the Danish spring Ravnkilde. [From Lindegaard et al. (1975).]

which the mosses are attached. Most of these were dipterans, although the plecopteran *Leuctra hippopus* also occurred in this lowermost zone.

In emergent moss at the source of a springbrook in Colorado (USA), Ward and Dufford (1979) identified 27 taxa of aquatic insects, most of which were dipterans (16) or coleopterans (7). The most abundant organisms were the dipterans *Euparyphus* (Stratiomyidae) and *Limonia* (Tipulidae), and the trichopteran *Hydroptila. Limonia* and *Pericoma* (Psychodidae) were not collected from any of several other substrates examined in this year-round study.

Insects were the numerically predominant faunal component on submerged *Fontinalis* in a small humic Danish lake (Berg and Petersen 1956). The mayfly *Leptophlebia* was by far the most abundant insect, and along with the caddisflies *Holocentropus* and *Cyrnus*, a dragonfly *(Cordulia)*, and chironomids, comprised the great majority of the entomofauna on moss.

Insects of Other Macrophytes Macrophytes other than mosses with which aquatic insects may be associated are primarily angiosperms, but also include charophytes (stoneworts) and some lower tracheophytes, such as aquatic ferns.

Soszka (1975a) used laboratory and field experiments in concert with field observations to examine several ways that phytophilous insects utilized submerged angiosperms in a Polish lake (Table 7.7). Only the lepidopteran larvae depended on the macrophytes for food, although some other species ingested living vascular plant tissue. Most insects fed on the epiphytic algae and detritus associated with the plant surfaces and readily colonized artifical (plastic) vegetation that had accumulated an organic film. Only lepidopteran larvae failed to

Table 7.7 Type and intensity of use of submerged macrophytes by aquatic insects in Mikolajskie Lake (A) and the extent to which the different plant species are used by the invertebrate community (B)

	Type and Intensity of Use[a]					
	Food	Oviposition Site	Overwintering Site	Mining	Case Material	Substrate
A. Insects						
Lepidoptera[b]	+++	++	+	+	++	+++
Phryganea grandis	+++	++	++		+++	++
Limnephilus sp.	+++	++	+++		+++	++
Chironomidae[c]	++		+	+++		+++
Cloeon dipterum	+		++			+
Hydroptilidae			+			+++
Mystacides sp.			+			+
Enallagma cyathigerum			+++			+
Caenis sp.			+			+
B. Macrophytes						
Potamogeton lucens	+++	+++		+++	++	++
P. perfoliatus	+++	+++		+++	++	++
Elodea canadensis	+	+	+++	+	++	++
Myriophyllum spicatum	+	+	+	+	++	++

[a]Intensity of use: +++ very strong, ++ medium, + weak.
[b]*Paraponyx stratiotata* and *Acentropus niveus* combined.
[c]Nonpredatory species.
Source: Modified from Soszka (1975a).

colonize the artificial plants. *Elodea* was most heavily utilized by invertebrates during the winter, when the other plant species had largely died back. The pondweeds (*Potamogeton* spp.) were generally utilized by aquatic insects to a greater extent than the other submerged angiosperms.

Insects associated with *Potamogeton*, the most diverse genus of freshwater angiosperms, have received intensive study in both Europe and North America. Müller-Liebenau (1956) identified 112 species of insects associated with five species of *Potamogeton* in the lakes of eastern Holstein, including 52 chironomids and 24 trichopterans. Berg (1949) listed 32 species of insects as directly related to one or more of the 17 species of *Potamogeton* occurring in Michigan (USA), water bodies. Only species that fed directly on *Potamogeton*, lived within the tissues, or tapped the plant's air stores were considered. In addition, Berg identified 10 species of parasitic Hymenoptera that were reared from the insects directly related to *Potamogeton*.

The composition and distribution of the phytophilous fauna of a given water body is, of course, partly a function of the composition and distribution of macrophytes. In Hodson's Tarn, England, Macan (1965) delineated two groups of phytophilous species: those associated with plants occurring in deeper portions of the tarn (mainly *Potamogeton natans* and *Myriophyllum alternaeflorum*) and those occurring on plants growing in the shallow margins (mainly *Carex rostrata* and *Littorella uniflora*). Some of the abundant species such as *Leptophlebia* (Ephemeroptera), *Pyrrhosoma* (Odonata), and *Corixa castanea* (Hemiptera) were largely restricted to the vegetation in shallow water. However, those species abundant in deeper water also occurred in low numbers in marginal vegetation. The only exception was the lepidopteran N*ymphula,* the only species restricted to one kind of plant, which occurred only in the deeper portions of the tarn in association with *Potamogeton.*

The general life form of hydrophytes may determine their suitability for particular species of insects. For example, plants lacking structures that project above the water surface are unsuitable as oviposition substrates for certain insects, and this alone may account for the absence of some phytophilous species from submerged angiosperms. Insects of submerged vegetation generally exhibited less specificity for a particular plant species than insects associated with floating or emergent angiosperms (Berg 1949, McGaha 1952). However, Krull (1970), who examined the fauna associated with 12 species of submerged macrophytes, found that nearly 60% of the invertebrate taxa occurred on three or fewer plant species. Thirty-three taxa were associated with only one species of macrophyte, this despite considering chironomids as a single taxon.

In an attempt to examine the abundance of phytophilous animals on different plants without the confounding effects of environmental variables such as depth and wave action, Krecker (1939) sampled seven species of submerged angiosperms (Fig. 7.11) in areas where they were intermingled in common stands. There was a 52-fold difference between the numerical abundance of insects on the least sparsely populated and the most densely populated plant species (Table 7.8). *Vallisneria,* the plant very sparsely populated by insects, has smooth linear leaves that afford minimal spatial heterogeneity, whereas *Myriophyllum* leaves are finely divided and provide a large surface area (Fig. 7.11). Krecker attributed the relatively high faunal densities associated with *P. crispus* to the crenulated leaves characteristic of that species.

Harrod (1964) designated the following factors as major determinants of the suitability of macrophytes for phytophilous stream invertebrates: (1) the morphological form of the plant, (2) the position of the plant in the stream, (3) the epiphytes present on plant surfaces, and (4) the chemical nature of the plant. In lakes, at least, another factor, habitat permanence, may also be important (Hargeby 1990).

Plants with finely divided leaves or those that grow in dense clumps provide protection and shelter for a variety of animals (Rooke 1986). In contrast, Harrod found that *Carex,* a plant giving little protection from the current, was unsuitable for most animals although it served as an attachment surface for large numbers of *Simulium* larvae. Plants that grow in rapid water are colonized by a different assemblage of insects than those occurring in pools or quiet margins, and plants forming dense mats on the substrate contain different faunal assemblages than

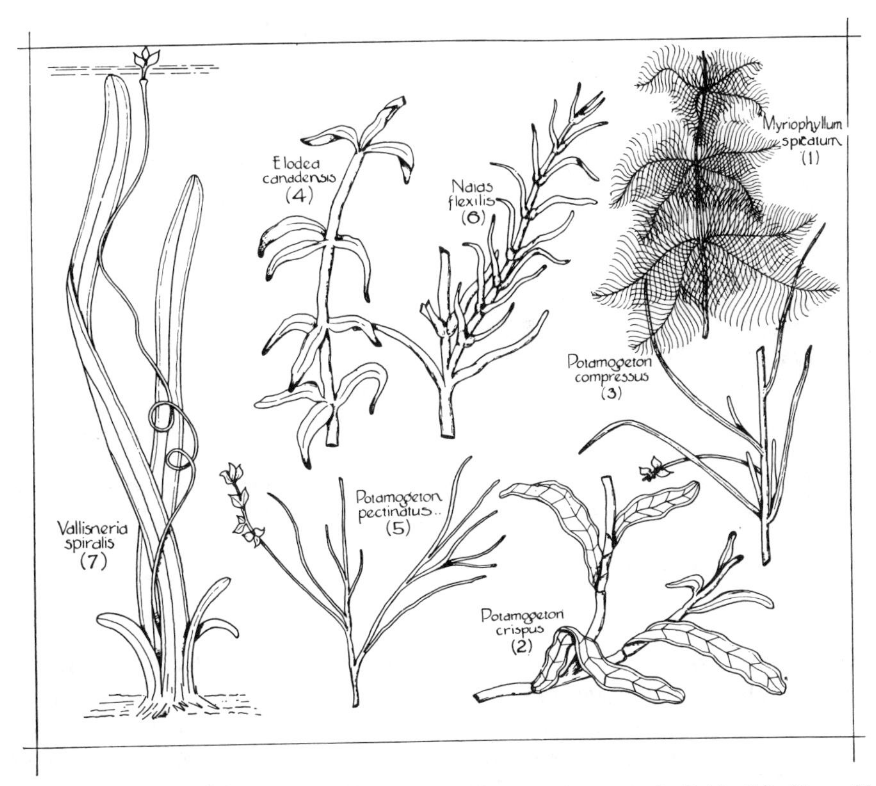

Figure 7.11 Morphological forms of the submerged angiosperms in Table 7.8. (From "A comparative study of the animal population of certain submerged aquatic plants" by F. H. Krecker, Ecology, 1939, 20, 553–562. Copyright © by The Ecological Society of America. Reprinted by permission.)

plants with a more erect growth form. Insects that graze on the periphyton growing on plant surfaces are undoubtedly influenced by the density and composition of epiphytes (Dudley 1988). However, while there may be considerable differences in epiphyte populations on different parts of the same plant, or on plants of different ages or physiological states, epiphytes do not exhibit a great deal of specificity for particular species of macrophytes (Hutchinson 1975, Fontaine and Nigh 1983). The chemical composition of aquatic plants does differ between species, sometimes strikingly so, making them more or less palatable to phytophagous insects. The stoneworts (Charophyta), for example, are aptly named since $CaCO_3$ may constitute over half of the dry weight of some species. Chemical defenses against herbivory are, however, apparently much less developed by aquatic plants than terrestrial species (Hutchinson 1975, but see Otto 1983).

In a newly created tropical impoundment, the appearance of macrophytes was an important stage in the development of the aquatic fauna (McLachlan 1969).

Table 7.8 Mean numbers of aquatic insects per sample unit[a] on the seven species of submerged angiosperms in Figure 7.11

	Plant Species[b]						
	V.s.	P.co.	E.c.	N.f.	P.p.	P.cr.	M.s.
Chironomidae	8	49	70	88	210	312	360
Anisoptera	0	0	1	0	0	1	1
Zygoptera	0	4	15	11	6	16	64
Ephemeroptera	0	0	1	21	3	3	63
Coleoptera	0	0	1	1	0	5	0
Hydroptila	2	2	1	2	9	16	26
Leptoceridae	0	0	1	0	0	3	4
Total	10	55	90	123	228	356	518

[a]Ten linear feet (305 cm) of plant.

[b]Key: V.s. = *Vallisneria spiralis*; P.co. = *Potamogeton compressus*; E.c. = *Elodea canadensis*; N.f. = *Naias flexilis*; P.p. = *Potamogeton pectinatus*; P.cr. = *Potamogeton crispus*; M.s. = *Myriophyllum spicatum*.

Source: Modified from Krecker (1939).

Even the herpobenthic fauna underwent changes that McLachlan attributed to the macrophytes. Following the invasion of rooted angiosperms (*Potamogeton* and *Ludwigia*), the number of mud-dwelling insect species increased from 25 to 39. However, there was a reduction or elimination of herpobenthos under floating mats of the fern *Salvinia.* This is exemplified by the following biomass data (mg dry weight m^{-2}) for the mud fauna (exclusively chironomids) at two sampling locations (A and B) in Lake Kariba with dense floating mats of *Salvinia:*

	A	B
100 m into mat	0	0
At edge of mat	30	45
100 m off edge of mat	180	390

Although McLachlan did not present an explanation for the absence of herpobenthic forms below mats of *Salvinia,* he did indicate that the floating plants were densely packed forming a nearly impenetrable layer. It is likely that the water below the mats is devoid of oxygen and that extended periods of anaerobic conditions account for the elimination of the mud fauna.

The best data on phytophilous insects in Lake Kariba are available for three species of macrophytes: *Potamogeton pusillus,* with a submerged growth form; *L. stolonifera,* a rooted species with floating leaves; and *S. auriculata,* a free-floating species. Mean faunal biomass (mg dry weight per 100 g dry vegetation) was 33 mg on the submerged species, 327 mg on the rooted plant with floating leaves, and 438 mg on the free-floating plants (McLachlan 1969).

Chironomids exhibited highest relative abundance on the submerged plant, where they constituted virtually the entire fauna. Chironomids were poorly represented on *Salvinia,* the free-floating plant; only 3 species occurred on *Salvinia,* whereas *Ludwigia* and *Potamogeton* contained 9 and 16 species, respectively. In contrast, Odonata were abundant on *Salvinia,* but rarely occurred on *Potamogeton.* Plants lacking structures that project above the water surface may not provide suitable emergence or oviposition substrates for Odonata, which may account for the paucity of this order on submerged plants.

After the invasion of Lake Kariba by macrophytes there was a several-fold increase in the total biomass of macroinvertebrates in areas where aquatic plants occurred (Table 7.9). In the case of *Ludwigia,* the increase included higher densities of the mud fauna in plant beds as well as the addition of phytophilous forms. Because *Salvinia* tends to colonize areas that had a less dense mud fauna, the increase in phytophilous forms more than compensates for the virtual elimination of the herpobenthos (Table 7.9).

Rzóska (1974) reported remarkable densities of insects on the free-floating angiosperm *Pistia stratiotes.* In a single plant 15 cm in diameter collected from a lagoon on the Upper Nile River system there were over 300 animals, including 166 dytiscid beetles, 66 corixid bugs, and 52 mayfly nymphs. Nineteen species of the beetle genus *Hydrovatus* are known to inhabit *Pistia,* with as many as 10 species occurring on a single plant. It is not known what mechanisms enable these congeneric species to coexist in so small an area.

Nymphs of phytophilous Odonata are characterized by a well-developed thigmotactic or clasping response (Corbet 1962). Among plant-dwelling odonates there is a tendency for a larger number of ommatidia, compared to bottom-dwellers, apparently as a means to extend their range of prey perception. Phytophilous species have also developed various mechanisms of protective coloration. For example, *Aeshna grandis* nymphs are green in summer and brown during the winter (Wesenberg-Lund 1913). Odonates that have moved away from the bottom to reside in vegetation near the water surface include some of the libellulids, aeshnids, coenagrionids, and lestids, many of which lay their eggs in plants near the surface. By remaining in plant beds near the water surface, growing nymphs are exposed to higher temperatures than are bottom-

Table 7.9 Mean areal biomass (mg dry weight m^{-2}) of fauna in vegetation and in mud associated with vegetation before and after the invasion of Lake Kariba by macrophytes

Macrophyte	Mud Fauna		Plant Fauna	Percent Increase[a]
	Before	After		
Ludwigia stolonifera	655	2273	1395	560
Salvinia auriculata	106	1	734	693

[a]Mud fauna (after) + plant fauna/mud fauna (before) x 100. McLachlan (1969) gives different percent increase values in his Table 7.

Source: Modified from McLachlan (1969).

dwelling species. The high growth potential thus attained has preadapted some phytophilous species to conditions in temporary ponds (Corbet 1962).

STANDING CROP AND DIVERSITY

The preceding section on insect-substrate relationships deals primarily with the composition of the entomofauna on different substrate types. This section presents a comparative analysis of the density, biomass, and diversity of insects on various substrates.

Diversity Comparisons of insect diversity as a function of substrate type (Table 7.10), while of considerable interest, should be interpreted cautiously. The results of such studies are influenced by sampling methods, level of taxonomic resolution, organic content of the substrate, proximity to other substrate types, stability, embeddedness, and numerous other variables.

The relatively large number of insect taxa reported for cobble relates to the stability and microhabitat diversity of this substrate type. While some substrates are relatively homogeneous, cobble is often actually a composite of many particle sizes because of the presence of finer materials beneath and between the stones. Cobble provides surfaces that are suitable for attachment by mosses and algae and has interstices that entrap organic detritus. Cobble thus contains abundant and diverse food resources for insects. In the small forested stream studied by Mackay and Kalff (1969), however, fewer species of insects were associated with stones than with either gravel or leaf litter. The authors attribute the differences between their findings and the results of most other studies to several factors. The general paucity of stony areas in the Quebec stream may be a major factor accounting for the poor development of a lithophilous fauna. In addition, the surfaces of the stones contained neither attached mosses nor algae, apparently because of the low incident radiation reaching the water of this well-canopied stream. The fact that leaf litter contained the most diverse assemblage of aquatic insects (92 species) should not be surprising given the predominance of this substrate type. Mackay and Kalff's findings are probably applicable to many other small, well-shaded deciduous forest streams.

The diversity and abundance of insects on substrates also varies over time. Substrate associations are less clearly manifest during periods of high flow when faunal spillover onto adjacent substrate types may occur, a phenomenon that likely accounts for the similar values on sand, gravel, and cobble for the Rocky Mountain stream listed in Table 7.10. However, a certain current velocity is necessary to maintain the integrity of substrate types. During periods when the sediments of a Canadian river were well sorted, the number of macrobenthic species was directly proportional to the mean particle size (de March 1976). At lesser velocities, fine sediment was deposited in the substrate interstices and the particle size-species diversity relationship became less distinct. Allan (1975) established a positive relationship between insect diversity and substrate hetero-

Table 7.10 Number of insect taxa associated with various substrates

	Number of Insect Taxa								
Habitat	Mud	Sand	Gravel	Cobble	Boulder	Moss	Angio.	Leaves	Ref.[a]
Lake littoral	10	31	37	39	—	—	27	—	(1)
Ontario stream	35	30	38	54	—	—	—	—	(2)
Quebec stream	—	61	82	76[b]	—	—	—	92	(3)
Rocky Mtn. stream	—	34	34	35	18	—	—	—	(4)
Spring source	—	19	—	26	—	27	17	—	(5)

[a]References: (1) Muttkowski (1918); (2) Sprules (1947) (based on Plecoptera, Trichoptera, and Ephemeroptera emergence only); (3) Mackay and Kalff (1969); (4) Ward (1975); (5) Ward and Dufford (1979).
[b]Cobble and pebble.

geneity by experimentally manipulating the particle size composition in colonization trays, but was unable to detect a similar relationship for the natural substrate.

When median size of particles was held constant, however, mineral substrate heterogeneity did not influence the number of invertebrate taxa colonizing experimental trays (Erman and Erman 1984). Variations in median particle size, current, and sedimentary detritus accounted for differences in the number of taxa.

Standing Crop Different substrates not only harbor different assemblages of aquatic insects, but the density and biomass of the fauna may vary over several orders of magnitude. In a study of the fauna on numerous substrate types in Michigan (USA) streams, Tarzwell (1936) found sand to harbor the poorest fauna. Tarzwell gave sand a population rating of 1.0 to provide a scale for comparing the relative standing crops of organisms on other substrate types, some of which follow:

Sand	1.0	Gravel and rubble	53
Marl	6	Moss on gravel	111
Fine gravel	9	Moss on gravel and rubble	140
Sand and silt	10.5	*Ranunculus*	194
Rubble	29	*Rorippa*	301
Chara	35	*Elodea*	452

The general paucity of aquatic insects in sand relates to the instability of this substrate and to the typically low organic content of sand. Tarzwell (1936) found that in areas where silt (presumably containing fine organic matter) was mixed with sand, there was a nearly 11-fold increase in faunal densities compared to "clean" sand. In the littoral of Oneida Lake, New York, sand actually contained higher densities of insects than several other substrate types (Table 7.11).

Faunal densities in sand may be much higher than previously thought. Using very fine mesh (0.06 mm), Soluk (1985) recorded macroinvertebrate densities of 12,000–78,000 organisms m^{-2} in shifting-sand areas of a river. Because most species inhabiting shifting sand are so small, it is likely that investigators sampling with "normal" mesh sizes fail to capture the majority of the fauna.

Table 7.11 Densities of aquatic insects on various substrates in lentic and lotic waters

	Insects m^{-2}					
	Mud	Sand	Gravel	Cobble	Boulder	Reference
Lake littoral	495	843	350	—	377	Baker 1918
Trout stream	—	85	244	265	136	Ward 1975
Large river[a]	371	17	—	399	—	Russev 1974

[a] Includes noninsects.

The addition of filamentous algae, mosses, or angiosperms to mineral substrate dramatically increases faunal densities, as the following data from a Yorkshire stream (Percival and Whitehead 1929) amply illustrate.

	Mean No. Insects dm^{-2}
Loose stones	32.82
Cladophora on stones	270.46
Loose moss	708.53
Thick moss	2139.61
Potamogeton on stones	2405.06

In a Canadian stream, however, more animals per unit surface area occurred on the adjacent stony bottom than on any of four species of macrophytes (Rooke 1984). The large standing crop of detritus associated with the stony substrate of this stream may at least partly account for these results, which are contrary to the findings of other workers.

In a Rocky Mountain trout stream (Table 7.11) lacking aquatic macrophytes or even appreciable algae, five orders of insects made up the entire entomofauna. The percentage composition of the insect orders on the predominant substrate types, based on numerical abundance data, is illustrated in Figure 7.12. Mayflies occurred at greater densities than all other insects combined on all substrates except boulders where, although abundant, they were superseded by dipterans (largely *Simulium* and *Deuterophlebia*).

The lower reaches of the River Endrick, Scotland, contain a diversity of substrate types (Table 7.12). More orders are present, dipterans are better represented, and insects, although predominant, represented a lower proportion of the total fauna than in the Rocky Mountain stream. In addition, Maitland (1966) was able to distinguish differences in the insect assemblages of pool and current habitats even within a given substrate type. Insects invariably comprised a greater proportion of the total fauna in current compared to pool habitats. In contrast, data from studies of primarily upland lotic systems in North America and the British Isles showed little difference in overall faunal composition between riffles and pools, although riffles contained significantly higher macroinvertebrate densities than pools (Logan and Brooker 1983).

The standing crop of the xylophilous fauna associated with flooded trees (Table 7.13) greatly exceeded that of the herpobenthos in a tropical impoundment (Petr 1970b). In temperate reservoirs, submerged wood may support much greater densities of chironomids, the main xylophilous insects of temperate artificial lakes, than the bottom substrate (Cowell and Hudson 1967).

Despite the ubiquity of wood in undisturbed forested streams, the biomass of insects on submerged wood is but a fraction of the biomass associated with leaf litter (Table 7.14). Anderson et al. (1978) attribute this difference largely to the higher amounts of refractory components (cellulose and lignin) in wood compared to leaves. As a result of the refractory nature of wood and the low surface to volume ratio of wood debris, microbial colonization is slow and decomposi-

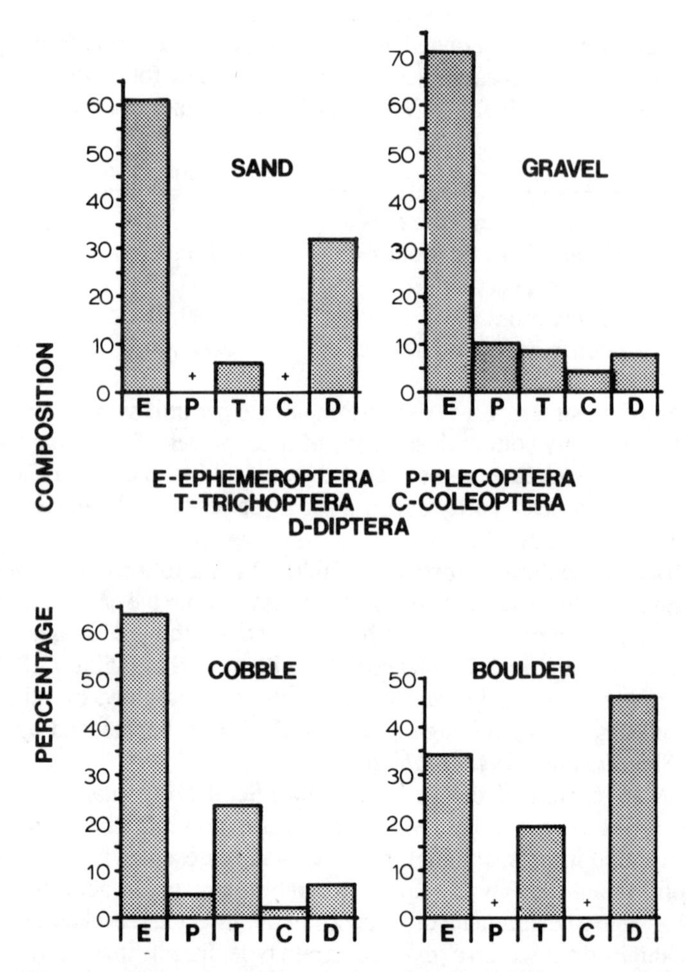

Figure 7.12 Relative contributions of orders to the numerical abundance of total insects on the mineral substrates of a Rocky Mountain trout stream. [From Ward (1975).]

tion requires years or even decades compared with months for leaf litter. Submerged wood thus provides a temporally stable substrate for aquatic insects in the relatively constant food resource for xylophagous species. Leaf litter, although a higher-quality food than wood, is a relatively transient habitat.

Table 7.12 Percentage composition (numbers) of the insect components of the invertebrate fauna in the main habitats of the lower reaches of the River Endrick, Scotland

	Stones		Gravel		Macrophytes[a]		
	Current	Pool	Current	Pool	Current	Pool	Silt Pool
Ephemeroptera	12.1	3.4	16.7	3.9	7.4	9.9	2.8
Plecoptera	1.2	0.1	5.1	0.3	0.4	0.2	0.1
Odonata	0	0	0	0	0	0.9	0
Hemiptera	0	0.1	0	0.1	0	14.4	14.1
Neuroptera	0	0.1	0	0	0	0	0
Trichoptera	0.6	5.1	0.1	0.1	0.4	2.0	1.7
Lepidoptera	0	0	0	0	0	0.1	0
Diptera	58.2	41.3	48.0	49.7	59.8	26.1	38.1
Coleoptera	7.5	18.1	14.2	2.1	3.6	1.4	0.7
Total insects (%)	79.6	68.2	84.1	56.2	71.6	55.0	57.5

[a]Largely *Potamogeton* and *Myriophyllum*.

Source: Modified from Maitland (1966).

Table 7.13 Density (organisms m^{-2}) and biomass (grams wet weight m^{-2}) of the predominant insect taxa on flooded trees at four sampling stations in Volta Lake

	Kpandu		Kete K.		Yeji		Ampem	
	Org. m^{-2}	g m^{-2}	Org. m^{-2}	g m^{-2}	Org. m^{-2}	g m^{-2}	Org. m^{-2}	g m^{-2}
Povilla	3752	55.05	1226	24.10	197	5.93	5896	63.33
Amphipsyche	1420	11.16	0	0	0	0	0	0
Ecnomus	469	1.09	113	0.23	133	0.08	133	0.48
Chironomidae	517	0.70	61	0.05	15	0.03	671	1.15

Source: Modified from Petr (1970b).

Table 7.14 Biomass (dry weight) of insects on submerged wood and leaf packs in Cascade Range streams of Oregon (USA)

	mg insects m^{-2} wood			mg insects kg^{-1}	
	Lara	*Heteroplectron*	Other Insects	Wood	Leaf packs
Devil's Club Creek	100.0	0	113.9	—	—
Mack Creek	104.5	27.9	356.4	63.3	8802
Lookout Creek	117.1	34.2	845.9	—	—
McKenzie River	148.7	0.9	478.3	—	—

Source: Modified from Anderson et al. (1978).

8

WATER LEVEL, CURRENT, AND DISCHARGE

Temporal and spatial variations in water level, current, and discharge play major roles in determining the structure of aquatic insect communities. This chapter considers the direct effects of these factors on insects and their influence on habitat conditions.

WATER-LEVEL FLUCTUATIONS

The amplitude and pattern of water-level fluctuations exhibited by different lentic and lotic waters are highly variable and subject to human alteration. Some lakes and spring-fed streams remain constant over long periods, whereas other natural water bodies exhibit considerable short-term variations in water level. Artificial lakes and regulated streams may exhibit fluctuations the extent of which exceeds the most extreme natural variations.

Lentic Waters In the context of aquatic insect ecology, changes in water levels are of special concern in lentic water bodies because it is the littoral zone, where most aquatic insects reside, that is directly affected. As a general rule, however, natural lakes, at least in mesic regions, only infrequently undergo large short-term changes in water level. Changes that do occur tend to be gradual and predictable in nature. Rapid, though relatively small (<1 m), fluctuations may occasionally occur in lakes that are particularly susceptible to flooding, but without any major adverse effects on the littoral fauna (Moon 1935). Some alpine lakes undergo extreme natural fluctuations in water level, and in such cases, the littoral fauna is poorly developed (Zschokke 1894, Schmassmann 1920); but the most definitive data on the effects of water-level fluctuations on lentic insects come from studies of regulated lakes and reservoirs.

The ways by which aquatic insect communities are altered by changing water levels depend on a complex of interrelated variables including the extent, duration, and rapidity of the fluctuations; the season during which drawdown occurs; the morphometry and trophic status of the water body; and the climate of the region. Temperature extremes and desiccation directly affect the aquatic insects remaining within the exposed zone. Water-level fluctuations also influence the entomofauna indirectly by reducing or eliminating aquatic plants, altering chemical conditions, and increasing erosion, thereby altering the substrate and water clarity.

The regulation of Lake Blåsjön in northern Sweden resulted in annual water level fluctuations of six vertical meters (Grimås 1961). Data collected prior to damming (Brundin 1949, Brinck 1949, Nilsson 1955) and investigations in nearby Lake Ankarvattnet, an unregulated lake, revealed major changes in the bottom fauna of Lake Blåsjön attributable to water-level fluctuations. The density of benthic animals was reduced 70% in the regulated zone, that portion of the lake bottom subjected to drawdown during winter subsequent to ice formation. Even in the deeper waters faunal densities were reduced by 25%.

The most notable change in faunal composition following regulation of Lake Blåsjön was an increase in the relative abundance of chironomids, which was accompanied by a reduction in the numerical importance of other insects as follows:

	Unregulated Lake Ankarvattnet	Lake Blåsjön	
		Before	After
Chironomids	34.9%	31.9%	54.1%
Other insects	2.4%	—	0.6%
Noninsects	62.7%	—	45.3%

The enhanced numerical importance of chironomids relates at least in part to their generally superior ability to survive being frozen, compared with insects from other orders. The insulating effect of an established ice cover, however, somewhat reduced the severity of winter temperatures in the regulated zone of Lake Blåsjön.

In Lake Ankarvattnet, as in Lake Blåsjön before regulation, there are four major emergence periods for chironomids (spring, early, mid-, and late summer). Following regulation, spring emergence was eliminated from the entire littoral zone, including the lower littoral that remained submerged. In the upper littoral, which lies entirely within the regulated zone, a retarded early-summer emergence is all that remained after regulation. About 80% of the chironomid species have been eliminated from the regulated zone of Lake Blåsjön, and 60% of all littoral species of chironomids no longer occurred in this zone after regulation. Although all the main groups of chironomids exhibited reductions in species following regulation, the relative abundance of Tanytarsini and Orthocladiinae increased, whereas Chironomini were severely reduced. Most species of chironomids remaining after regulation exhibited a more stenobathic distribution and there was a tendency for each depth zone within the littoral to be dominated by a very few species or even a single species. *Parakiefferiella bathophilus* was predominant in the 2–4-m-depth zone; *Tanytarsus gregarius* from 4–6 m; *Acricotopus thienemanni* from 6–10 m; and *Micropsectra groenlandica* predominated from 10–13 m.

Insects other than chironomids occur at mean densities of 78 organisms m^{-2} and 119 organisms m^{-2}, respectively, in rocky areas and soft-bottom areas of Lake Ankarvattnet. Those few mayflies, stoneflies, and caddisflies occurring in the lake following regulation exhibited constricted depth distributions (Fig. 8.1), as did the remaining chironomids.

Ten years after regulation of Lake Blåsjön, the annual fluctuations in water level were increased from 6 m to 13 m (Grimås 1962). The additional amplitude further reduced the bottom fauna, altered their bathymetric distribution patterns, and eliminated the sparse stands of *Nitella*, the only macrophyte remaining in the lake.

Grimås (1961, 1962) considered temperature extremes and desiccation to be

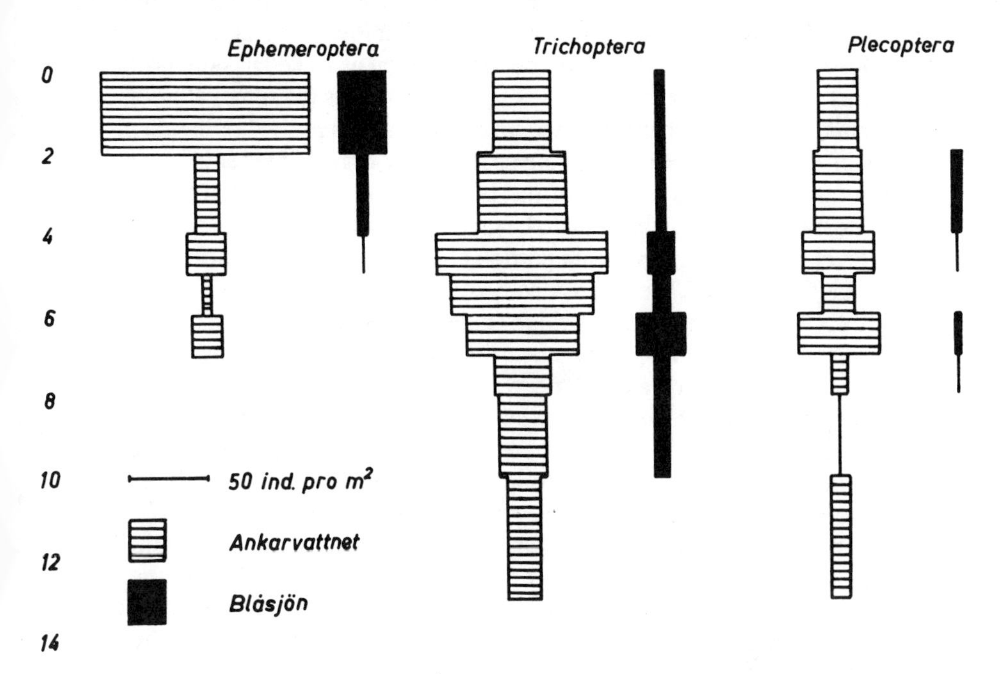

Figure 8.1 Depth distribution of Ephemeroptera, Trichoptera, and Plecoptera in regulated Lake Blåsjön and unregulated Lake Ankarvattnet, northern Sweden. The vertical scale is depth in meters. [From Grimås (1961).]

the major factors resulting from water-level fluctuations to directly influence the bottom fauna of Lake Blåsjön. Because drawdown occurs during winter, freezing of the organisms and littoral substrate is very probably the most important effect of regulation. The extreme lowering of the lake level also reduced winter water temperatures below the drawdown limit at depths that would normally be within the profundal zone. The severity of thermal conditions apparently accounts for the shift toward an arctic species assemblage of chironomids (Brundin 1949). In a Canadian reservoir with drawdown occurring during spring, Chironomini rather than Orthocladiinae were the dominant forms in the regulated zone, and many species exhibited wide bathymetric limits in contrast to the tendency toward stenobathy in Lake Blåsjön (Fillion 1967).

The elimination of aquatic macrophytes and a concomitant reduction in benthic microphytes is a general result of severe water level fluctuations (Quennerstedt 1958, Stube 1958). The demise of certain mayflies, caddisflies, and chironomids following regulation likely relates to their phytophilous nature. Aquatic vegetation contributes organic matter to littoral and deep sediments. The reduction of this source of detritus adversely affects faunal components that utilize sedimentary organic matter as a food source. The absence of rooted macrophytes exacerbates erosion in the regulated zone, which further reduces the organic content of the substrate. The littoral zones of regulated lakes typically consist of largely mineral sediments. Fine mineral particles may, however, be entrained by eroding forces and deposited in deeper water, thereby blanketing organic sediments and thus reducing food availability for profundal benthos. Lake Jormsjön, a regulated lake in the same region as Lake Blåsjön, possesses a peculiar morphometry that to some extent ameliorates the removal of organic sediments despite fluctuating water levels. A submerged ridge dampens the effects of erosion so that much of the sedimentary organic matter is retained in the littoral zone in one portion of the lake. Faunal components (e.g., certain mayflies and caddisflies) that tend to be greatly reduced in lakes with fluctuating water levels represent a significant portion of the fauna in the "protected" littoral area of Lake Jormsjön. The mean density of bottom animals in the portion of the lake protected by the submerged ridge is about 6000 organisms m^{-2}, compared to only 700 organisms m^{-2} in the upper littoral zone in other areas of the lake.

Complete drainage of an Ontario reservoir allowed an investigation of the winter survival of benthic animals exposed to atmospheric conditions (Paterson and Fernando 1969). Air and substrate temperatures remained above freezing during the first 50 days of exposure in autumn. During the next 100 days the substrate froze to depths of >20 cm. Air temperature dropped to -22°C during this period. The reservoir was refilled in spring, inundating the substrate 168–176 days (depending on the sampling station) after the initial exposure. The majority of benthic species did not survive the entire period of exposure. Among insects, only small numbers of a few species of chironomids and the hydroptilid caddisfly *Agraylea multipunctata* survived. The only exception was the chironomid *Glyphotendipes barbipes,* which exhibited 12.5–46.4% survival at the various stations. Successful pupation and emergence from larvae exposed during the entire period of drawdown was demonstrated for *A. multipunctata, G.*

barbipes, and *Cricotopus trifasciatus* (Paterson and Fernando 1969). Although there was no evidence of vertical migration by chironomids, the probability of survival was greater for individuals deeper within the substrate as shown below.

	Length of Exposure (Days)		
	<50	50–100	>100
>3-cm depth	92%	85%	68%
<3-cm depth	8%	15%	32%

These data are based on the depth distribution of surviving chironomids. Absolute numbers of surviving chironomids below the 3-cm level were relatively constant throughout the period of exposure. Because survival of larvae in the surficial substrate layer declined markedly over time, the relative contribution of deeper-dwelling larvae to total surviving chironomids quadrupled over the period of exposure.

The benthic invertebrates of three habitats were studied in a fluctuating reservoir in Wisconsin (Kaster and Jacobi 1978). In areas that remained inundated during maximum drawdown, annual density and wet biomass of benthos (largely chironomids and oligochaetes) averaged 8558 organisms m^{-2} and 16.0 g m^{-2}. Areas of substrate covered with ice prior to drawdown exhibited mean values of 4311 organisms m^{-2} and 4.5 g m^{-2}. Areas experiencing drawdown before ice formation, that were therefore exposed to desiccation prior to freezing, contained an average of 3025 organisms m^{-2} with a biomass of 1.8 g m^{-2}. At one location exposed to air, density of total benthos exceeded 5000 organisms m^{-2} prior to drawdown but declined to zero after 35 days of exposure. *Chironomus plumosus* exhibited a downward migration within the exposed substrate, and some surviving individuals may have penetrated beyond the sampling depth of ~10 cm. As much as 21% of the benthic fauna survived in drawdown areas that were covered by ice. Total density, number of taxa, and biomass were greatest immediately below the drawdown limit.

The maximum values of benthic animals frequently recorded just below the drawdown limit of fluctuating reservoirs may be explained in part by migratory movements. Cowell and Hudson (1967) studied the migration of chironomids along transects in the regulated zone of a Missouri River reservoir during summer. On a gently sloping shore, large numbers of chironomids became stranded and died as the water receded. However, a portion of the population migrated with the receding water level. On a steeper shore stranding was lessened presumably because larvae were required to traverse a shorter distance to keep up with the receding water.

Studies of Llyn Tegid (Lake Bala), Wales, offer an excellent opportunity to examine the short-term effects of water level fluctuations and long-term recovery of the littoral insect fauna following restabilization (Dunn 1961, Hynes 1961a, Hunt and Jones 1972, Hynes and Yadav 1985). In 1955 a 3-m drop in mean lake level occurred and annual fluctuations of 5 m commenced as a controlled outflow scheme was implemented. The maximum permissible rate of change (15 cm/day) was often exceeded, resulting in rapid and irregular fluctuations for several years after dam construction. However, because of improved management and construction of an additional reservoir, fluctuations lessened, and by 1967 the water-level amplitude was similar to pre-1955 levels.

Prior to 1955 a rich macrobenthos, including a diverse and varied aquatic entomofauna, occupied the littoral zone of Llyn Tegid. The initial drop in water level stranded and killed great numbers of littoral organisms. Littoral sampling in 1957 revealed no Plecoptera, Ephemeroptera, Megaloptera, or Coleoptera (Table 8.1). The rich trichopteran fauna had been reduced to a small population of a formerly abundant species, *Anabolia nervosa.* The exposed stony shores were silt-laden. Sheltered shores, formerly composed of silty sand, gravel, and stones with extensive growths of the submerged angiosperm *Littorella,* were covered with sublittoral mud. Oligochaetes proliferated, and some species of sublittoral chironomids established populations in the littoral zone. Hynes (1961a) attributed these major changes in the fauna to stranding losses associated with rapid drops in lake level and to alterations in the substrate and the demise of macrophytes.

Some faunal recovery was evident by 1959, apparently reflecting incipient

Table 8.1 Number of species of aquatic insects (excluding Diptera) collected from littoral zone of Llyn Tegid, Wales, prior to (1951/52) and following the installation of a regulating dam (see text)

	Number of Species				
	1951/52	1957	1959	1968/69	1968/69 Fauna
Plecoptera	3	0	1	3	Abundance and distribution similar to 1951/52
Ephemeroptera	3	0	3	3	Dominance has shifted from *Leptophlebia marginata* to *Caenis moesta–Ephemera danica*
Trichoptera	15	1	2	23	Caddis fauna has clearly reestablished, but not to former abundance levels
Megaloptera	1	0	1	1	Increased abundance of *Sialis lutaria* attributed to pelophilous nature
Hemiptera	4	1	1	4	*Micronecta poweri* abundance unaffected by regulation
Coleoptera	9	0	2	10	Only one species of those recorded in 1951/52

Source: Hunt and Jones (1972).

restabilization of the littoral zone. At least some species from all the usual major groups of insects were again present.

Beginning in 1967, the amplitude of fluctuation in Llyn Tegid was reduced and was similar to that extant prior to 1955. Wave action had by this time removed much of the silt from the rocky shores, and aquatic macrophytes were reestablished on the sheltered shores. All major groups of insects and many species present before 1955 were again recorded in 1968/69, although recovery was far from complete. *Leptophlebia marginata,* the most abundant mayfly prior to regulation had not attained preimpoundment population levels by 1968/69. *Caenis moesta,* a silt-adapted species, and the burrowing mayfly *Ephemera danica,* although eliminated initially, increased in numbers under the new regime. Increased abundance of *Sialis lutaria* was attributed to the pelophilous nature of this megalopteran. The caddisfly fauna had clearly reestablished, although the abundance of most species was still low in 1968/69. The distribution and abundance of plecopterans was similar to 1951/52 values. Only the corixid bug *Micronecta poweri* was not visibly affected by regulation of Llyn Tegid, as their recorded distribution and abundance were similar for all surveys. The main change in the dipteran fauna involved the frequently reported increase in chironomids. Hunt and Jones (1972) predicted that, even if water level fluctuations are not increased, it would be years before the littoral community reverts to its original composition and abundance. This prediction was substantiated by additional sampling in 1978/79 (Hynes and Yadav 1985). Several species of insects recorded in 1951/52 had not appeared by 1978/79. Although sponges had reestablished, spongillaflies *(Sisyra fuscata)* had not.

Chironomids comprise virtually the entire mud fauna in the regulated zones of Lake Kariba and other African impoundments with highly fluctuating water levels (McLachlan 1974). By the end of the relatively stable period at the maximum water-retention level, the macrobenthos in the shallow water of Lake Kariba consisted of an assemblage of six species of chironomids. Drawdown inflicted heavy stranding losses, but such losses were largely compensated by oviposition at the receding water margin. Grasses that developed on the exposed mud flats attracted grazing animals, including elephants, buffalo, and zebra. When the water again rose, conditions were different in the aquatic environment. Dung from the grazing animals and inundated terrestrial plants supplied large amounts of organic matter. Oxygen levels were reduced and nutrient levels increased. The original six species of chironomids were eliminated by the altered conditions and replaced by enormous numbers of *Chironomus transvaalensis.* McLachlan concluded that the water level fluctuations maintain an artificial subclimax in the regulated zone. According to this viewpoint, *C. transvaalensis* is a pioneer species that is replaced by the next seral stage consisting of the assemblage of six species that develop following a period of stabilized water level. The next drawdown, however, induces a regression to the pioneer "community."

In some instances, water-level fluctuations have a generally stimulating effect on aquatic insects. Woodland inundated by tropical impoundments provides suitable substrate for large populations of xylophilous insects, most notably *Povilla*, the wood-burrowing mayfly (McLachlan 1970). However, *Povilla* nymphs are unable to penetrate hardwood trees such as those inundated in some areas of Lake Kariba. Periods of drawdown expose the dead trees to terrestrial wood-boring beetles, thereby creating suitable burrows for colonization by *Povilla* and other cryptic aquatic insects when the trees are resubmerged (see Fig. 7.8). McLachlan (1970) reported a nearly sixfold greater faunal biomass on resubmerged compared to submerged hardwood, with the increase almost entirely due to *Povilla*.

Drawdown has in some cases been utilized as a managerial strategy to increase fish production and waterfowl utilization of water bodies. In contrast to most operational schemes, regulation of Lake Tohopekaliga in Florida resulted in a more stabilized water level (Wegener et al. 1975). Because the large natural fluctuations were thought responsible for maintaining the biological integrity of this lake–wetland complex, an extreme drawdown (exposing 50% of the lake bottom) was undertaken to stimulate aquatic productivity of the regulated lake. Drawdown stimulated the areal expansion of submerged and emergent hydrophytes. Macroinvertebrates of the littoral zone (the dewatered area) rapidly increased in abundance following reflooding. The mud fauna increased 63% and the phytophilous fauna increased 350% compared to predrawdown values, an enhancement attributed to improved substrate conditions and to increases in the density and diversity of vegetation with associated epiphytes. Chironomids and oligochaetes largely accounted for increases in the mud fauna. Chironomids also increased in plant beds, but even greater (at least order-of-magnitude) increases in the phytophilous fauna were exhibited by baetid mayflies, dragonflies, damselflies, corixid bugs, and hydrophilid beetles. Within 2 years, however, macroinvertebrate standing crops had declined to approximate predrawdown values.

Summer drawdown of a shallow impoundment in Michigan (USA) was undertaken to increase waterfowl utilization (Kadlec 1962). Alteration of aquatic and semiaquatic plants resulting from drawdown increased the attractiveness of the impoundment to the waterfowl. Aquatic insects, including chironomids, were, however, severely depleted by the drawdown (Table 8.2).

Water-level management has been used as a method of mosquito control for impounded waters. To control *Anopheles quadrimaculatus*, four phases of water-level management were proposed by Hess and Kiker (1944). The first phase involves raising the water level to flood stage for a brief period during late winter or spring to strand floating debris, which if it remained in the water would provide suitable breeding conditions for *A. quadrimaculatus*. During the second phase, a high constant water level is maintained to limit the invasion of marginal vegetation, which, if present, would enhance breeding conditions for mosquitoes. Seeds of plants adapted to conditions in a fluctuating littoral zone may require a period of dewatering before they germinate. Considerable reductions in the effective growing season of such species is thus possible if their seeds are continually covered with water during the spring. The third phase of control

Table 8.2 Average number of insects per sample (929 cm^2) before and after summer drawdown of a Michigan (USA) waterfowl impoundment

	Station I 122 cm deep Sparse Vegetation		Station II 31 cm deep Dense Vegetation	
	Before	After	Before	After
Trichoptera	6.0	0.0	30.4	0.4
Ephemeroptera	4.0	0.0	45.6	0.0
Odonata	9.6	0.0	12.0	2.0
Diptera[a]	785.6	51.2	540.0	206.8

[a] Largely Chironomidae.

Source: Modified from Kadlec (1962).

involving weekly cyclical water-level fluctuations is initiated when mosquito production reaches moderate levels in late spring. The water level is dropped below the band of marginal hydrophytes once each week, thus creating unfavorable oviposition conditions, reducing larval food supply, exposing larvae and pupae to predators, and increasing stranding losses of eggs and larvae. The final phase of water level management combines the weekly fluctuations with a gradual drop in water level. This phase, which coincides with the period of maximum production of *A. quadrimaculatus* ensures that the low amplitude of the weekly cycle keeps ahead of the advancing band of vegetation. It is seldom, however, possible or even desirable to attain the regime of water level fluctuations as outlined by Hess and Kiker.

Before concluding this section, further mention should be made of lentic waters that undergo relatively large natural fluctuations in water level. The lake-wetland complex in Florida has already been discussed. Additional examples may be found in north temperate ponds and lakes in the arid tropics.

The Nearctic caddisfly *Glyphopsyche irrorata* has evolved a life-cycle strategy enabling it to survive in permanent ponds with fluctuating water levels (Berté and Pritchard 1983). Caddisflies adapted to temporary vernal ponds emerge during spring, pass the summer dry period as adults, and lay desiccation-resistant eggs on the dry pond basin in early autumn (Wiggins et al. 1980). *Glyphopsyche irrorata*, the eggs of which are apparently not resistant to desiccation, emerges in autumn as the water level of the habitat is receding. Autumn oviposition would risk exposure of eggs to desiccation. Instead, it is the adult stage that overwinters. (Adults have been collected in winter as far north as Alaska.) Oviposition is thereby postponed until spring when the water of the ponds is at the highest level.

Lake Chilwa, a large, shallow lake in central Africa, exhibits annual fluctuations associated with the wet and dry seasons, superimposed on minor recessions in mean lake level that occur at approximately 6-year intervals (Kalk et al. 1979, Cantrell 1988). Two major recessions, resulting in complete desiccation of the lake bed, have occurred in the past century. McLachlan (1979) has reconstructed the responses of the benthos to the drying (Fig. 8.2) and subsequent refilling (Fig. 8.3) of Lake Chilwa during a major recession. As the lake contracts, the shoreline recedes from the extensive littoral vegetation. This is particularly significant because the marginal aquatic plants contain a much more diverse entomofauna than the mud habitat. The number of species of benthic invertebrates (largely insects) recorded in shallow water at the lake's edge declined from 45 species at high water level to only 11 species shortly after the shoreline had receded just below the level of the marginal aquatic vegetation. The demise of the chironomid *Nilodorum brevipalpis* is particularly striking because this species normally dominates the benthos of the marginal vegetation. Further reduction in lake volume resulted in high salinity and generally extreme physicochemical conditions, which were associated with the loss of another previously abundant chironomid, *Nilodorum brevibucca,* and the corixid *Micronecta scutellaris.* Along windward shores the bodies of *M. scutellaris* accumulated in windrows about 30 cm deep that extended for many kilometers. Well before the lake is completely dry, virtually the only macroinvertebrates to remain are larvae of the beetle *Berosus vitticollis.* No aquatic insects are known to survive the rare periods of complete desiccation.

Faunal recovery from the 1968 drying of Lake Chilwa occurred rapidly as

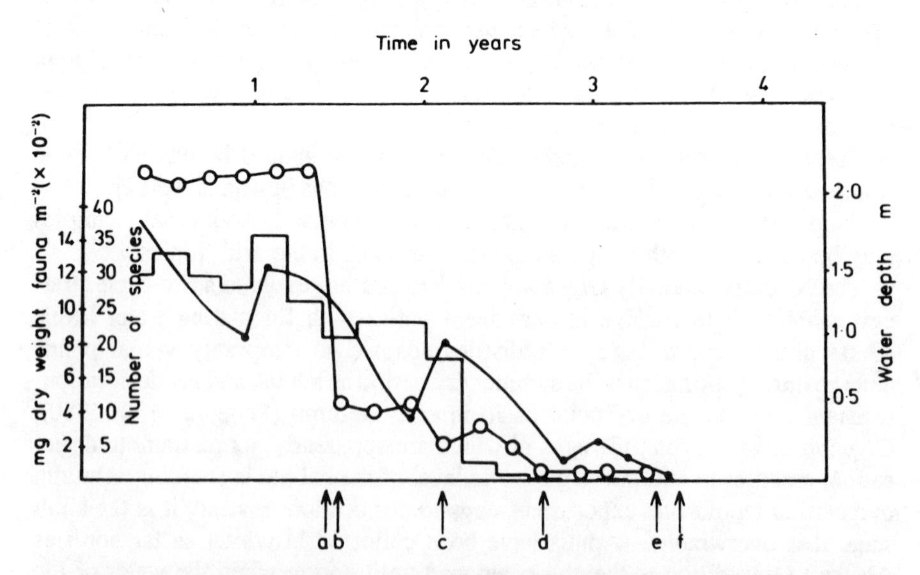

Figure 8.2 The probable sequence of events in a typical drying phase of Lake Chilwa in central Africa. The number of species surviving (open circles) and the total weight of the fauna (histograms) are shown in relation to decreasing water depth (solid circles) at the lake center. The letters and arrows indicate when the marginal swamps become stranded (*a*), the last record of the chironomid *Nilodorum brevipalpis* (*b*), the last record of live snails (*c*), the last record of *Nilodorum brevibucca* (*d*), death of the corixid *Micronecta scutellaris* (*e*), and lake dry (*f*). [From McLachlan (1979). Reprinted by permission of Kluwer Academic Publishers.]

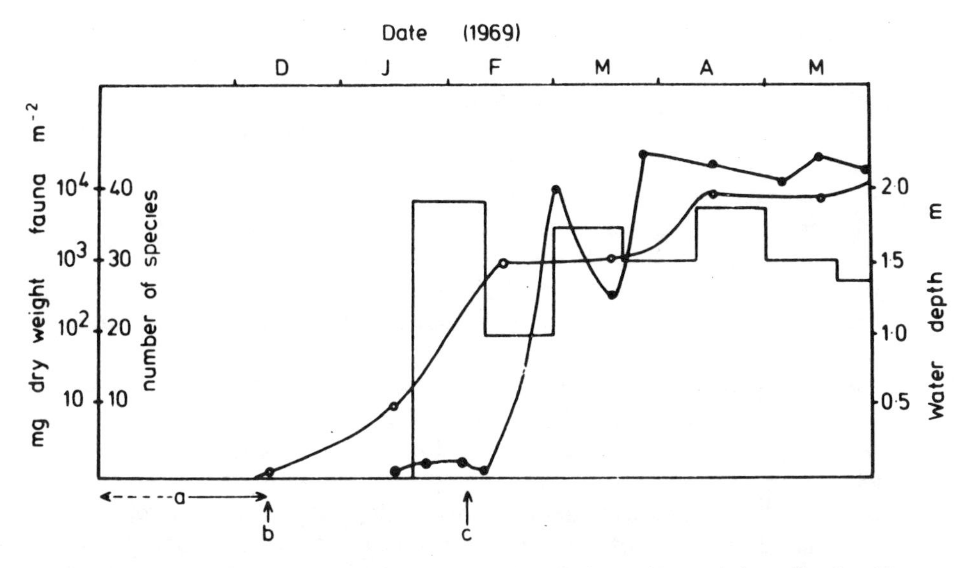

Figure 8.3 Colonization of Lake Chilwa, Africa, by benthic invertebrates during reflooding. The total weight of the fauna (histograms) and number of species (solid circles) present on the available substrate are shown in relation to the rising water level (open circles). The letters and arrows indicate when the lake is dry (*a*), reflooding begins (*b*), and the reflooding of marginal swamp habitats (*c*). [From McLachlan (1979). Reprinted by permission of Kluwer Academic Publishers.]

reflooding began during the next rainy season. Dense populations of chironomid larvae followed the advancing shoreline as the lake filled. *Chironomus transvaalensis,* the most common species in the newly inundated mud, was restricted to a narrow band behind the rising water. *Chironomus transvaalensis,* a pioneer species, was eventually replaced by *N. brevibucca. Nilodorum brevipalpas* attained high densities on the remains of terrestrial plants that had colonized the dry lake bed (Cantrell 1988). Shortly after the rising water reached the littoral zone, there was a sharp increase in the number of invertebrate species (Fig. 8.3). The recovery of the aquatic entomofauna is thought to occur primarily from oviposition by females from nearby water bodies.

Lotic Waters Considerable fluctuations in water level may occur during spates and as a natural consequence of the seasonal variations in flow characterizing many lotic habitats. Some aquatic insects are in fact dependent on natural water-level fluctuations. For example, the elmid beetle *Stenelmis sexlineata* depends on changes in water level to complete the life cycle (White 1978). Larvae do not actively leave the water to seek a terrestrial pupation site. Rather, mature larvae migrate up the shoreline during a spate, the recession of which strands them above the water's edge, whereupon they construct the pupal chamber. The wood-dwelling crane fly *Lipsothrix nigrilinea* uses receding water level as a proximate cue for pupation (Dudley and Anderson 1987). In the absence of this cue, the larval stage is extended for an additional year or more.

In most cases, however, insects residing in habitats exhibiting wide fluctuations in water level must avoid or tolerate stranding. This is accomplished in various ways. The eggs of many lotic insects tolerate at least brief periods of exposure. Larvae may avoid stranding by releasing their hold on the substrate and temporarily entering the current, or by swimming or crawling to deeper water as the water level recedes. Although aquatic arthropods generally contain much less fat than their terrestrial relatives, "many of the species that live in environments that are liable to dry up have far more fat than those that live in more stable aquatic environments" (Hinton 1953, p. 219). According to Hinton, combustion of fat helps maintain the moisture content of larval tissues during periods of exposure.

Immobile pupae of aquatic insects are especially vulnerable to recessions in the water level of the habitat. Mature larvae of some caddisflies move to deeper water to pupate (Scott 1958). The heavy stone cases of many stream-dwelling Trichoptera may reduce the chances of pupae being transported by the current to marginal areas more likely to be exposed during periods of low discharge. It may be that the addition of large mineral particles to the cases of some species of caddisflies has its primary function as ballast in the pupal rather than the larval stage. The pupal stages of some Diptera and the few Coleoptera with truly aquatic pupae have evolved spiracular gills (discussed in detail in Chapter 2). Plastron-bearing spiracular gills are dual organs possessed primarily by species inhabiting environments that are alternately dry and flooded (Hinton 1953). If the water level drops and exposes the pupa to air, the insect breathes through functional spiracles that provide but a small surface area and excessive water loss is avoided. When immersed, a large air-water interface for respiratory exchange is provided for pupae possessing plastron-bearing spiracular gills. Pupae of *Tabanus autumnalis,* although lacking spiracular gills, are highly tolerant of alternating periods of exposure and immersion (Hinton 1953). On immersion, uptake of water by the pupae fully exposes the intersegmental membranes, whereas a diminished volume when exposed to air results in a telescoping of the segments into one another so that no intersegmental membranes are exposed. The intersegmental membranes are highly permeable and serve as the primary respiratory organs for the pupae when a large respiratory surface is required during immersion. During exposure to air the relatively impermeable areas of the segments overly the intersegmental membranes, thus greatly reducing water loss.

Dramatic short-term fluctuations in water level commonly occur in streams below hydroelectric dams. Large portions of the shoreline are alternately exposed and inundated, often with a diel periodicity. The extensive dewatered zone below hydroelectric dams has been referred to as the "freshwater intertidal" by Fisher and LaVoy (1972), who examined benthic invertebrate communities along a sand–gravel bar below a hydroelectric dam on the Connecticut River. The bar was submerged during high flow but was largely exposed at low discharge. Samples were collected from four zones along a transect from the high water mark to below the low water mark. The zones were exposed 70, 40,

13, and 0% of the time during summer (Fig. 8.4). Number of taxa, diversity index values, and standing crop of the macroinvertebrate assemblages increased from zone 1 (70% exposure) to zone 3 (13% exposure). However, the macroinvertebrates of zone 4, which remained submerged, did not differ significantly from those of zone 3, suggesting that the prevailing benthic community was not greatly affected by brief periods of exposure. Chironomids were virtually the only insects collected from zones 1 and 2 (Table 8.3). Inexplicably, all species of tipulids were restricted to zone 2.

Rapid lowering of the water level may strand large numbers of benthic organisms. When the water released through Jackson Lake Dam was reduced from 2.8 to 0.3 m^3/sec in less than 5 minutes, large areas of the bottom of the Snake River, Wyoming, were exposed (Kroger 1973). On the basis of samples taken from the exposed portions of the river bed, Kroger estimated that over 3 billion macroinvertebrates, largely insects, were killed by stranding in the first 3 km below the dam.

Stranding susceptibility, tolerance to exposure, and ability of aquatic insects to migrate with the receding water depend on a variety of factors, including the rate of recession, the substrate type, the slope of the shoreline, and atmospheric conditions, and also varies along taxonomic lines.

The predominance of chironomids in the dewatered zone apparently is attributable to their tolerance of exposure (Fisher and LaVoy 1972, Brusven et al. 1974), since several investigators comment on the failure of lotic chironomid larvae to actively migrate to deeper water with rapidly receding water levels (Denham 1938, Brusven et al. 1974, Corrarino and Brusven 1983). Brusven et al. (1974) reported negligible mortality of stranded chironomid larvae after 24

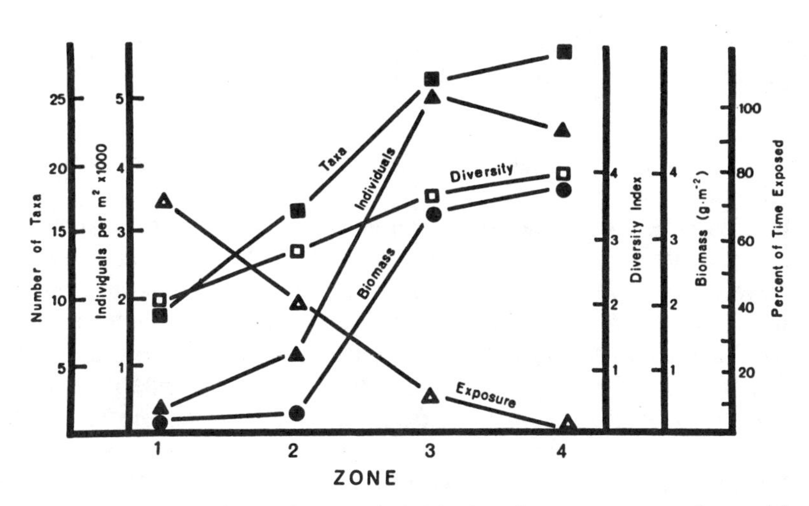

Figure 8.4 Distribution of benthic macroinvertebrates along an exposure transect in the Connecticut River (USA). [From Fisher and LaVoy (1972).]

Table 8.3 Mean numbers of benthic insects m^{-2} along a transect running from the high water mark to below the low water mark of the Connecticut River below a hydroelectric dam

	Zone[a]			
	1	2	3	4
Ephemeroptera				
Stenonema sp.	0	0	0	12
Ephemerella sp.	0	0	20	24
Tricorythodes sp.	7	0	76	156
Caenis sp.	0	0	12	4
Ephoron sp.	0	0	0	4
Trichoptera				
Limnephilus sp.	0	4	4	20
Molanna sp.	0	0	0	4
Lepidostoma sp.	0	0	0	4
Cheumatopsyche sp.	0	0	12	48
Tascobia sp.	0	0	4	0
Orthotrichia sp.	0	0	4	4
Odonata				
Gomphus sp.	0	0	4	0
Coleoptera				
Anchycteis sp.	0	4	8	20
Agabus sp.	0	4	0	0
Diptera				
Erioptera sp.	0	4	0	0
Pedicia sp.	0	8	0	0
Helius sp.	0	8	0	0
Antocha sp.	0	12	0	0
Chironomidae	27	556	764	792

[a]Exposure times during summer: zone 1, 70%; zone 2, 41%; zone 3, 13%; zone 4, 0%.
Source: Modified from Fisher and LaVoy (1972).

hours and remarkably high survival after 96 and 120 hours' exposure under the prevailing cool springtime conditions.

Mayflies as a group tend to be particularly susceptible to stranding and relatively intolerant of exposure (Brusven et al. 1974). Whereas mats of the filamentous alga *Cladophora* provided protection from exposure for most groups of insects, certain mayflies became entangled in the algal filaments at the edge of the mat and desiccated. However, *Baetis*, a widespread and often predominant mayfly in lotic habitats, is an exception (Corrarino and Brusven 1983). *Baetis* nymphs avoid stranding by drifting in response to receding water and by actively crawling or swimming toward deeper water as flows are reduced. Experimental dewatering of stream channels during periods of high air temperatures resulted in nearly 100% stranding of nearshore insects; only *Baetis* successfully avoided stranding under such conditions (Corrarino and Brusven

1983). Denham (1938) found that mayfly nymphs from several genera migrated with receding water levels under ideal laboratory test conditions. However, under field conditions, aquatic insects tended to congregate in small depressions that contained water for a short time, but eventually dried.

A propensity toward stranding has been reported for hydropsychid caddisfly larvae in field tests (Brusven et al. 1974, Corrarino and Brusven 1983), although Denham (1938) found active migration with receding water levels under ideal laboratory conditions. Under prevailing cool springtime field conditions both trichopterans and lepidopterans exhibited high survival after 24 and 48 hours' exposure (Brusven et al. 1974). Larvae of these insects sought shelter under rocks and in algal mats during the period of exposure. *Amphipsyche* larvae employ a special adaptive strategy in a widely fluctuating river below a tropical impoundment (Boon 1979). By inhabiting the cavities in vesicular volcanic rocks, larvae are better able to survive exposure during low water and are protected against dislodgement and scouring during high water. The vesicular refugia probably also confer protection from predators and provide a competitive advantage for the species utilizing this special microhabitat.

Nymphs of the plecopteran *Pteronarcys californica*, although relatively intolerant of exposure, were superior to the trichopterans, and especially ephemeropterans, tested in their ability to migrate with receding water levels (MacPhee and Brusven 1974). The greater tolerance of cased caddisflies to exposure and high temperatures as compared to *P. californica* was attributed to the protection afforded by the case.

There is a surprising dearth of definitive information on the migratory behavior and exposure tolerance of the aquatic stages of insects in response to receding water levels, and any generalizations developed in the preceding text must be regarded as preliminary. Entire groups of aquatic insects remain uninvestigated in this regard. Data are available for only a tiny fraction of the species of aquatic insects; congeneric (and certainly confamilial) differences in stranding susceptibility and exposure tolerance are to be expected within each order. This is a research area where data derived from carefully designed field experiments, conducted in conjunction with rigorous laboratory tests, could significantly increase our understanding of aquatic insect ecology.

DROUGHT AND FLOOD

Severe droughts and floods have devastating effects on the aquatic insects of permanent waters. Because of their infrequency, and because not all stream segments are affected, such events may have little influence on the long-term structure of many natural stream communities. In some natural lotic habitats, however, droughts and floods are primary determinants of the community structure and life-cycle patterns of the inhabitants. The ensuing material examines the effects of floods and droughts on aquatic insects in both types of streams, those for which such phenomena are rare events and those for which they are of frequent occurrence.

Drought Most studies of the effects of drought on aquatic insects of normally perennial running waters have been conducted in small streams where extended dry periods result in cessation of surface flow (Engelhardt 1951, Hynes 1958, Larimore et al. 1959, Kamler and Riedel 1960, Iversen et al. 1978, Ladle and Bass 1981, Canton et al. 1984). Complete drying of the bed rarely, if ever, occurs in most large streams or rivers except by human intervention. However, damage may be inflicted on riverine insects without cessation of surface flow as when the water level of the Moselle River dropped severely during a drought year, stranding large numbers of invertebrates, and markedly reducing the population of *Heptagenia sulphurea*, a formerly abundant mayfly (Mauch 1963).

Although surface flow may cease during drought, isolated pools are often present and may retain water for the duration of the dry period. Conditions in the pools are, however, hardly reminiscent of the normal lotic habitat, especially of high-gradient streams. The pools may develop oxygen deficits and other adverse conditions, and water temperature may reach extreme levels. Competitive and predator-prey interactions are altered as organisms are concentrated into the diminishing habitat space.

A few species of aquatic insects of normally perennial streams are able to survive periods of drought in an active state. This may involve residing in isolated pools until flow resumes. As the pools stagnate, some species move to the water surface, where they rest only partly submerged, whereas others leave the pool proper to seek moist marginal microhabitats such as accumulations of debris (Larimore et al. 1959). The insects of a Danish stream survived a 2-week period without flow that lowered the water level and resulted in stagnant conditions (Iversen et al. 1978). The populations of most species were not greatly reduced by the short period of stagnant water, although *Baetis rhodani* was nearly eliminated. Many members of the fauna of high-gradient streams, however, have inherent current requirements and are much less tolerant of conditions in isolated pools than are the inhabitants of slow-flowing lowland streams such as those studied by Iversen and Larimore and their colleagues.

The active forms of a very few perennial steam insects may, however, survive drought even if no surface water remains. Some seek the damp microhabitats under rocks or survive at least for short periods in moist sand or gravel.

Larimore et al. (1959) reported an active downward migration of several orders of aquatic insects as the surficial sediments of a lowland stream began to dry during an extended drought. At least some species from each of the orders, except perhaps mayflies, were apparently able to tolerate short periods without surface water (Table 8.4). Chironomids, elmid beetles, and certain drought-adapted stoneflies were least affected by the extended dry period. It is, however, difficult to ascertain whether an organism that reappears had tolerated the drought conditions or had moved into the affected area from upstream reaches as flow resumed.

When an unusual dry period disrupted the flow of a normally perennial Welsh mountain stream, a special set of circumstances enabled a fairly precise examination of the effects of drought on the invertebrate fauna (Hynes 1958). The

Table 8.4 Number of insects per sample (929 cm²) collected from the riffle of a small lowland stream during a period of extended drought[a]

	April 2, 1954	July 13, 1954	Nov. 11, 1954	Apr. 14-15, 1955
Diptera	145	2	0	54
Coleoptera	2	5	0	8
Ephemeroptera	0	19	0	2
Trichoptera	1	7	0	166
Plecoptera	5	<1	0	10
Total	(153)	(33)	(0)	(240)

[a]April 2—flow had recently resumed; July 13—substrate surface moist, saturated with water at depth of 10 cm, flow became discontinous just prior to sampling; November 1—substrate surface dry after several months without flow; April 14-15—bankful conditions.

Source: Modified from Larimore et al. (1959).

extended drought resulted in complete drying of a 1.5-km segment of the stream for 10 weeks. Although Hynes was unable to find living metazoans within the substrate during the drought, macroinvertebrates reappeared when the water table rose above the gravel stream bed, even before flow had resumed. Aquatic insects that survived as active forms, presumably deep within the hyporheic zone, included certain chironomids and elmid beetles. The high tolerance to drought exhibited by elmid beetles [also reported by Larimore et al. (1959) and Iversen et al. (1978)] is significant because adults lose the ability to fly on entering the water following an initial dispersal flight.

A number of insects of the Welsh stream, including stoneflies (7 spp.), mayflies (2 spp.), and dipterans (several species of chironomids, simuliids, and tipulids), survived the drought in the egg stage. Those species whose egg hatching period began after the drought ended exhibited a high rate of survival, whereas species whose hatching periods coincided with the drought were eliminated or greatly reduced in population size. *Nemoura cinerea*, a stonefly not formerly present in the Welsh stream, appeared in considerable numbers following the drought, completed one generation and disappeared. Because the flight period of *N. cinerea* coincided with the end of the drought, it is possible that the newly formed pools, prior to resumption of normal flow, provided suitable oviposition sites normally lacking in this stream.

Insects that had been important members of the fauna of the Welsh stream, but that failed to reappear following drought, included four species of stoneflies, three mayflies, and four caddisflies. All of these, except *Rhithrogena semicolorata*, which lacks drought resistant eggs, perished because they were in active aquatic stages during the drought.

Species that emerge prior to the dry period would also be eliminated from the fauna if the drought prevents oviposition in the affected stream reach. The seasonal timing of the dry phase is therefore an important determinant of the influence that drought has on the perennial stream fauna.

By altering the composition and abundance of species, drought may shift the trophic relationships of the stream fauna. Drought of a Colorado montane stream reduced the relative abundance of deposit and filter-feeding invertebrates, while increasing the shredders and predators (Canton et al. 1984). The predaceous nymphs of the dragonfly *Ophiogomphus severus* appeared to be favored by the low flow conditions that concentrated their prey species and eliminated fish predators.

Floods The susceptibility of running waters to flood varies greatly between streams, and is largely a function of the climatic regime interacting with the geology and terrestrial vegetation of the drainage basin. Streams in undisturbed forested watersheds with well-developed aquifer systems are considerably buffered against dramatic short-term increases in discharge. In contrast, streams traversing impervious bedrock in arid regions are much more prone to flash-flooding. Mountain streams fed largely by snowmelt may be viewed as flooding each year because of the large proportion of the annual discharge that passes down the channel during spring runoff. In tropical regions with a monsoon climate, lotic habitats are also subjected to annual flooding of considerable magnitude.

Allen (1951) established a general relationship between the rainfall of single storm events and the rise in level of the Horokiwi Stream, New Zealand, and designated three categories of floods as follows:

	Rainfall (cm)	Rise in Level (cm)	Extent of Disturbance
Slight flooding	2.5–5	30–61	Some movement of surface gravel
Moderate flooding	5–7.6	61–91	Extensive movement of surface gravel; some movement of deeper gravel in regions of low stability; some changes in stream course; occasional bank erosion
Severe flooding	>7.6	>91	Widespread disturbance of gravel to a considerable depth; many changes in stream course; frequent bank erosion

Over a 3-year period, the Horokiwi Stream averaged 7.3 slight floods, 1.3 moderate floods, and 0.7 severe floods per year.

During a year with a series of major floods the annual mean faunal density in most portions of the Horokiwi Stream was between 40 and 50% of the values for the preceding year (Allen 1951). An area with unstable substrate was most severely affected; faunal densities were only 20–25% of preflood values. Maitland (1964) reported greater reductions in the fauna of sandy than stony areas during flooding, apparently because of the lesser stability of the former substrate.

Severe flooding of a Welsh mountain stream transported large quantities of gravel and reduced aquatic mosses by 80% (Hynes 1968). The fauna residing in gravel was considerably reduced, although the density of animals in the remaining moss did not decrease appreciably. Annual production of invertebrates was reduced to 25% of preflood estimates. Three species of mayflies *(Baetis rhodani, Ecdyonurus venosus, Rhithrogena semicolorata)* were eliminated by the flood. These species did not reappear until the next generation. Other aquatic insects that were severely reduced by the flood reappeared in increasing numbers before there was any chance of their having reproduced. This group included stoneflies, caddisflies, and elmid beetles. Analysis of size classes showed that hatching of eggs laid before the flood would not account for the reappearance of these species. Because the entire valley was affected by flooding, it is improbable that these insects recolonized the study reach by downstream drift. Hynes feels that those individuals were in the hyporheic zone deep within the stream bed during the flood and that they began to appear in samples as the populations became redistributed following decimation of the individuals inhabiting the surficial layer of gravel.

Other studies of the influence of floods on lotic habitats not particularly liable to spate include Jones' (1951) study of a river in Wales, Pomeisl's (1953) examination of an Austrian woodland stream, Maitland's (1964) study of a Scottish river, and Hoopes' (1974) study of a small forested stream in Pennsylvania. The latter case is a most dramatic example of a short-term flood. The flooding occurred in June as a result of Hurricane Agnes. Discharge increased 150-fold over several days. After the water subsided the devastation was described as follows: "Large amounts of rubble, gravel, and small boulders were heaped through the valley bottom. The stream substrate appeared to have been literally sand-blasted with major portions being scoured" (Hoopes 1974, p. 855). Yet by October of the same year a remarkable degree of faunal recovery was apparent. As stated by Thorup, who studied the effects of a short-term flood on a springbrook community, ". . . a very serious disturbance of the substrate does not change the community bound to this substrate, not even when the biotope in question is characterized by a high degree of constancy" (Thorup 1970, p. 456). Within 2 or 3 months of the flood event, the springbrook populations had recovered to normal levels with the exception of one caddisfly. *Wormaldia occipitalis* was in the pupal stage when the flood occurred and many of the fixed pupae, attached to rocks, were crushed as the rocks rolled downstream. *Wormaldia* was scarce the year after the flood and did not attain normal abundance until the following year.

Streams that are liable to flooding are characterized by invertebrate communities that are less diverse and abundant than those not regularly subjected to dramatic changes in discharge (Hynes 1970a, Siegfried and Knight 1977). In such streams, fluctuations in the density of benthos have been attributed largely to reduced abundances during spates, followed by a gradual buildup in numbers until the next flood event (e.g., Bishop 1973a, Turcotte and Harper 1982). If severe flooding is of frequent occurrence, the benthic fauna tends to be dominated by pioneer species (Fisher 1983). In a Utah stream chironomids and simuliids, which have short life cycles, were the predominant organisms during the initial stages of recovery following a severe flood (Moffett 1936).

Disturbance from flooding may, however, facilitate coexistence of species. In a small California stream, areas of substrate scoured by winter floods are initially recolonized by *Simulium virgatum*, the abundance of which gradually declines concomitant with increases in *Hydropsyche oslari* (Hemphill and Cooper 1983). By experimentally disturbing areas of substrate at different intervals, Hemphill and Cooper showed that periodic disturbance allows *Simulium*, a pioneer, to coexist with *Hydropsyche*, a superior competitor. Their data suggest that in the absence of flooding, *Hydropsyche* would competitively exclude *Simulium*, whereas frequent flooding would favor *Simulium*, thus supporting the applicability of the intermediate disturbance hypothesis to stream insect communities (Ward and Stanford 1983).

A detailed study of the aquatic insect community of a lake outlet stream (McAuliffe 1984) provides another demonstration that moderate disturbance resulting from changes in discharge may facilitate the coexistence of species. On the upper surfaces of large stones in deep water the caddisfly *Leucotrichia*, a sedentary herbivore limited to the upper surfaces of sunlit stones, accounted for the vast majority (70–80%) of the benthic animals. *Leucotrichia* competitively monopolized the space and excluded other species or kept their populations at low levels. However, in shallow water where stone surfaces are exposed during periods of low flow and on small stones that have a high probability of overturning during high flow, *Leucotrichia* (one generation per year) is unable to maintain competitive dominance, allowing colonization by insects with shorter generation times and resulting in a more equitable distribution of species.

An annual spring flood resulting from snowmelt runoff typifies the flow regimes of many high mountain streams. The aquatic insects residing in such habitats, although adapted to withstand this period of high discharge, are nonetheless influenced by it. Figure 8.5 shows the density patterns of macroinvertebrates (≈95% insects) in a high mountain stream during the ice-free period of a typical year (1975), and during a year (1976) of reduced spring runoff but with severe flooding in early August (Short and Ward 1980a). Density values are typically depressed during peak discharge in June, the reduced values partly reflecting emergence losses of winter stoneflies and reduced sampling efficiency during high water. During a typical year (1975), fauna density rapidly increases as runoff subsides, reaches maximum values in August-September,

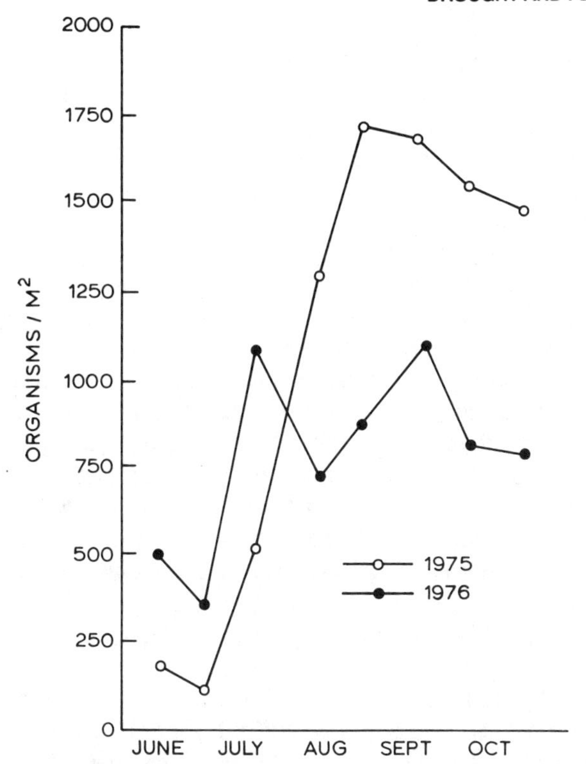

Figure 8.5 Total macroinvertebrate density during two consecutive open seasons in a high-elevation Rocky Mountain stream. [From Short and Ward (1980a).]

and remains at relatively high levels for the remainder of the ice-free period. In years with reduced snowpack (1976), spring runoff is much less severe and does not depress the fauna to as great an extent. It was therefore anticipated that density values would be greater in 1976. This was indeed the case until an exceptionally heavy summer rainstorm increased discharge 400% over a several-hour period. Following an initial drop in density, the fauna exhibited some recovery in the remainder of August and in September, but did not attain anywhere near the expected levels. Mean density of total macroinvertebrates during the ice-free period of 1975 (1467 organisms m^{-2}) was nearly twice the 1976 value (774 organisms m^{-2}), despite the less severe spring runoff the second year.

In many low-elevation streams in California, annual flooding occurs during winter. In a Sierra foothills stream a series of winter storms led to repeated flooding that reduced the standing crop of macroinvertebrates to less than 4% of preflood values (Siegfried and Knight 1977). Net-spinning caddisflies and stoneflies suffered the greatest depressions during floods. Baetid mayflies and aquatic moths (that construct shelters in depressions on rock surfaces) were less affected. Black flies increased during the winter floods and chironomid larvae dominated the riffle community after repeated washout. Siegfried and Knight (1977) concluded that streams subjected to annual flooding maintain a less diverse assemblage of benthic organisms than do streams for which flooding is not an annual event.

Flash-flooding exerts an especially important selective pressure on the fauna of desert streams (Bruns and Minckley 1980, Collins et al. 1981, Gray and Fisher 1981, Fisher et al. 1982, Meffe and Minckley 1987, Grimm and Fisher 1989). "Because the permanent desert stream ecosystem is a small trickle sustained by an enormous watershed, rare isolated cloudburst events in the drainage subject various portions of the stream to massive destruction. Recession of floods is rapid and the physical structure is restored in hours or (rarely) days, but with a much reduced biota" (Fisher et al. 1982, p. 94). Discharge increased from 0.23 m^3/sec to 53.3 m^3/sec during flash flooding of Sycamore Creek, Arizona (Fig. 8.6). This flood event was, however, atypical in that the rate of recession occurred slowly (Fisher and Minckley 1978). Such floods greatly increase suspended matter (to 55 g/liter in Sycamore Creek) and alter chemical conditions. The substrate may be scoured to considerable depths even in small streams (Fig. 8.7).

A single flash-flood removed 98% of the macroinvertebrate standing crop from the Sonoran Desert stream studied by Fisher et al. (1982). Gray (1981) calculated the depletion of aquatic insect populations (mean percentage losses) resulting from nine flood events that occurred over a 27-month period as follows:

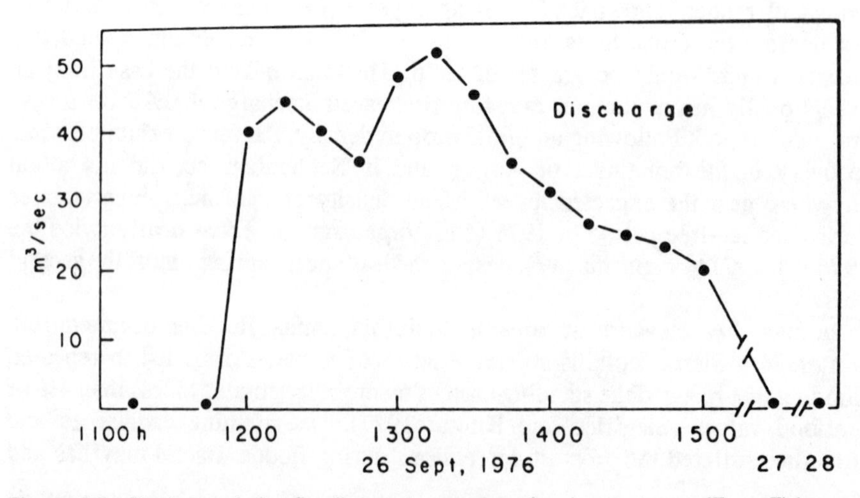

Figure 8.6 Discharge during flooding and recession of a desert stream. [From Fisher and Minckley (1978).]

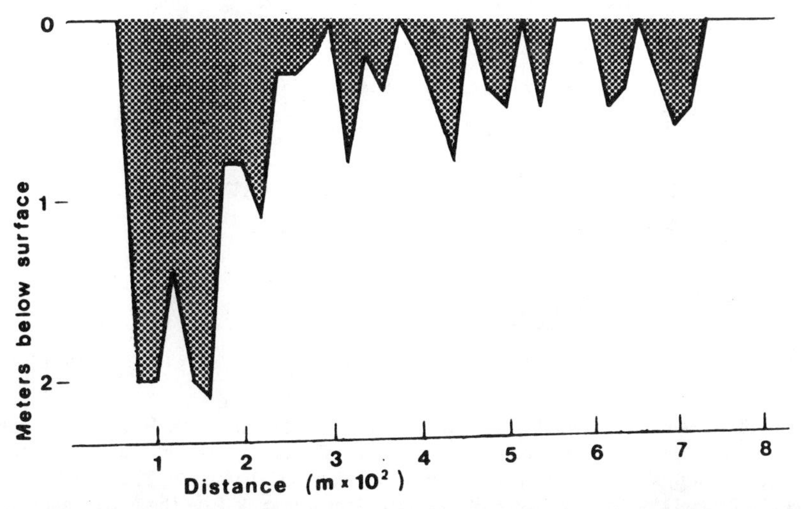

Figure 8.7 Net change in channel depth (zero represents the former channel level) at 20-m intervals along an 800-m reach of a desert stream, resulting from entrainment of sediments by winter flooding. [Modified from Collins et al. (1981).]

	Percent Loss
Ephemeroptera (all taxa)	96
Baetis quilleri	80
Leptohyphes packeri	96
Tricorythodes dimorphus	96
Trichoptera (all taxa)	74
Helicopsyche mexicana	74
Cheumatopsyche arizonensis	68
Diptera	
Chironomidae	78
Ceratopogonidae	77
Coleoptera	
Adults	0
Larvae	89
Hemiptera	
Graptocorixa serrulata	28
Abedus herberti	0
All taxa	86

Taxa with aquatic adults (beetles and bugs) exhibited behavioral avoidance by flying or swimming to protected areas during floods and were less severely depleted than amphibiotic groups such as mayflies and midges. Recovery was rapid even when flooding was widespread in the catchment (Fig. 8.8), and for most taxa, occurred primarily via aerial pathways (i.e., oviposition by aerial

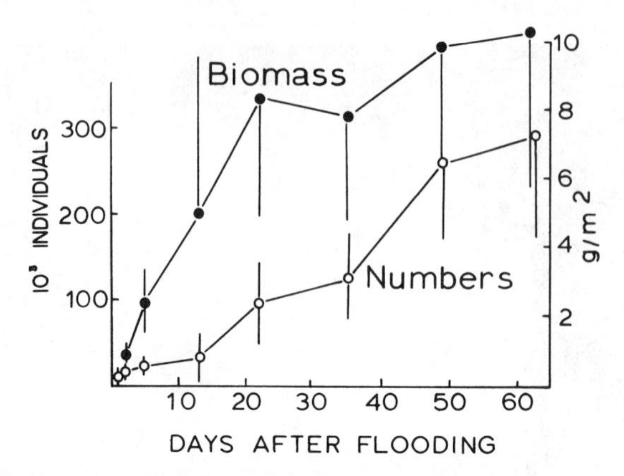

Figure 8.8 Changes in mean invertebrate numbers and biomass after flooding of a desert stream. Vertical bars show 95% confidence intervals. (From "Temporal succesion in a desert stream ecosystem following flash flooding" by S. G. Fisher, L. J. Gray, N. B. Grimm and D. E. Busch, Ecological Monographs, 1982, 52, 93–110. Copyright © 1982 by The Ecological Society of America. Reprinted by permission.)

adults that had emerged prior to flooding and recolonization by aquatic adults capable of flight). When flooding was more localized and upstream reaches remained undisturbed, downstream drift served as the principal recolonization mechanism of a few species. Within one month following flash flooding, the predominant species of mayflies and midges had completed several generations. Extremely rapid development, continuous reproduction, and the absence of dormant stages are common life history strategies exhibited by the aquatic insects of desert streams (Gray 1981). Gray estimated the probability of successful reproduction for species with 10-, 21-, 35-, and 50-day development times based on the historical discharge records (flood probability) of a desert stream (Fig. 8.9). The curves were constructed from calculations for each day of an annual cycle. For example, a species with a 50-day development time that oviposited in the stream on January 1 is considered to have reproduced successfully if no floods occur from January 1 to February 19. Because floods did not occur during that period in 7 of the 19 years of record, the probability of successful reproduction on January 1 is 7/19 or 0.37 (Fig. 8.9). Probability of successful reproduction so calculated is greater than 0.5 throughout the year for species with 10- or 21-day development times (Ephemeroptera, Chironomidae, Ceratopogonidae, and the corixid *Graptocorixa serrulata*). No selection for seasonal reproduction was exhibited by such species which reproduced continuously throughout the year. For species with longer development times (e.g., Coleoptera, Trichoptera) the probability of successful reproduction varied greatly with the season, being highest in spring and late summer. Indeed, the aquatic beetles reproduced only in spring and after the summer rains. The predominant caddisflies, however, reproduced continuously and were successful only during years with average precipitation. During years of extreme flooding, caddisflies were absent from main stream sites (Gray 1981).

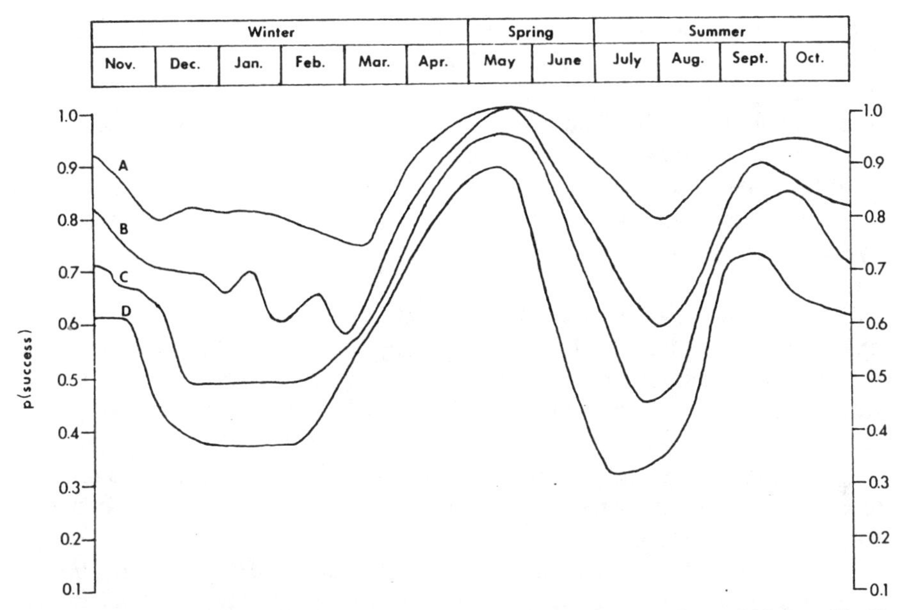

Figure 8.9 Annual flood probability curves for projected development times of 10 days (*A*), 21 days (*B*), 35 days (*C*), and 50 days (*D*) based on 19 years of discharge records for a Sonoran desert stream. Probability of successful reproduction, *p* (success), equals the probability of no floods during larval development if oviposition occurred on that day. See text for further explanation. [From Gray (1981).]

Perhaps the most impressive aspect of the effects of droughts and floods on aquatic insects is not the devastation of the fauna that results, but rather the remarkable ability of species to recover from such severe perturbations. The resiliency of the community is not, however, really so surprising if it is remembered that floods and droughts are natural, if infrequent, phenomena that have occurred throughout the evolutionary history of aquatic insects. Let us now examine in more detail the recolonization mechanisms that are responsible, in part, for the recovery of the fauna.

Recolonization Mechanisms Colonization dynamics encompass a range of spatiotemporal scales from the immediate recolonization of an overturned stone to long-term intercontinental faunal exchanges (Sheldon 1984). In the context of this chapter, consideration is restricted to the recolonization of stream substrates following drought or flood.

Aquatic insects recolonize denuded stream substrate via four major pathways (Williams and Hynes 1976b): (1) downstream drift, (2) upstream migration within the water, (3) upward vertical migration from the hyporheic zone, and (4) aerial migration by winged adults. Aerial colonization may occur from oviposition by terrestrial adults or from immigration by aquatic adults capable of flight (i.e., some Coleoptera and Hemiptera). On the basis of numbers of individuals, drift served as the most important colonization pathway for mayflies, caddisflies, and dipterans during summer in a Canadian stream (Table 8.5). Townsend and Hildrew (1976) also implicated drift as the major colonization mechanism of stream insects. Coleopterans utilized drift, upstream, and vertical pathways about equally. The few plecopterans collected depended exclusively on aerial colonization. Most taxa, however, utilized all four pathways. (This would probably also be true of plecopterans over an annual cycle.) Williams and Hynes concluded that excluding any of these pathways (e.g., truncation of drift by a dam) would disrupt normal recolonization, thus altering community structure.

Aerial pathways constituted the principal recolonization mechanism for the majority of insect taxa, and the sole pathway for some groups, after summer flooding of a desert stream (Table 8.6). Numerically, drift was the major pathway because of large numbers of individuals of a few species. Following recession of flood waters, a greater proportion of individuals moved upstream than during nonflood periods, suggesting a behavioral component to recolonization by this pathway (Gray and Fisher 1981). The vertical pathway was utilized by few taxa or individuals (Table 8.6).

In temperate streams of mesic regions with suitable substrate, the hyporheic zone may be a refuge for many species of insects and therefore contains an important reservoir of organisms to repopulate the surface benthos following drought (Hynes 1958, Williams 1977) or flood (Hynes 1968). However, such is not the case for tropical streams where oviposition by aerial adults serves as the most important recolonization pathway following drought (Harrison 1966, Hynes 1975).

Minshall et al. (1983a) applied the species equilibrium model (MacArthur and Wilson 1963) to macroinvertebrate recolonization of a riverine habitat subjected to severe short-term flooding resulting from dam failure. When the dam failed, a wall of water up to 23 m high was recorded in the river below. Discharge increased from 30 to 3500 m^3/sec and current velocities exceeded 12 m/sec. Over one year was required before recolonization was complete. Species

Table 8.5 Recolonization pathways utilized by aquatic insects during summer in a Canadian stream

	Relative Importance (%)[a]			
	Drift	Upstream	Vertical	Aerial
Ephemeroptera	67.9	21.4	7.1	3.6
Trichoptera	53.9	23.1	0	23.1
Plecoptera	0	0	0	100.0
Coleoptera	29.3	29.3	34.2	7.3
Chironomidae	41.3	8.9	14.9	34.8
Other Diptera	76.9	7.7	0	15.4

[a]Within-taxon comparisons based on numbers of colonizers utilizing each of the four pathways.
Source: Modified from Williams and Hynes (1976b).

Table 8.6 Recolonization pathways utilized by aquatic insects after summer flooding of a Sonoran Desert stream

	Drift	Upstream	Vertical	Aerial
Percent of numbers[a]				
Ephemeroptera	78.9	13.0	0.2	7.9
Trichoptera	26.3	12.6	tr	61.1
Coleoptera	0	0	0	100.0
Odonata	60.0	0	tr	40.0
Lepidoptera	0	0	0	100.0
Chironomidae	65.6	27.6	4.8	1.9
Other Diptera	8.6	42.1	1.4	47.9
Total (No. m^{-2} day^{-1})	3329	717	38	434
Percent of taxa	23	4	8	65

[a]Within-taxon comparisons based on numbers of colonizers utilizing each of the four pathways.
Source: Modified from Gray and Fisher (1981).

that feed by grazing tended to predominate during the first few months of recovery, whereupon they were replaced by collectors. There was a considerable lag phase (>250 days) before predators became established in substantial numbers. The trophic structure of the community had apparently not returned to preflood conditions by the end of the study (3.5 years following dam failure). Estimates of complete recovery of the benthic community vary from one year to over 3.5 years, depending on the measures employed (Minshall et al. 1983) as follows:

	Recovery Time (days)
Total numbers	≈375
Colonization rate	≥200
Number of taxa per collection	≈440
Shannon diversity	≈440
Cumulative number of taxa	≈700
Trophic equilibrium	>1275
Species equilibrium (immigration = extinction)	≈625

CURRENT AND DISCHARGE

"Current" refers to the velocity of moving water. It is a distance/time ratio usually measured in cm/sec. Discharge or flow refers to the amount of water moving down a channel. It is a volume/time function usually measured in cubic meters per second (m^3/sec). See John (1978) for a detailed review of the various devices and techniques used for measuring current and discharge. The ensuing material addresses spatial and temporal variations in current and discharge in lotic habitats and the influence of such variation on aquatic insects.

The Flow Regime The flow regime, by influencing mineral and organic substrate, suspended matter, and aquatic flora, plays a major role in structuring habitat conditions for stream insects. Basin geology, relief, climate, and terrestrial vegetation are primary determinants of the variations in discharge patterns of different lotic systems.

The first attempt at a global classification of river regimes (Haines et al. 1988), identified 15 types based on the shape of the annual hydrograph and named them according to the season of peak flow (mid-late spring, late spring-early summer, etc.). Although useful from geographic and hydrologic perspectives, a good deal of ecologically pertinent data were lost because of the reliance on monthly mean discharge values.

With daily flow values it is possible to simultaneously analyze a range of temporal components of the discharge regime (i.e., frequency, timing, and duration of floods and intermittency; flow variability; predictability of defined flow events), all of which may play a role in structuring aquatic insect communities. Statistical analysis of long-term (17–81 years) daily discharge records of gauged streams located throughout the conterminous United States distinguished several types of annual flow regimes (Fig. 8.10). The combined values of four hydrologic variables—intermittency, flood frequency, flood predictability, and overall flow predictability—were largely responsible for separation of the different types (Poff and Ward 1989). Predictability was separated into two components, constancy and contingency (Colwell 1974). A spring-fed stream with uniform flow (high constancy) exhibits highly predictable discharge, but so does a stream with highly variable flow, if the variance has a fixed periodicity (high contingency). Evolution of life history traits to accommodate even extreme flow variability is possible if such variability exhibits a predictable pattern. Three types of intermittent streams (not illustrated here) were distinguished, with reference to the duration of zero flow and the frequency and predictability of flooding (Poff and Ward 1989, 1990). Six types of perennial streams were identified, four of which are illustrated in Figure 8.10.

Current Velocity Maximum water velocity in a straight section of stream channel is normally greatest at or near the surface of the water in the center of the stream with greatly diminished values along the edges and near the bottom. Ideally, velocity declines exponentially with depth, with the mean column velocity 0.6 of the distance from the surface to the bed. Variations of depth and irregularities of the stream bed result in complex current patterns even in low-gradient stream reaches (Fig. 8.11).

It is the microcurrent regime near the surface of the stream bed, where many aquatic insects reside, that is of most interest to stream ecologists (Statzner et al. 1988, Davis and Barmuta 1989, Wetmore et al. 1990). Near-bed conditions may be hydraulically smooth (e.g., over mud bottoms, flat sheets of bedrock, or flat blades of macrophytes) or hydraulically rough (over irregular stream bottoms). In hydraulically smooth situations there is a "laminar sublayer" consisting of a microzone of nonturbulent water above the bottom surface (Davis and Barmuta

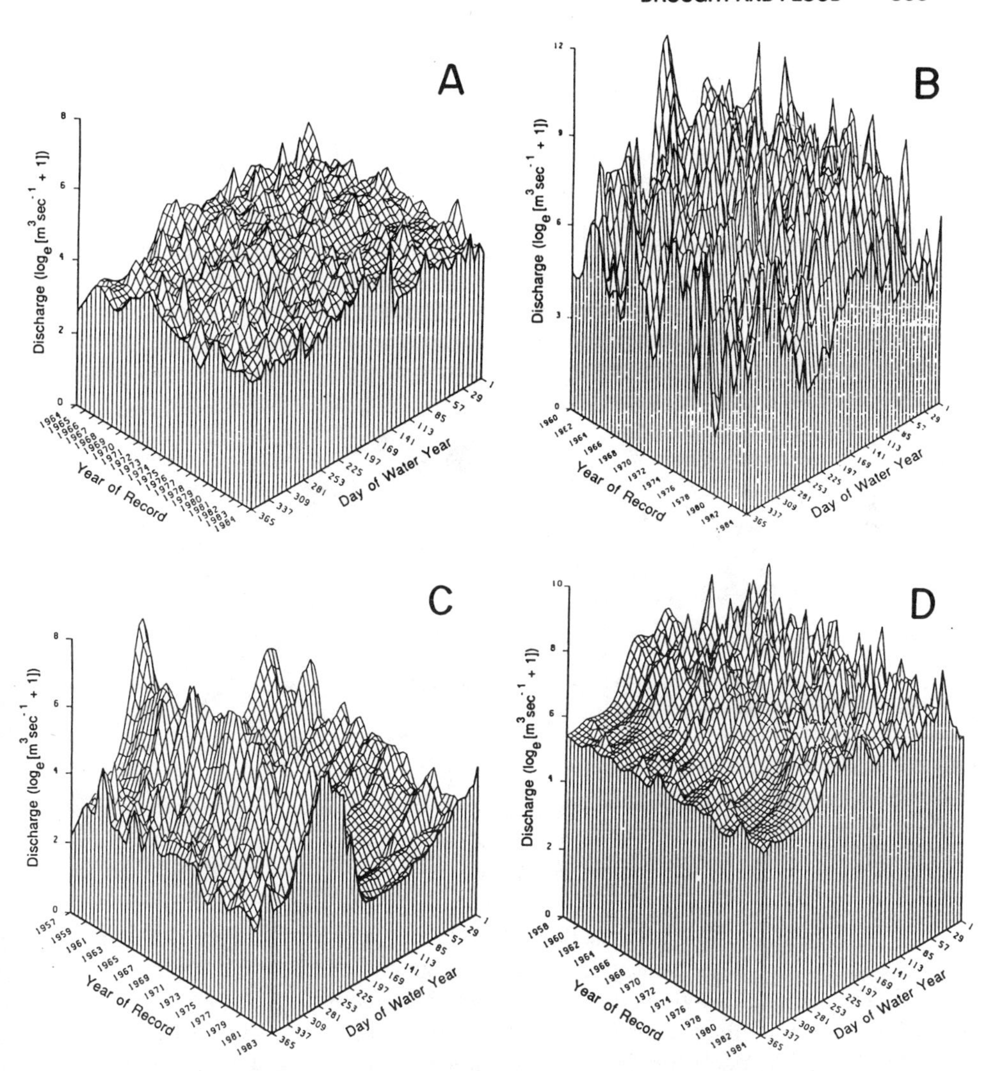

Figure 8.10 Streamflow histories based on long-term, daily mean discharge records for four gauged stream stations showing intra- and interyear temporal flow variability. Year of record is water year, October 1–September 30. Stream category designations from Poff and Ward (1989). (*A*) "mesic groundwater" Augusta Creek, Michigan; (*B*) "perennial flashy" Satilla River, Georgia; (*C*) "snowmelt" Upper Colorado River, Colorado; (*D*) "winter rain" South Fork of the McKenzie River, Oregon. [Modified from Poff and Ward (1990).]

1989). In hydraulically rough situations even the near-bed current micro-environment is normally turbulent. Fluid dynamics depend on (1) the spacing of individual substrate elements, (2) the height of individual elements, and (3) "relative roughness," the height of the individual elements in relation to water depth. When widely spaced, the vortex formed downstream of each substrate

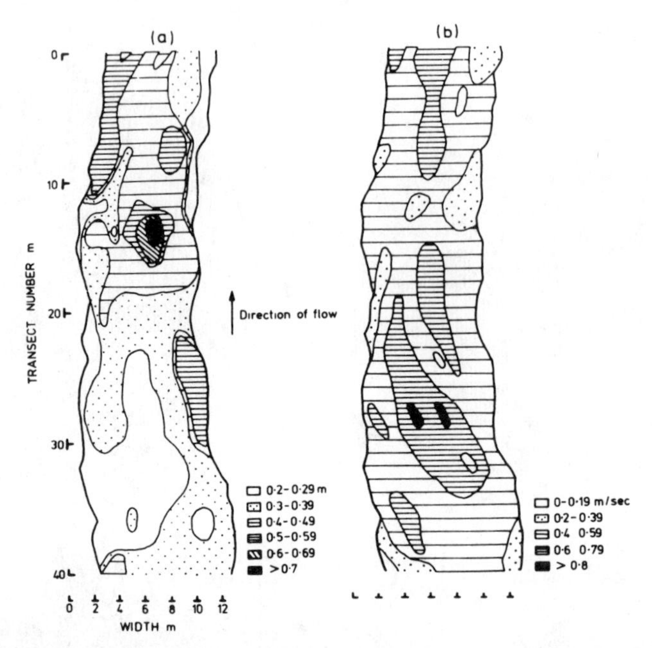

Figure 8.11 Depth (*a*) and current velocity (*b*) contours for a low-gradient reach of the River Coln, England. [From Mackey et al. (1982).]

element dissipates, whereas closer spacing creates overlapping vortices, high turbulence, and high velocities. Even closer spacing results in stable eddies between elements as the current skims over the top. Shallow rapids with rocks projecting above the water surface, such as in high-gradient riffles at low water, exhibit near-bed current patterns that are extremely complex with high velocities close to the bed. Nonetheless, highly heterogeneous microcurrent regimes prevail, which include low-velocity microdepositional habitats. Refuges from rapid current occur within clumps of moss and detrital aggregates, and in dead-water zones behind boulders and in cracks and crevices. Within beds of a submerged angiosperm, for example, the current velocity was reduced by 58–92%, compared to current in the open water (Madsen and Warncke 1983). Water velocities rapidly approach zero with increasing depth within the interstitial spaces of the hyporheic zone (Fig. 8.12).

Aquatic Insect Response In addition to their direct effects, current and discharge interact with other ecological variables (e.g., substrate, food supply, dissolved oxygen) to determine habitat conditions for aquatic insects. The complexity of the interactions often confounds attempts to assess the relative importance of the variables involved, especially under field conditions.

In streams with highly variable and/or unpredictable flow regimes, abiotic factors are presumed to play a more important role in structuring aquatic insect

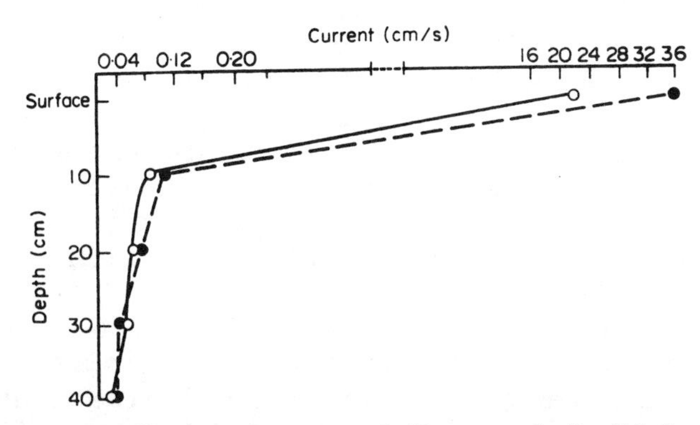

Figure 8.12 Two plots of intragravelar current velocities versus depth within the substrate in a Canadian river. [From Williams and Hynes (1974).]

communities than in streams with benign or predictable discharge patterns, where biotic interactions such as predation and competition become more important (Peckarsky 1983, Ward and Stanford 1983, Resh et al. 1988, Poff and Ward 1989).

For lotic habitats subject to prolonged periods of intermittency, other hydrologic variables (flood frequency and predictability, overall flow predictability) are of secondary importance and biotic interactions are thought to have little influence on community structure. The aquatic insects of harsh intermittent streams are under strong selective pressures for special physiological and life history traits enabling them to persist in the demanding environment. (The details of such adaptations are considered in Chapter 4.)

In perennial streams (and intermittent streams with less prolonged drying), flood frequency and predictability assume greater roles in structuring aquatic insect communities as discussed earlier in this chapter. The constant predictability of mesic groundwater streams (Fig. 8.10*A*), for example, is thought to increase the importance of biotic interactions, whereas in perennial flashy streams (Fig. 8.10*B*) the putative importance of biotic interactions is constrained by the high frequency of nonseasonal flooding. Moderate disturbance from flooding may facilitate coexistence by preventing monopolization of resources by one or a few species as might occur if disturbance was either more or less frequent (Ward and Stanford 1983).

Much of the research on water current-insect relationships is based on mean column velocities, rather than the near-bed microcurrents experienced by the animals. In an experimental trough, Dorier and Vaillant (1954) measured the minimum current velocities at which various species of macroinvertebrates were swept away and, the maximum velocities against which they will ascend, and compared these with the range of current occupied in nature. For species normally inhabiting smooth rock surfaces exposed to the current, the results roughly correspond to what is known of their natural occurrences. For example, larvae of the blephariceridd *Liponeura cinerascens,* which bear ventral suckers and inhabit the upper surfaces of rocks in torrential streams, were able to maintain their positions in the trough and ascend against a velocity of 240 cm/sec, the highest value tested. The absence of current refugia in the trough makes this approach of limited value for more cryptic stream insects. In addition, the microdistribution of some aquatic insects relates more to the "pattern and turbulence of flow rather than the velocity of flow" (Osborne and Herricks 1987).

Stream riffles contain areas of rapid current as well as low velocity microhabitats, whereas only slowly moving water normally occurs in pools. So-called pool species may, therefore, find suitable current microhabitats within riffles, but "riffle species" rarely make incursions into pools (Edington 1968). Daily fluctuations in discharge below a hydroelectric dam changed the Kennebec River, Maine (USA) from "one long uninterrupted stretch of swift flowing water" at high flow to a series of pools connected by slowly flowing channels at low flow (Trotzky and Gregory 1974). Conditions were not suitable for either rheostenic insects or pool species. The diversity and abundance of aquatic insects was most severely depressed at sampling stations having the lowest current velocity during periods of low flow.

Changes in discharge increase the number of aquatic insects drifting downstream with the current (Anderson and Lehmkuhl 1968, Minshall and Winger 1968, Scullion and Sinton 1983). An artifical release of water over a 2-day period resulted in substantially greater densities of drifting insects in a Welsh river (Brooker and Hemsworth 1978). The two most abundant insects exhibited quite different responses to the increased flow (Fig. 8.13). The chironomid *Rheotanytarsus* exhibited an immediate and dramatic increase in drift rate as discharge increased, followed by a rapid decline in drift density although discharge remained high. Nymphs of the mayfly *Ephemerella ignita* did not show an immediate response. The natural diel pattern with a nighttime drift peak was maintained, but the number of nymphs drifting was greatly increased during the period of elevated flow.

Much of the ecological information on current-insect interactions deals with the relationship between species distribution patterns and the corresponding spatial variations in water velocity (Scott 1958, Minshall and Minshall 1977, Edington 1968, Kamler 1967, Haddock 1977, Jaag and Ambühl 1964, Petr 1970a, Phillipson 1956, Needham and Usinger 1956, Gore 1978). For example, Needham and Usinger (1956) examined the distribution of aquatic insects across the width (30 m) of a single riffle. Depth and current measurements were taken at 10 points across the riffle (Fig. 8.14). Nymphs of the mayfly *Rhithrogena* exhibited a pattern similar to the current/depth profile (i.e., were most abundant in deep, fast water), but most aquatic insects either exhibited an inverse relationship to depth and current (*Ephemerella,* Fig. 8.14) or did not show consistent trends in their distribution across the riffle. All parts of the riffle were fully exposed to the sun, so differential shading had no influence on the distribution patterns. There was no apparent correlation between bottom type and faunal

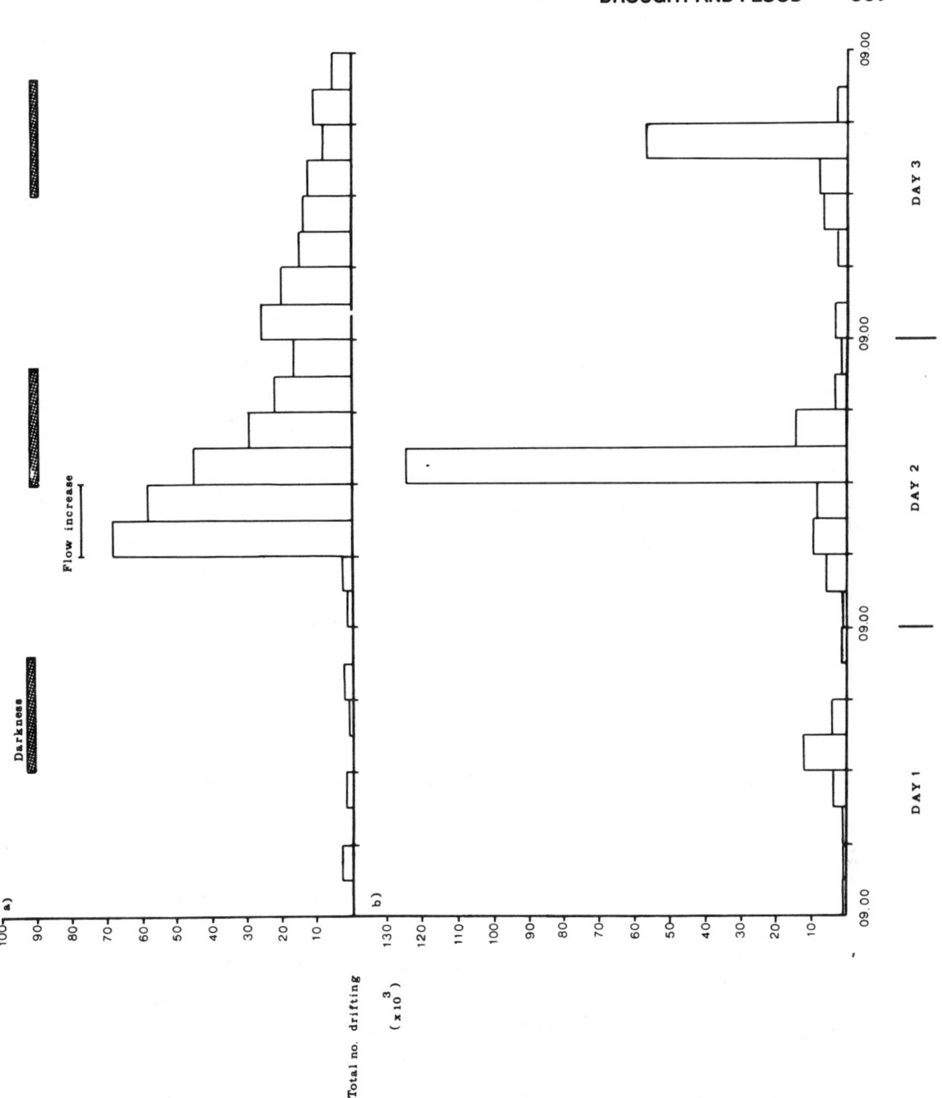

Figure 8.13 Differential drift responses of *Rheotanytarsus* (*a*) and *Ephemerella ignita* (*b*) to increased flow in a Welsh river. [From Brooker and Hemsworth (1978). Reprinted by permission of Kluwer Academic Publishers.]

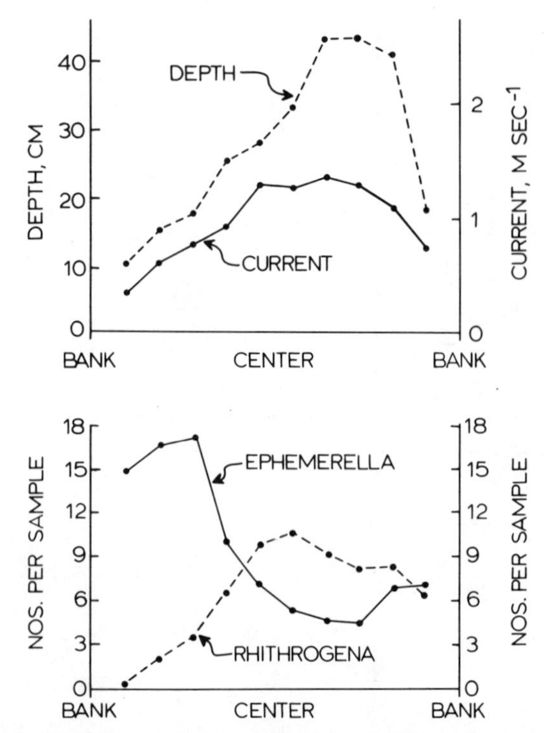

Figure 8.14 Distribution of two mayflies across the width of a riffle in relation to the current–depth profile. [Modified from Needham and Usinger (1956).]

distribution in this rather uniform riffle. Many of the aquatic insects, therefore, appeared to be responding to current or depth, or both of these factors.

Chutter (1969b) used Kendall partial correlation coefficients (Kendall 1948) to separate the effects of depth and current on species distribution patterns. Only *Cheumatopsyche thomasseti* and *Simulium* spp., of the 10 insect taxa whose distributions were correlated with depth and current in a South African river, exhibited a significant correlation between density and current when the effect of depth was eliminated. *Cheumatopsyche,* net-spinning caddisflies, and *Simulium,* filter-feeding black flies, both have feeding mechanisms that depend on current. However, the fact that the distribution pattern of a species is significantly correlated with depth when the effects of current were statistically eliminated does not negate their requirements for a particular range of current velocity.

Gersabeck and Merritt (1979) examined the current preferenda of black fly larvae in an artifical stream designed so that depth would remain constant over a range of current velocities. They constructed velocity preference curves (Fig. 8.15), assuming that larvae release their hold from the substrate as current velocities become unfavorable.

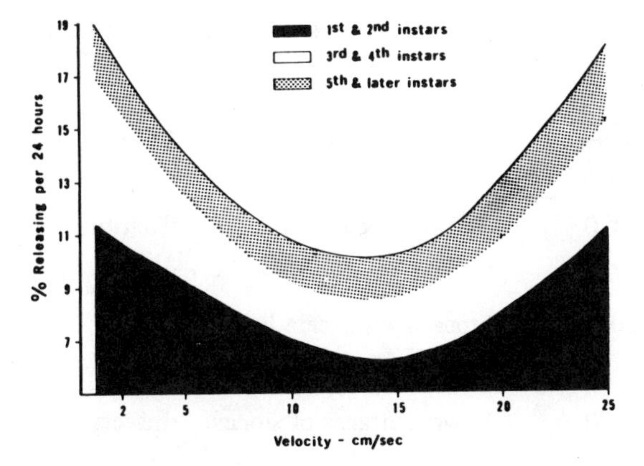

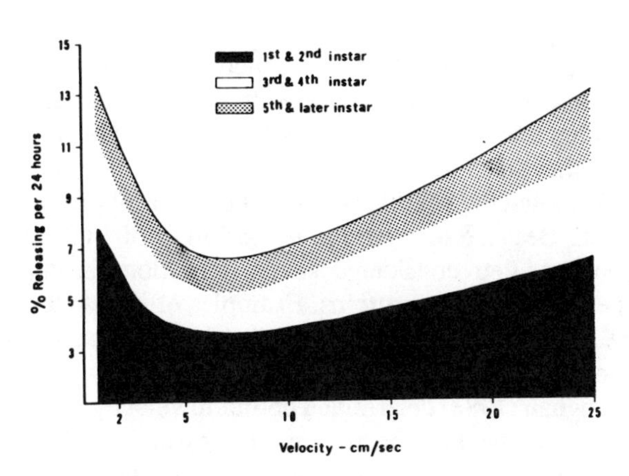

Figure 8.15 Velocity preference curves for *Prosimulium mixtum/fuscum* (*A*) and *Cnephia dacotensis* (*B*) (Simuliidae) based on the percentage of larval instars that released their hold from the substrate with changes in current velocity in laboratory channels. [From Gersabeck and Merritt (1979). Reprinted by permission of the Entomological Society of America.]

Scott (1958) grouped the caddisflies of an English river into three categories based on their distribution patterns relative to the surface current velocity (Table 8.7). He found that current preferenda corresponded with microhabitat preferences and food habits. Scott concluded that the larvae were largely responding to food supply rather than current per se.

By varying the placement of colonization trays with regard to current, Minshall and Minshall (1977) determined that the distribution patterns of some

Table 8.7 Current and microhabitat preferences and food habits of the larval caddisflies of the River Dean, England

Category	Current Preferenda[a] (cm/sec)	Microhabitat Preferenda	Predominant Food	Examples
Group 1	<20	Substrate below stones	Detritus	*Stenophylax stellatus*
Group 2	20-40	Stone faces except lower	Algae	*Glossosoma boltoni*
Group 3	>40	Lower surfaces of stones	Insects	*Hydropsyche fulvipes*

[a] Surface current velocity.

Source: From Scott (1958).

species of stream insects were largely a function of substrate type. However, some species were responding primarily or solely to current, while current operating in conjunction with substrate appeared responsible for the distribution patterns of yet other species.

Petr (1970a) studied the benthic fauna of the rapids of the Black Volta River in central Africa. Because the water was shallow (<30 cm) and the substrate quite homogeneous, Petr considered spatial variations in current to largely account for species abundance patterns. Examples of insects favoring rapid and slow areas are shown in Figure 8.16. The distribution of some species did not indicate distinct velocity preferences.

Orth and Maughan (1983) determined optimum velocity, depth, and substrate for the major taxa of benthic macroinvertebrates of a warm-water woodland stream. The combination of a current velocity of 60 cm/sec, a depth of 34 cm, and rubble-boulder substrate resulted in optimal diversity of benthic assemblages. Recognizing that habitat selection by benthos may be based on factor combinations, the investigators derived “joint preference factors” using the product of the individual preference factors, ranging from 0 (unsuitable) to 1 (optimum), for velocity, depth, and substrate. They found significant correlations between biomass and “joint preference factors” for the 10 species of aquatic insects examined. In addition, they found that the Froude number, an index of turbulence, was an important factor for some species.

A major problem in assessing current preferenda for aquatic insects is that the velocity measured in the water column may not accurately represent the current to which the organism is subjected. Ambühl (1959, 1962, Jaag and Ambühl 1964) developed an optical technique to measure microcurrent patterns in experimental channels containing artificial stones. Using the optical measurements, the channel bottom was divided into regions of known velocities and the distribution of various species of aquatic insects placed in the channel was

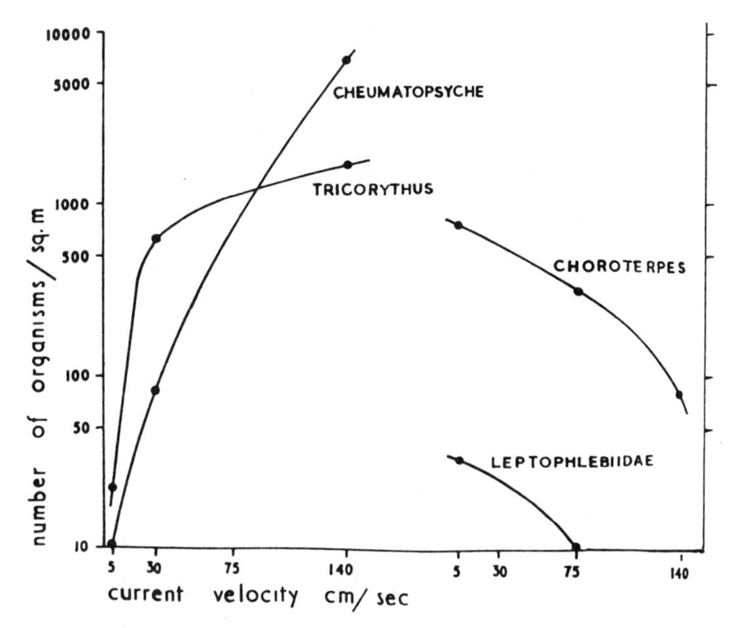

Figure 8.16 Relationship between surface current velocity and abundance of a caddisfly *(Cheumatopsyche)* and mayflies in a rapids on the Black Volta River in Ghana. [From Petr (1970a). Reprinted by permission of Kluwer Academic Publishers.]

plotted, resulting in patterns such as those shown in Figure 8.17. The current preferenda determined by this method generally correspond to the known occurrences of the species in nature. *Baetis, Rhyacophila,* and *Rhithrogena,* for example, are known to inhabit fast water areas, whereas *Ephemerella ignita* avoids areas of rapid current in the experimental channel and in natural streams.

Using a miniature current meter, Edington (1968) was able to measure accurately water velocity over the range of 5-200 cm/sec alongside the nets of caddisfly larvae. He found that the nets of the various species were concentrated in certain velocity ranges. Further research concentrated on *Hydropsyche instabilis,* a "rapids species" that constructs nets in fast water (15-100 cm/sec), and *Plectrocnemia conspersa,* a "pool species" that resides in stream pools or sheltered sites in rapids (0-20 cm/sec). Net form apparently accounts for the absence of *P. conspersa* from rapid water since the nets have a large surface area and in laboratory tests they disintegrated when velocity was increased to 25 cm/sec. In contrast, the nets of *H. instabilis* are streamlined and rigidly supported. At low velocities the nets appear to function normally. However, because the nets of *H. instabilis* have a small surface area, it may be necessary for this species to inhabit higher velocity water to enable the larvae to filter a sufficiently large volume of water to meet their food requirements. Under experimental conditions Edington found that the percent of larvae constructing nets varied as a function of the water velocity to which larvae were exposed as follows:

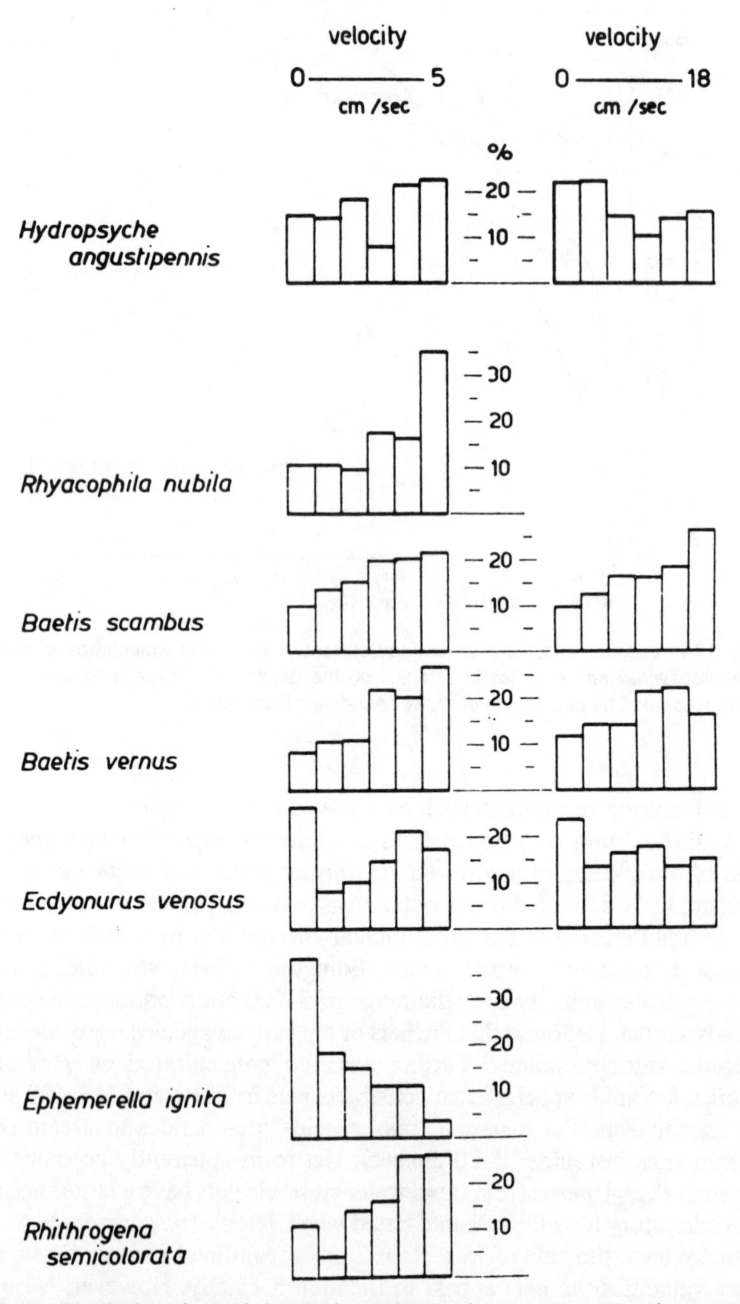

Figure 8.17 Distribution of aquatic insects in relation to microcurrent patterns (measured with an optical technique) in experimental channels. [From Jaag and Ambühl (1964). Reprinted with permission of Pergammon Press plc.]

	Current (cm/sec)		
	10	15	20
Hydropsyche instabilis	20%	48%	73%
Plectrocnemia conspersa	72%	50%	4%

Hydropsyche instabilis produced complete nets at velocities of 100 cm/sec, the highest velocity tested. Nets were only rarely produced at velocities of <10 cm/sec; no nets were constructed by larvae kept in still water. The proportion of aberrant nets increased with decreasing current. Field distribution patterns and the results of net construction in the laboratory tests are in general concordance.

Dense populations of *H. instabilis* on flat bedrock in a lake outflow enabled Edington to analyze the reactions of the larvae to changes in current velocity in a field situation. When baffles were installed to divert the current, larvae finding themselves in an area of low velocity moved to areas of higher velocity. When water was diverted from a well populated area to a previously dry portion of the stream bed, larvae readily moved into the newly created area of rapid current.

One of the earliest studies of the microdistribution of stream insects dealt with the positions of mayflies on boulders in torrential reaches of Rattlesnake Creek, Montana (Linduska 1942). At a surface velocity of slightly over 240 cm/sec, the heptageniid *Epeorus grandis* and two species of *Baetis* were the only mayflies to occur on the upper surfaces of boulders. These three species also occurred on other boulder faces. *Epeorus longimanus, Drunella doddsi,* and a third species of *Baetis* colonized the sides and the downstream surfaces of the boulders. *Rhithrogena doddsi* nymphs, when they occurred on boulders, only colonized the underside. Other species did not occur on the boulder faces, at least during periods of high discharge, but were found only in microhabitats with reduced current such as between or under stones or in accumulations of detritus.

In the stony-bottomed reach of an English river, Scott (1958) recorded the microhabitat distributions of the common species of caddisflies (Table 8.8). Scott felt that these spatial patterns largely reflected the interactions of current and substrate on the distribution of food. However, current appears to have some direct effects on the fauna. *Glossosoma* larvae, for example, normally inhabit exposed microhabitats, but tend to move to sheltered locations as current velocity increases (Fig. 8.18). Moving into sheltered microhabitats during periods of high discharge has been documented in other groups of aquatic insects (Phillipson 1956, Zahar 1951) and is undoubtedly a common phenomenon among rheophilic species. Pupal distribution patterns closely resembled larval distributions for some species of Trichoptera (Scott 1958). Pupae of *G. boltoni* and *S. pallipes* occurred in areas of higher velocity than the larvae, suggesting that final instar larvae move into faster water to pupate, which may be an adaptation to reduce the chances of stranding if the water level drops.

Maitland and Penney (1967) examined the microdistribution of black flies (Simuliidae) in a Scottish river. Most larvae occurred near the upstream end of the upper surface of boulders (Fig. 8.19), a finding in concordance with the

Table 8.8 Distribution patterns of Trichoptera in an English river based on the percent of total larvae of the important species occupying various microhabitats

	Exposed			Sheltered			
	Stone Faces			Stone Faces		Stream Bed	
	Upstream	Top	Sides	Underside	Downstream	Under Stones	Gravel Bottom
Glossosoma boltoni	23.9	27.9	21.2	7.9	13.8	1.5	3.8
Rhyacophila dorsalis	3.8	33.1	9.1	42.6	0.8	9.5	1.1
Hydropsyche fulvipes	1.3	5.2	3.5	73.5	0.9	12.1	3.5
Stenophylax spp.[a]	0.0	1.7	0.0	2.4	0.0	80.6	15.3
Ecclisopteryx guttulata	3.0	21.0	15.0	2.0	9.0	36.0	14.0
Odontocerum albiocorne	0.6	0.6	1.2	0.6	0.0	77.2	19.8
Silo pallipes	5.1	15.2	10.9	5.9	7.6	42.0	13.3

[a] *S. latipennis* and *S. stellatus* exhibited similar microdistribution patterns.

Source: Modified from Scott (1958).

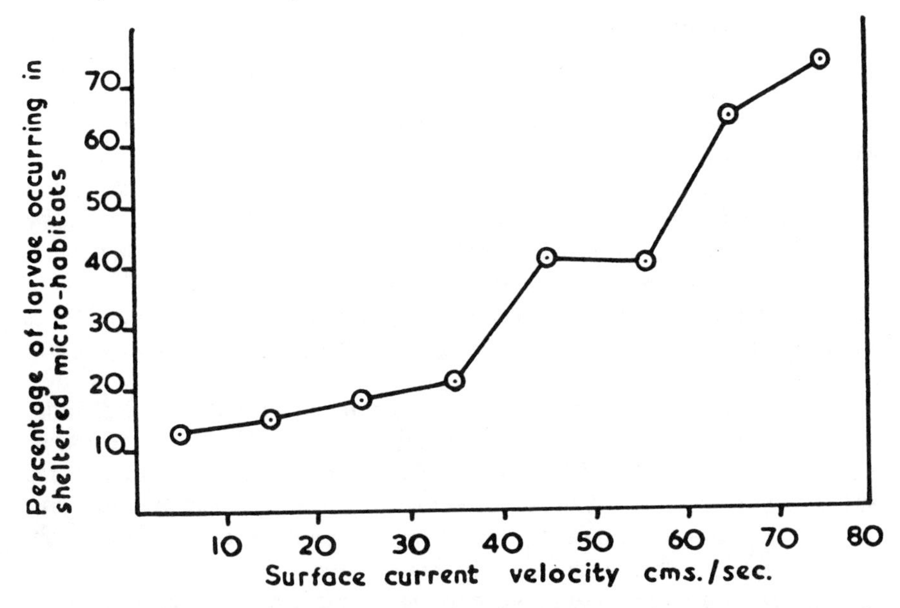

Figure 8.18 Microdistribution of *Glossosoma boltoni* (Trichoptera) showing the tendency for larvae to move from exposed microhabitats (the upstream, top, and sides of stones) to sheltered microhabitats behind and under stones as surface current velocity increases. [From Scott (1958).]

investigations of Grenier (1949). Final instar larvae move to the downstream face, or to crevices on more exposed surfaces, to pupate. The predominant larval attachment site at the upper front surface of stones enables larvae to filter large volumes of water in an area with little turbulence. The cephalic fans of simuliid larvae, used to filter food particles from the water, are held open by the force of

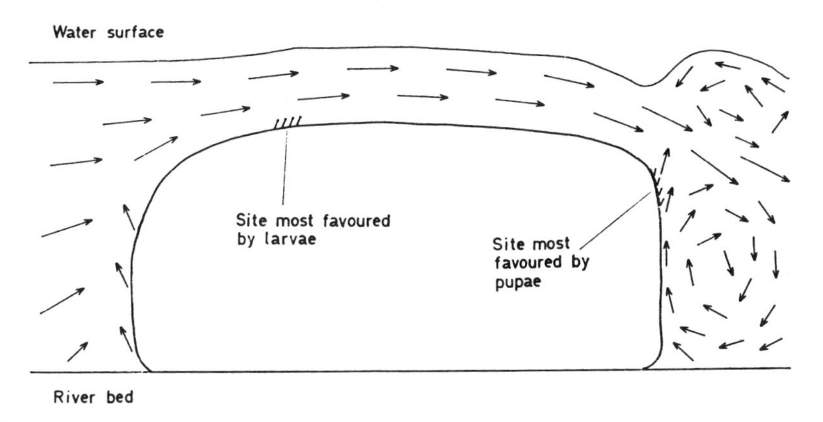

Figure 8.19 Typical microcurrent patterns in the vicinity of a boulder and site preferences of larval and pupal Simuliidae. [From Maitland and Penney (1967).]

the current. The cephalic fans of *Simulium ornatum* var. *nitidifrons,* for example, close and lose their feeding function at velocities <19 cm/sec (Harrod 1965). It would, however, be disadvantageous for immobile pupae to be exposed to high current velocities, especially during the formation of the cocoon when they are especially liable to dislodgement.

Pupal cases tend to all be oriented in the same direction in areas where flow is more or less laminar, but pupation typically occurs in locations of turbulent flow where cases are oriented in many directions (Fig. 8.20). Although the pattern may reflect the actual microcurrent patterns to which each individual is subjected, biotic interactions may also play a role. Under crowded conditions, the open end of each pupal case tends to be nearest the closed end of the adjacent case.

These microdistribution patterns generally apply to all species of Simuliidae in the river system investigated by Maitland and Penney (1967). There was no sharp spatial segregation of species, but where more than one species occurred on the same boulder, small areas tended to be dominated by larvae of the same species. This was not, however, true of pupae; the cases of one species frequently were in contact with cases of other species.

Results of much of the published work on aquatic insect responses to current should be critically examined, and in some cases reinterpreted, in light of our increasing understanding of near-bed microcurrent regimes. Because conventional sampling devices for riffles enclose areas of substrate containing a diverse range of microcurrent habitats (Davis and Barmuta 1989), such devices do not provide the appropriate spatial scale for addressing certain types of questions regarding insect distribution patterns. For example, late instar larvae of the filter-feeding caddisfly *Brachycentrus occidentalis* occur in locations that can be characterized by defined hydraulic variables (Wetmore et al. 1990). Individual larvae in reaches with very different depths, mean current, and substrate, were positioned in identical hydraulic regimes.

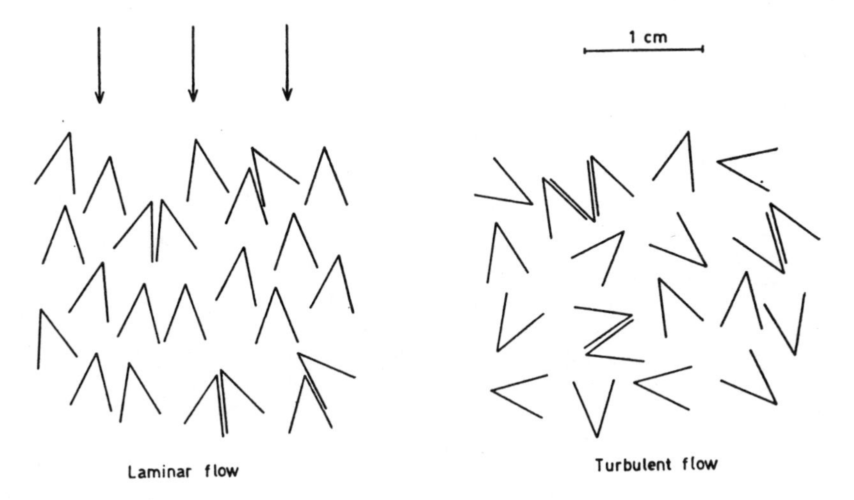

Figure 8.20 The orientation of simuliid pupae in relation to the type of flow present. [From Maitland and Penney (1967).]

9

OTHER ABIOTIC VARIABLES

Physical and chemical factors not covered in the previous chapters are included here. Topics considered include the influence of suspended sediment, light, oxygen, and other selected chemical variables.

SUSPENDED SEDIMENT

Turbidity, an optical property of water, is a measure of light attenuation by suspended particles. Highly turbid waters have larger concentrations of particulate matter than clear water; however, because light transmission is influenced by the sizes, shapes, and refractive indices of the particles (Sorensen et al. 1977), there is no absolute relationship between turbidity and the concentration (mg/liter) of suspended solids in a water sample. Suspended solids include both inorganic sediments and particulate organic matter (seston). This chapter deals largely with the former component.

It is estimated that at least 13.5×10^9 tons of sediment are transported to the sea each year via the world's water courses (Milliman and Meade 1983). The amount of suspended sediment in inland waters is highly variable in space and time, ranging from nearly imperceptible levels in the clearest lakes and streams to concentrations approaching and, in some cases, exceeding 10,000 mg/liter in the most turbid rivers. Detailed accounts of the physical and chemical effects of suspended particles on inland water bodies, especially in relation to optical properties of water, are contained in limnology books (e.g., Hutchinson 1957a, Dussart 1966, Bayly and Williams 1973, Cole 1983, Margalef 1983, Wetzel

1983). Wilber (1983) presents a general review of turbidity as an environmental factor in fresh and oceanic waters.

There is a surprising dearth of definitive data on the effects of suspended inorganic sediments on aquatic insects; most studies deal only with sediment deposition, or at least do not separate the effects of the particles while in suspension from those resulting from their deposition (Tebo 1955; Cordone and Kelley 1961; Chutter 1969a; Rosenberg and Wiens 1975, 1978; but see Culp et al. 1986).

Semiarid regions of the world contain argillotrophic (Gr. = clay-nourished) lakes and ponds, water bodies that are characterized by persistent and extreme turbidity (Daborn 1975). Wind-induced circulation is largely responsible for the continuous suspension of fine mineral particles in these shallow water bodies. Aquatic macrophytes and phytoplankton are rare or absent, presumably because of the rapid light attentuation by the turbid water. As the name suggests, the food base of argillotrophic lakes consists largely of bacteria and fine organic detritus, both of which may be intimately associated with the suspended clay particles. In Fleeinghorse Lake, an argillotrophic system in western Canada, the macrofauna consists of a few species of planktonic crustaceans and benthic insects (Fig. 9.1). Although some data are available for the crustaceans, the ecological roles of insects in argillotrophic lakes remain largely unexplored.

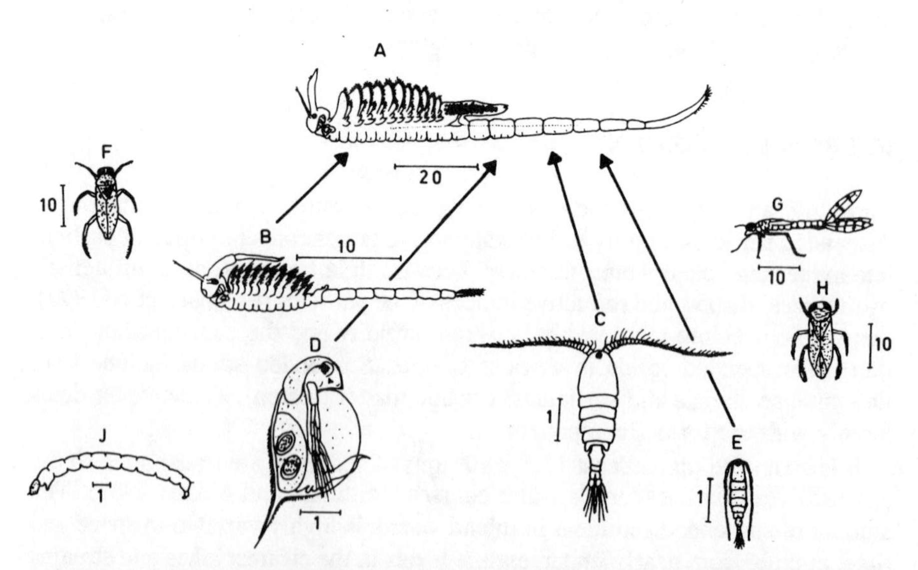

Figure 9.1 The fauna of Fleeinghorse Lake, an argillotrophic lake in Canada, consists of a few species of planktonic crustaceans (A-E) and benthic insects (F-J). Trophic pathways are indicated by arrows for the crustaceans; the ecological role of insects in such systems remains largely unexplored. (*A*) *Branchinecta gigas;* (*B*) *B. mackini;* (C) *Diaptomus nevadensis;* (*D*) *Daphnia similis;* (*E*) *Diaptomus sicilis;* (*F*) *Notonecta kirbyi;* (G) *Lestes congener;* (*H*) *Cenocorixa dakotensis;* (*J*) *Cricotopus* sp. Figure scales in millimeters. [From Daborn (1975).]

The net-spinning caddisfly *Hydropsyche separata* appears to be favored by turbid riverine conditions. It is abundant in turbid rivers but does not occur in clear streams (Smith 1979). Whereas *H. separata* is the predominant net-spinning caddisfly in the turbid Saskatchewan River, Canada, another hydropsychid (*H. recurvata*) replaces it as the dominant species in the clarified river segment below a reservoir. Although the sublethal effects of the altered temperature regime below dams preclude many species of aquatic insects (Ward 1976), this does not appear to be the case for *H. separata* in the Saskatchewan River since a small population occurs below the dam.

Several investigations of lotic waters have attempted to assess the effects of suspended inorganic sediment per se on aquatic insects. Gammon (1970) studied a small, low-gradient stream receiving sediment input from a stone quarry. Based on four years of research, he found that suspended solids loads <40 mg/liter above normal levels, with no appreciable deposition, resulted in a 25% reduction of macrobenthos of riffles. Densities were reduced 60% when suspended solids were 120 mg/liter or more above normal levels (usually some deposition occurred at these high inputs). Population reductions were attributed solely to emigration by drift since there was no evidence of mortality. Experimental inputs of sediment resulted in a linear increase of drift rates as suspended sediment concentrations increased from 18 to 160 mg/liter above background levels (10–28 mg/liter). Most taxa were reduced in approximately the same proportions as their abundance in the benthos. Nymphs of the mayfly *Tricorythodes* and adults of the beetle *Berosus*, taxa normally occurring in silted habitats, both increased during periods of heavy sediment input.

Based on studies of a Scottish stream with a sustained load of suspended inorganic matter from sand pit washings, Hamilton (1961) concluded that high turbidity alone does not adversely affect the macroinvertebrate fauna of a shallow lotic environment. During the study, very little sediment was deposited; the substrate was comparable, and algae and mosses occurred above and below the discharge. During an earlier period when sediment input was greater, the normal stream fauna was absent for a short distance below the point of sediment input (Badcock 1952), and a similar situation was noted during Hamilton's study when a breach formed in one of the settling tanks. In both instances, thick deposits of sediment covered the substrate. Nuttall and Bielby (1973) concluded that the decimation of bottom fauna in English streams receiving china clay wastes resulted not from turbidity or the abrasive action of suspended clay but rather from the deposition of fine particles.

A short-term release of sediment from a reservoir increased suspended solids 20-fold in the downstream river (Gray and Ward 1982). The current was sufficient to maintain the particles (75% silt, 25% clay) in suspension so that the mean particle size of the substrate was unaltered and faunal changes could be attributed largely to the effects of increased turbidity. Densities of some of the numerically dominant benthic species decreased (chironomids), others increased (*Tricorythodes* and oligochaetes), while yet others exhibited little change (baetid mayflies). The absence of recruitment by tubiferous chironomids during the high-turbidity period suggests that suspended particles may have interfered with larval feeding or respiratory activities by scouring or clogging the silk tubes. Despite high turbidities, growth of the filamentous alga *Cladophora* was stimulated by the sediment release, presumably because of nutrients associated with the sediment particles.

Volcanic ash deposition from the 1980 eruption of Mt. St. Helens affected large areas of the Pacific Northwest (USA). The caddisfly fauna of the Klickitat, a river moderately impacted by ash fall, did not appear to be adversely affected (Cushing and Smith 1982). In the Crispus River, located in a drainage basin that received up to 50 mm of ash, caddisfly numbers were only about one-half those of the Klickitat River. The net-spinning *Hydropsyche* species, previously common in the Crispus River, were nearly eradicated, whereas all size classes of *Arctopsyche,* a net-spinner with coarser mesh than *Hydropsyche,* were present. *Glossosoma,* a caddisfly that feeds by scraping algae from rock surfaces, also decreased significantly in the Crispus River following ash deposition. Gersich and Brusven (1982) subjected four species of stream insects to suspended Mt. St. Helens ash concentrations of 2000 mg/liter for 48 hours without discernible adverse effects despite substantial accumulations of particles on exoskeletal surfaces (Fig. 9.2). Of the 10 species of stream insects exposed to 2000 mg/liter of suspended Mt. St. Helens ash for 14 days, four exhibited higher daily

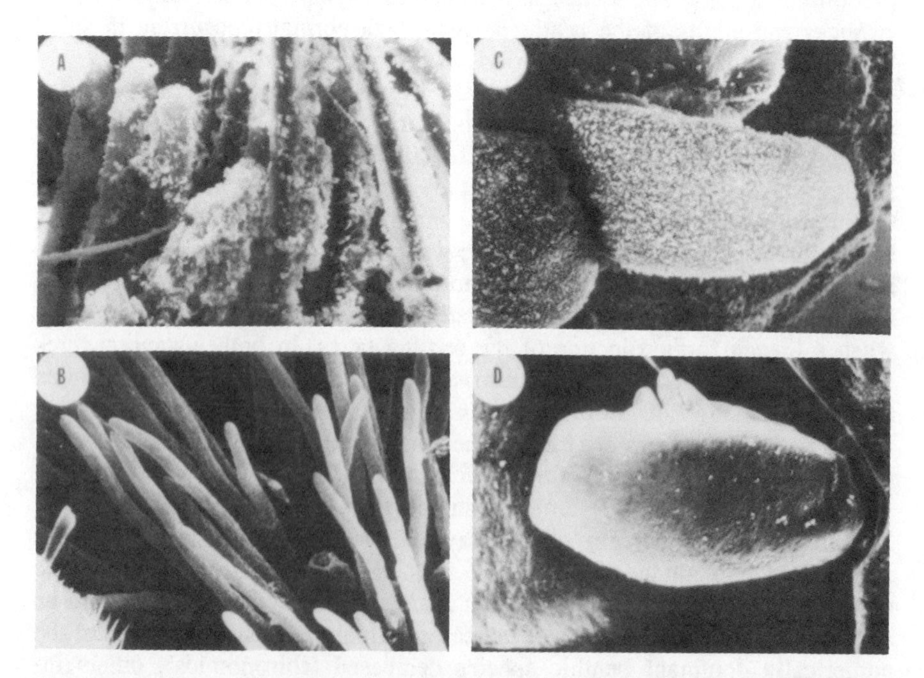

Figure 9.2 Scanning electron micrographs of gills of the stonefly *Hesperoperla pacifica* (*A* = ash-impacted; *B* = control 360X) and the mayfly *Drunella doddsi* (*C* = ash-impacted; *D* = control 220X). [From Gersich and Brusven (1982).]

mortality rates (Fig. 9.3) compared with controls, but by the end of 14 days, mortality was not significantly different from the controls for any of the species tested (Brusven and Hornig 1984).

It is perhaps not really surprising that stream insects as a group are relatively tolerant of elevated levels of inert suspended solids. Suspended particles occur to a lesser or greater degree in all water bodies, and relatively high concentrations periodically occur naturally in many undisturbed lotic habitats. The ability to withstand high levels of turbidity, at least for short periods, is an integral part of the evolutionary heritage of stream insects. Even insects of clean rocky streams may be adapted to withstand periods with high suspended solids loads, especially if such episodes occur during runoff, when elevated levels of suspended matter occur naturally and when deposition is minimal (Cline et al. 1982).

In summary, high levels of suspended solids may exert various direct and indirect effects on aquatic organisms. The potential effects of suspended particles on aquatic insects include (1) abrasion of respiratory epithelia, (2) clogging of respiratory structures, (3) reduced feeding rates, (4) reduced feeding

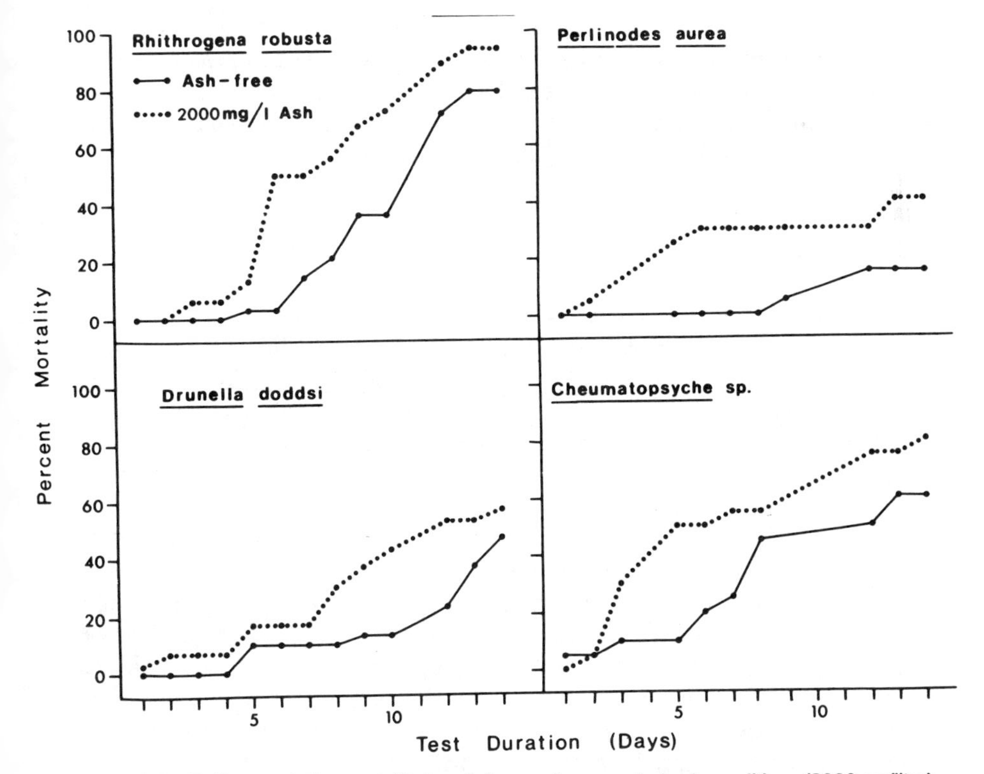

Figure 9.3 Daily cumulative mortality in ash-free and suspended ash conditions (2000 mg/liter) for four species of stream insects. [From Brusven and Hornig (1984).]

efficiency, (5) exposure to toxins (sediment particles provide adsorption sites where toxins may concentrate), (6) reduced vision, and (7) induction of drift. In addition to these direct effects, high levels of suspended solids may significantly modify the aquatic habitat by (1) reducing light penetration, (2) reducing primary production, (3) altering trophic structure (e.g., favoring detritivores over herbivores), (4) altering nutrient dynamics, (5) changing thermal conditions, (6) reducing oxygen levels, (7) altering behavioral patterns, (8) altering predator-prey interactions, and (9) changing the outcomes of competition.

It should, however, be stressed that few empirical data are available to document the extent to which these alterations, and their interactions, influence the aquatic insect fauna. The limited database suggests that the most dramatic effects of inert suspended sediment on aquatic insects occur subsequent to the deposition of the particles.

Major alterations of lotic insect communities may occur if the combination of mineral particle size and the stream's tractive force results in neither suspension nor deposition. The abrasive action of sand (0.5–2.0 mm) bouncing along the bottom (saltation) induced catastrophic drift of aquatic insects in a Canadian stream (Culp et al. 1986). The experimental addition of sand induced an immediate drift response in some insects (*Baetis*, *Cinygmula*, *Zapada*), whereas the enhanced drift rates of others (*Alloperla*, *Paraleptophlebia*) did not occur for 6–9 hours. Taxa exhibiting an immediate response tended to reside at the substrate surface, where the effects were instantaneous. *Alloperla* and *Paraleptophlebia* both reside within substrate interstices during the day (the time when sediment was introduced), and thus were not exposed to the scouring effects until later when noctural activities occurred at the substrate surface.

LIGHT

Light, in addition to influencing habitat conditions (e.g., food type, oxygen levels), may exert direct effects on aquatic insects. Various life-history components may be under photoperiodic control and light plays a major role in the diel drift patterns of stream insects. This chapter considers the influence of light on activity patterns (other than those related to life-cycle phenomena and drift) and on the distribution of aquatic insects. In addition, the special topic of lunar periodicity is briefly examined.

Many aquatic insects are believed to be negatively phototactic during most of their lives. For example, *Stenonema* mayflies that occur on the rocky shores of Douglas Lake, Michigan, reside under rocks during the day, but move to the tops of rocks at night, at least during calm weather (Lyman 1945). If a light is shone on the tops of rocks during the night, the nymphs move to the lower surfaces. When glass culture dishes covered with dark cloth are illuminated from below, most nymphs move to the tops of the rocks (where light intensity is lowest). Elliott (1968) demonstrated that the nocturnal positioning of five lotic mayfly species on the tops of rocks is an exogenous rhythm controlled by light intensity.

Aquatic insects that are normally negatively phototactic may, however, move toward light under conditions of respiratory stress. In aquatic habitats, moving toward light often means moving toward the air-water interface, where oxygen concentrations and current velocities (in running waters) tend to be highest. Because of this association, aquatic insects may use light rather than dissolved oxygen concentration as a cue for behavioral respiratory regulation [i.e., microhabitat positioning; see Wiley and Kohler (1984)].

In Norwegian ponds, the mayfly *Leptophlebia vespertina* is negatively phototactic until ice cover forms and oxygen levels decline, whereupon the phototactic response reverses resulting in migration of nymphs to microhabitats (shallow shoreline areas) with sufficient oxygen for winter survival (Brittain and Nagell 1981). A similar reversal of phototaxis was observed in another mayfly (*Cloeon dipterum*) in small ponds in Sweden (Nagell 1977). About 10 days after ice cover, when the entire water mass became anoxic, nymphs migrated en masse to the underside of the ice, where they remained throughout the winter. Percolation of meltwater through cracks in the ice formed a thin layer of oxygenated water in otherwise anoxic ponds. Laboratory experiments demonstrated that whereas both species of mayflies mentioned above were negatively phototactic in aerated water, the response to light reversed under anoxic conditions. As the concentration of carbon dioxide was experimentally elevated, nymphs of the mayfly *Baetis rhodani* moved from shaded to illuminated sections of an artificial stream (Scherer 1965).

The interaction of light and substrate may also influence the distribution patterns of aquatic insects. In Douglas Lake, Michigan, light plays a major role in influencing the distribution of the mayfly fauna (Lyman 1956). According to Lyman, "It seems that the factor of light alone is sufficient to eliminate these mayflies from open, sandy, or marly areas since these situations offer non-burrowing forms no means of protection from light" (Lyman 1956, p. 572). Rocks that are embedded in finer substrate are unsuitable since nearly all mayfly species of rocky shores are negatively phototactic and reside under stones during the day.

First instars of many lentic chironomids are positively phototactic, resulting in their appearance in surface strata of water bodies (Davies 1974). This apparently functions to disperse the larvae during a time when, because of their small size and transparency, they are least vulnerable to visual predators. Within a few days larvae become negatively phototactic and assume a benthic mode of existence.

Luferov (1966) considers light as the factor largely responsible for the differential depth distribution of lithophilous chironomid larvae. He identifies photophilous and photophobic species based on the illumination levels they inhabit. The greater depth (lower illumination) occupied by later instars of a species is explained as changes in photoresponsiveness as larvae mature.

Faunal distribution patterns in running waters may also exhibit a concordance with illumination. The mayfly *Baetis rhodani*, several caddisfly larvae, and the beetle *Helodes minuta* occurred at higher frequencies in illuminated as opposed to shaded sections of a Danish springbrook (Thorup 1966). Not surprisingly, these species are mainly algal feeders. A net-spinning caddisfly, in contrast, occurred with greater frequencies in the shaded sections of the brook. Production of aquatic insects, except black flies, was higher at an open site (4% canopy) than a shaded site (61% canopy) in a Utah (USA) stream (Behmer and Hawkins 1986).

Hughes (1966a), who examined three shaded-unshaded site pairs in mountain streams of the eastern Transvaal, Africa, found no significant differences in population densities of the total stream fauna in shaded versus unshaded reaches. Some species of aquatic insects were clearly more abundant in either shaded or illuminated stream segments, although, based on their distribution patterns, many taxa appeared unaffected by light. There was no overall density effect on any major taxonomic category. Any differences were discernible mainly at the species level. One congener of the mayfly *Centroptilum* was significantly more abundant in open reaches, whereas another species occurred with greater frequency in shaded segments.

With a black polythene canopy over a riffle on a New Zealand stream, daytime light intensity was reduced to 6% that at a control site (Towns 1981). All visible attached periphyton rapidly disappeared from the artifically shaded segment, where algal growth had been extensive during most of the previous year. There was no significant difference in the total densities of stream invertebrates between experimental (shaded) and control (open) sites, although differences in the abundance of certain taxa were apparent within a few weeks following canopy placement. A simuliid and a hydropsychid caddisfly were more abundant beneath the canopy, whereas certain caddisflies from three other families and all chironomids were more abundant at the control site. Species that feed by filtering organic particles from the current (simuliids and hydropsychid caddisflies) exhibited greater relative abundance values beneath the canopy; algal piercers (hydroptilid caddisflies) and those feeding by a collector-gatherer mode (chironomids) constituted a greater proportion of the total fauna at the control site (Fig. 9.4). Towns attributes these differences in small-scale spatial distribution patterns largely to the demise of algae beneath the artificial canopy, a conclusion also reached by Fuller et al. (1986), who artificially shaded a riffle of a stream in the United States. Although algal feeders were reduced, sessile filter feeders were favored as clean rock surfaces suitable for attachment sites appeared (see also Dudley et al. 1986).

Because several major environmental factors (e.g., temperature, autotrophic production, allochthonous inputs, attractiveness to ovipositing adults) may vary concomitantly with changes in illumination levels along watercourses, it is difficult to identify the causal mechanisms responsible for any correlations between faunal distribution and light. However, it does appear that illumination per se plays a major role in the microhabitat preferences of some aquatic insects.

Hughes (1966b) designed a series of experiments to examine the light responses of the mayflies *Baetis harrisoni,* a species occurring predominantly in illuminated stream reaches, and *Tricorythus discolor,* a species largely restricted to reaches shaded by dense riparian vegetation (Hughes 1966a). *Baetis*

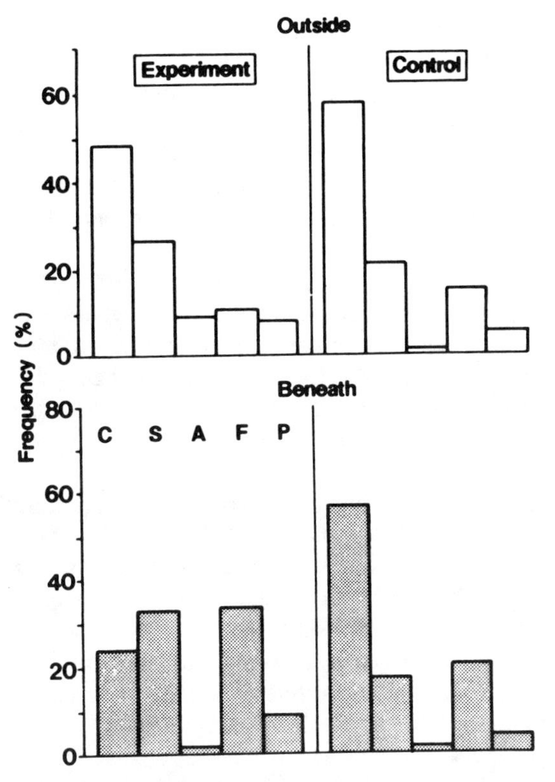

Figure 9.4 Percentage contribution by feeding categories outside and beneath canopy during experimental (canopy in place) and control (canopy absent) periods. C, collector-gatherers; S, scrapers; A, algal piercers; F, filterers; P, predators. Control samples, $n = 6$ outside and 6 beneath; experimental samples, $n = 27, 27$). (From Figure 3 in Towns, D. R. 1981: Effects of artificial shading on periphyton and invertebrates in a New Zealand stream. New Zealand Journal of Marine and Freshwater Research 15: 185–192.)

harrisoni, a streamlined swimming mayfly highly adapted to life in rapid current, typically occupies the upper surfaces of stones. *Tricorythus discolor,* a crawling nymph poorly adapted to current, occurs beneath stones.

In the case of *T. discolor,* a negative phototactic response directs nymphs away from regions of high light intensity and skototaxis, an attraction toward dark objects, guides nymphs to the microhabitat beneath stones. In addition, *T. discolor* exhibits a strong orthokinetic response to light intensity. This means that nymphs move continuously (although randomly) under high light intensities, whereas locomotion practically ceases under low light conditions.

The results of laboratory experiments are less conclusive for *B. harrisoni.* The field distribution of nymphs, on the tops of rocks in open reaches of streams, appears to result from the interaction of responses to light and current. The strong, positive phototaxis of this species accounts for its predominance in illuminated stream reaches. However, *Baetis* nymphs exhibit an orthokinetic response to light similar to that shown by *Tricorythus* (Fig. 9.5). Orthokinesis would, therefore, counter the tendency of nymphs to inhabit well-illuminated areas of the stream. Hughes (1966b) feels that some "as yet unidentified" aspect of current suppresses the orthokinetic response, thus explaining the distribution of *B. harrisoni* on the tops of rocks in illuminated reaches. Observations in the natural habitat show that nymphs are continuously active on the upper surfaces of stones until a suitable position is attained when "all movement suddenly ceases," suggesting that microcurrent conditions supersede the orthokinetic response (Hughes 1966b).

Baetis nymphs also exhibit a dorsal light response that maintains the organism's normal primarily dorsoventral body orientation (Hughes 1966c). This response is especially effective in maintaining an upright position in swimming nymphs. A ventral light reflex has been demonstrated for backswimmers (Notonectidae), aquatic bugs that swim with their ventral surface upward (Rabe 1953).

Vertical Migration Larvae of some species of the phantom midge *Chaoborus* exhibit extensive vertical migrations (see Table 3.2). The first larval instars are nonmigratory and remain near the surface; as larvae mature, they tend to move progressively deeper during the day, resulting in increasingly greater vertical migrations (Swift and Forward 1980). Light plays a major role in the diel vertical migration of *Chaoborus* (e.g., Teraguchi and Northcote 1966, Chaston 1969). First-instar larvae exhibit positive phototaxis, whereas mature larvae are negatively phototactic (Berg 1937, Cook and Connors 1964).

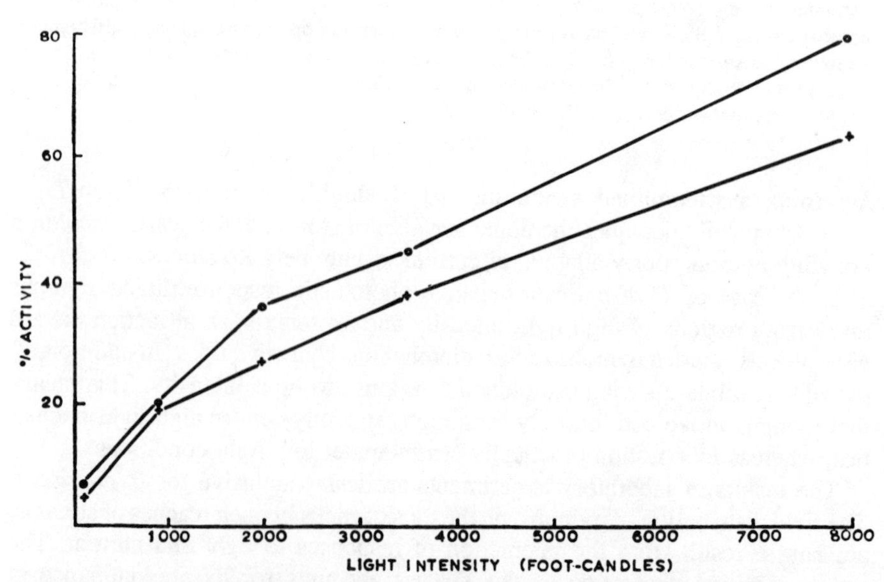

Figure 9.5 Effect of different light intensities on degree of locomotory activity shown by nymphs of the mayflies *Baetis harrisoni* (O) and *Tricorythus discolor* (+). [From Hughes (1966b).]

Swift and Forward (1980) examined the photoresponses of mature *C. punctipennis* larvae under laboratory conditions. On the basis of the light intensity required to evoke a negative phototactic response, larvae showed maximum sensitivity to wavelengths of 400 nm (the short end of the visible spectrum). Larvae typically occur in lentic habitats stained with humic acid colloids. Such waters rapidly attenuate light of all wavelengths, but are most transparent to light near the long end of the visible spectrum (Hutchinson 1957a). Swift and Forward showed that the phototactic response of mature larvae followed a circadian rhythm with an increase in negative photoresponsiveness in early morning to a daytime maximum and a decrease in early evening to a minimum at night. LaRow (1968, 1969) demonstrated an endogenous component in the vertical migration of *Chaoborus,* the timing of which is modified by light, to accommodate seasonal changes in day length. The circadian rhythm of photoresponsiveness may fine-tune the seasonal adjustments of vertical migrations to maximize the time spent feeding in the water column while ensuring that larvae are not exposed to visual predators in surface strata during the day.

Lunar Periodicity Lunar periodicity of emergence is known for several marine insects and has been intensively studied in the chironomid *Clunio marinus* (Caspers 1951, Neumann 1976). Moonlight is the synchronizing agent for the semilunar rhythm of emergence exhibited by intertidal populations of *C. marinus* (Fig. 5.3). Emergence is thus synchronized with periods of low tide when the habitat is not subject to wave action and the transformation from aquatic pupa to terrestrial adult is least hazardous.

Lunar patterns of emergence also occur among insects residing in lakes (Hartland-Rowe 1955, 1958; Macdonald 1956; Corbet 1958, 1964; Fryer 1959b; Corbet et al. 1974). In equatorial Lake Victoria, Africa, Hartland-Rowe (1955) found that the wood burrowing mayfly *Povilla adusta* exhibits a striking periodicity of emergence despite the constancy of environmental conditions. Corbet (1958) examined the emergence patterns of 37 species of aquatic insects in Lake Victoria and reported lunar periodicities for the following:

	Emergence Peak (days after new moon)
Ephemeroptera	
Povilla adusta	17–18
Trichoptera	
Athripsodes stigma	23–29
A. ugandanus	8–9
Diptera	
Clinotanypus claripennis	7–8
Tanytarsus balteatus	2–5
Chaoborus edulis	24–27

The time of peak emergence varies considerably between species. *Povilla,* for example, emerges shortly after full moon, whereas the emergence of *Chaoborus* peaks shortly before new moon. Emergence of three species of *Chaoborus* in a western African lake also peaked at or shortly before new moon (Hare and Carter 1986). At least one nonmarine aquatic insect is reported to have two emergence peaks during a lunar cycle [Tjønneland (1962) cited in Corbet (1964)].

Differential responses to the lunar cycle have been documented for different populations of the same species. The mayfly *Povilla* exhibits a striking lunar emergence periodicity in some lakes (Fig. 9.6), yet some populations fail to show a lunar rhythm (Corbet et al. 1974). The African midge *Chironomus brevibucca* emerges at full moon in one location (Fryer 1959b), but at new moon in another (Corbet 1964).

Temporal changes in lunar illumination may provide the only distinct environmental cue for aquatic insects in the inland waters of equatorial regions; this may be responsible for maintaining the synchronous emergence patterns of some tropical species with the selective advantages conferred by such synchrony (Hartland-Rowe 1984). As postulated by Fryer (1959b), it may be that some species respond to the lunar photoperiod in a manner similar to the well-known response of many temperate organisms to the solar photoperiod. Indeed, for aquatic insects such as *Povilla* that reside in burrows by day and forage at night, moonlight may provide the only phased photic information.

Studies of *Povilla* in western African lakes showed that the most distinct lunar emergence periodicity occurred in clear lakes (Corbet et al. 1974). Populations in the more turbid lakes exhibited less synchronous emergence patterns or failed to show any lunar periodicity. Studies by Hartland-Rowe (1958) suggest that the synchronous emergence of *Povilla* may be controlled by an endogenous rhythm initiated early in larval life by the lunar photoperiod. Changes in lunar illumination would be readily detectable by aquatic organisms only in waters of high clarity.

Among stream-dwelling insects, moonlight may effect drift rates. Species that normally exhibit nocturnal peaks in numbers of drifting organisms may not show a nighttime increase during clear nights with a full moon (Anderson 1966).

DISSOLVED OXYGEN

Dissolved oxygen is an environmental variable of considerable importance to those aquatic insects that depend on oxygen in solution to meet their respiratory needs. Respiratory adaptations are considered from an evolutionary perspective in Chapter 2. This chapter begins with a brief and general description of the factors influencing temporal and spatial variations of dissolved oxygen in aquatic systems. The interactions of oxygen with current, substrate, and temperature are then considered in the context of aquatic insect ecology. This is followed by an examination of the extent to which oxygen distribution patterns influence the distribution of aquatic insects. Finally, adaptations to low-oxygen waters are considered.

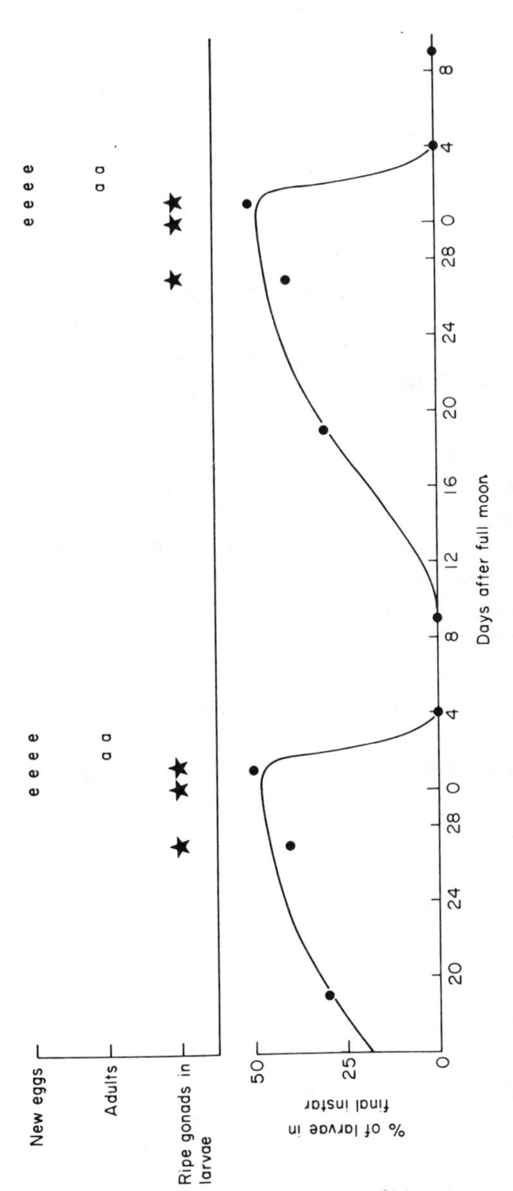

Figure 9.6 Periodicity of development of the mayfly *Povilla* in Barombi Mbo, a crater lake in Africa, in relation to the phases of the moon. [From Corbet et al. (1974). Reprinted with permission of the Zoological Society of London.]

The main sources of dissolved oxygen are the atmosphere and photosynthesis by aquatic plants. The solubility of oxygen is a function of water temperature, pressure, and salinity. Cold water holds more oxygen at saturation than warm water. Aquatic habitats at high altitudes (low barometric pressure) hold less oxygen than do those near sea level. However, the lower temperatures characterizing high altitude aquatic habitats often counters the influence of the lower barometric pressure on oxygen solubility. The decrease in oxygen solubility with increasing salinity, while not appreciably affecting oxygen concentrations in freshwaters, significantly lowers saturation values of marine and inland saline waters. Oxygen solubility coefficients for freshwaters are available in limnology textbooks (e.g., Goltermann 1975, Cole 1983, Wetzel 1983) and in the review by Mortimer (1981). Green and Carritt (1967) present oxygen saturation values for saline waters.

Respiratory uptake, especially bacterial decomposition of organic matter, is largely responsible for oxygen depletion in water bodies. Autumn leaf fall into forest pools typically results in oxygen deficits as the leaves decompose. Likewise, oxygen uptake from the decomposition of organic lake sediments reduces dissolved oxygen levels in the overlying waters, which may result in anaerobic conditions in the hypolimnion of eutrophic lakes by the end of summer stratification.

Because molecular diffusion is a very slow process, turbulent mixing is largely responsible for maintaining saturation in well-oxygenated habitats (e.g., high-gradient mountain streams) and for restoring oxygen to under-saturated waters. The turbulence of high-gradient streams ensures that dissolved oxygen levels are normally near or slightly above saturation. Even slow-flowing lotic habitats are generally well mixed and exhibit oxygen deficits only under special circumstances (Hynes 1970a).

Many lentic water bodies exhibit large spatial and temporal variations in oxygen levels. During periods of thermal stratification only the epilimnion is well mixed. Oxygen carried into the depths during the previous overturn of the lake is depleted by decomposition processes, but is not renewed because of the stagnant conditions and because photosynthesis is confined to the photic zone.

While epilimnetic waters remain saturated with oxygen, by the end of summer stratification the bottom waters of highly productive lakes may be anaerobic. At fall overturn the lake again becomes saturated with oxygen from top to bottom. Ice cover isolates the entire water body from the atmosphere and prevents wind-induced mixing. Clear ice without snow cover transmits sufficient light for considerable photosynthesis by phytoplankton that may result in oxygen supersaturation. More typically the lack of exchange of atmospheric gases and the screening effect of snow cover on the ice causes a reduction in oxygen beneath the ice. In cold climates the entire water mass of shallow, highly organic ponds may become anaerobic before ice breakup in the spring (Nagell and Brittain 1977).

In tropical regions dense mats of floating hydrophytes such as water hyacinth (*Eichhornia*) (which add oxygen to the air rather than the water) may create anaerobic conditions in the underlying water by impeding circulation and shading submerged plants.

Water bodies containing well-developed submerged plant communities may exhibit diel fluctuations in oxygen content as the balance between photosynthesis and respiration shifts over the dark-light cycle (e.g., Eriksen 1984). As an extreme example, Kushlan (1979) recorded fluctations from 4% to 200% oxygen saturation over a single diel cycle in a subtropical pond (Fig. 9.7).

Oxygen-Current Interactions

In the introduction of their important paper on the effects of current on aquatic organisms, Jaag and Ambühl (1964, p. 32) state, in reference to respiratory physiology, that "the movement of the water may be of greater importance than its oxygen content." They further state that "it is fundamentally impossible to discuss the significance of one component without at the same time taking account of the other." It is this interaction between oxygen and current that serves as the focus of the following discussion.

Insects inhabiting running waters typically differ in certain respiratory characteristics from species residing in standing-water bodies. Some of these basic differences are presented in relative terms in Table 9.1. Lentic species are characterized by independent-type respiration, since oxygen consumption re-

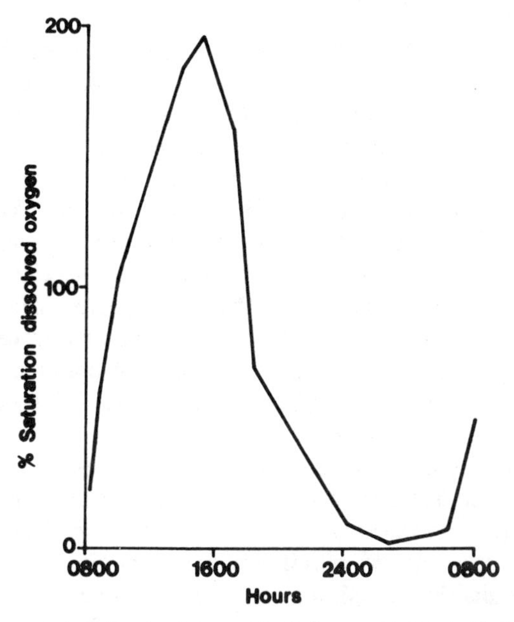

Figure 9.7 Diel fluctuation of dissolved oxygen in Cottonmouth Pond, Florida, during a period of low water (April 29–30). [From Kushlan (1979). Reprinted by permission of Kluwer Academic Publishers.]

Table 9.1 Contrasting features in some general respiratory characteristics of insects in running and standing waters

Characteristic	Standing Waters	Running Waters
Type of respiration	Independent	Dependent
Metabolic rate	Low	High
Tolerance of low O_2	High	Low
Ventilatory movements	Well developed	Poorly developed[a]

[a] Well developed in case- and tube-dwelling forms.

mains fairly constant over a wide range of environmental oxygen concentrations. In contrast, many lotic insects exhibit dependent respiration and are unable to regulate oxygen consumption, which varies directly with the oxygen concentration of the surrounding medium (Fox et al. 1937, Knight and Gaufin 1964, Hynes 1970a). Nagell (1974) cautions against interpreting or predicting distribution patterns in nature based solely on the results of laboratory experiments. The damselfly *Lestes disjunctus,* for example, is a poor regulator yet resides in temporary ponds exhibiting large diel fluctuations in oxygen (Eriksen 1984). Eriksen postulates that the reduction in metabolic rate induced by low oxygen levels during the night is energetically efficient for a visual predator and enables this species to complete development during the wet phase of the temporary habitat.

It was demonstrated some time ago that certain rheophilic insects exhibit higher metabolic rates than related species of similar size from standing waters (Fox and Simmonds 1933, Fox et al. 1935, Walshe 1948). Tests conducted with both anesthetized and unanesthetized animals produced consistent results, indicating that differences were not merely artifacts caused by higher activity rates of lotic than lentic species under experimental conditions. For example, using anesthetized mayfly nymphs of the same size and from the same family (Baetidae), Fox et al. (1935) demonstrated that the stream species *(Baetis rhodani)* consumed 4 times more oxygen than the pond species *(Cloeon dipterum)* at 10°C and nearly 3 times more at 16°C. All nymphs recovered fully from the anesthetic. The higher metabolic rates (oxygen consumption) of species from running than standing waters, measured under identical laboratory conditions, appear to be a general phenomenon.

The high oxygen requirements of lotic species, in association with a dependent-type respiration and a limited ability to increase oxygen uptake by ventilatory movements, undoubtedly account for their general intolerance of low-oxygen conditions, compared with insects from standing waters. Because most undisturbed streams do not exhibit spatial oxygen gradients or severe oxygen deficits, insects with an evolutionary history tied to running waters have had little or no selective pressures to develop mechanisms to deal with low-oxygen conditions.

Even different populations of a single species may vary in their respiratory abilities. For example, Wichard (1978) found that last instar larvae of the caddisfly *Molanna angustata* from lakes possessed a significantly larger number of tracheal gills than nearby populations inhabiting slow-moving creeks. He suggests that the intraspecific variation in gill numbers is an adaptive morphological response to the oxygen concentration of the environment to which individuals are exposed during larval development. It is likely that species able to persist over a wide range of currents possess adaptive responses similar to that of *M. angustata.*

In a series of experiments, summarized by Jaag and Ambühl (1964), respiration of various stream insects was examined over a range of current velocities. For some species oxygen consumption was directly and highly correlated with current speed, whereas for other species, oxygen consumption remained relatively constant as current velocity varied. Those insects whose respiratory rates varied as a function of current velocity were able to tolerate lower oxygen concentrations at higher current velocities (i.e., lethal levels also varied as a function of current).

Stonefly nymphs (*Hesperoperla pacifica)* exhibited 100% mortality in 24 hours when subjected to a low dissolved-oxygen level with a current velocity of 1.5 cm/sec, while a second group of nymphs exposed to a current velocity of 7.6 cm/sec exhibited no mortality under otherwise identical conditions (Knight and Gaufin 1963). Using another species, *Pteronarcys californica,* Knight and Gaufin (1964) demonstrated that the combined action of current and temperature "has a significantly greater effect upon the stonefly survival than either one of the factors when used alone."

The most extreme cases are species that, while able to survive fairly low levels of oxygen in moving water, are unable to acquire sufficient oxygen to meet their respiratory needs in still waters even at high concentrations of oxygen. Such obligatory rheophiles have been reported among black flies (Wu 1931), caddisflies (Philipson 1954), and damselflies (Zahner 1959), and representatives undoubtedly occur in other orders of aquatic insects as well.

The oxygen available to an aquatic organism is determined by three major factors (Jaag and Ambühl 1964): (1) the amount of oxygen in the surrounding water, (2) the rate of oxygen uptake by the animal, and (3) the rate at which the oxygen contained within the boundary layer enveloping the animal is renewed. An oxygen concentration gradient (diffusion gradient) occurs within the boundary layer because the oxygen consumed at respiratory surfaces can be replenished only by diffusion through that thin layer of water (Eriksen et al. 1984).

The ventilatory movements performed by virtually all lentic insects, and to a lesser extent by many stream species, enhance oxygen uptake by creating currents across respiratory exchange surfaces. Lotic insects with an inherent current requirements (i.e., obligate rheophiles) are unable to perform ventilatory movements, thereby relying entirely on the natural flow of water to shorten the diffusion path so that oxygen renewal within the boundary layer is sufficiently rapid to satisfy respiratory requirements.

Oxygen–Substrate Interactions

The relationship between substrate type and respiration by aquatic insects has received little study, but the results of a few investigations indicate that the interrelationship warrants further examination. Working with European mayfly nymphs, Wautier and Pattée (1955) found large differences in respiration rates when the substrate in the experimental chamber was altered. The respiratory rate of a species from rapid streams (*Ecdyonurus venosus*) declined as much as 40% when stones were placed in the glass chamber with the nymphs. A burrowing species (*Ephemera danica*) exhibited the highest respiratory rate in the glass chamber without substrate, a lower rate when pebbles were added to the chamber, and the lowest rate (45% that on glass) when sand, a substrate in which nymphs occur in nature, was added.

Eriksen (1964, 1968) extended this line of research using North American species of burrowing mayflies. In addition to using a wide range of substrate particle sizes in respiration experiments, he independently examined nymphal substrate preferences, in both the laboratory and the field. Curves of oxygen consumption as a function of substrate particle size and substrate selection (determined independently for *Ephemera simulans*) are superimposed in Figure 9.8. The inverse relationship between the two curves shows that oxygen consumption is highest in the least preferred substrate and lowest in the substrate type most frequently selected by the nymphs, thereby confirming the general findings and interpretations of Wautier and Pattée (1955). The higher oxygen consumption of insects on unsuitable substrate may result from elevated respi-

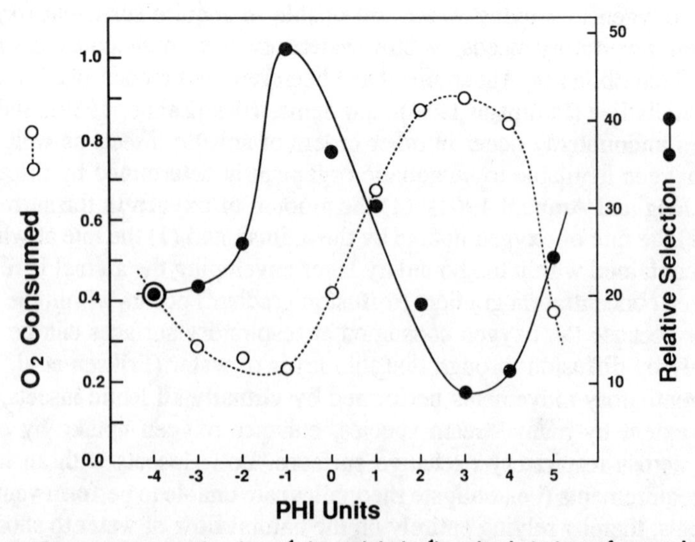

Figure 9.8 Oxygen consumption (cc g^{-1} dry weight hr^{-1}) and substrate preference by the burrowing mayfly *Ephemera simulans* in relation to substrate particle size in phi (Ø) units. [Modified from Eriksen (1964).]

ration caused by stress or increased activity. In the case of burrowing mayflies, ease of penetration provides an additional explanation for the relationship between respiratory rate and substrate type (Eriksen 1968). For example, while *Ephemera simulans* is able to burrow in a range of substrate types, lowest oxygen consumption occurs when nymphs are exposed to the particle size most easily penetrated.

The interactions of respiration and substrate type by aquatic insects may call for a reinterpretation of data derived from tests conducted in glass chambers without added substrate. For example, the two species of burrowing mayflies studied by Eriksen (1968) appear to exhibit dependent-type respiration if no substrate is added to the respiratory chamber, but in suitable substrate, both species are respiratory regulators over a wide range of oxygen concentrations.

Oxygen–Temperature Interactions

Basic and ecologically important interactions between oxygen and temperature involve the inverse relationship between them regarding oxygen solubility and the positive relationship between them regarding an organism's oxygen requirements. Stated more simply, while aquatic organisms require more oxygen at higher temperatures, less is available. However, interactions are apparent even when temperature and oxygen are controlled independently in laboratory experiments. For example, onset of mortality occurred at an oxygen concentration 2.4 times higher when nymphs of the stonefly *Pteronarcys californica* were tested at 15.6°C than when in 10°C water (Knight and Gaufin 1964).

The Q_{10}, the factor by which oxygen consumption is increased by a 10°C increase in temperature, varied from 0.6 to 3.2 for the stoneflies studied by Knight and Gaufin (1966b) (Table 9.2). The highest Q_{10} values usually occur at lower temperatures with values <1.0 (declining metabolic rate with rising temperature) sometimes occurring over the highest range examined (20–30°C).

The microdistribution of a stream caddisfly, described in more detail in the following section, is apparently determined by changes in the respiratory requirements induced by seasonal changes in water temperature (Kovalak 1976).

Increasing the water temperature will increase ventilatory movements of

Table 9.2 Q_{10} values for oxygen consumption over various temperature ranges for year class 1 nymphs of four species of stoneflies

	Q_{10} Values				
Species	5-10°C	5-15°C	10-20°C	15-25°C	20-30°C
Pteronarcys californica	2.1	3.2	1.5	2.0	0.6
Hesperoperla pacifica	2.2	1.9	1.4	1.3	0.9
Claassenia sabulosa	—	—	1.4	—	—
Megarcys signata	—	—	1.7	—	—

Source: Modified from Knight and Gaufin (1966b).

aquatic insects in the same general manner as decreasing ambient oxygen concentration (e.g., Philipson 1978). In air-saturated water the proportion of time devoted to continuous abdominal undulation by the two species of *Hydropsyche* (Trichoptera) examined by Philipson increased as water temperature increased. *Hydropsyche pellucidula* undulated 25% of the time at 10°C, 45% at 18°C, and 55% at 25°C. *Hydropsyche siltalai* undulated 50% of the time at 10°C and 60% of the time at 18°C and 25°C. The rate of undulation, in addition to the proportion of time spent in continuous undulation, also increases with rising temperature.

Hynes (1970a) feels that the cold stenothermy of many stream animals relates as much to high requirements for oxygen as it does to temperature per se. For example, the young nymphs of the mayfly *Habroleptoides modesta* are able to withstand high summer temperatures but mature nymphs must emerge before the maximum temperatures of the following summer are attained (Pleskot 1953). The intolerance of large nymphs to high temperatures apparently relates to their greater respiratory demands, caused by smaller surface/volume ratios (Istenic 1963), rather than any direct effect of warm water. Some caddisflies exhibit intraspecific differences in gill numbers related to the temperature regime (Badcock et al. 1987). Both larvae and pupae from warmer waters had significantly more gill filaments than conspecifics residing in cooler habitats.

Insect Distribution and Oxygen

Dissolved oxygen plays a major role in the spatiotemporal distribution patterns of aquatic insects. Responses include, but are not restricted to, microspatial positioning, depth distribution, migratory behavior, and predator–prey interactions.

Madsen (1968) describes a method of measuring oxygen availability at the microhabitat level, which he designates oxygen diffusion rate (ODR). ODR, measured with a platinum electrode, is a function of (1) microhabitat oxygen concentration, (2) mixing by water movement, and (3) diffusion. Figure 9.9 shows ODR measurements made in several microhabitats in a small Danish brook. The values at time zero are indicative of the oxygen content of the microhabitats, whereas the slopes of the graphs reflect mixing and diffusion. Madsen likens the reducing electrode to an oxygen consuming organism. In well-mixed microhabitats (free current and exposed surfaces of stones) there is no slope to the graphs because oxygen consumed is immediately replenished. Beneath stones and in the interstices of sand and gravel, initial ODR values are low and decline rapidly over time, indicating that diffusion is largely responsible for replenishing oxygen reduced (utilized) by the cathode.

Many species of stream insects exhibit predictable distribution patterns on individual stones, which have generally been considered direct responses to microcurrent regimes or feeding activities (see Chapter 8). A few investigators have suggested that microhabitat positioning may represent behavioral regulation of oxygen consumption (Hynes 1970a, Kovalak 1976, Wiley and Kohler

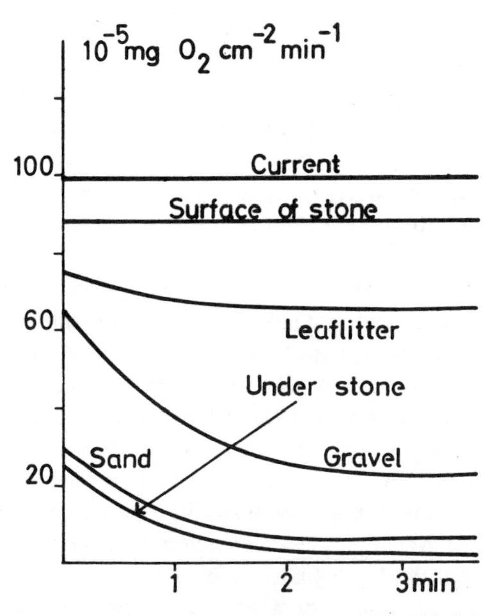

Figure 9.9 Oxygen diffusion rate over time in the various microhabitats of a small Danish brook, March 18, 1967 (see text for further explanation). [From Madsen (1968).]

1980). Kovalak (1976) recorded diel and seasonal changes in the positioning of *Glossosoma nigrior* larvae (Trichoptera) on bricks placed in a stream, which he attributed to temporal changes in respiratory requirements. Seasonal changes in positioning were associated with the thermal regime, with a larger proportion of larvae occupying current-exposed surfaces, where oxygen availability was greatest, at higher temperatures. At lower temperatures respiratory requirements can be met at lower current velocities and a higher proportion of larvae occupied less exposed positions. Kovalak hypothesized that diel position changes are a response to changing respiratory needs induced by diel rhythms of oxygen consumption by the larvae.

In an artificial stream mayfly nymphs moved to current-exposed positions on stones as dissolved-oxygen levels declined, a response Wiley and Kohler (1980) interpreted as a mechanism to enhance oxygen renewal at respiratory exchange surfaces. The species most tolerant of low oxygen levels, *Stenacron interpunctatum,* remained on the undersides of stones except at very low oxygen concentrations. Most individuals of the species least tolerant of low oxygen levels, *Pseudocloeon* sp. (a taxon incapable of gill movements), remained in current-exposed positions even in air-saturated water. Mayfly nymphs unable to satisfy their respiratory needs by gill movements and behavioral regulation (positioning), emigrate by actively entering the drift.

Nymphs of the mayfly *Leptophlebia vespertina* overwinter in humic forest ponds in Norway (Brittain and Nagell 1981). The main bodies of these ponds are anoxic for several months, yet *L. vespertina* nymphs are unable to tolerate long-term anoxia in laboratory experiments. Brittain and Nagell (1981) found that low oxygen conditions induce positive phototaxis and negative thermotaxis resulting in nymphal migrations to the underside of the ice along the shore where temperatures are near zero (in contrast to 4°C in deeper water) and oxygen levels are adequate for winter survival. A similar migratory response induced by anoxic conditions in winter, was demonstrated for the mayfly *Cloeon dipterum* (Nagell 1977).

Some benthic insects of lakes may engage in migratory movements involving incursions from the sediment into the limnetic zone (Mundie 1959, Cowell and Hudson 1967, Hudson and Lorenzen 1981). In the case of first-instar chironomid larvae, the planktonic phase may serve largely for dispersal (Davies 1976). However, vertical movements may also enable benthic organisms to move into aerated strata to avoid periods when oxygen is depleted from bottom waters.

In *Chaoborus* (phantom midges) the tracheal system has been modified to serve a hydrostatic function (as detailed in Chapter 3), enabling larvae to regulate their own buoyancy. Typically, mature larvae reside in bottom sediments during the day and are planktonic in surface strata at night. Oxygen concentration is a major regulatory factor in the vertical migration of *Chaoborus punctipennis* (LaRow 1970). Light functions as the entraining agent (Zeitgeber) that results in larvae moving to the sediment-water interface (from deeper within the sediment) at sunset. However, it is the oxygen content of the bottom water that determines whether larvae undergo vertical migration within the water column. When water overlying sediment was 81% saturated with oxygen, only about 1% of the population migrated to upper waters, whereas 30% of the population migrated to the surface when water above the sediment was only 3% saturated. Although factors other than oxygen may modify the migratory behavior of *C. punctipennis,* neither temperature nor food abundance induced migration when oxygen levels were high.

Depth distribution patterns of benthic insects, discussed in some detail in Chapter 3, are determined partly by differences in the oxygen requirements of the various species (Jónasson 1978). This becomes apparent when comparing the benthic fauna of lakes of small and moderate size with that of very large lakes where major currents carry oxygenated water to greater depths. Insects restricted to rather shallow water in small, productive lakes occur at much greater depths in large, well-circulated lakes (Barton and Smith 1984).

Jónasson (1978) related the depth distribution of oxygen and benthos with the respiratory adaptations of the predominant species in each major depth zone of Lake Esrom, Denmark. The species of the surf zone characteristically exhibit a positive linear relationship between oxygen consumption and oxygen concentration (i.e., a linear decline in respiratory rate as oxygen in the surrounding water declines). This dependent-type respiration, typical of animals residing in well-aerated habitats, stresses how similar are the respiratory environments of wave-swept shores and rapid streams (see Table 9.1). In the littoral macrophyte zone, oxygen content may range from supersaturation to <50% saturation over a diel cycle. Species inhabiting this zone exhibit independent respiration (i.e., oxygen consumption remains constant as oxygen content declines) over an intermediate range of oxygen concentrations. These species are thus adapted to the diel changes in dissolved oxygen, but are unable to survive the prolonged periods of low oxygen that occur below the littoral zone. Sublittoral species exhibit a wider range of independent respiration. For example, the respiration of *Chironomus plumosus* is independent from full air saturation to 56% saturation, below which consumption gradually declines. *Chironomus anthracinus* of the profundal zone exhibits independent-type respiration down to 25% air saturation. However, even at 2.5% saturation, the respiratory rate is still high (75% of the rate at full air saturation). Nonetheless, severe and prolonged oxygen deficits retard growth and may result in longer life cycles (Jónasson 1984). *Chironomus anthracinus* populations occurring in deep water in Esrom Lake have a 2-year life cycle, whereas populations residing in shallower water complete development in 1 year, apparently a result of the extended period of low oxygen at greater depths. A more detailed account of adaptations to low-oxygen conditions, such as occur in the profundal zone of productive lakes, will be presented in the next section of this chapter.

To the extent that oxygen content influences the distribution of predator and prey populations, so will it influence predator–prey interactions. Cockrell (1984) examined predator–prey interactions in laboratory experiments using the aquatic bug *Notonecta glauca* as the predator. She analyzed prey choice under various combinations of oxygen and temperature to determine how these factors may act to constrain predation in nature. Three potential prey species were used: adult houseflies *(Musca)* trapped on the water surface, mosquito larvae *(Culex)* that remain largely suspended from the underside of the surface film, and benthic crustaceans *(Asellus)*. Tests were conducted at five constant temperatures (5, 10, 15, 20, and 25°C) in air-saturated water. In addition, tests were conducted at 15°C with three oxygen levels: air-saturated, supersaturated to the concentration of 5°C air-saturated water, and deoxygenated to the concentration of 25°C air-saturated water. Both oxygen and temperature (through its influence on respiratory rate and oxygen content of water) influenced the vertical position of the predator, a notonectid that respires underwater using a physical gill. *Notonecta glauca* spent more time submerged at lower temperatures and at higher oxygen levels. Oxygen and temperature thus influenced the proportion of time that the predator encountered the three prey types. Because *Culex* larvae also spent more time submerged at low temperature and high oxygen levels, spatial overlap between the predator and this prey item was maintained over a range of conditions. This is reflected in a greater proportion of submerged attacks on *Culex* at low temperatures or high oxygenation, and a greater proportion of surface attacks at high temperature or low oxygenation. At high oxygen levels, *N. glauca* less frequently attacked floating *Musca* and more frequently attacked the benthic prey *Asellus*. Therefore, oxygen—and temperature, through its influence on oxygen—modified the spatiotemporal distribution of a predaceous water bug and certain prey types in the water column, thereby determining the vulnerability of different prey species.

For species able to tolerate low-oxygen conditions, the profundal zone of eutrophic lakes provides a habitat with reduced predation pressures. For example, the predaceous chironomid *Procladius* is less well adapted to low- oxygen conditions than some of its prey (other species of chironomids) and is unable to colonize the profundal zone of Lake Esrom during summer stratification (Berg et al. 1962).

Adaptations to Low-Oxygen Waters

The respiratory mechanisms of some aquatic insects enable them to reside in low-oxygen habitats without other special adaptations. Species that rely solely on atmospheric oxygen, either directly using tubes that extend to the surface as *Eristalis* or indirectly by tapping the air spaces of underwater plants as *Donacia,* are able to reside in biotopes where oxygen is deficient or even absent. In the case of *Donacia* the special respiratory mechanism enables the insect to colonize a microhabitat (anaerobic mud around plant roots) virtually devoid of competitors or predators. In the case of *Eristalis* and a few other dipterans, the telescopic respiratory siphon enables larvae to lie concealed from terrestrial predators while inhabiting shallow water biotopes with few aquatic predators or competitors.

Both intra- and interspecies differences in tracheal gills have been related to environmental oxygen levels. Dodds and Hisaw (1924) reported a negative relationship between the gill surface area of different species of mayflies and the oxygen tension characterizing the nymphal habitat. Although the conclusions of Dodds and Hisaw have been questioned (Hynes 1970a, Macan 1974a), differences in the number of gills of different species of limnephilid caddisflies correspond to the oxygen conditions of the habitat (Wichard 1978). Wichard also reported intraspecific differences in gill numbers of last-instar larvae of *Molanna angustata* as a function of habitat oxygen levels.

Some aquatic insects can respire anaerobically, and at least some members of the genus *Chironomus* are able to excrete the lactic acid that is produced during anaerobic metabolism, thus avoiding an oxygen debt (Walshe 1947). Some chironomids are adapted to survive long periods without oxygen (e.g., Nagell and Landahl 1978), but no species of macroinvertebrates are able to survive indefinitely in waters devoid of oxygen, at least in an active state. A dense, albeit species-limited, community of benthic invertebrates characterizes the deep waters of lakes that are anaerobic for several months each year. In contrast, bottom waters permanently devoid of oxygen lack insects, except chaoborids, which make daily migrations to near surface waters.

Ventilatory Movements The aquatic stages of many insects increase respiratory movements in response to low ambient oxygen levels. Some underwater bugs use their legs to create currents over the ventral air stores. The respiratory regulation of *Molanna* is achieved partly by varying the frequency of abdominal undulations. Such ventilatory movements enhance oxygen uptake by creating currents that supply "new" water to respiratory epithelia while reducing the diffusion barrier.

Many, but not all, mayfly nymphs exhibit gill movements. The gills of burrowing species beat with a frequency inverslely proportional to the oxygen concentration in the burrow (Wingfield 1939, Eriksen 1968). In these large burrowing mayflies the gills, in addition to maintaining current through the burrow, function directly in oxygen uptake. However, in *Cloeon* evidence from experiments with gilled and gillless nymphs suggests that the primary, perhaps sole, function of gill movement is to reduce the thickness of the diffusion barrier (Wingfield 1939).

Eastham (1932, 1934, 1936, 1937, 1939) made detailed studies of the micro-current pathways produced by ventilatory movements in mayfly nymphs. Despite quite diverse species-specific current patterns, in all mayflies examined the gills beat rhythmically, not in unison, but with each gill lamella beating slightly in advance of the one posterior to it. *Caenis horaria* lives in silty habitats. The second pair of gills have evolved as enlarged opercular plates that cover the more delicate third through sixth pairs of gills. When ventilating, the opercular plates are raised and the remaining gills beat to create a current that is exceptional because it moves laterally, pulsating back and forth across the dorsum of the nymphs. Each gill is about one-third of an oscillation out of phase with the other member of that gill pair (Eastham 1934). The gills of some running-water species of Ephemeroptera (and lotic representatives from other orders) have lost the ability to engage in ventilatory movements.

Aquatic insects living within cases would be especially prone to encounter oxygen-depleted waters were it not for ventilatory movements. Cased larval caddisflies have evolved efficient respiratory adaptations in this regard (Fig. 9.10). Spacing humps on the first abdominal segment ensures that free areas remain between the animal's body and the wall of the case; lateral hair fringes increase the efficiency of undulatory movements; the hole in the posterior end of the case provides for the unidirectional flow of water through the case. The respiratory advantages possible for case-dwellers are in part responsible for the remarkable adaptive radiation exhibited by larval caddisflies (Wiggins and Mackay 1978), although cases do not universally confer a respiratory advantage (Williams et al. 1987). The larvae of chironomids and aquatic moths that reside in tubes or cases also create currents by undulations of the body that vary as a function of environmental conditions. In both cased and caseless aquatic insects the frequency of undulatory movements generally varies inversely with ambient-oxygen levels, and directly with temperature; in caseless species, ventilatory movements also exhibit an inverse relationship with water movement.

Rectal ventilation in dragonfly (Anisoptera) nymphs is accomplished by various muscles, two valves, and a diaphragm, the structure and functioning of which have been studied in some detail (see Hughes and Mill 1966, Mill and Pickard 1972, Mill 1974). The frequency of ventilation is directly related to temperature and inversely related to the oxygen content of the water. When

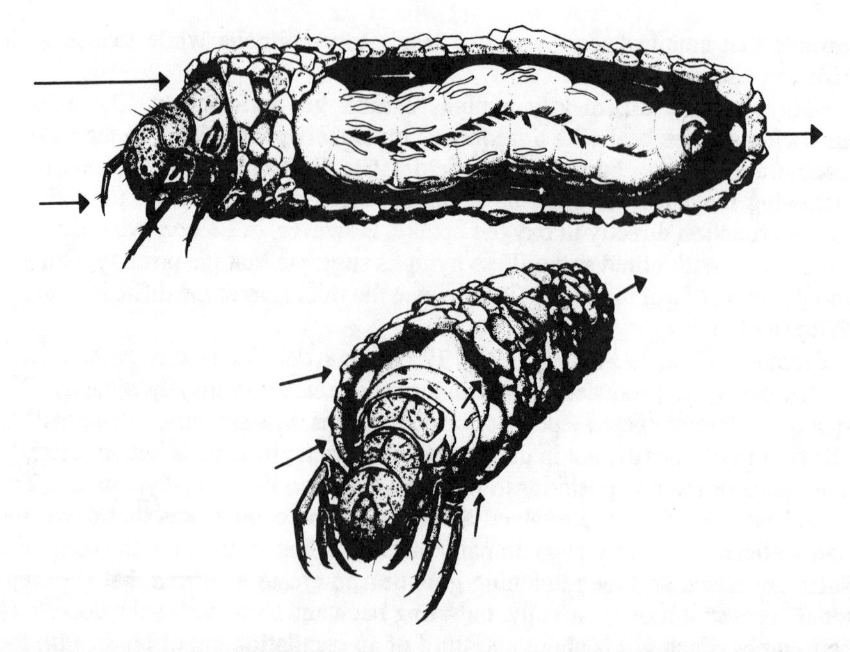

Figure 9.10 Circulation of water through the case of a typical caddisfly larva induced by abdominal undulations. [From Wiggins (1977). Reprinted by permission of The Royal Ontario Museum.]

oxygen reaches a critically low level (2.5 cm^3/liter at 17°C for *Aeshna grandis*), nymphs project their abdomen above the water surface and pump air instead of water (Mill 1974). An alternate response to low-oxygen conditions for intermediate- and final-instar larvae with functional thoracic spiracles, is to raise their thorax above the water (Hinton 1947). Zygopteran nymphs undulate the body from side to side in low-oxygen waters, and may move to the water surface as oxygen declines further (Robert 1958, Zahner 1959).

Some Plecoptera nymphs do "pushups" if ambient-oxygen levels decline (Knight and Gaufin 1966b). Most stoneflies, however, reside in well-aerated waters and have little or no need for adaptations to low-oxygen conditions in their natural habitat.

Chironomids, on the other hand, are often found in oxygen-deficient waters; several adaptations are involved, including undulations of the body. Walshe (1950) observed the activities of *Chironomus plumosus* in U-shaped glass tubes that simulated the mud tubes constructed by last instar larvae. When the water was 80–100% saturated with oxygen, *Chironomus* larvae devoted an average of 36% of active time to feeding activities and 52% to respiratory movements. In water only 5–7% saturated with oxygen, feeding activites ceased and all active time (88% of total time) was devoted to respiratory movements.

Hemoglobin of Aquatic Insects Only a few chironomid midges and water bugs possess hemoglobin. Among aquatic insects, hemoglobin functions in buoyancy control, as an oxygen transport pigment during hypoxia, and as a short-term oxygen store (Weber 1980).

In a few genera of chironomids hemoglobin is an extracellular component of the hemolymph. Members of the *Chironomus plumosus* group, the so-called blood worms, have received the most intensive study. These species are among the most characteristic inhabitants of the low-oxygen waters and are often the only insects to permanently reside in such habitats. *Chironomus* larvae have the ability to synthesize hemoglobin in response to low levels of environmental oxygen (Fox 1955). The temporally limited oxygen storage capacity [9 minutes for *C. plumosus* (Walshe 1950)] negates a significant storage function for hemoglobin during extended periods of anoxia. However, the short-term oxygen storage properties of hemoglobin allow *Chironomus* to alternate periods of feeding with ventilatory movements under low-oxygen conditions (Walshe 1950). For example, larvae ceased feeding when ambient-oxygen saturation levels fell below 26% when carbon monoxide was used to prevent oxygen uptake by hemoglobin. In contrast, normal larvae of *C. plumosus* engaged in feeding activities down to 10% oxygen saturation. Under conditions of hypoxia, the time spent filter-feeding by the leaf-mining species *Endochironomus dispar*, was also dramatically reduced when carbon monoxide was used to deactivate the hemoglobin (Walshe 1951). The presence of hemoglobin also allows chironomid larvae to more rapidly repay oxygen debts incurred during anaerobic metabolism (Walshe 1947).

The hemoglobin of chironomids becomes 95% saturated at a partial pressure of oxygen of 10 mm Hg, a pressure at which mammalian hemoglobin is only 1% saturated (Clarke 1954). However, *Chironomus* hemoglobin is a high-affinity pigment; only at environmental partial pressures of <7 mm Hg at 17°C does the hemoglobin become partially reduced and thus make oxygen available to meet the needs of the larvae (Wigglesworth 1972). The advantages that hemoglobin confer have enabled *Chironomus* to exploit aquatic habitats where low oxygen levels are commonly encountered, and where anoxic conditions periodically occur.

Hemoglobin is also possessed by the notonectid bugs *Anisops* and *Buenoa*, but is used for buoyancy control and not as an adaptation to low-oxygen waters (see Chapter 2).

OTHER CHEMICAL FACTORS

Apart from salinity and dissolved oxygen, the influence of normal ranges of chemical variables on aquatic insects is often difficult to discern, especially under field conditions. Part of the difficulty arises from problems of isolating the effects of single chemical variables. However, while much more research is needed, it appears that many species of aquatic insects are capable of tolerating relatively wide natural ranges of many chemical variables. This section examines the potential influence of natural variations in selected chemical factors on aquatic insect ecology. Comprehensive accounts of chemical conditions in natural waters are readily available elsewhere (e.g., Hutchinson 1957a, Hem 1970, Stumm and Morgan 1970, Faust and Aly 1981, Drever 1982).

Acidity The pH of natural waters ranges from <3.0 to >12.0. Most unpolluted waters, however, exhibit pH values in the range 6.0–9.0 (Cole 1983). Nonetheless, in certain regions there may be numerous water bodies with natural pH values around 5.0. Waters receiving drainage from peat bogs, for example, are typically acidic. Generally pH increases from the headwaters to the lower reaches of river systems and from the bottom to the surface of lakes. Habitats with dense growths of aquatic plants may exhibit large diel fluctuations in pH as photosynthetic uptake of CO_2 varies (altering carbonic acid levels), with corresponding changes in the acidity of the water.

Acid waters are characterized by low species diversity and low productivity (Welch 1952). Major components are typically missing from the fauna of the most acid water bodies, although factors other than pH may contribute to their absence. There is a surprising dearth of information on aquatic insects in naturally acidic waters, especially lentic habitats.

The five lentic waters listed in Table 9.3 are arranged in order of decreasing pH. Two are lakes (Store Gribsø, Denmark, and Lake Trestickeln, Sweden), two are ponds [Blaxter Lough, England, and Lake Whitney, Georgia (USA)], and the sphagnum pools are in a bog mat associated with Mud Lake, Michigan. Although there is a general diminution in insect species richness associated with the trend of increasing acidity in Table 9.3, the species richness of the Coleoptera exhibits a bimodal pattern. The richest beetle fauna occurs in Lake Whitney, which possesses diverse and abundant aquatic macrophytes and in the acid bog pools lined with sphagnum moss. Blaxter Lough, lacking aquatic macrophytes, also lacks a beetle fauna. This suggests that, over the pH range of the five lentic habitats, macrophytes may be of greater direct importance than acidity for some species of aquatic beetles.

While aquatic insects exhibit an overall impoverishment in acid biotopes, their relative abundance may actually increase (Table 9.3) because of a more rapid decline of certain noninsect taxa with decreasing pH. Lake Whitney contains a diverse assemblage of noninsect benthos, including crustaceans and mollusks. Store Gribsø has isopod crustaceans but lacks mollusks. Neither benthic crustaceans nor molluscs occur in the remaining lentic habitats and no macrobenthic forms, but insects occur in the highly acidic sphagnum pools.

The impoverished fauna of Blaxter Lough is especially interesting because this water body lacks some of the adverse conditions often associated with acid lakes and implicated as contributing to faunal depletions. For example, this pond is shallow and well mixed with adequate dissolved oxygen at all times (McLachlan and McLachlan 1975). Blaxter Lough has a firm bottom without overlying flocculent deposits of dy often found in acid bog lakes. It appears that low pH, a virtual absence of macrophytes, and the general lack of habitat heterogeneity account for the poorly developed zoobenthos in this pond.

Somewhat more data are available for the aquatic insects of acidic lotic

Table 9.3 Number of species in major insect groups and percentage contributed by insects to total benthos in some acidic lentic water bodies

	Lake Whitney[a] (pH 3.5–5.9)	Store Gribsø[b] (pH≈5.0)	Lake Trestickeln[c] (pH 3.9–4.6)	Blaxter Lough[d] (pH≈4.0)	Sphagnum pools[e] (pH 3.3–3.6)
Ephemeroptera	2	1	1	1	—
Odonata	16+	9	2	2	2
Hemiptera	7	1	3	—	—
Lepidoptera	3	—	—	—	—
Coleoptera	14	6	3	—	16
Megaloptera	—	1	1	1	—
Trichoptera	7+	9	5	2	—
Diptera					
Chaoboridae	1	1	1	—	—
Chironomidae[f]	+	+	+	+	+
Others	11	1	1	—	1
% Insects[g]	48–77	51–79	90–95	96	100

[a] Smock et al. (1981).
[b] Berg and Petersen (1956).
[c] Wiederholm and Eriksson (1977).
[d] McLachlan and McLachlan (1975).
[e] Jewell and Brown (1929).
[f] Only presence shown for chironomids, since not taken beyond family in some cases.
[g] Percent based on numbers except for Blaxter Lough (weight). Ranges include values for different habitat types.

waters. Although some species of stream insects can tolerate and even thrive in waters of quite low pH (Wright and Shoup 1945, Pomeisl 1961, Otto and Svensson 1983; Winterbourn and Collier 1987) (Fig. 9.11), many other species or higher taxa may be eliminated. In a preliminary survey of a naturally acid stream in Illinois, Jewell (1922) commented on the absence of Ephemeroptera. A comparative examination of the chironomid faunas of a chalk stream (pH 7.9–8.3) and slightly acid streams (pH 6.1–7.0) in England found numerous species restricted to the chalk stream, others occurring only in the acid streams, and yet other species that inhabited both stream types (Hall 1951). Harrison and Agnew (1962) identified species of stream invertebrates that were restricted to the soft acid streams of Table Mountain, South Africa. Ziemann (1975) conducted a comparative study of two adjacent European mountain brooks that differed in hydrogen ion content. Below pH 6.0 certain species of Ephemeroptera and Plecoptera were eliminated, and a distinct acidophilic diatom assemblage was present. Ziemann characterized the acid stream by its small number of species, the predominance of one or a few species in contrast to the balanced frequency distribution of species above pH 6.5, and by its generally oligotrophic character.

Naturally acidic streams are common in areas of New Zealand, where they may occur in proximity to alkaline streams (Winterbourn and Collier 1987). Benthic invertebrate assemblages (largely insects) were investigated in 34 New Zealand streams that were acidobiontic (pH 3.5–5.5), acidophilic (pH 5.6–6.9), or alkaline (pH 7.0–8.1). Similar numbers of taxa occurred in the three groups of streams, except at the five most acidic sites (pH <4.6) where the taxonomic richness of mayflies, stoneflies, and caddisflies was about 50% of the richness at the other acidobiontic sites.

The invertebrate fauna of the River Duddon system in the English Lake District is comprised of two distinctive communities (Minshall and Kuehne 1969). In the upper basin Plecoptera predominate and the faunal assemblage has been designated the "Plecoptera community." Numerous additional taxa, plus most members of the "Plecoptera community," occur in the lower basin; because Ephemeroptera, rare in the upper basin, are predominant, the invertebrate assemblage of the lower basin is designated the "Ephemeroptera community." Minshall and Kuehne (1969) ruled out temperature, substrate, and several other factors as responsible and suggested that differences in chemical conditions may account for the faunal discontinuity. Sutcliffe and Carrick (1973) conducted a detailed analysis of water chemistry in the Duddon basin and concluded that pH was responsible for the differences in the insect fauna between the upper and lower drainage basins. The water of the upper basin ("Plecoptera community") is characterized by pH <6.0, whereas higher pH values generally occur in the lower basin ("Ephemeroptera community"). However, five small tributaries were discovered in the upper basin with pH >6.0; these tributaries contained nearly all of the major faunal components of the "Ephemeroptera community." Likewise, two lower-basin tributaries contained only the "Plecoptera community" in their acidic upper reaches. It was proposed that pH influences the fauna indirectly through changes in the food base. Nearly all taxa excluded from acidic

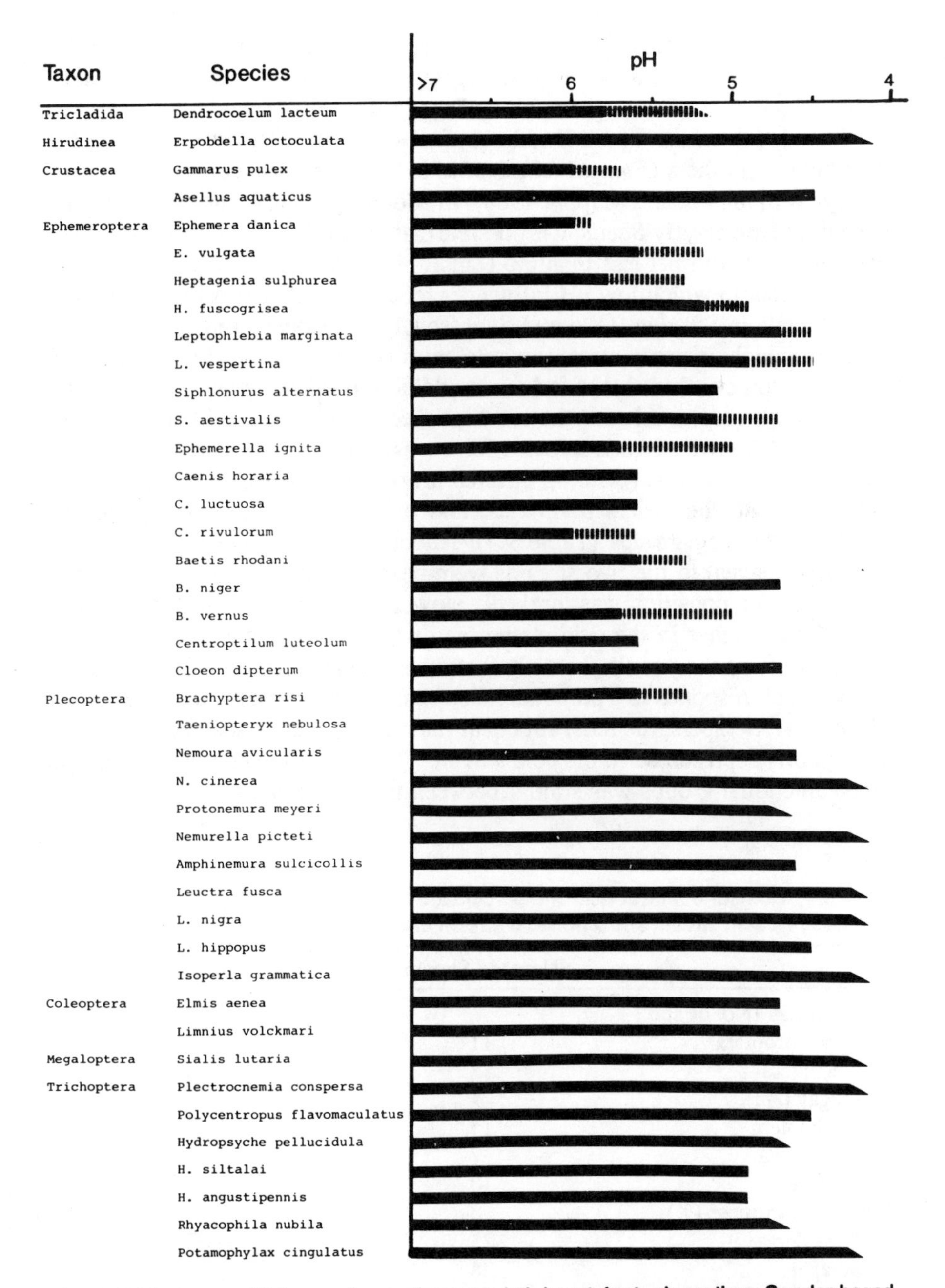

Figure 9.11 Lower pH tolerance levels of common lotic invertebrates in southern Sweden based on field occurrences. Continuous line, continuous occurrence; hatched line, occasional occurrence; obliquely cut lines, tolerance levels probably lower. [From Otto and Svensson (1983).]

reaches of the Duddon system are herbivores, which corresponds to the paucity of epilithic algae and a slower decomposition of leaf litter in acidic waters. Laboratory experiments suggest that some mayflies *(Baetis rhodani, B. muticus)* are excluded from the Upper Duddon because of water chemistry, whereas the distribution of others *(Ephemerella ignita)* is largely determined by food availability (Willoughby and Mappin 1988). In addition, considering the egg-laying behavior of the mayfly *Baetis,* it is possible that ovipositing females avoid laying eggs in water with pH less than 6.0 (Sutcliffe and Carrick 1973).

The extent to which low pH influences food availability was investigated in acid streams in Sweden (Otto and Svensson 1983). Considering the lower pH levels at which common stream invertebrates occur in nature (Fig. 9.11), Otto and Svensson concluded that factors in addition to the direct effects of acidity must also be responsible for the impoverished fauna of the acid streams they examined. The two streams selected for intensive study, one acid (pH 4.8) and the other circumneutral (pH 6.8), differed in species richness, the distribution of species among the orders of insects, and in the percentage composition of functional feeding groups (Table 9.4). Allochthonous detrital inputs (largely terrestrial leaves) to the two streams were nearly identical. However, because leaf litter decomposition was markedly slower at low pH, the standing crop of CPOM was higher in the acid stream and was never depleted as occurred at certain times of the year in the neutral stream. The abundance and permanence of CPOM corresponds to a preponderance of shredder species in the acid stream (Table 9.4). Scrapers, the most abundant functional group in the neutral stream, were poorly represented in the acid stream. The percentage composition of the other functional groups was similar between the two streams.

Table 9.4 Number of insect taxa and percentage composition of functional feeding groups in an acid stream and a nonacid stream in Sweden

	Nonacid Stream	Acid Stream
Insect orders (No. of spp.)		
Ephemeroptera	11	0
Plecoptera	12	9
Megaloptera	1	1
Trichoptera	15	7
Coleoptera	7	2
Diptera	10	4
Functional groups[a] (%)		
Shredders	25	48
Scrapers	28	4
Deposit feeders	15	11
Filter feeders	11	15
Predators	21	22

[a]Based on number of taxa of insect and noninsect benthos.

Source: Modified from Otto and Svensson (1983).

Pupae of the shredding caddisfly *Potamophylax cingulatus* were largest in an acid stream, intermediate in size in a neutral stream, and smallest in an alkaline stream. Otto and Svensson concluded that the abundance and availability of food accounted for the increased growth of *P. cingulatus* in the acid stream, more than compensating for the lower food quality of detritus (lower microbial biomass) or any physiological stress imposed by low pH.

Analysis of data generated from a survey of 34 stream sites in England revealed that patterns of species distribution and community structure were strongly associated with differences in acidity, even though the mean pH of all sites examined was <7.0 (Townsend et al. 1983). The mean numbers of invertebrate taxa per site (shaded histograms) increased with increasing pH over the entire range of acidity examined (Fig. 9.12). The total number of available taxa at all sites within each pH range (open histograms) increased up to pH 6.0 and remained essentially constant with further increase in pH. The benthic community of a given pH range contains 40–50% of the pool of available species, suggesting that constraints to diversity based on the limited number of suitably adapted species only partly explains the low diversity of invertebrates in acid streams, a conclusion also reached by Otto and Svensson (1983). Large-bodied invertebrate predators were more abundant at acidic stream sites where fish were rare or absent, suggesting that shifts in predation strategies and pressures may also contribute to differences in the diversity and composition of stream insects in waters varying in pH.

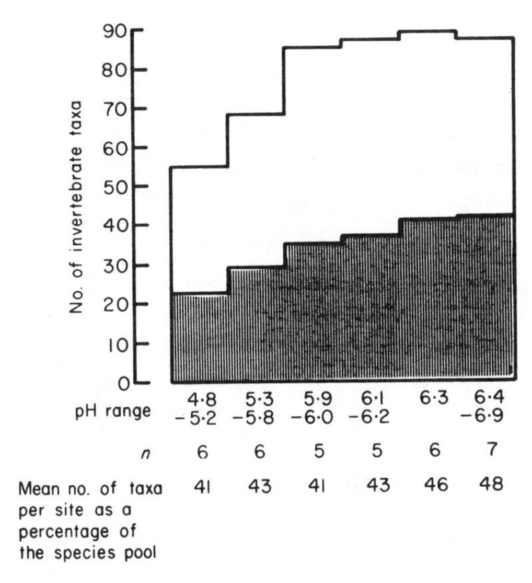

Figure 9.12 The number of invertebrate taxa found in six mean pH categories based on a survey of 34 stream sites in southern England. Shaded histograms indicate mean number of taxa per site for each pH category; open histograms indicate the total number of taxa (species pool) collected. [From Hildrew et al. (1984a).]

Nearly all of those who have studied the fauna of acid waters recognize that factors in addition to low pH may contribute to the elimination of species and attendant changes in community structure. Table 9.5 summarizes the major factors associated with acid waters that have been suggested as responsible, in part, for alterations in aquatic insect population and community parameters.

Water Hardness and Related Factors Hardness values exhibit considerable variation in freshwaters, ranging from the extremely soft waters of alpine lakes and streams situated on insoluble bedrock, to the hard waters of calcareous aquatic habitats located in limestone regions. Calcium and magnesium normally account for most of the hardness of freshwaters, although other ions such as chloride and sulfates also contribute to total hardness values (Cole 1983). It is difficult to separate the potential ecological role of hardness in aquatic systems from the effects of pH, total ionic content, or alkalinity (capacity to neutralize acids), since soft waters tend to be more acidic, less alkaline, and lower in conductivity (an index of total ionic content) than hard waters. For example, Costello et al. (1984), who examined the influence of water chemistry on the distribution of Plecoptera, selected total hardness as the representative chemical

Table 9.5 Factors other than the direct physiological effects of pH that may contribute to alterations of the entomofauna of acid water bodies

Factors	Selected References
Physicochemical Factors	
Oxygen deficits	Jewell and Brown 1929
Calcium deficiency[a]	Hall 1951
Low transparency	Otto and Svensson 1983
Flocculent false bottom (lentic)	Jewell and Brown 1929
Ferric hydroxide deposition (lotic)	Otto and Svennson 1983
Metal solubility	Townsend et al. 1983
Food Resources	
Slow decomposition	Hildrew et al. 1984a
Larger CPOM pool	Otto and Svennson 1983
Permanent CPOM pool	Otto and Svennson 1983
Lower food quality	Otto and Svennson 1983
Reduced algae	Sutcliffe and Carrick 1973
Reduced macrophytes	McLachlan and McLachlan 1975
Other Biotic Factors	
Altered competitive interactions	Wiederholm and Eriksson 1977
Altered predation pressure	Hildrew et al. 1984b
Oviposition avoidance	Sutcliffe and Carrick 1973
Reduced resource heterogeneity	Hildrew et al. 1984b

[a] Apparently more applicable to mollusks and crustaceans than insects.

parameter for their analyses because of its correlation with calcium hardness, pH, and conductivity.

While some aquatic animals occur over a wide range of water hardness, others are largely limited to hard (calciphiles) or soft waters (calciphobes), although the underlying causal mechanisms for such "preferences" are poorly understood (Macan 1961). As Hynes (1970a) points out, to reside in soft water of low total ionic content demands extremely well-developed osmoregulatory mechanisms, and this alone must account for the absence of some species from dilute aquatic habitats. Some invertebrates, such as mollusks and to a lesser extent crustaceans, contain a preponderance of species with a high physiological demand for calcium, which accounts for their restriction to waters with a certain amount of hardness. Aquatic insects, in contrast, as a group appear rather indifferent to normal ranges of water hardness (Macan 1974a). Because of this, noninsect invertebrates are often numerically dominant in the benthos of calcareous waters, whereas insects tend to predominate in most soft-water habitats. Those aquatic insects that are influenced by water hardness, or interrelated chemical variables, form the basis for the following discussion.

A very few species of aquatic insects live only in habitats where calcium carbonate deposits occur and are thus restricted to very hard waters (Geijskes 1935). Such is the case for the chironomid *Lithotanytarsus,* the larvae of which inhabit tubes in travertine deposits (Thienemann 1954, Symoens 1955, Sourie 1962). Certain species of the elmid beetle *Riolous* and the psychodid *Pericoma* also inhabit these areas of calcium carbonate deposition and are apparently restricted to them (Geijskes 1935, Beier 1948). Travertine deposits may, however, adversely alter the substrate for other aquatic insects.

In some cases congeneric species exhibit contrasting occurrences in relation to the calcium content of the habitat. Macan (1961) reports that the mayfly *Heptagenia sulfurea* inhabits the wave-swept shores of calcareous lakes in Ireland, whereas *H. lateralis* is found under otherwise similar conditions in lime-poor lakes. Similar hard and soft-water species pairs occur in the mosquito (Macan 1951) and corixid bug (Macan 1954) fauna of Britain.

Despite the numerical predominance by noninsects in a hard-water pond studied by Palmer (1973), the entomofauna was much more diverse than that of a soft-water pond where insects clearly dominated. Although a few species (the mayflies *Leptophlebia vespertina* and *L. marginata,* and the corixid bug *Sigara scotti)* occurred only in the soft-water habitat, numerous species (especially among bugs and beetles), absent from the calcium poor habitat were present in the calcareous pond.

A survey of streams in the southeastern United States revealed two distinct lotic types based on chemical characteristics (Neel 1973). One group of streams contained soft waters of low alkalinity, whereas hard, highly alkaline waters occurred in another set of streams. In addition to mollusks, mayflies, beetles, and dipterans were better developed in the hard, highly alkaline streams; stoneflies and caddisflies were better developed in the soft waters of low alkalinity. The aquatic flora was much more abundant in the hard-water, highly alkaline streams (although concentrations of plant nutrients were no higher), which undoubtedly accounts for some of the differences in the insect communities of the two stream types.

Egglishaw and Morgan (1965) classified the streams of the Scottish Highlands into five classes on the basis of total cation concentrations, which corresponded closely with calcium and bicarbonate levels (Table 9.6). Stream type I contained significantly fewer organisms with a significantly lower mean weight than stream type II, but above 800 μeq total cations per liter the differences are nonsignificant. On two rock types, granite and schist, both chemically poor (<400 μeq total cations/liter) and chemically rich (>700 μeq total cations/liter) stream types were present, allowing faunal comparisons within the same rock type. The chemically rich streams of both rock types contained a more diverse and abundant fauna than did the chemically poor streams. Other comparisons also supported the contention that faunal differences resulted from differences in water chemistry rather than through the influence of rock type on substrate characteristics. Differences in altitude were also ruled out as responsible for the observed faunal changes.

In a subsequent study, Egglishaw (1968) established a positive correlation between the calcium content of stream water and the rate of decomposition of plant detritus. He postulates that water chemistry may exert its primary control on the standing stock of benthos by regulating decomposition processes.

Differences in alkalinity were more highly correlated with changes in the aquatic insect fauna of the Firehole River, Yellowstone National Park (USA) than was temperature (Armitage 1958). A series of sampling stations, established above and below major influxes of water from hot springs, exhibited progressive increases in temperature and alkalinity. The mean standing crop of insects also increased downstream; statistical analyses indicated a stronger relationship between insect biomass and alkalinity than between biomass and temperature. The densities of Trichoptera and Ephemeroptera exhibited significant positive and negative correlations, respectively, with alkalinity levels, whereas densities of Diptera and Coleoptera were not significantly correlated with alkalinity.

Table 9.6 Mean numbers and weights (mg) of benthos per sample[a] in stream types differing in chemical "richness" (concentrations in μeq/liter)

	Stream Type				
	I	II	III	IV	V
Total cations	224-400	401-800	801-1200	1201-1600	>1600
Ca^{2+}	40-115	94-394	186-667	656-1067	492-1457
HCO_3^-	26-136	20-404	204-748	328-1076	340-1232
$\bar{X}$ numbers	79	261	239	260	273
$\bar{X}$ weights	158	588	524	585	716

[a] Based on samples collected during spring.

Source: Modified from Egglishaw and Morgan (1965).

Savage (1982) proposes using the corixid bug fauna in lake classification because of species-specific correlations with the conductivity of lentic habitats. For example, whereas *Sigara scotti* is restricted to water bodies with low conductivity levels, its congener *S. concinna* is confined to high-conductivity waters. Because conductivity levels generally increase as lake succession proceeds, it is not surprising that there is a general concordance between the conclusions of Savage and the earlier work of Dr. T. T. Macan (referred to in Chapter 3), who examined corixid assemblages in relationship to the trophic status of lakes.

A further word of caution is appropriate. In addition to the chemical interrelationships already mentioned (e.g., the generally positive relationship between hardness and pH), many other chemical and nonchemical factors may also covary with a parameter such as water hardness. For example, soft waters tend to be more prevalent near the headwaters, where summer temperatures are often lower and oxygen levels higher than in the harder waters lower in the catchment. Special care must therefore be taken when drawing conclusions on factors controlling the distribution and abundance of aquatic insects. Egglishaw and Morgan (1965) realized this and analyzed the data from 52 stream sites in several ways before suggesting that chemical factors, rather than substrate characteristics, stream gradient, or altitude (and presumably temperature), were primarily responsible for the observed faunal patterns.

REFERENCES

Aagaard, K. 1978. "The Chironomidae of the exposed zone of Øvre Heimdalsvatn." *Holarct. Ecol.* **1**: 261–265.

Abeille de Perrin, E. 1904. "Description d'un Coléoptère hypogé francais." *Bull. Soc. Ent. France* **1904**: 226–228.

Abell, D. A. 1959. "Observations on mosquito populations of an intermittent foothills stream in California." *Ecology* **40**: 186-193.

Adams, A. 1984. "Cryptobiosis in Chironomidae (Diptera)—two decades on." *Antenna* **8**: 58–61.

Adamstone, F. B. 1924. "The distribution and economic importance of the bottom fauna of Lake Nipigon with an appendix on the bottom fauna of Lake Ontario." *Univ. Toronto Studies Biology, Publ. Ontario Fish. Res. Lab.* **24**: 35–100.

Alba-Tercedor, J. and A. Sanchez-Ortega (eds.). 1991. *Overview and Strategies of Ephemeroptera and Plecoptera.* Sandhill Crane Press, Gainesville, Fla.

Alekseyev, N. K. 1955. "Concerning the chironomid dissemination in a water basin." *Vop. Ikhtoil.* **5**: 145–149 (in Russian).

Alexander, R. D. and W. L. Brown. 1963. "Mating behavior and the origin of insect wings." *Occas. Pap. Mus. Zool. Univ. Mich.* **628**: 1–19.

Alimov, A. F. and G. G. Winberg. 1972. "Biological productivity of two northern lakes." *Verh. Internat. Verein. Limnol.* **18**: 65–70.

Allan, J. D. 1975. "The distributional ecology and diversity of benthic insects in Cement Creek, Colorado." *Ecology* **56**: 1040-1053.

Allen, K. R. 1951. "The Horokiwi Stream. A study of a trout population." *Fish. Bull. N. Z.* **10**: 1-238.

Ambühl, H. 1959. "Die Bedeutung der Strömung als ökologischer Faktor." *Schweiz. Z. Hydrol.* **21**: 133–264.

Ambühl, H. 1962. "Die Besonderheiten der Wasserströmung in physikalischer, chemischer und biologischer Hinsicht." *Schweiz. Z. Hydrol.* **24**: 367-382.

Andersen, N. M. 1976. "A comparative study of locomotion on the water surface in semiaquatic bugs." *Vidensk. Meddel. Dan. Naturhist. Foren.* **139**: 337-396.

Andersen, N. M. 1982. *The Semiaquatic Bugs (Hemiptera, Gerromorpha). Phylogeny, Adaptations, Biogeography and Classification. Entomonograph,* Vol. 3. Scandinavian Science Press, Ltd., Malov, Denmark.

Andersen, N. M. and J. T. Polhemus. 1976. "Water-striders (Hemiptera: Gerridae, Veliidae, etc.)," pp. 187-224 in L. Cheng (ed.), *Marine Insects.* North-Holland Publ. Co., Amsterdam.

Anderson, N. H. 1966. "Depressant effect of moonlight on activity of aquatic insects." *Nature* **209**: 319-320.

Anderson, N. H. 1982. "A survey of aquatic insects associated with wood debris in New Zealand streams." *Mauri Ora* **10**: 21-33.

Anderson, N. H. and K. W. Cummins. 1979. "Influences of diet on the life histories of aquatic insects." *J. Fish. Res. Board Can.* **36**: 335-342.

Anderson, N. H. and D. M Lehmkuhl. 1968. "Catastrophic drift of insects in a woodland stream." *Ecology* **49**: 198-206.

Anderson, N. H. and J. R. Sedell. 1979. "Detritus processing by macroinvertebrates in stream ecosystems." *Ann. Rev. Ent.* **24**: 351-377.

Anderson, N. H., J. R. Sedell, L. M. Roberts, and F. J. Triska. 1978. "The role of aquatic invertebrates in professing of wood debris in coniferous forest streams." *Am. Midl. Nat.* **100**: 64-82.

Anderson, N. H., R. J. Steedman, and T. Dudley. 1984. "Patterns of exploitation by stream invertebrates of wood debris (xylophagy)." *Vern. Internat. Verein. Limnol.* **22**: 1847-1852.

Andrews, D. and F. H. Rigler. 1985. "The effects of an Arctic winter on benthic invertebrates in the littoral zone of Char Lake, Northwest Territories." *Can. J. Zool.* **63**: 2825-2834.

Andrews, D. A. and G. W. Minshall. 1979. "Longitudinal and seasonal distribution of benthic invertebrates in the Little Lost River, Idaho." *Am. Midl. Nat.* **102**: 225-236.

Angelier, E. 1962. "Remarques sur la répartition de la faune dans le milieu interstitiel hyporhéique." *Zool. Anzg.* **166**: 351-356.

Armitage, K. B. 1958. "Ecology of the riffle insects of the Firehole River, Wyoming." *Ecology* **39**: 571-580.

Armitage, P. D. 1976. "A quantitative study of the invertebrate fauna of the River Tees below Cow Green Reservoir." *Freshwat. Biol.* **6**: 229-240.

Arvy, L. and W. L. Peters. 1975. "L'Ephemeri Vita' A Trois Siècles." *Eatonia,* Suppl. No. 1: 1-16.

Ashton, G. D. 1979. "River ice." *Am. Sci.* **67**: 38-45.

Aubert, J. 1963. "Les Plécoptères des cours d'eau temporaires de la péninsule ibérique." *Mitt. Schweiz. Ent. Ges.* **35**: 301-315.

Axtell, R. C. 1976. "Coastal horse flies and deer flies (Diptera: Tabanidae)," pp. 415-445 in L. Cheng. (ed.), *Marine Insects.* North-Holland Publ. Co., Amsterdam.

Badcock, R. M. 1952. "Notes on the effect of sandy inflows on the fauna of the Fruin Water." M.S. thesis, Glasgow University, Glasgow, Scotland.

Badcock, R. M. 1953. "Observation on oviposition underwater of the aerial insect *Hydropsyche angustipennis* (Curtis) (Trichoptera)." *Hydrobiologia* **5**: 222–225.

Badcock, R. M., M. T. Bales, and J. D. Harrison. 1987. "Observations on gill number and respiratory adaptation in caddis larvae," pp. 175–178 in M. Bournaud and H. Tachet (eds.), *Proc. Fifth Internat. Symp. Trichoptera.* Dr. W. Junk, Dordrecht.

Baker, F. C. 1918. "The productivity of invertebrate fish food on the bottom of Oneida Lake, with special reference to mollusks." *New York State College of Forestry Tech. Publ. No. 9.* **18**: 1-264.

Balciunas, J. K. and T. D. Center. 1981. "Preliminary host specificity tests of a Panamanian *Parapoynx* as a potential biological control agent for *Hydrilla.*" *Envir. Ent.* **10**: 462–467.

Banse, K. 1975. "Pleuston and Neuston: On the categories of organisms in the uppermost pelagial." *Internat. Revue ges. Hydrobiol.* **60**: 439–447.

Bärlocher, F., R. J. Mackay, and G. B. Wiggins. 1978. "Detritus processing in a temporary vernal pool in southern Ontario." *Arch. Hydrobiol.* **81**: 269–295.

Barmuta, L. A. and P. S. Lake. 1982. "On the value of the river continuum concept." *N. Z. J. Mar. Freshwat. Res.* **16**: 227-229.

Barnes, J. R. and G. W. Minshall (eds.). 1983. *Stream Ecology—Application and Testing of General Ecological Theory.* Plenum, New York.

Barnes, T. C. and T. L. Jahn. 1934. "Properties of water of biological significance." *Q. Rev. Biol.* **9**: 292–341.

Barton, D. R. 1980. "Benthic macroinvertebrate communities of the Athabasca River near Ft. Mackay, Alberta." *Hydrobiologia* **74**: 151–160.

Barton, D. R. and H. B. N. Hynes. 1978. "Wave zone macrobenthos of the exposed Canadian shores of the St. Lawrence Great Lakes." *J. Great Lakes Res.* **4**: 27–45.

Barton, D. R. and M. A. Lock. 1979. "Numerical abundance and biomass of bacteria, algae and macrobenthos of a large northern river, the Athabasca." *Internat. Revue ges. Hydrobiol.* **64**: 345–359.

Barton, D. R. and S. M. Smith. 1984. "Insects of extremely small and extremely large aquatic habitats," pp. 456–483 in V. H. Resh and D. M. Rosenberg (eds.), *The Ecology of Aquatic Insects.* Praeger, New York.

Bar-Zeev, M. 1958. "The effect of temperature on the growth rate and survival of the immature stages of *Aedes aegyptii* (L.)." *Bull. Ent. Res.* **49**: 157–163.

Baskerville, G. L. and P. Emin. 1969. "Rapid estimation of heat accumulation from maximum and minimum temperatures." *Ecology* **50**: 514–517.

Baumann, R. W. 1979. "Nearctic stonefly genera as indicators of ecological parameters (Plecoptera: Insecta)." *Great Basin Naturalist* **39**: 241–244.

Bayly, I. A. E. 1972. "Salinity tolerance and osmotic behaviour of animals in athalassic saline and marine hypersaline waters." *Ann. Rev. Ecol. Syst.* **3**: 233–268.

Bayly, I. A. E. and W. D. Williams. 1966. "Chemical and biological studies on some saline lakes of south-east Australia." *Austral. J. Mar. Freshwat. Res.* **17**: 177–228.

Bayly, I. A. E. and W. D. Williams. 1973. *Inland Waters and Their Ecology.* Longman Australia Pty Ltd., Victoria.

Beadle, L. C. 1943. "Osmotic regulation and the faunas of inland waters." *Biol. Rev.* **18**: 172–183.

Beadle, L. C. 1957. "Comparative physiology: Osmotic and ionic regulation in aquatic animals." *Ann. Rev. Physiol.* **19**: 329-358.

Beadle, L. C. 1959. "Osmotic and ionic regulation in relation to the classification of brackish and inland saline waters." *Arch. Oceanogr. Limnol.* **11** (Suppl.): 143-151.

Beadle, L. C. 1969. "Osmotic regulation and the adaptation of freshwater animals to inland saline waters." *Verh. Internat. Verein. Limnol.* **17**: 421-429.

Beadle, L. C. 1974. *The Inland Waters of Tropical Africa.* Longman, London.

Bebutova, I. M. 1941. "Biology and systematics of the Baikalian caddisflies." *Izvestiya Akad. Nauk. USSR, Biol. Ser.* **1**: 82-104 (in Russian).

Behmer, D. J. and C. P. Hawkins. 1986. "Effects of overhead canopy on macroinvertebrate production in a Utah stream." *Freshwat. Biol.* **16**: 287-300.

Behning, A. 1928. "Das Leben der Wolga, zugleich eine Einführung in die Flussbiologie." *Die Binnengewässer* **5**: 1-162.

Beier, M. 1948. "Zur Kenntnis von Körperbau und Lebensweise der Helminen (Col. Dryopidae)." Eos, Wien **24**: 123-211.

Beier, M. and E. Pomeisl. 1959. "Einiges über Körperbau und Lebensweise von *Ochthebius exsculpus* Germ. und seiner Larve (Col. Hydroph. Hydraen.)." *Z. Morph. Ökol. Tiere* **48**: 72-88.

Belkin, J. N. 1962. *The Mosquitoes of the South Pacific (Diptera, Culicidae),* Vol. 1. Univ. Calif. Press, Berkeley.

Benech, V. 1972. "Etude expérimentale de l'incubation des oeufs de *Baetis rhodani* Pictet." *Freshwat. Biol.* **2**: 243-252.

Bengtsson, B.-E. 1981. "The growth of some ephemeropteran nymphs during winter in a North Swedish river." *Aquatic Insects* **3**: 199-208.

Benke, A. C., T. C. Van Arsdall, D. M. Gillespie, and F. K. Parrish. 1984. "Invertebrate productivity in a subtropical blackwater river: The importance of habitat and life history." *Ecol. Monogr.* **54**: 25-63.

Berg, C. O. 1949. "Limnological relations of insects to plants of the genus *Potamogeton.*" *Trans. Am. Microsc. Soc.* **68**: 279-291.

Berg, C. O. 1950. "Biology of certain aquatic caterpillars (Pyralidae: *Nymphula* spp.) which feed on *Potamogeton.*" *Trans. Am. Microsc. Soc.* **69**: 254-266.

Berg, K. 1937. "Contributions to the biology of *Corethra* Meigen (*Chaoborus* Lichtenstein)." *Det. Kgl. Danske Vidensk. Selskab. Biolog. Medd.* **13**: 1-101.

Berg, K. 1938. "Studies on the bottom animals of Esrom Lake." *Kgl. Danske Vidensk. Selskab Skr., Nat. Mat. Afd. 9 Rk.* **8**: 1-255.

Berg, K. 1941. "Contributions to the biology of the aquatic moth *Acentropus niveus* (Oliv.)." *Vidensk. Medd. fra Danske Naturh. Foren.* **105**: 60-139.

Berg, K., S. A. Boisen-Bennike, P. M. Jónasson, J. Keiding, and A. Nielsen. 1948. "Biological studies on the River Susaa." *Folia Limnol. Scand.* **4**: 1-318.

Berg, K., P. M. Jónasson, and K. W. Ockelmann. 1962. "The respiration of some animals from the profundal zone of a lake." *Hydrobiologia* **19**: 1-39.

Berg, K. and I. C. Petersen. 1956. "Studies on the humic, acid Lake Gribsø." *Folia Limnol. Scand.* **8**: 1-273.

Berner, L. 1940. "*Baetisca rogersi*, a new mayfly from northern Florida." *Can. Ent.* **72**: 156-160.

Berner, L. and W. C. Sloan. 1954. "The occurrence of a mayfly nymph in brackish water." *Ecology* **35**: 98.

Berté, S. B. and G. Pritchard. 1983. "The life history of *Glyphopsyche irrorata* (Trichoptera, Limnephilidae): A caddisfly that overwinters as an adult." *Holarctic Ecol.* **6**: 69–73.

Berthélemy, C. 1966. "Recherches écologiques et biogéographiques sur les plécoptères et coléoptères d'eau courante (Hydraena et Elminthidae) des Pyrénées." *Ann. Limnol.* **2**: 227–458.

Berthélemy, C. (ed.). 1984. "Eighth International Symposium on Plecoptera." *Ann. Limnol.* **20**(1–2): 1–145.

Bertrand, H. 1954. *Les Insectes Aquatiques d'Europe*. 2 vols. Lechevalier, Paris.

Bertrand, H. P. I. 1972. *Larves et nymphs des coléoptères aquatiques du globe.* Centre National de la Recherche Scientifique, Paris.

Bick, G. H. and R. H. Penn. 1947. "Resistance of mosquito larvae and pupae to experimental drought." *Ann. Ent. Soc. Am.* **25**: 91–94.

Bilby, R. E. and G. E. Likens. 1980. "Importance of organic debris dams in the structure and function of stream ecosystems." *Ecology* **61**: 1107–1113.

Birge, E. A., C. Juday, and H. W. March. 1928. "The temperature of the bottom deposits of Lake Mendota: A chapter in the heat exchanges in the Lake." *Trans. Wisc. Acad. Sci. Arts Lett.* **23**: 187–231.

Bishop, J. E. 1973a. *Limnology of a Small Malayan River Sungai Gombak.* Dr. W. Junk, The Hague.

Bishop, J. E. 1973b. "Observations on the vertical distribution of the benthos in a Malaysian Stream." *Freshwat. Biol.* **3**: 147–156.

Blake, R. W. 1986. "Hydrodynamics of swimming in the water boatman, *Cenocorixa bifida*." *Can. J. Zool.* **64**: 1606–1613.

Blinn, D. W., C. Pinney, and M. W. Sanderson. 1982. "Nocturnal planktonic behavior of *Ranatra montezuma* Polhemus (Nepidae: Hemiptera) in Montezuma Well, Arizona." *J. Kans. Ent. Soc.* **55**: 481–484.

Bohle, H. W. 1969. "Untersuchungen über die Embryonalentwicklung und die embryonale Diapause bei *Baetis vernus* Curtis und *Baetis rhodani* (Pictet) (Baetidae, Ephemeroptera)." *Zool. Jb. Abt. Anat. Ökol.* **86**: 493–575.

Bohle, H. W. 1972. "Die Temperaturabhängigkeit der Embryogenese und der embryonalen Diapause von *Ephemerella ignita* (Poda) (Insecta, Ephemeroptera)." *Oecologia* **10**: 253–268.

Boling, R. H., E. D. Goodman, J. V. Zimmer, K. W. Cummins, S. R. Reice, R. C. Petersen, and J. A. Van Sickle. 1975. "Toward a model of detritus processing in a woodland stream." *Ecology* **56**: 141–151.

Bonazzi, G. and P. F. Ghetti. 1977. "I macroinvertebrati del torrente Parma. Risultati di un ciclo annuale di ricerche." *Ateneo Parmense, Acta Nat.* **13**: 351–395.

Boon, P. J. 1979. "Adaptive strategies of *Amphipsyche* larvae (Trichoptera: Hydropsychidae) downstream of a tropical impoundment," pp. 237–255 in J. V. Ward and J. A. Stanford (eds.), *The Ecology of Regulated Streams.* Plenum, New York.

Boon, P. J. and S. W. Shires. 1976. "Temperature studies on a river system in north-east England." *Freshwat. Biol.* **6**: 23–32.

Borkent, A. 1981. "The distribution and habitat preferences of the Chaoboridae (Culicomorpha: Diptera) of the Holarctic Region." *Can. J. Zool.* **59**: 122–133.

Borutsky, E. V. 1939. "Dynamics of the total benthic biomass in the profundal of Lake Beloye." *Works Limnol. Sta. Kossino* **22**: 196-218.

Botosaneanu, L. 1959. "Recherches sur les trichoptères du massif de Retezat et des Monts du Banat." *Bibl. Biol. Animala* (Bucarest) **1**: 1-165.

Botosaneanu, L. 1988. "Zonation et classification, biologiques des cours d'eau: Développements récents, alternatives, perspectives." *Atti XV Congr. naz. ital. Ent., L'Aquila* **1988**: 33-61.

Botosaneanu, L. 1989. "Three western Palaearctic caddisflies (Trichoptera) from the British Museum (Natural History) collection." *Ent. Gaz.* **40**: 165–169.

Botosaneanu, L. and J. R. Holsinger, 1991. "Some aspects concerning colonization of the subterranean realm—especially of subterranean waters: a response to Rouch and Danielopol, 1987." *Stygologia* **6**: 11-39.

Bott, H. R. 1928. "Beiträge zur Kenntnis von *Gyrinus natator substriatus* Steph. I. Lebensweise und Entwicklung. II. Der Sehapparat." *Z. Morphol. Ökol. Tiere* **10**: 207-306.

Bou, C. 1979. "Etude de la faune interstitielle des alluvions du Tarn mise en place d'une station d'etude et resultats preliminaires." *Bull. Féd. Tarn Spéleo.-Archéol.* **16**: 117-129.

Boudreaux, H. B. 1979. *Arthropod Phylogeny with Special Reference to Insects.* Wiley, New York.

Boulton, A. J. and P. S. Lake. 1988. "Australian temporary streams—some ecological characteristics." *Verh. Internat. Verein. Limnol.* **23**: 1380–1383.

Bournaud, M. 1971. "Observations biologiques sur les Trichoptères cavernicoles." *Bull. Mens. Soc. Linn. Lyon* **7**: 196–211.

Bournaud, M., G. Keck, and P. Richoux. 1980. "Les prelevements de macroinvertebres benthiques en tant que revelateurs de la physionomie d'une riviere." *Ann. Limnol.* **16**: 55–75.

Bournaud, M. and H. Tachet (eds.). 1987. *Proceedings of the Fifth International Symposium on Trichoptera.* Dr. W. Junk, Dordrecht.

Bradley, J. C. 1942. "The origin and significance of metamorphosis and wings among insects." *Proc. VIII Pan. Am. Congr., Biol. Sect.* **3**: 303-309.

Bradshaw, W. E. 1973. "Homeostasis and polymorphism in vernal development of *Chaoborus americanus*." *Ecology* **54**: 1247-1259.

Bradshaw, W. E. 1980. "Thermoperiodism and the thermal environment of the pitcher plant mosquito, *Wyeomyia smithii*." *Oecologia* **46**: 13-17.

Bradshaw, W. E. and R. A. Creelman. 1984. "Mutualism between the carnivorous purple pitcher plant and its inhabitants." *Am. Midl. Nat.* **112**: 294–303.

Braukmann, U. 1987. "Zoozönologische und saprobiologische Beiträge zu einer allgemeinen regionalen Bachtypologie." *Ergebnisse Limnol.* **26**: 1-355.

Bretschko, G. 1981. "Vertical distribution of zoobenthos in an alpine brook of the RITRODAT-LUNZ study area." *Verh. Internat. Verein. Limnol.* **21**: 873–876.

Bretschko, G. 1985. "Quantitative sampling of the fauna of gravel streams." *Verh. Internat. Verein. Limnol.* **22**: 2049-2052.

Brewer, M. C. 1958. "The thermal regime of an arctic lake." *Trans. Am. Geophys. Union* **39**: 278–284.

Brinck, P. 1949. "Studies on Swedish stoneflies (Plecoptera)." *Opuscula Ent.* (Suppl. 11): 1-250.

Brinkhurst, R. O. 1959. "Alary polymorphism in the Gerroidea (Hemiptera-Heteroptera)." *J. An. Ecol.* **28**: 211–230.

Brinkhurst, R. O. 1974. *The Benthos of Lakes.* St. Martin's Press, New York.

Britt, N. W. 1962. "Biology of 2 species of Lake Erie Mayflies, *Ephoron album* (Say) and *Ephemera simulans* Walker." *Bull. Ohio Biol. Surv.* **5**: 1–70.

Brittain, J. E. 1974. "Studies on the lentic Ephemeroptera and Plecoptera of southern Norway." *Nor. Tiddsskr.* **21**: 135–154.

Brittain, J. E. 1975. "The life cycle of *Baetis macani* Kimmins (Ephemerida) in a Norwegian mountain biotope." *Ent. Scand.* **6**: 47–51.

Brittain, J. E. 1976a. "The temperature of two Welsh Lakes and its effect on the distribution of two freshwater Insects." *Hydrobiologia* **48**: 37–49.

Brittain, J. E. 1976b. "Experimental studies on nymphal growth in *Leptophlebia vespertina* (L.) (Ephemeroptera)." *Freshwat. Biol.* **6**: 445–449.

Brittain, J. E. 1978a. "The Ephemeroptera of Øvre Heimdalsvatn." *Holarct. Ecol.* **1**: 239–254.

Brittain, J. E. 1978b. "The aquatic Coleoptera of Øvre Heimdalsvatn." *Holarct. Ecol.* **1**: 266–270.

Brittain, J. E. 1979. "Emergence of Ephemeroptera from Øvre Heimdalsvatn, a Norwegian subalpine lake," pp. 115–123 in K. Pasternak and R. Sowa (eds.), *Proc. Second Internat. Conf. Ephemeroptera.* Panstwowe Wydawnictwo Naukowe, Warsaw.

Brittain, J. E. 1982. "Biology of mayflies." *Ann. Rev. Ent.* **27**: 119-147.

Brittain, J. E. 1983. "The influence of temperature on nymphal growth rates in mountain stoneflies (Plecoptera)." *Ecology* **64**: 440–446.

Brittain, J. E. 1990. "Life history strategies in Ephemeroptera and Plecoptera," pp. 1–12 in I. C. Campbell (ed.), *Mayflies and Stoneflies.* Kluwer, Dordrecht.

Brittain, J. E. and A. Lillehammer. 1978. "The fauna of the exposed zone of Øvre Heimdalsvatn: Methods, sampling stations and general results." *Holarct. Ecol.* **1**: 221–228.

Brittain, J. E. and B. Nagell. 1981. "Overwintering at low oxygen concentrations in the mayfly *Leptophlebia vespertina.*" *Oikos* **36**: 45–50.

Bro Larsen, E. 1952. "On subsocial beetles from the salt-marsh, their care of progeny and adaptation to salt and tide." *Proc. 9th Internat. Congr. Ent., Amsterdam* **I**: 502–506.

Brock, T. D. 1985. "Life at high temperatures." *Science* **230**: 132-138.

Brodsky, K. A. 1980. *Mountain Torrent of the Tien Shan.* Dr. W. Junk, The Hague.

Brooker, M. P. and R. J. Hemsworth. 1978. "The effect of the release of an artificial discharge of water on invertebrate drift in the R. Wye, Wales." *Hydrobiologia* **59**: 155–163.

Brooks, J. L. 1950. "Speciation in ancient lakes." *Q. Rev. Biol.* **25**: 30-60, 131–176.

Brown, A. V. and L. C. Fitzpatrick. 1978. "Life history and population energetics of the dobson fly, *Corydalus cornutus.*" *Ecology* **59**: 1091–1109.

Brown, C. J. D., W. D. Clothier, and W. Alvord. 1953. "Observations on ice conditions and bottom organisms in the West Gallatin River, Montana." *Proc. Mont. Acad. Sci.* **13**: 21–27.

Brown, H. P. 1987. "Biology of riffle beetles." *Ann. Rev. Ent.* **32**: 253–273.

Brown, J. H. and C. R. Feldmeth. 1971. "Evolution in constant and fluctuating environments: Thermal tolerances of desert pupfish *(Cyprinodon)*." *Evolution* **25**: 390–398.

Brues, C. T. 1924. "Observations on animal life in the thermal waters of Yellowstone Park, with a consideration of the thermal environments." *Proc. Am. Acad. Arts Sci.* **59**: 371–437.

Brues, C. T. 1927. "Animal Life in Hot Springs." *Rev. Biol.* **2**: 181-203.

Brues, C. T. 1928. "Studies of the fauna of hot springs in the western United States and the biology of thermophilous animals." *Proc. Am. Acad. Arts Sci.* **63**: 139–228.

Brues, C. T. 1932. "Further studies on the fauna of North American hot springs." *Proc. Am. Acad. Arts Sci.* **67**: 185–303.

Brundin, L. 1949. "Chironomiden und andere Bodentiere der Südschwedischen Urgebirgsseen." *Rep. Instit. Freshwat. Res. Drottningholm* **30**: 1–914.

Brundin, L. 1956. "Die bodenfaunistischen Seetypen und ihre Anwendbarkheit auf die Südhalkugel Zugleich ein Theorie der produktionbiologischen Bedeutung der glazialen Erosion." *Rep. Instit. Freshwat. Res. Drottningholm* **37**: 186–235.

Bruns, D. A. and W. L. Minckley. 1980. "Distribution and abundance of benthic invertebrates in a Sonoran Desert stream." *J. Arid Envir.* **3**: 117–131.

Bruns, D. A., G. W. Minshall, J. T. Brock, C. E. Cushing, K. W. Cummins, and R. L. Vannote. 1982. "Ordination of functional groups and organic matter parameters from the Middle Fork of the Salmon River, Idaho." *Freshwat. Invert. Biol.* **1**: 2–12.

Brusven, M. A. and C. E. Hornig. 1984. "Effects of suspended and deposited volcanic ash on survival and behavior of stream insects." *J. Kans. Ent. Soc.* **57**: 55–62.

Brusven, M. A., C. MacPhee, and R. Biggam. 1974. "Effects of water fluctuation on benthic insects," pp. 67–79 in *Anatomy of a River*. Pacific Northwest River Basins Commission Report, Vancouver, Washington.

Brusven, M. A., W. R. Meehan, and R. C. Biggam. 1990. "The role of aquatic moss on community composition and drift of fish-food organisms." *Hydrobiologia* **196**: 39–50.

Bullock, T. H. 1955. "Compensation for temperature in the metabolism and activity of poikilotherms." *Biol. Rev.* **30**: 311–342.

Burger, J. F., J. R. Anderson, and M. F. Knudsen. 1980. "The habits and life history of *Oedoparena glauca* (Diptera: Dryomyzidae), a predator of barnacles." *Proc. Ent. Soc. Wash.* **82**: 360–377.

Burmeister, H. 1839. "Plecoptera," pp. 863–881 in *Handbuch der Entomologie,* Vol. II, part 2. T. C. F. Enslin, Berlin.

Butler, M. G. 1982. "A 7-year life cycle for two *Chironomus* species in arctic Alaskan tundra ponds (Diptera: Chironomidae)." *Can. J. Zool.* **60**: 58–70.

Butler, M., M. C. Miller, and S. Mozley. 1980. "Macrobenthos," pp. 297–339 in J. E. Hobbie (ed.), *Limnology of Tundra Ponds.* Dowden, Hutchinson and Ross, Stroudsburg, Pa.

Butler, P. M. and E. J. Popham. 1958. "The effects of the floods of 1953 on the aquatic insect fauna of Spurn (Yorkshire)." *Proc. Royal Ent. Soc. London, A.* **33**: 149–158.

Buxton, P. A. 1926. "The colonization of the sea by insects: With an account of the habits of *Pontomyia*, the only known submarine insect." *Proc. Zool. Soc. London* **1926**: 807-814.

Campbell, B. C. and R. F. Denno. 1978. "The structure of the aquatic insect community associated with intertidal pools on a New Jersey salt marsh." *Ecol. Ent.* **3**: 181-187.

Campbell, I. C. (ed.). 1990. *Mayflies and Stoneflies.* Kluwer, Dordrecht.

Cannings, R. A., S. G. Cannings, and R. J. Cannings. 1980. "The distribution of the genus *Lestes* in a saline lake series in central British Columbia, Canada (Zygoptera: Lestidae)." *Odonatologica* **9**: 19-28.

Canton, S. P., L. D. Cline, R. A. Short, and J. V. Ward. 1984. "The macroinvertebrates and fishes of a Colorado stream during a period of fluctuating discharge." *Freshwat. Biol.* **14**: 311-316.

Canton, S. P. and J. V. Ward. 1981. "Emergence of Trichoptera from Trout Creek, Colorado, USA," pp. 39-45 in G. P. Moretti (ed.), *Proc. Third Internat. Symp. Trichoptera.* Dr. W. Junk, The Hague.

Cantrall, I. J. 1984. "Semiaquatic Orthoptera," pp. 177-181 in R. W. Merritt and K. W. Cummins (eds.), *An Introduction to the Aquatic Insects of North America.* Kendall/Hunt, Dubuque, Iowa.

Cantrell, M. A. 1979. "Invertebrate communities in the Lake Chilwa swamp in years of high level," pp. 161-173 in M. Kalk, A. J. McLachan, and C. Howard-Williams (eds.), *Lake Chilwa. Studies of Change in a Tropical Ecosystem.* Dr. W. Junk, The Hague.

Cantrell, M. A. 1988. "Effect of lake fluctuations on the habitats of benthic invertebrates in a shallow tropical lake." *Hydrobiologia* **158**: 125-131.

Cantrell, M. A. and A. J. McLachlan. 1982. "Habitat duration and dipteran larvae in tropical rain pools." *Oikos* **38**: 343-348.

Carlsson, G. F. 1962. "Studies on Scandinavian black flies (Fam. Simuliidae Latr.)." *Opuscula Ent. Suppl.* **21**: 1-280.

Carlsson, M., L. M. Nilsson, Bj. Svensson, S. Ulfstrand, and R. S. Wooten. 1977. "Lacustrine seston and other factors influencing the blackflies (Diptera: Simuliidae) inhabiting lake outlets in Swedish Lapland." *Oikos* **29**: 229-238.

Carpenter, F. M. 1953. "The Geological history and evolution of insects." *Am. Sci.* **41**: 256-270.

Carpenter, F. M. 1977. "Geological history and the evolution of the insects." *Proc. 15th Internat. Congr. Ent., Washington, D.C.*, pp. 63-70.

Carpenter, K. E. 1927. "Faunistic ecology of some Cardiganshire streams." *J. Ecol.* **15**: 33-53.

Caspers, H. 1951. "Rhythmische Erscheinungen in der Fortpflanzung von *Clunio marinus* (Dipt. Chiron.) und das problem der lunaren Periodizitat bei organismen." *Arch. Hydrobiol. Suppl.* **18**: 415-594.

Chance, M. M. and D. A. Craig. 1986. "Hydrodynamics and behaviour of Simuliidae larvae (Diptera)." *Can. J. Zool.* **64**: 1295-1309.

Chaston, I. 1969. "The light threshold controlling the vertical migration of *Chaoborus punctipennis* in a Georgia impoundment." *Ecology* **50**: 916-920.

Cheng, L. 1973a. "Can *Halobates* dodge nets? I: By daylight?" *Limnol. Oceanogr.* **13**: 564-570.

Cheng, L. 1973b. "Marine and freshwater skaters: Differences in surface fine structures." *Nature* **242**: 132–133.

Cheng, L. 1974. "Notes on the ecology of the oceanic insect *Halobates*." *Mar. Fish. Rev.* **36**: 1-7.

Cheng, L. (ed.). 1976. *Marine Insects.* North-Holland Publ. Co., Amsterdam.

Cheng, L. 1985. "Biology of *Halobates* (Heteroptera: Gerridae)." *Ann. Rev. Ent.* **30**: 111–135.

Cheng, L. and M. C. Birch. 1978. "Insect flotsam: An unstudied marine resource." *Ecol. Ent.* **3**: 87–97.

Cheng, L. and J. T. Enright. 1973. "Can *Halobates* dodge nets? II: By moonlight?" *Limnol. Oceanogr.* **18**: 666–669.

Cheng, L. and H. Hashimoto. 1978. "The marine midge *Pontomyia* (Chironomidae) with a description of females of *P. oceana* Tokunaga." *Syst. Ent.* **3**: 189–196.

Chessman, B. C. 1986. "Dietary studies of aquatic insects from two Victorian rivers." *Austral. J. Mar. Freshwat. Res.* **37**: 129–146.

China, W. E. 1955. "The evolution of the water bugs." *Bull. Nat. Instit. Sci. India* **7**: 91-103.

Christophers, S. R. 1960. *Aedes aegypti (L.) The Yellow Fever Mosquito.* Cambridge Univ. Press, Cambridge.

Chutter, R. M. 1969a. "The effects of silt and sand on the invertebrate fauna of streams and rivers." *Hydrobiologia* **34**: 57-76.

Chutter, F. M. 1969b. "The distribution of some stream invertebrates in relation to current speed." *Internat. Revue ges. Hydrobiol.* **54**: 413–422.

Chutter, F. M. 1970. "Hydrobiological studies in the catchment of Vaal Dam, South Africa. Part 1. River Zonation and the benthic fauna." *Internat. Revue ges. Hydrobiol.* **55**: 445–494.

Claassen, P. W. 1940. "A catalog of the Plecoptera of the world." *Cornell Univ. Agric. Exp. Sta. Mem.* **232**: 1–235.

Claflin, T. O. 1968. "Reservoir aufwuchs on inundated trees." *Trans. Am. Microsc. Soc.* **87**: 97–104.

Claire, E. and R. Phillips. 1968. "The stonefly *Acroneuria pacifica* as a potential predator on salmonid embryos." *Trans. Am. Fish. Soc.* **97**: 50–52.

Clarke, G. L. 1954. *Elements of Ecology.* Wiley, New York.

Clausen, C. P. 1950. "Respiratory adaptations in the immature stages of parasitic insects." *Arthropoda* **1**: 198–224.

Clegg, J. 1959. *The Freshwater Life of the British Isles,* 2nd ed. Warne, London.

Clifford, H. F. 1966. "The ecology of invertebrates in an intermittent stream." *Invest. Indiana Lakes Streams* **7**: 57–98.

Clifford, H. F. 1972. "A year's study of the drifting organisms in a brown-water stream of Alberta, Canada." *Can. J. Zool.* **50**: 975-983.

Clifford, H. F. 1978. "Descriptive phenology and seasonality of a Canadian brown-water stream." *Hydrobiologia* **58**: 213–231.

Clifford, H. F., M. R. Robertson, and K. A. Zelt. 1973. "Life cycle patterns of mayflies (Ephemeroptera) from some streams of Alberta, Canada," pp. 122–131 in W. L. Peters and J. G. Peters (eds.), *Proc. First Internat. Conference Ephemeroptera.* E. J. Brill, Leiden.

Clifford, H. F., H. Hamilton, and B. A. Killins. 1979. "Biology of the mayfly *Leptophelbia cupida* (Say) (Ephemeroptera: Leptophlebiidae)." *Can. J. Zool.* 57: 1026-1045.

Cline, L. D., R. A. Short, and J. V. Ward. 1982. "The influence of highway construction on the macroinvertebrates and epilithic algae of a high mountain stream." *Hydrobiologia* **96**: 149-159.

Cockrell, B. J. 1984. "Effects of temperature and oxygenation on predator-prey overlap and prey choice of *Notonecta glauca*." *J. An. Ecol.* **53**: 519-532.

Coker, R. E., V. Millsaps, and R. Rice. 1936. "Swimming plume and claws of the broad-shouldered water-strider *Rhagovelia flavicinta* Bueno (Hemiptera)." *Bull. Brooklyn Ent. Soc.* **31**: 81-85.

Colbo, M. H. 1979. "Distribution of winter-developing Simuliidae (Diptera), in eastern Newfoundland." *Can. J. Zool.* **57**: 2143-2152.

Cole, G. A. 1968. "Desert limnology," pp. 423-486 in G. W. Brown, Jr. (ed.), *Desert Biology*, Vol. 1. Academic Press, New York.

Cole, G. A. 1983. *Textbook of Limnology*. Mosby, St. Louis.

Coleman, M. J. and H. B. N. Hynes. 1970. "The vertical distribution of the invertebrate fauna in the bed of a stream." *Limnol. Oceanogr.* **15**: 31-40.

Collins, J. P., C. Young, J. Howell, and W. L. Minckley. 1981. "Impact of flooding in a Sonoran Desert stream, including elimination of an endangered fish population (*Poeciliopsis o. occidentalis*, Poeciliidae)." *Southwest. Nat.* **26**: 415-423.

Colwell, R. K. 1974. "Predictability, constancy and contingency of periodic phenomena." *Ecology* **55**: 1148-1153.

Comstock, J. H. 1887. "Note on the respiration of aquatic bugs." *Am. Nat.* **21**: 577-578.

Cook, D. G. and M. G. Johnson. 1974. "Benthic macroinvertebrates of the St. Lawrence Great Lakes." *J. Fish. Res. Board Can.* **31**: 763-782.

Cook, S. F., Jr. and J. D. Connors. 1964. "The direct response of *Chaoborus astictopus* to light (Diptera: Culicidae)." *Ann. Ent. Soc. Am.* **57**: 387-388.

Corbet, P. S. 1957a. "The life-histories of two summer species of dragonfly (Odonata: Coenagriidae)." *Proc. Zool. Soc. London* **128**: 403-418.

Corbet, P. S. 1957b. "Larvae of East African Odonata." *Entomologist* **90**: 28-34.

Corbet, P. S. 1958. "Lunar periodicity of aquatic insects in Lake Victoria." *Nature* **182**: 330-331.

Corbet, P. S. 1962. *A Biology of Dragonflies*. Witherby, London.

Corbet, P. S. 1964. "Temporal patterns of emergence in aquatic insects." *Can. Ent.* **96**: 264-279.

Corbet, P. S. 1980. "Biology of Odonata." *Ann. Rev. Ent.* **25**: 189-217.

Corbet, P. S., C. Longfield, and N. W. Moore. 1960. *Dragonflies*. New Naturalist Series. Collins, London.

Corbet, S. A., R. D. Sellick, and N. G. Willoughby. 1974. "Notes on the biology of the mayfly *Povilla adusta* in West Africa." *J. Zool.*, (London) **172**: 491-502.

Cordone, A. J. and D. W. Kelley. 1961. "The influences of inorganic sediment on the aquatic life of streams." *Calif. Fish Game* **47**: 189-228.

Corkum, L. D. and H. F. Clifford. 1981. "Function of caudal filaments and correlated structures in mayfly nymphs, with special reference to *Baetis* (Ephemeroptera)." *Quaest. Ent.* **17**: 129-146.

Corrarino, C. A. and M. A. Brusven. 1983. "The effects of reduced stream discharge on insect drift and stranding of near shore insects." *Freshwat. Invert.* **2**: 88–98.

Costello, M. J., T. K. McCarthy, and M. M. O'Farrell. 1984. "The stoneflies (Plecoptera) of the Corrib catchment area, Ireland." *Ann. Limnol.* **20**: 25–34.

Courtney, G. W. 1986. "Discovery of the immature stages of *Parasimulium crosskeyi* Peterson (Diptera: Simuliidae), with a discussion of a unique black fly habitat." *Proc. Ent. Soc. Wash.* **88**: 280–286.

Coutant, C. C. 1982. "Positive phototaxis in first instar *Hydropsyche cockerelli* Banks (Trichoptera)." *Aquatic Insects* **4**: 55–59.

Cowell, B. C. and W. C. Carew. 1976. "Seasonal and diel periodicity in the drift of aquatic insects in a subtropical Florida stream." *Freshwat. Biol.* **6**: 587–594.

Cowell, B. C. and P. L. Hudson. 1967. "Some environmental factors influencing benthic invertebrates in two Missouri River reservoirs," pp. 541–555 in *Reservoir Fishery Resources Symposium*. Am. Fish. Soc., Washington, D.C.

Cowie, B. and M. J. Winterbourn. 1979. "Biota of a subalpine springbrook in the southern alps." *N. Z. J. Mar. Freshwat. Res.* **13**: 295–302.

Craig, D. A. 1990. "Behavioural hydrodynamics of *Cloeon dipterum* larvae (Ephemeroptera: Baetidae)." *J. N. Am. Benthol. Soc.* **9**: 346–357.

Crawford, C. S. 1981. *Biology of Desert Invertebrates.* Springer-Verlag, Berlin.

Crichton, M. I. (ed.). 1978. *Proceedings of the Second International Symposium on Trichoptera, Reading, England, 1977.* Dr. W. Junk, The Hague.

Crisp, D. T. and E. D. LeCren. 1970. "The temperature of three different small streams in northwest England." *Hydrobiologia* **35**: 305–323.

Crowson, R. A. 1960. "The phylogeny of Coleoptera." *Ann. Rev. Ent.* **5**: 111–134.

Crowson, R. A. 1981. *The Biology of the Coleoptera.* Academic Press, London.

Cudney, M. D. and J. B. Wallace. 1980. "Life cycles, microdistribution and production dynamics of six species of net-spinning caddisflies in a large southeastern (U.S.A.) river." *Holarct. Ecol.* **3**: 169-182.

Cullen, M. J. 1969. "The biology of giant water bugs in Trinidad." *Proc. Roy. Ent. Soc. London, A* **44**: 123–136.

Culp, J. M. and R. W. Davies. 1982. "Analysis of longitudinal zonation and the river continuum concept in the Oldman-South Saskatchewan River system." *Can. J. Fish. Aquat. Sci.* **39**: 1258–1266.

Culp, J. M., S. J. Walde, and R. W. Davies. 1983. "Relative importance of substrate particle size and detritus to stream benthic macroinvertebrate microdistribution." *Can. J. Fish. Aquatic Sci.* **40**: 1568–1574.

Culp, J. M., F. J. Wrona, and R. W. Davies. 1986. "Response of stream benthos and drift to fine sediment deposition versus transport." *Can. J. Zool.* **64**: 1345–1351.

Cummins, K. W. 1962. "An evaluation of some techniques for the collection and analysis of benthic samples with special emphasis on lotic waters." *Am. Midl. Nat.* **67**: 477–504.

Cummins, K. W. 1964. "Factors limiting the microdistribution of the caddisflies *Pycnopsyche lepida* (Hagen) and *Pycnopsyche guttifer* (Walker) in a Michigan stream (Trichoptera: Limnephilidae)." *Ecol. Monogr.* **34**: 271–295.

Cummins, K. W. 1974. "Structure and function of stream ecosystems." *BioScience* **24**: 631–641.

Cummins, K. W. 1979. "The natural stream ecosystem," pp. 7–24 in J. V. Ward and J. A. Stanford (eds.), *The Ecology of Regulated Streams.* Plenum, New York.

Cummins, K. W. and M. J. Klug. 1979. "Feeding ecology of stream invertebrates." *Ann. Rev. Ecol. Syst.* **10**: 147–172.

Cummins, K. W., M. J. Klug, G. M. Ward, G. L. Spengler, R. W. Speaker, R. W. Ovink, D. C. Mahan, and R. C. Petersen. 1981. "Trends in particulate organic matter fluxes, community processes and macroinvertebrate functional groups along a Great Lakes Drainage Basin river continuum." *Verh. Internat. Verein. Limnol.* **21**: 841-849.

Cummins, K. W., G. W. Minshall, J. R. Sedell, C. E. Cushing, and R. C. Petersen. 1984. "Stream ecosystem theory." *Verh. Internat. Verein. Limnol.* **22**: 1818–1827.

Cushing, C. E. 1964. "Plankton and water chemistry in the Montreal River lake-stream system, Saskatchewan." *Ecology* **45**: 306–313.

Cushing, C. E. and S. D. Smith. 1982. "Effects of Mt. St. Helens ashfall on lotic algae and caddisflies." *J. Freshwat. Ecol.* **1**: 527–538.

Daborn, G. R. 1971. "Survival and mortality of coenagrionid nymphs (Odonata: Zygoptera) from the ice of an aestival pond." *Can. J. Zool.* **49**: 569–571.

Daborn, G. R. 1974. "Biological features of an aestival pond in Western Canada." *Hydrobiologia* **44**: 287-299.

Daborn, G. R. 1975. "The argillotrophic lake system." *Verh. Internat. Verein. Limnol.* **19**: 580–588.

Daborn, G. R. 1976. "Physical and chemical features of a vernal temporary pond in western Canada." *Hydrobiologia* **51**: 33–38.

Daborn, G. R. and H. F. Clifford. 1974. "Physical and chemical features of an aestival pond in western Canada." *Hydrobiologia* **44**: 43–59.

Dahl, E. 1956. "Ecological salinity boundaries in poikilohaline waters." *Oikos* **7**: 1–21.

Dale, H. M. and T. Gillespie. 1977a. "Diurnal fluctuations of temperature near the bottom of shallow water bodies as affected by solar radiation, bottom color and water circulation." *Hydrobiologia* **55**: 87–92.

Dale, H. M. and T. J. Gillespie. 1977b. "The influence of submersed aquatic plants on temperature gradients in shallow water bodies." *Can. J. Bot.* **55**: 2216–2225.

Dall, P. C., C. Lindegaard, and P. M. Jónasson. 1990. "In-lake variations in the compositions of zoobenthos in the littoral of Lake Esrom, Denmark." *Verh. Internat. Verein. Limnol.* **24**: 613-620.

Daly, H. V. 1984. "General classification and key to the orders of aquatic and semiaquatic insects," pp. 76–81 in R. W. Merritt and K. W. Cummins (eds.), *An Introduction to the Aquatic Insects of North America.* Kendall/Hunt, Dubuque, Iowa.

Danecker, E. 1961. "Studien zur hygropetrischen Fauna. Biologie und Ökologie von *Stactobia* und *Tinodes* (Insecta: Trichoptera)." *Internat. Revue ges. Hydrobiol.* **46**: 214–254.

Danielopol, D. L. 1976. "The distribution of the fauna in the interstitial habitats of riverine sediments of the Danube and the Piesting (Austria)." *Int. J. Speleol.* **8**: 23–51.

Danks, H. V. 1971a. "Overwintering of some north temperate and arctic Chironomidae: I. The winter environment." *Can. Ent.* **103**: 589-604.

Danks, H. V. 1971b. "Overwintering of some north temperate and arctic Chironomidae. II. Chironomid biology." *Can. Ent.* **103**: 1875–1910.

Danks, H. V. 1978. "Modes of seasonal adaptation in the insects. I. Winter survival." *Can. Ent.* **110**: 1167–1205.

Danks, H. V. and D. R. Oliver. 1972a. "Diel periodicities of emergence of some high Arctic Chironomidae (Diptera)." *Can. Ent.* **104**: 903–916.

Danks, H. V. and D. R. Oliver. 1972b. "Seasonal emergence of some high arctic Chironomidae (Diptera)." *Can. Ent.* **104**: 661–686.

Davies, B. R. 1974. "The planktonic activity of larval Chironomidae in Loch Leven, Kinross." *Proc. Roy. Soc. Edinburgh, B* **74**: 275–283.

Davies, B. R. 1976. "The dispersal of Chironomidae: A review." *J. Ent. Soc. South Africa* **39**: 39–62.

Davies, B. R. and K. F. Walker (eds.). 1986. *The Ecology of River Systems*. Dr. W. Junk, Dordrecht.

Davies, L. and C. D. Smith. 1958. "The distribution and growth of *Prosimulium* larvae (Diptera: Simuliidae) in hill streams in northern England." *J. Anim. Ecol.* **27**: 335–348.

Davis, J. A. 1986. "Boundary layers, flow microenvironments and stream benthos," pp. 293–312 in P. DeDekker and W. D. Williams (eds.), *Limnology in Australia*. Dr. W. Junk, Dordrecht.

Davis, J. A. and L. A. Barmuta. 1989. "An ecologically useful classification of mean and near-bed flows in streams and rivers." *Freshwat. Biol.* **21**: 271–282.

Davis, L. V. and I. E. Gray. 1966. "Zonal and seasonal distribution of insects in North Carolina salt marshes." *Ecol. Monogr.* **36**: 275-295.

Day, J. H. 1951. "The ecology of South African estuaries. I. A review of estuarine conditions in general." *Trans. Roy. Soc. S. Afr.* **33**: 53–91.

Décamps, H. 1967. "Écologie des Trichoptères de la vallée d'Aure (Hautes-Pyrénées)." *Ann. Limnol.* **3**: 399–577.

Décamps, H. 1968. "Vicariances écologiques chez les Trichoptères des Pyrénées." *Ann. Limnol.* **4**: 1-50.

Decksbach, N. K. 1929. "Zur klassifikation der Gewässer vom astatischen Typus." *Arch. Hydrobiol.* **20**: 399–406.

Deevey, E. S., Jr. 1955. "Limnological studies in Guatemala and El Salvador." *Verh. Internat. Verein. Limnol.* **12**: 278–283.

Delucchi, C. M. 1988. "Comparison of community structure among streams with different temporal flow regimes." *Can. J. Zool.* **66**: 579-586.

de March, B. G. E. 1976. "Spatial and temporal patterns in macrobenthic stream diversity." *J. Fish. Res. Board Can.* **33**: 1261–1270.

Denham, S. C. 1938. "A limnological investigation of the West Fork and Common Branch of White River." *Invest. Indiana Lakes Streams* **1**: 17–71.

Dodds, G. S. and F. L. Hisaw. 1924. "Ecological studies of aquatic insects. II. Size of respiratory organs in relation to environmental conditions." *Ecology* **5**: 262–271.

Dodds, G. S. and F. L. Hisaw. 1925. "Ecological studies on aquatic insects. IV. Altitudinal Range and zonation of mayflies, stoneflies and caddisflies in the Colorado Rockies." *Ecology* **6**: 380-390.

Dodson, V. E. 1975. "Life histories of three species of Corixidae (Hemiptera: Heteroptera) from Western Colorado." *Am. Midl. Nat.* **94**: 257–266.

Dole, M.-J. 1983. "Le domaine aquatique souterraine de la plaine alluviale du Rhône à l'est de Lyon 1. Diversité hydrologique et biocénotique de trois stations représentatives de la dynamique fluviale." *Vie Milieu* **33**: 219–229.

Donald, D. B. and R. S. Anderson. 1980. "The lentic stoneflies (Plecoptera) from the Continental Divide region of southwestern Canada." *Can. Ent.* **112**: 753–758.

Dorier, A. 1933. "Sur la biologie et les metamorphosis de *Psectrocladius obvius* Walk." *Trav. Lab. Hydrobiol. Piscic. Univ. Grenoble* **25**: 205–215.

Dorier, A. and F. Vaillant. 1954. "Observations et expériences relatives à la resistance au courant de divers invertébrés aquatiques." *Trav. Lab. Hydrobiol. Piscic. Univ. Grenoble* **45/46**: 9–31.

Dosdall, L. and D. M. Lehmkuhl. 1979. "Stoneflies (Plecoptera) of Saskatchewan." *Quaest. Ent.* **15**: 3–116.

Douglas, M. M. 1981. "Thermoregulatory significance of thoracic lobes in the evolution of insect wings." *Science* **211**: 84–86.

Downes, J. A. 1964. "Arctic insects and their environment." *Can. Ent.* **96**: 279–307.

Doyen, J. T. 1976. "Marine beetles (Coleoptera excluding Staphylinidae)," pp. 497–519 in L. Cheng (ed.), *Marine Insects*. North-Holland Publ. Co., Amsterdam.

Drever, J. I. 1982. *The Geochemistry of Natural Waters.* Prentice-Hall, Englewood Cliffs, N.J.

Drummond, H. and G. W. Wolfe. 1981. "An observation of a diving beetle larva (Insecta: Coleoptera: Dytiscidae) attacking and killing a garter snake *Thamnophis elegans* (Reptilia: Serpentes: Colubridae)." *Coleopt. Bull.* **35**: 121–124.

Dudgeon, D. 1982a. "Spatial and temporal changes in the sediment characteristics of Tai Po Kau Forest Stream, New Territories, Hong Kong, with some preliminary observations upon within-reach variations in current velocity." *Arch. Hydrobiol. Suppl.* **64**: 36-64.

Dudgeon, D. 1982b. "Spatial and temporal variations in the standing crop of periphyton and allochthonous detritus in a forest stream in Hong Kong, with notes on the magnitude and fate of riparian leaf fall." *Arch. Hydrobiol. Suppl.* **64**: 189–220.

Dudley, T. L. 1988. "The roles of plant complexity and epiphyton in colonization of macrophytes by stream insects." *Verh. Internat. Verein. Limnol.* **23**: 1153–1158.

Dudley, T. and N. H. Anderson. 1982. "A survey of invertebrates associated with wood debris in aquatic habitats." *Melanderia* **39**: 1–21.

Dudley, T. L. and N. H. Anderson. 1987. "The biology and life cycles of *Lipsothrix* spp. (Diptera: Tipulidae) inhabiting wood in western Oregon streams." *Freshwat. Biol.* **17**: 437–451.

Dudley, T. L., S. D. Cooper, and N. Hemphill. 1986. "Effects of macroalgae on a stream invertebrate community." *J. N. Am. Benthol. Soc.* **5**: 93–106.

Dumont, H. J. 1981. "Kratergöl, a deep hypersaline crater-lake in the steppic zone of western-Australia (Turkey), subject to occasional limno-meteorological perturbations." *Hydrobiologia* **82**: 271–279.

Dunn, D. R. 1961. "The bottom fauna of Llyn Tegid (Bala Lake)." *J. An. Ecol.* **30**: 267–281.

Dussart, B. 1966. *Limnologie; l'étude des eaux continentales.* Gauthier-Villars, Paris.

Eastham, L. E. S. 1932. "Currents Produced by the Gills of Mayfly Nymphs." *Nature* **130**: 58.

Eastham, L. E. S. 1934. "Metachronal rhythms and gill movements of the nymph of

Caenis horaria (Ephemeroptera) in relation to water flow." *Proc. Roy. Soc. London, B* **115**: 30–48.

Eastham, L. E. S. 1936. "The rhymical movements of the gills of nymphal *Leptophlebia marginata* (Ephemeroptera) and the currents produced by them in water." *J. Exp. Biol.* **13**: 443–449.

Eastham, L. E. S. 1937. "The gill movements of nymphal *Ecdyonurus venosus* (Ephemeroptera) and the currents produced by them in the water." *J. Exp. Biol.* **14**: 219–228.

Eastham, L. E. S. 1939. "Gill movements of nymphal *Ephemera danica* (Ephemeroptera) and the water currents caused by them." *J. Exp. Biol.* **16**: 18–33.

Eddy, S. 1931. "The plankton of the Sangamon River in the summer of 1929." *Bull. Illinois Nat. Hist. Surv.* **19**: 469–486.

Edington, J. M. 1968. "Habitat preferences in net-spinning caddis larvae with special reference to the influence of water velocity." *J. An. Ecol.* **37**: 675–692.

Edmonds, J. S. and J. V. Ward. 1979. "Profundal benthos of a multibasin foothills reservoir in Colorado, U.S.A." *Hydrobiologia* **63**: 199–208.

Edmunds, G. F. Jr. 1972. "Biogeography and evolution of Ephemeroptera." *Ann. Rev. Ent.* **17**: 21–42.

Edmunds, G. F., Jr. 1973. "Trends and priorities in mayfly research," pp. 7–11 in W. L. Peters and J. G. Peters (eds.), *Proc. First Internat. Conf. Ephemeroptera.* Brill, Leiden.

Edmunds, G. F., Jr., S. L. Jensen, and L. Berner. 1976. *The Mayflies of North and Central America.* Univ. Minnesota Press, Minneapolis.

Edmunds, G. F., Jr. and G. G. Musser. 1960. "The mayfly fauna of Green River in the Flaming Gorge Reservoir basin of Wyoming and Utah." *Univ. Utah Anthropol. Pap.* **48**: 11–23.

Ege, R. 1915. "On the respiratory function of the air stores carried by some aquatic insects." *Z. Allg. Physiol.* **17**: 81–124.

Eggleton, F. E. 1937. "Productivity of the profundal benthic zone in Lake Michigan." *Pap. Mich. Acad. Sci. Arts Lett.* **22**: 593–611.

Egglishaw, H. J. 1964. "The distributional relationship between the bottom fauna and plant detritus in streams." *J. An. Ecol.* **33**: 463–476.

Egglishaw, H. J. 1968. "The quantitative relationship between bottom fauna and plant detritus in streams of different calcium concentrations." *J. Appl. Ecol.* **5**: 731–740.

Egglishaw, H. J. 1969. "The distribution of benthic invertebrates on substrata in fast-flowing streams." *J. An. Ecol.* **38**: 19–33.

Egglishaw, H. J. and N. C. Morgan. 1965. "A survey of the bottom fauna of streams in the Scottish Highlands. Part II. The relationship of the fauna to the chemical and geological conditions." *Hydrobiologia* **26**: 173–183.

Ehrenberg, H. 1957. "Die Steinfauna der Brandungsufer ostholsteinischer Seen." *Arch. Hydrobiol.* **53**: 87–159.

Einsele, W. 1960. "Die Strömungsgeschwindigkeit als beherrschender Faktor bei den limnologischen Gestaltung der Gewässer." *Österreichs Fischei,* Suppl. 1, 2: 1–40.

Elgmork, K. and O. R. Saether. 1970. "Distribution of invertebrates in a high mountain brook in the Colorado Rocky Mountains." *Univ. Colo. Stud. Ser. Biol.* **31**: 1–55.

Elliott, J. M. 1968. "The daily activity patterns of mayfly nymphs (Ephemeroptera)." *J. Zool.* **155**: 201-222.

Elliott, J. M. 1972. "Effects of temperature on the time of hatching in *Baetis rhodani* (Ephemeroptera: Baetidae)." *Oecologia* **9**: 47-51.

Elliott, J. M. 1978. "Effect of temperature on the hatching time of eggs of *Ephemerella ignita* (Poda) (Ephemeroptera: Ephemerellidae)." *Freshwat. Biol.* **8**: 51-58.

Elliott, J. M. 1982. "The life cycle and spatial distribution of the aquatic parasitoid *Agriotypus armatus* (Hymenoptera: Agriotypidae) and its caddis host *Silo pallipes* (Trichoptera: Goeridae)." *J. An. Ecol.* **51**: 923-942.

Elliott, J. M. 1983. "The responses of the aquatic parasitoid *Agriotypus armatus* (Hymenoptera: Agriotypidae) to the spatial distribution and density of its caddis host *Silo pallipes* (Trichoptera: Goeridae)." *J. An. Ecol.* **52**: 315-330.

Elliott, J. M. and U. H. Humpesch. 1980. "Eggs of Ephemeroptera." *Ann. Rep. Freshwat. Biol. Assoc.* **48**: 41-52.

Ellis, R. A. and J. Y. Borden. 1970. "Predation by *Notonecta undulata* on larvae of the yellow fever mosquito." *Ann. Ent. Soc. Am.* **63**: 963-973.

Elvang, O. and B. L. Madsen. 1973. "Biologiske undersøgelser over *Taeniopteryx nebulosa* (L.) (Plecoptera) med bemaerkning over vaekst." *Ent. Medd.* **41**: 49-59.

Enaceanu, V. 1964. "Das Donauplankton auf rumänischen Gebiet." *Arch. Hydrobiol. Suppl.* **27**: 442-456.

Engelhardt, W. 1951. "Faunistische-ökologische untersuchungen über Wasserinsekten an den südlichen Zuflüssen des Ammersees." *Mitt. Münch. Ent. Ges.* **41**: 1-135.

Eriksen, C. H. 1964. "The influence of respiration and substrate upon the distribution of burrowing mayfly naiads." *Verh. Internat. Verein. Limnol.* **15**: 903-911.

Eriksen, C. H. 1966. "Diurnal limnology of two highly turbid puddles." *Verh. Internat. Verein. Limnol.* **16**: 507-514.

Eriksen, C. H. 1968. "Ecological significance of respiration and substrate for burrowing Ephemeroptera." *Can J. Zool.* **46**: 93-103.

Eriksen, C. H. 1984. "The physiological ecology of larval *Lestes disjunctus* Selys (Zygoptera: Odonata)." *Freshwat. Invert. Biol.* **3**: 105-117.

Eriksen, C. H., V. H. Resh, S. S. Balling, and G. A. Lamberti. 1984. "Aquatic insect respiration," pp. 27-37 in R. W. Merritt and K. W. Cummins (eds.), *An Introduction to the Aquatic Insects of North America.* Kendall/Hunt, Dubuque, Iowa.

Erman, D. C. and N. A. Erman. 1984. "The response of stream macroinvertebrates to substrate size and heterogeneity." *Hydrobiologia* **108**: 75-82.

Erman, N. A. 1981. "Terrestrial feeding migration and life history of the stream-dwelling caddisfly, *Desmona bethula* (Trichoptera: Limnephilidae)." *Can. J. Zool.* **59**: 1658-1665.

Fahy, E. 1973. "Observations on the growth of Ephemeroptera in fluctuating and constant temperature conditions." *Proc. Roy. Irish Acad.* **73**: 133-149.

Faust, S. D. and O. M. Aly. 1981. *Chemistry of Natural Waters.* Ann Arbor Science, Ann Arbor, Michigan.

Fedorenko, A. Y. and M. C. Swift. 1972. "Comparative biology of *Chaoborus americanus* and *Chaoborus trivittatus* in Eunice Lake, British Columbia." *Limnol. Oceanogr.* **17**: 721-730.

Fernando, C. H. 1959. "The colonization of small freshwater habitats by aquatic insects. 2. Hemiptera (the waterbugs)." *Ceylon J. Sci. (Biol. Sci.)* **2**: 5-32.

Fernando, C. H. and D. Galbraith. 1973. "Seasonality and dynamics of aquatic insects colonizing small habitats." *Verh. Internat. Verein. Limnol.* **18**: 1564-1575.

Fey, J. M. 1977. "Die Aufheizung eines Mittelgebirgsflusses und ihre Auswirkungen auf die Zoozönose-dargestellt an de Lenne (Sauerland)." *Arch. Hydrobiol. Suppl.* **53**: 307-363.

Fey, J. M. and H. Schuhmacher. 1978. "Zum Einfluss wechselnder Temperatur auf den Netzbau von Larven der Köcherfliegen-Art *Hydropsyche pellucidula* (Trichoptera: Hydropsychidae)." *Ent. Germ.* **4**: 1-11.

Fillion, D. B. 1967. "The abundance and distribution of benthic fauna of three mountain reservoirs on the Kananaskis River in Alberta." *J. Appl. Ecol.* **4**: 1-11.

Fischer, F. C. J. 1960-73. *Trichopterorum Catalogus* (in 15 volumes). Nederlandse Entomolgische Vereeniging, Amsterdam.

Fish, D. and D. W. Hall. 1978. "Succession and stratification of aquatic insects inhabiting the leaves of the insectivorous pitcher plant, *Sarracenia purpurea*." *Am. Midl. Nat.* **99**: 172-183.

Fisher, J. B. 1982. "Effects of macrobenthos on the chemical diagenesis of freshwater sediments," pp. 177-218 in P. L. McCall and M. J. S. Tevesz (eds.), *Animal-Sediment Relations.* Plenum, New York.

Fisher, K. 1932. "*Agriotypus armatus* (Walk) (Hymenoptera) and its relations with its hosts." *Proc. Zool. Soc. London* 451-461.

Fisher, S. G. 1983. "Succession in streams," pp. 7-27 in J. R. Barnes and G. W. Minshall (eds.), *Stream Ecology—Application and Testing of General Ecological Theory.* Plenum, New York.

Fisher, S. G., L. J. Gray, N. B. Grimm, and D. E. Busch. 1982. "Temporal succession in a desert stream ecosystem following flash flooding." *Ecol. Monogr.* **52**: 93-110.

Fisher, S. G. and A. LaVoy. 1972. "Differences in littoral fauna due to fluctuating water levels below a hydroelectric dam." *J. Fish. Res. Board Can.* **29**: 1472-1476.

Fisher, S. G. and G. E. Likens. 1973. "Energy flow in Bear Brook, New Hampshire: An integrative approach to stream ecosystem metabolism." *Ecol. Monogr.* **43**: 421-439.

Fisher, S. G. and W. L. Minckley. 1978. "Chemical characteristics of a desert stream in flash flood." *J. Arid. Envir.* **1**: 25-33.

Fittkau, E. -J. 1967. "On the ecology of Amazonian rain-forest streams." *Atlas do Simpósio sôbre a Biota Amazônica* **3** *(Limnologia)*: 97-108.

Fittkau, E.-J. 1976. "Kinal und kinon, Lebensraum und Lebensgemeinschaft der Oberflächen-drift am Beispiel amazonischer Fliessgewässer." *Biogeographica* **7**: 101-113.

Flannagan, J. F. 1979. "The burrowing mayflies of Lake Winnipeg, Manitoba, Canada," pp. 103-113 in K. Pasternak and R. Sowa (eds.), *Proc. Second Internat. Conf. Ephemeroptera.* Panstwowe Wydawnictwo Naukowe, Warsaw.

Flannagan, J. F. and G. H. Lawler. 1972. "Emergence of caddisflies (Trichoptera) and mayflies (Ephemeroptera) from Heming Lake, Manitoba." *Can. Ent.* **104**: 173-183.

Flannagan, J. F. and K. E. Marshall (eds.). 1980. *Advances in Ephemeroptera Biology.* Plenum Press, New York.

Fontaine, T. D. and S. M. Bartell (eds.). 1983. *Dynamics of Lotic Ecosystems.* Ann Arbor Science Publ., Ann Arbor, Mich.

Fontaine, T. D. and D. G. Nigh. 1983. "Characteristics of epiphyte communities on natural and artificial submersed lotic plants: Substrate effects." *Arch. Hydrobiol.* **96**: 293–301.

Forel, F. A. 1895. *Le Léman: monographie limnologique.* Tome 2, *Mécanique, Chimie, Thermique, Optique, Acoustique.* F. Rouge, Lausanne.

Foster, W. A. and J. E. Treherne. 1976. "Insects of marine salt marshes: Problems and adaptations," pp. 5–42 in L. Cheng (ed.), *Marine Insects.* North-Holland Publ. Co., Amsterdam.

Fox, H. M. 1955. "The effect of oxygen on the concentration of haem in invertebrates." *Proc. Roy. Soc. London, B* **135**: 195–212.

Fox, H. M. and B. G. Simmonds. 1933. "Metabolic rates of aquatic arthropods from different habitats." *J. Exp. Biol.* **10**: 67–74.

Fox, H. M., B. G. Simmonds, and R. Washbourn. 1935. "Metabolic rates of ephemerid nymphs from swiftly flowing and from still waters." *J. Exp. Biol.* **12**: 179–184.

Fox, H. M., C. A. Wingfield, and B. G. Simmonds. 1937. "The oxygen consumption of ephemerid nymphs from flowing and from still waters in relation to the concentration of oxygen in the water." *J. Exp. Biol.* **14**: 210–218.

Franciscola, M. E. 1979. "On a new Dytiscidae from a Mexican cave, a preliminary description (Coleoptera)." *Fragm. Ent.* (Rome) **15**: 233–241.

Frank, J. H. and L. P. Lounibos (eds.). 1983. *Phytotelmata: Terrestrial Plants as Hosts for Aquatic Insect Communities.* Plexus, Medford, N.J.

Franke, C. 1987. "Detection of transversal migration of larvae of *Chaoborus flavicans* (Diptera, Chaoboridae) by the use of a sonar system." *Arch. Hydrobiol.* **109**: 355–366.

Fraser, F. C. 1957. "A reclassification of the Order Odonata." *Royal Zool. Soc. New South Wales, Handbook No. 12*: 1–133.

Fremling, C. R. and D. K. Johnson. 1990. "Recurrence of *Hexagenia* mayflies demonstrates improved water quality in Pool 2 and Lake Pepin, upper Mississippi River," pp. 243–248 in I. C. Campbell (ed.), *Mayflies and Stoneflies.* Kluwer, Dordrecht.

Frey, D. G. 1964. "Remains of animals in Quaternary lake and bog sediments and their interpretation." *Arch. Hydrobiol. Beih.* **2**: 1–114.

Friesen, M. K., J. F. Flannagan, and S. G. Lawrence. 1979. "Effects of temperature and cold storage on development time and viability of eggs of the burrowing mayfly *Hexagenia rigida* (Ephemeroptera: Ephemeridae)." *Can. Ent.* **111**: 665–673.

Frison, T. H. 1935. "The stoneflies, or Plecoptera, of Illinois." *Bull. Illinois Nat. Hist. Surv.* **20**: 281–471.

Frost, W. E. 1942. "River Liffey, Survey IV. The fauna of the submerged mosses in an acid and an alkaline water." *Proc. Roy. Irish Acad.* **47B**: 293–369.

Fryer, G. 1959a. "The trophic relationships and ecology of some littoral communities of Lake Nyasa with special reference to the fishes; and an account of the evolution and ecology of rock-frequenting cichlidae." *Proc. Zool. Soc. London* **132**: 153-281.

Fryer, G. 1959b. "Lunar rhythm of emergence, differential behaviour of sexes and other phenomena in the African midge *Chironomus brevibucca* (Kiet)." *Bull. Ent. Res.* **50**: 1–8.

Fuller, R. L., J. L. Roelofs, and T. J. Fry. 1986. "The importance of algae to stream invertebrates." *J. N. Am. Benthol. Soc.* **5**: 290-296.

Furneaux, W. 1897. *Life in Ponds and Streams.* Longmans, Green, and Co., London.

Galat, D. L., E. L. Lider, S. Vigg, and S. R. Robertson. 1981. "Limnology of a large, deep, North American terminal lake, Pyramid Lake, Nevada." *Hydrobiologia* **82**: 281-317.

Galewski, K. 1971. "A study on morphobiotic adaptations of European species of the Dytiscidae (Coleoptera)." *Polskie Pismo Entomolgiczne* **41**: 488-702.

Gallepp, G. W. 1977. "Responses of Caddisfly larvae (*Brachycentrus* spp.) to temperature, food availability and current velocity." *Am. Midl. Nat.* **98**: 59-84.

Gammon, J. R. 1970. "The effect of inorganic sediment on stream biota." *Water Poll. Control Res. Ser. No. 18050 DWC.* U.S. Environmental Protection Agency, Washington, D.C.

Gaufin, A. 1959. "Production of bottom fauna in the Provo River, Utah." *Iowa State Coll. J. Sci.* **33**: 395-419.

Geijskes, D. C. 1935. "Faunistisch-okologische Untersuchungen am Rõserenbach bei Liestal im Basler Tafeljura." *Tijdschr. Ent.* **78**: 249-382.

Geijskes, D. C. 1942. "Observations on temperature in a tropical river." *Ecology* **23**: 106-110.

Gersabeck, E. F., Jr. and R. W. Merritt. 1979. "The effect of physical factors on the colonization and relocation behavior of immature black flies (Diptera: Simuliidae)." *Envir. Ent.* **8**: 34-39.

Gersich, R. M. and M. A. Brusven. 1982. "Volcanic ash accumulation and ash-voiding mechanisms of aquatic insects." *J. Kans. Ent. Soc.* **55**: 290-296.

Gibert, J., M.-J. Dole-Olivier, P. Marmonier, and P. Vervier. 1990. "Surface water-groundwater ecotones," pp. 199-225 in R. J. Naiman and H. Décamps (eds.), *The Ecology and Management of Aquatic-Terrestrial Ecotones.* Parthenon, Casterton Hall, England.

Gieysztor, M. 1961. "Les températures et la chimie du littoral des lacs." *Verh. Internat. Verein. Limnol.* **14**: 84-86.

Giguère, L. A. and L. M. Dill. 1980. "Seasonal patterns of vertical migration: A model for *Chaoborus trivittatus*," pp. 122-128 in W. C. Kerfoot (ed.), *Evolution and Ecology of Zooplankton Communities.* Univ. Press New England, Hanover, N.H.

Gisin, H. 1978. "Collembola," pp. 254-255 in J. Illies (ed.), *Limnofauna Europaea.* Fischer, Stuttgart.

Gíslason, G. M. 1978. "Life cycle of *Limnephilus affinis* Curt. (Trichoptera: Limnephilidae) in Iceland and in Northumberland, England." *Verh. Internat. Verein. Limnol.* **20**: 2622-2629.

Gíslason, G. M. 1981. "Distribution and habitat preferences of Icelandic Trichoptera," pp. 99-109 in G. P. Moretti (ed.). *Proc. Third Internat. Symp. Trichoptera.* Dr. W. Junk, The Hague.

Gittelman, S. H. 1974. "Locomotion and predatory strategy in back swimmers." *Am. Midl. Nat.* **92**: 496-500.

Gittelman, S. H. 1975. "Physical gill efficiency and winter dormancy in the pigmy backswimmer, *Neoplea striola* (Hemiptera: Pleidae)." *Ann. Ent. Soc. Am.* **68**: 1011-1017.

Glazier, D. S. and J. L. Gooch. 1987. "Macroinvertebrate assemblages in Pennsylvania (U.S.A.) springs." *Hydrobiologia* **150**: 33–43.

Glime, J. M. and R. M. Clemons. 1972. "Species diversity of stream insects on *Fontinalis* spp. compared to diversity on artificial substrates." *Ecology* **53**: 458–464.

Golterman, H. L. 1975. *Physiological Limnology.* Elsevier, Amsterdam.

Goma, L. K. H. 1960. "The swamp breeding mosquitos of Uganda: Records of larvae and their Habitats." *Bull. Ent. Res.* **51**: 77–94.

Gore, J. A. 1978. "A technique for predicting in-stream flow requirements of benthic macroinvertebrates." *Freshwat. Biol.* **8**: 141–151.

Gose, K. 1970. "Life history and instar analysis of *Stenopsyche griseipennis* (Trichoptera)." *Jap. J. Limnol.* **31**: 96–106.

Grafius, E. and N. H. Anderson. 1980. "Population dynamics and role of two species of *Lepidostoma* (Trichoptera: Lepidostomatidae) in an Oregon coniferous forest stream." *Ecology* **61**: 808–816.

Grant, P. M. and K. W. Stewart. 1980. "The life history of *Isonychia sicca* (Ephemeroptera: Oligoneuriidae) in an intermittent stream in north central Texas." *Ann. Ent. Soc. Am.* **73**: 747–755.

Gray, L. J. 1981. "Species composition and life histories of aquatic insects in a lowland Sonoran Desert Stream." *Am. Midl. Nat.* **106**: 229–242.

Gray, L. J. and S. G. Fisher. 1981. "Postflood recolonization pathways of macroinvertebrates in a lowland Sonoran Desert stream." *Am. Midl. Nat.* **106**: 249–257.

Gray, L. J. and J. V. Ward. 1982. "Effects of sediment releases from a reservoir on stream macroinvertebrates." *Hydrobiologia* **96**: 177-184.

Gray, L. J., J. V. Ward, R. Martinson, and E. Bergey. 1983. "Aquatic macroinvertebrates of the Piceance Basin, Colorado: Community response along spatial and temporal gradients of environmental conditions." *Southwest. Nat.* **28**: 125–135.

Green, E. J. and D. E. Carritt. 1967. "New tables for oxygen saturation of seawater." *J. Mar. Res.* **25**: 140–147.

Green, J. 1968. *The Biology of Estuarine Animals.* Sidgwick and Jackson, London.

Gregg, W. W. and F. L. Rose. 1985. "Influence of aquatic macrophytes on invertebrate community structure guild structure, and microdistribution in streams." *Hydrobiologia* **128**: 45-56.

Grenier, P. 1949. "Contribution à l'étude biologique des simuliides de France." *Physiol. Comp. Oecol.* **1**: 165-330.

Grenier, S. 1970. "Biologie d'*Agriotypus armatus* Curtis (Hymenoptera: Agriotypidae) parasite de nymphes de trichopteres." *Ann. Limnol.* **6**: 317-361.

Greze, I. I. 1953. "Hydrobiology of the lower part of the River Angara." *Trudy vses. gidrobiol. Obshch.* **5**: 203-311 (in Russian).

Griffiths, D. 1973. "The structure of an acid moorland pond community." *J. An. Ecol.* **42**: 263-283.

Grimås, U. 1961. "The bottom fauna of natural and impounded lakes in northern Sweden." *Rep. Instit. Freshwat. Res. Drottningholm* **42**: 183–237.

Grimås, U. 1962. "The effect of increased water level fluctuation upon the bottom fauna in Lake Blåsjön, Northern Sweden." *Rep. Instit. Freshwat. Res. Drottningholm* **44**: 14–41.

Grimm, N. B. and S. G. Fisher. 1989. "Stability of periphyton and macroinvertebrates to disturbance by flash floods in a desert stream." *J. N. Am. Benthol. Soc.* **8**: 293–307.

Guignot, F. 1925. "Description d'un *Siettitia* noveau du Midi de la France. (Col. Dytiscidae)." *Bull. Soc. Ent. France* **1925**: 23-24.

Gullefors, Bo and K. Müller. 1990. "Seasonal and diurnal occurrence of adult caddisflies (Trichoptera) from the brackish water of the Bothnia Sea." *Aquatic Insects* **12**: 227–239.

Gunter, G. and J. Y. Christmas. 1959. "Corixid insects as part of the offshore fauna of the sea." *Ecology* **40**: 724–725.

Günther, K. 1934. "Die Dermapteren der Deutschen Limnologischen Sunda-Expedition." *Arch. Hydrobiol. Suppl.* **12**: 503–517.

Gupta, A. P. (ed.). 1979. *Arthropod Phylogeny.* Van Nostrand Reinhold, New York.

Haddock, J. D. 1977. "The effect of stream current velocity on the habitat preference of a net-spinning caddisfly larva, *Hydropsyche oslari* Banks." *Pan-Pac. Ent.* **53**: 169–174.

Haeckel, E. 1890. *Plankton-studien.* Fischer, Jena.

Hagen, K. S. 1984. "Aquatic Hymenoptera," pp. 438–447 in R. W. Merritt and K. W. Cummins (eds.), *An Introduction to the Aquatic Insects of North America,* 2nd ed. Kendall/Hunt, Dubuque, Iowa.

Haines, A. T., B. L. Finlayson, and T. A. McMahon. 1988. "A global classification of river regimes." *Appl. Geogr.* **8**: 255–272.

Håkanson, L. 1982. "Bottom dynamics in lakes." *Hydrobiologia* **91**: 9-22.

Hall, R. E. 1951. "Comparative observations on the chironomid fauna of a chalk stream and a system of acid streams." *J. Soc. Br. Ent.* **3**: 253–262.

Halse, S. A. 1981. "Faunal assemblages of some saline lakes near Marchagee, western Australia." *Austral. J. Mar. Freshwat. Res.* **32**: 133–142.

Hamilton, A. 1940. "The New Zealand Dobson-Fly (*Archichauliodes diversus* Walk.): Life-history and bionomics." *N. Z. J. Sci. Tech.* **22**: 44a–55a.

Hamilton, J. D. 1961. "The effect of sand-pit washings on a stream fauna." *Verh. Internat. Verein. Limnol.* **14**: 435–439.

Hamilton, K. G. A. 1971. "The insect wing, Part I. Origin and development of wings from notal lobes." *J. Kans. Ent. Soc.* **44**: 421-433.

Hammer, U.T., R. C. Haynes, J. M. Heseltine, and S. M. Swanson. 1975. "The saline lakes of Saskatchewan." *Verh. Internat. Verein. Limnol.* **19**: 589–598.

Hansen, K. 1959. "The terms gyttja and dy." *Hydrobiologia* **13**: 309-315.

Hare, L. and J. C. H. Carter. 1986. "The benthos of a natural West African lake, with emphasis on the diel migrations and lunar and seasonal periodicities of the *Chaoborus* populations (Diptera, Chaoboridae)." *Freshwat. Biol.* **16**: 759–780.

Hargeby, A. 1990. "Macrophyte associated invertebrates and the effect of habitat permanence." *Oikos* **57**: 338–346.

Harker, J. E. 1952. "A study of the life cycles and growth-rates of four species of mayflies." *Proc. Roy. Ent. Soc. London, A* **27**: 77-85.

Harmon, M. E., J. F. Franklin, F. J. Swanson, P. Sollins, S. V. Gregory, J. D. Lattin, N. H. Anderson, S. P. Cline, N. G. Aumen, J. R. Sedell, G. W. Lienkaemper, K. Cromack, Jr., and K. W. Cummins. 1986. "Ecology of coarse woody debris in temperate ecosystems." *Adv. Ecol. Res.* **15**: 133–302.

Harper, P. P. 1973. "Life histories of Nemouridae and Leuctridae in Southern Ontario (Plecoptera)." *Hydrobiologia* **41**: 309-356.

Harper, P. P. 1986. "Relations entre les macrophytes et les insectes dan les milieux d'eau douce." *Rev. Ent. Québec* **31**: 76-86.

Harper, P. P. and H. B. N. Hynes. 1970. "Diapause in the nymphs of Canadian winter stoneflies." *Ecology* **51**: 925-927.

Harper, P. P. and J.-G. Pilon. 1970. "Annual patterns of emergence of some Quebec stoneflies (Insecta: Plecoptera)." *Can. J. Zool.* **48**: 681-694.

Harrison, A. D. 1964. "An ecological survey of the Great Berg River," pp. 143-158 in D. H. S. Davis (ed.), *Ecological Studies in Southern Africa*, Dr. W. Junk, The Hague.

Harrison, A. D. 1965a. "River zonation in southern Africa." *Arch. Hydrobiol.* **61**: 380-386.

Harrison, A. D. 1965b. "Geographical distribution of riverine invertebrates in Southern Africa." *Arch. Hydrobiol.* **61**: 387-394.

Harrison, A. D. 1966. "Recolonization of a Rhodesian stream after drought." *Arch. Hydrobiol.* **62**: 405-421.

Harrison, A. D. and J. D. Agnew. 1962. "The distribution of invertebrates endemic to acid streams in the Western and Southern Cape Province." *Ann. Cape Province Museum* **2**: 273-291.

Harrison, A. D. and H. B. N. Hynes. 1988. "Benthic fauna of Ethiopian mountain streams and rivers." *Arch. Hydrobiol. Suppl.* **81**: 1-36.

Harrison, A. D. and J. J. Rankin. 1976. "Hydrobiological studies of eastern lesser Antillean Islands II. St. Vincent: Freshwater Fauna—its distribution, tropical river zonation and biogeography." *Arch. Hydrobiol. Suppl.* **50**: 275-311.

Harrod, J. J. 1964. "The distribution of invertebrates on submerged aquatic plants in a chalk stream." *J. An. Ecol.* **33**: 335-348.

Harrod, J. J. 1965. "Effect of current speed on the cephalic fans of the larvae of *Simulium ornatum* var. *nitidifrons* Edwards." *Hydrobiologia* **26**: 8-12.

Hartland-Rowe, R. 1953. "Feeding mechanisms of an ephemeropteran nymph." *Nature* **172**: 1109-1110.

Hartland-Rowe, R. 1955. "Lunar rhythm in the emergence of an ephemeropteran." *Nature* **176**: 657.

Hartland-Rowe, R. 1958. "The biology of a tropical mayfly *Povilla adusta* Navás (Ephemeroptera, Polymitarcidae) with special reference to the lunar rhythm of emergence." *Rev. Zool. Bot. Afr.* **58**: 185-202.

Hartland-Rowe, R. 1964. "Factors influencing the life histories of some stream insects in Alberta." *Verh. Internat. Verein. Limnol.* **15**: 917-925.

Hartland-Rowe, R. 1966. "The fauna and ecology of temporary pools in western Canada." *Verh. Internat. Verein. Limnol.* **16**: 577-584.

Hartland-Rowe, R. 1972. "The limnology of temporary waters and the ecology of Euphyllopoda," pp. 15-31 in R. B. Clark and R. Wootton (eds.), *Essays in Hydrobiology*. Univ. Exeter.

Hartland-Rowe, R. 1984. "The adaptive value of synchronous emergence in the tropical African mayfly *Povilla adusta*: A preliminary investigation," pp. 283-289 in V. Landa, T. Soldán, and M. Tonner (eds.), *Proc. Fourth Internat. Conf. Ephemeroptera.* Czechoslovak Academy of Sciences, Ceské Budejovice.

Hashimoto, H. 1976. "Non-biting midges of marine habitats (Diptera: Chironomidae)," pp. 377–414 in L. Cheng (ed.), *Marine Insects*. North-Holland Publ. Co., Amsterdam.

Hatch, M. H. 1925. "An outline of the ecology of Gyrinidae." *Bull. Brooklyn Ent. Soc.* **20**: 101–114.

Hawkes, H. A. 1975. "River zonation and classification," pp. 312–374 in B. A. Whitton (ed.), *River Ecology*. Blackwell Sci. Publ., Oxford.

Hawkins, C. P. and J. R. Sedell. 1981. "Longitudinal and seasonal changes in functional organization of macroinvertebrate communities in four Oregon streams." *Ecology* **62**: 387–397.

Hayden, W. and H. F. Clifford, 1974. "Seasonal movements of the mayfly *Leptophlebia cupida* (Say), in a brown-water stream of Alberta, Canada." *Am. Midl. Nat.* **91**: 90–102.

Heiman, D. R. and A. W. Knight. 1975. "The influence of temperature on the bioenergetics of the carnivorous stonefly nymph, *Acroneuria californica* Banks (Plecoptera: Perlidae)." *Ecology* **56**: 105–116.

Hem, J. D. 1970. *Study and Interpretation of the Chemical Characteristics of Natural Water*. U.S. Geol. Surv. Water-Supply Paper 1473.

Hemphill, N. and S. D. Cooper. 1983. "The effect of physical disturbance on the relative abundances of two filter-feeding insects in a small stream." *Oecologia* **58**: 378–383.

Henderson, L. J. 1913. *The Fitness of the Environment; an Inquiry into the Biological Significance of the Properties of Matter*. Macmillan, New York.

Hennig, W. 1948–1952. *Die larvenformen der Dipteren* (three parts). Akademie Verlag, Berlin.

Hennig, W. 1981. *Insect Phylogeny* (English ed. transl., ed. A. C. Pont). Wiley, New York.

Henriksen, K. L. 1918. "De europaeiske Vandsynltehvespe og deres Biologi." *Ent. Medd.* **12**: 137–251.

Henson, E. B. 1966. "Aquatic insects as inhalent allergens: A review of American literature." *Ohio J. Sci.* **66**: 529–532.

Henson, E. B. 1988. "Macro-invertebrate associations in a marsh ecosystem." *Verh. Internat. Verein. Limnol.* **23**: 1049–1056.

Herbst, D. B. 1988. "Comparative population ecology of *Ephydra hians* Say (Diptera: Ephydridae) at Mono Lake (California) and Abert Lake (Oregon)." *Hydrobiologia* **158**: 145–166.

Herbst, G. N. and H. J. Bromley. 1984. "Relationships between habitat stability, ionic composition, and the distribution of aquatic invertebrates in the desert regions of Israel." *Limnol. Oceanogr.* **29**: 495–503.

Herring, J. L. 1961. "The genus *Halobates* (Hemiptera: Gerridae)." *Pacific Insects* **3**: 223–305.

Hershey, A. E. 1985. "Littoral chironomid communities in an arctic Alaskan lake." *Holarct. Ecol.* **8**: 39–48.

Hess, A. D. and C. C. Kiker. 1944. "Water level management for malaria control on impounded waters." *J. Nat. Malaria Soc.* **3**: 181–196.

Hickin, N. E. 1967. *Caddis Larvae*. Hutchinson, London.

Hildrew, A. G. and J. M. Edington. 1979. "Factors facilitating the coexistence of hydropsychid caddis larvae (Trichoptera) in the same river system." *J. An. Ecol.* **48**: 557-576.

Hildrew, A. G., C. R. Townsend, and J. Francis. 1984a. "Community structure in some southern English streams: The influence of species interactions." *Freshwat. Biol.* **14**: 297-310.

Hildrew, A. G., C. R. Townsend, J. Francis, and K. Finch. 1984b. "Cellulolytic decomposition in streams of contrasting pH and its relationship with invertebrate community structure." *Freshwat. Biol.* **14**: 323-328.

Hinton, H. E. 1947. "On the reduction of functional spiracles in the aquatic larvae of the Holometabola, with notes on the moulting process of spiracles." *Trans. Roy. Ent. Soc. London* **98**: 449-473.

Hinton, H. E. 1953. "Some adaptations of insects to environments that are alternately dry and flooded, with some notes on the habits of the Stratiomyidae." *Trans. Soc. Br. Ent.* **11**: 209-227.

Hinton, H. E. 1955. "On the respiratory adaptations, biology, and taxonomy of the Psephenidae, with notes on some related families (Coleoptera)." *Proc. Zool. Soc. London* **125**: 543-568.

Hinton, H. E. 1960. "Cryptobiosis in the larva of *Polypedilum vanderplanki* Hint. (Chironomidae)." *J. Insect Physiol.* **5**: 286-300.

Hinton, H. E. 1966. "Respiratory adaptations of the pupae of beetles of the family Psephenidae." *Phil. Trans. Roy. Soc.* **B251**: 211-245.

Hinton, H. E. 1967a. "Structure of the plastron in *Lipsothrix*, and the polyphyletic origin of plastron respiration in the Tipulidae." *Proc. Roy. Ent. Soc. London, A* **42**: 35-38.

Hinton, H. E. 1967b. "On the spiracles of the larvae of the suborder Myxophaga (Coleoptera)." *Austral. J. Zool.* **15**: 955-959.

Hinton, H. E. 1968. "Spiracular gills." *Adv. Insect Physiol.* **5**: 65-162.

Hinton, H. E. 1969. "Plastron respiration in adult beetles of the suborder Myxophaga." *J. Zool. London* **159**: 131-137.

Hinton, H. E. 1976a. "Respiratory adaptations of marine insects," pp. 43-78 in L. Cheng (ed.), *Marine Insects*. North-Holland Publ. Co., Amsterdam.

Hinton, H. E. 1976b. "Plastron respiration in bugs and beetles." *J. Insect Physiol.* **22**: 1529-1550.

Hinton, H.E. 1977. "Enabling mechanisms." *Proc. 15th Internat. Congress Ent.* Washington, D.C., pp. 71-83.

Hippa, H., S. Koponen, and R. Mannila. 1985. "Invertebrates of Scandinavian caves III. Ephemeroptera, Plecoptera, Trichoptera, and Lepidoptera." *Notulae Ent.* **65**: 25-28.

Hocking, B. and C. D. Sharplin. 1965. "Flower basking by Arctic insects." *Nature* **206**: 215.

Holsinger, J. R. 1988. "Troglobites: The evolution of cave-dwelling organisms." *Am. Sci.* **76**: 146-153.

Hoopes, R. L. 1974. "Flooding, as the result of Hurricane Agnes, and its effect on a macrobenthic community in an infertile headwater stream in central Pennsylvania." *Limnol. Oceanogr.* **19**: 853-857.

Hora, S. L. 1928. "Animal life in torrential streams." *J. Bombay Nat. Hist. Soc.* **32**: 111-126.

Hora, S. L. 1930. "Ecology, bionomics and evolution of the torrential fauna with special reference to the organs of attachment." *Phil. Trans. Roy. Soc. London, B.* **218**: 171-282.

Horsfall, W. R. 1955. *Mosquitoes. Their Bionomics and Relation to Disease.* Ronald Press, New York.

Horton, R. E. 1945. "Erosional development of streams and their drainage basins; hydrophysical approach to quantitative morphology." *Bull. Geol. Soc. Am.* **56**: 275-370.

Houlihan, D. F. 1969. "Respiratory physiology of the larva of *Donacia simplex*, a root-piercing beetle." *J. Insect Physiol.* **15**: 1517-1536.

Houlihan, D. F. 1970. "Respiration in low oxygen partial pressures: The adults of *Donacia simplex* that respire from the roots of aquatic plants." *J. Insect Physiol.* **16**: 1607-1622.

Howarth, F. G. 1983. "Ecology of cave arthropods." *Ann. Rev. Ent.* **28**: 365-389.

Howe, R. W. 1967. "Temperature effects on embryonic development in insects." *Ann. Rev. Ent.* **12**: 15-42.

Howmiller, R. 1969. "Studies on some inland waters of the Galapagos." *Ecology* **50**: 73-80.

Hrbácek, J. 1950. "On the morphology and function of the antennae of the central European Hydrophilidae." *Trans. Roy. Ent. Soc.* **101**: 239-256.

Hubault, E. 1927. "Contribution à l'étude des invertébrés torrenticoles." *Bull. Biol. France Beligique, Suppl.* **9**: 1-390.

Hubbard, M. D. 1979. "A nomenclatural problem in Ephemeroptera: *Prosopistoma* or *Binoculus*?," pp. 73-77 in K. Pasternak and R. Sowa (eds.), *Proc. Second Internat. Conf. Ephemeroptera.* Panstwowe Wydawnictwo Naukowe, Warsaw.

Hubbard, M. D. 1990. *Mayflies of the World. A Catalog of the Family and Genus Group Taxa (Insecta: Ephemeroptera).* Sandhill Crane Press, Gainesville, Fla.

Hudson, P. L. and W. E. Lorenzen. 1981. "Manipulation of reservoir discharge to enhance tailwater fisheries," pp. 568-579 in R. M. North, L. B. Dworsky, and D. J. Allee (eds.), *Unified River Basin Management.* Am. Wat. Res. Assoc., Minneapolis, Minn.

Hughes, D. A. 1966a. "Mountain streams of the Barberton area, eastern Transvaal. Part II, the effect of vegetational shading and direct illumination on the distribution of stream fauna." *Hydrobiologia* **27**: 439-459.

Hughes, D. A. 1966b. "The role of responses to light in the selection and maintenance of microhabitat by the nymphs of two species of mayfly." *An. Behav.* **14**: 17-33.

Hughes, D. A. 1966c. "On the dorsal light response in a mayfly nymph." *An. Behav.* **14**: 13-16.

Hughes, G. M. 1958. "The co-ordination of insect movements. III. Swimming in *Dytiscus, Hydrophilus*, and a dragonfly nymph." *J. Exper. Biol.* **35**: 567-583.

Hughes, G. M. and P. J. Mill. 1966. "Patterns of ventilation in dragonfly larvae." *J. Exper. Biol.* **44**: 317-333.

Humpesch, U. 1971. "Zur Faktorenanalyse des Schlüpfrhythmus der Flugstadien von *Baetis alpinus* Pict. (Baetidae, Ephemeroptera)." *Oecologia* **7**: 328-341.

Humpesch, U. 1978. "Preliminary notes on the effect of temperature and light-condition on the time of hatching in some Heptageniidae (Ephemeroptera)." *Verh. Internat. Verein. Limnol.* **20**: 2605–2611.

Humpesch, U. H. 1979. "Life cycles and growth rates of *Baetis* spp. (Ephemeroptera: Baetidae) in the laboratory and in two stony streams in Austria." *Freshwat. Biol.* **9**: 467–479.

Humpesch, U. 1980. "Effect of temperature on the hatching time of eggs of five *Ecdyonurus* spp. (Ephemeroptera) from Austrian streams and English streams, rivers and lakes." *J. An. Ecol.* **49**: 317–333.

Humpesch, U. H. 1982. "Effect of fluctuating temperatures on the duration of embryonic development in two *Ecdyonurus* spp. and *Rhithrogena* cf. *hybrida* (Ephemeroptera) from Austrian streams." *Oecologia* **55**: 285–288.

Hungerford, H. B. 1919. "The biology and ecology of aquatic and semiaquatic Hemiptera." *Univ. Kans. Sci. Bull.* **21**: 1-341.

Hungerford, H. B. 1948. "The Corixidae of the Western Hemisphere (Hemiptera)." *Univ. Kans. Sci. Bull.* **32**: 1-827.

Hungerford, H. B. 1958. "Some interesting aspects of the world distribution and classification of aquatic and semiaquatic Hemiptera." *Proc. Tenth Internat. Congr. Ent.* **1**: 337–348.

Hunt, P. C., and J. W. Jones. 1972. "The effect of water level fluctuations on a littoral fauna." *J. Fish Biol.* **4**: 385–394.

Hurlbert, S. H., G. Rodríguez, and N. D. Santos (eds.). 1981. *Aquatic Biota of Tropical South America,* Part 1: *Arthropoda.* San Diego State Univ., San Diego, Calif.

Husmann, S. 1971. "Ecological studies on freshwater meiobenthon in layers of sand and gravel." *Smithsonian Contrib. Zool.* **76**: 161-169.

Hutchinson, G. E. 1931. "On the occurrence of *Trichocorixa* Kirkaldy (Corixidae, Hemiptera-Heteroptera) in salt water and its zoo-geographical significance." *Am. Nat.* **65**: 573–574.

Hutchinson, G. E. 1937a. "Limnological studies in Indian Tibet." *Internat. Revue ges. Hydrobiol. Hydrogr.* **35**: 134–176.

Hutchinson, G. E. 1937b. "A contribution to the limnology of arid regions." *Trans. Conn. Acad. Arts Sci.* **33**: 47–132.

Hutchinson, G. E. 1957a. *A Treatise on Limnology,* Vol. I. Wiley, New York.

Hutchinson, G. E. 1957b. "Concluding remarks." *Cold Spring Harbor Symp. Quant. Biol.* **22**: 415–427.

Hutchinson, G. E. 1967. *A Treatise on Limnology,* Vol. II. Wiley, New York.

Hutchinson, G. E. 1975. *A Treatise on Limnology,* Vol. III. Wiley, New York.

Hutchinson, G. E. 1981. "Thoughts on aquatic insects." *BioScience* **31**: 495-500.

Hutchinson, G. E. and H. Löffler. 1956. "The thermal classification of lakes." *Proc. Nat. Acad. Sci. U.S.A.* **42**: 84–86.

Hynes, H. B. N. 1941. "The taxonomy and ecology of the nymphs of British Plecoptera with notes on the adults and eggs." *Trans. Roy. Ent. Soc. London* **91**: 459–557.

Hynes, H. B. N. 1958. "The effect of drought on the fauna of a small mountain stream in Wales." *Verh. Internat. Verein. Limnol.* **13**: 826–833.

Hynes, H. B. N. 1961a. "The effect of water level fluctuation on littoral fauna." *Verh. Internat. Verein. Limnol.* **14**: 652–656.

Hynes, H. B. N. 1961b. "The invertebrate fauna of a Welsh mountain stream." *Arch. Hydrobiol.* **57**: 344–388.

Hynes, H. B. N. 1963. "Imported organic matter and secondary productivity in streams." *Internat. Congr. Zool.* **16**: 324–329.

Hynes, H. B. N. 1968. "Further studies on the invertebrate fauna of a Welsh mountain stream." *Arch. Hydrobiol.* **65**: 360–379.

Hynes, H. B. N. 1970a. *The Ecology of Running Waters.* Univ. Toronto Press, Toronto.

Hynes, H. B. N. 1970b. "The ecology of stream insects." *Ann. Rev. Ent.* **15**: 25–42.

Hynes, H. B. N. 1971. "Zonation of the invertebrate fauna in a West Indian stream." *Hydrobiologia* **38**: 1–18.

Hynes, H. B. N. 1976. "Biology of Plecoptera." *Ann. Rev. Ent.* **21**: 135–153.

Hynes, H. B. N. 1988. "Biogeography and origins of the North American stoneflies (Plecoptera)." *Mem. Ent. Soc. Can.* **144**: 31–37.

Hynes, H. B. N. and M. E. Hynes. 1975. "The life histories of many of the stoneflies (Plecoptera) of south-eastern mainland Australia." *Austral. J. Mar. Freshwat. Res.* **26**: 113–153.

Hynes, H. B. N. and U. R. Yadav. 1985. "Three decades of post-impoundment data on the littoral fauna of Llyn Tegid, North Wales." *Arch. Hydrobiol.* **104**: 39–48.

Hynes, J. D. 1975. "Annual cycles of macro-invertebrates of a river in southern Ghana." *Freshwat. Biol.* **5**: 71–83.

Ide, F. P. 1935. "The effect of temperature on the distribution of the mayfly fauna of a stream." *Publ. Ont. Fish. Res. Lab.* **50**: 1–76.

Ikeshoji, T. and M. S. Mulla. 1970. "Oviposition attractants for four species of mosquitoes in natural breeding waters." *Ann. Ent. Soc. Am.* **63**: 1322–1327.

Illies, J. 1952. "Die molle faunistischokologische untersuchungen an einem forellenbach im lipper bergland." *Arch. Hydrobiol.* **46**: 424–612.

Illies, J. 1953a. "Die Besiedlung der Fulda (insbes. das Benthos der Salmonidenregion) nach dem jetzigen Stand der Untersuchung." *Ber. Limnol. Flussstat. Freudenthal* **5**: 1–28.

Illies, J. 1953b. "Beitrag zur Verbreitungsgeschichte der europäischen Plecopteren." *Arch. Hydrobiol.* **48**: 35–74.

Illies, J. 1956. "Seeausfluss-Biozönosen lappländischer Waldbäche." *Ent. Tidskr.* **77**: 138–153.

Illies, J. 1961a. "Versuch einer allgemeinen biozönotischen Gliederung der Fliessgewässer." *Internat. Revue ges. Hydrobiol.* **46**: 205-213.

Illies, J. 1961b. "Gebirgsbäche in Europa und in Südamerika-ein limnologischer Vergleich." *Verh. Internat. Verein. Limnol.* **14**: 517–523.

Illies, J. 1964. "The invertebrate fauna of the Huallaga, A Peruvian tributary of the Amazon River, from the sources down to Tingo Maria." *Verh. Internat. Verein. Limnol.* **15**: 1077–1083.

Illies, J. 1965. "Phylogeny and zoogeography of the Plecoptera." *Ann. Rev. Ent.* **10**: 117–140.

Illies, J. 1966. "Katalog der rezenten Plecoptera." *Das Tierreich* **82**: 1–623.

Illies, J. 1968. "Ephemeroptera (Eintagsfliegen)." *Handb. Zool.* **4**(2) 2/5: 1–63.

Illies, J. 1969. "Biogeography and ecology of neotropical freshwater insects, especially those from running waters," pp. 685–708 in E. J. Fittkau, J. Illies, H. Klinge, G. H. Schwabe, and H. Sioli (eds.), *Biogeography and Ecology in South America,* Vol. 2. Dr. W. Junk, The Hague.

Illies, J. (ed.). 1978. *Limnofauna Europaea. A Checklist of the Animals Inhabiting European Inland Waters, with Account of Their Distribution and Ecology.* 2nd ed. Gustav Fischer Verlag, Stuttgart.

Illies, J. and L. Botosaneanu. 1963. "Problèmes et méthodes de la classification et de la zonation écologique des eaux courantes, considerées surtout du point de vue faunistique." *Mitt. Internat. Verein. Limnol.* **12**: 1–57.

Illies, J. and E. C. Masteller. 1977. "A possible explanation of emergence patterns of *Baetis vernus* Curtis (Ins.: Ephemeroptera) on the Breitenbach-Schlitz studies on productivity, Nr. 22." *Internat. Revue ges. Hydrobiol.* **62**: 315–321.

Inman, D. L. 1949. "Sorting of sediments in the light of fluid mechanics." *J. Sėdiment Petrol.* **19**: 51–70.

Irons, J. G., S. R. Ray, L. K. Miller, and M. W. Oswood. 1989. "Spatial and seasonal patterns of streambed water temperatures in an Alaskan subarctic stream," pp. 381–390 in W. W. Woessner and D. F. Potts (eds.), *Headwaters Hydrology.* Am. Water Res. Assoc., Bethesda, Md.

Issel, R. 1910. "La faune des sources thermales de Viterbo." *Internat. Revue ges. Hydrobiol.* **3**: 178–180.

Istenic, L. 1963. "Rate of oxygen consumption of larvae of *Perla marginata* Pz. in relation to body size and temperature." *Raspr. slov. Akad. Znan. Umet.* **4**(7): 201–236 (in Czech.).

Istock, C. A., S. A. Wasserman, and H. Zimmer. 1975. "Ecology and evolution of the pitcher-plant mosquito. 1. Population dynamics and laboratory responses to food and population density." *Evolution* **239**: 296–312.

Iversen, T. M. 1975. "Disappearance of autumn shed beech leaves placed in bags in small streams." *Verh. Internat. Verein. Limnol.* **19**: 1687–1692.

Iversen, T. M., J. Thorup, and J. Skriver. 1982. "Inputs and transformation of allochthonous particulate organic matter in a headwater stream." *Holarct. Ecol.* **5**: 10–19.

Iversen, T. M., P. Wiberg-Larsen, S. B. Hansen, and F. S. Hansen. 1978. "The effect of partial and total drought on the macroinvertebrate communities of three small Danish streams." *Hydrobiologia* **60**: 235–242.

Jaag, O. and H. Ambühl. 1964. "The effect of the current on the composition of biocoenoses in flowing water streams." *Adv. Water Poll. Res.* **1**: 31–49.

Jackson, D. J. 1958. "Observations on the biology of *Caraphractus cinctus* Walker (Hymenoptera: Mymaridae), a parasitoid of the eggs of Dytiscidae. I. Methods of rearing and numbers bred on different host eggs." *Trans. Roy. Ent. Soc. London* **110**: 533-566.

Jackson, D. J. 1964. "Observations on the life-history of *Mestocharis bimacularis* (Dalman) (Hym. Eulophidae), a parasitoid of the eggs of Dytiscidae." *Opusc. Ent.* **29**: 81–97.

Jaczewski, T. and A. S. Kostrowicki. 1969. "Number of species of aquatic and semiaquatic Heteroptera in the fauna of various parts of the Holarctic in relation to the world fauna." *Mem. Soc. Ent. Ital.* **48**: 153–156.

Jansson, A. 1977. "Distribution of Micronectae (Heteroptera, Corixidae) in Lake Päijänne, central Finland: Correlation with eutrophication and pollution." *Ann. Zool. Fennici* **14**: 105–117.

Jeannel, R. 1950. "Un Elmide cavernicole du Congo belge (Col. Dryopidae)." *Rev. Franc. Ent.* **17**: 168–172.

Jewell, M. E. 1922. "The fauna of an acid stream." *Ecology* **3**: 22–28.

Jewell, M. E. and H. W. Brown. 1929. "Studies on northern Michigan bog lakes." *Ecology* **10**: 427–475.

Jewett, S. G. 1963. "A stonefly aquatic in the adult stage." *Science* **139**: 484–485.

Johannsen, O. A. 1934–1937. "Aquatic Diptera." *Mem. Cornell Univ. Agric. Exp. Station* **164**: 1–71; **177**: 1–62; **205**: 3–84; **210**: 3–80.

Johannsson, O. E. 1980. "Energy dynamics of the eutrophic chironomid *Chironomus plumosus* f. *semireductus* from the Bay of Quinte, Lake Ontario." *Can. J. Fish. Aquat. Sci.* **37**: 1254–1265.

Johansen, A. C. (ed.) 1918. *Randers Fjords Naturhistorie.* Copenhagen.

John, P. H. 1978. "Discharge measurement in lower order streams." *Internat. Revue ges. Hydrobiol.* **63**: 731–755.

Jónasson, P. M. 1977. "Lake Esrom research 1867–1977." *Folia Limnol. Scand.* **17**: 67–90.

Jónasson, P. M. 1978. "Zoobenthos of Lakes." *Verh. Internat. Verein. Limnol.* **20**: 13–37.

Jónasson, P. M. 1984. "Oxygen demand and long term changes of profundal zoobenthos." *Hydrobiologia* **115**: 121–126.

Jones, J. R. E. 1949. "An ecological Study of the River Rheidol, North Cardiganshire, Wales." *J. An. Ecol.* **18**: 67–88.

Jones, J. R. E. 1951. "An ecological study of the River Towy." *J. An. Ecol.* **20**: 68–86.

Joossee, E. N. G. 1976. "Littoral apterygotes (Collembola and Thysanura)," pp. 151–186 in L. Cheng (ed.). *Marine Insects.* North-Holland Publ. Co., Amsterdam.

Juday, C. 1921. "Quantitative studies of the bottom fauna in the deeper waters of Lake Mendota." *Trans. Wisc. Acad. Sci. Arts Lett.* **20**: 461–493.

Judd, W. W. 1949. "Insects collected in the Dundas Marsh, Hamilton, Ontario, 1946–47, with observations on their periods of emergence." *Can. Ent.* **81**: 1–10.

Junk, W. J., P. B. Bayley, and R. E. Sparks. 1989. "The flood pulse concept in river-floodplain systems." *Can. Spec. Publ. Fish. Aquat. Sci.* **106**: 110–127.

Kadlec, J. A. 1962. "Effects of a drawdown on a waterfowl impoundment." *Ecology* **43**: 267–281.

Kalk, M., A. J. McLachlan, and C. Howard-Williams (eds.). 1979. *Lake Chilwa.* Dr. W. Junk, The Hague.

Kalpage, K. S. P. and R. A. Brust. 1973. "Oviposition attractant produced by immature *Aedes atropalpus.*" *Envir. Ent.* **2**: 729–730.

Kalugina, N. S. 1959. "Changes in morphology and biology of chironomid larvae in relation to growth (Diptera, Chironomidae)." *Trudy vses. gidriobiol. Obshch.* **9**: 85–107.

Kamler, E. 1964. "Recherches sur les Plécoptéres des Tatras." *Pol. Arch. Hydrobiol.* **12**: 145–184.

Kamler, E. 1965. "Thermal conditions in mountain waters and their influence on the distribution of Plecoptera and Ephemeroptera larvae." *Ekol. Pol. Ser. A* **13**: 377–414.

Kamler, E. 1967. "Distribution of Plecoptera and Ephemeroptera in relation to attitude above mean sea level and current speed in mountain waters." *Pol. Arch. Hydrobiol.* **14**(27): 29–42.

Kamler, E. and W. Riedel. 1960. "The effect of drought on the fauna, Ephemeroptera, Plecoptera and Trichoptera, of a mountain stream." *Pol. Arch. Hydrobiol.* **8**: 87–94.

Karny, H. H. 1934. *Biologie der Wasserinsekten.* Fritz Wagner, Vienna.

Kaster, J. L. and G. Z. Jacobi. 1978. "Benthic macroinvertebrates of a fluctuating reservoir." *Freshwat. Biol.* **8**: 283–290.

Kaushik, N. K., and H. B. N. Hynes. 1971. "The fate of the dead leaves that fall into streams." *Arch. Hydrobiol.* **64**: 465–515.

Kavaliers, M. 1981. "Rhythmic thermoregulation in larval cranefly (Diptera: Tipulidae)." *Can. J. Zool.* **59**: 555–558.

Keller, E. A. and F. J. Swanson. 1979. "Effects of large organic material on channel form and fluvial processes." *Earth Surf. Processes* **4**: 361–380.

Kendall, M. G. 1948. *Rank Correlation Methods.* Charles Griffin & Co., London.

Kenk, R. 1949. "The animal life of temporary and permanent ponds in southern Michigan." *Misc. Publ. Mus. Zool. Univ. Mich.* **71**: 1–66.

Kennedy, C. H. 1922. "The ecological relationships of the dragonflies of the Bass Islands of Lake Erie." *Ecology* **3**: 325–336.

Key, K. H. L. 1970. "Orthoptera," pp. 323–347 in I. M. Mackerras (ed.), *The Insects of Australia.* Melbourne Univ. Press, Melbourne, Australia.

Khoo, S. G. 1964. "Studies on the Biology of *Capnia bifrons* (Newman) and notes on the diapause in the nymphs of this species." *Gewäss. Abwäss* **34/35**: 23–30.

Khoo, S. G. 1968a. "Experimental studies on diapause in stoneflies. I. Nymphs of *Capnia bifrons* (Newman)." *Proc. Roy. Ent. Soc. London, A* **43**: 40–48.

Khoo, S. G. 1968b. "Experimental studies on diapause in stoneflies. II. Eggs of *Diura bicaudata* (L.)." *Proc. Roy. Ent. Soc. London, A* **43**: 49–56.

Kim, K. C. and R. W. Merritt (eds.). 1987. *Black Flies: Ecology, Population Management, and Annotated World List.* Penn. State Press, University Park, Pennsylvania.

King, J. M. 1981. "The distribution of invertebrate communities in a small South African river." *Hydrobiologia* **83**: 43–65.

King, J. M., J. A. Day, B. R. Davies, and M.-P. Henshall-Howard. 1987a. "Particulate organic matter in a mountain stream in the south-western Cape, South Africa." *Hydrobiologia* **154**: 165–187.

King, J. M., M.-P. Henshall-Howard, J. A. Day, and B. R. Davies. 1987b. "Leaf-pack dynamics in a southern African mountain stream." *Freshwat. Biol.* **18**: 325–340.

Kingsolver, J. G. 1979. "Thermal and hydric aspects of environmental heterogeneity in the pitcher plant mosquito." *Ecol. Monogr.* **49**: 357–376.

Kinne O. 1963. "The effects of temperature and salinity on marine and brackish water animals. I. Temperature." *Oceanogr. Mar. Biol. Ann. Rev.* **1**: 301–340.

Kitching, R. L. 1971. "An ecological study of water-filled tree holes and their position in the woodland ecosystem." *J. An. Ecol.* **40**: 281–302.

Kitching, R. L. 1987. "Spatial and temporal variation in food webs in water-filled treeholes." *Oikos* **48**: 280-288.

Kitching, R. L. and C. Callaghan. 1982. "The fauna of water-filled tree holes in box forest in south-east Queensland." *Austral. Ent. Mag.* **8**: 61-70.

Knight, A. W. and A. R. Gaufin. 1963. "The effect of water flow, temperature, and oxygen concentration on the Plecoptera nymph, *Acroneuria pacifica* Banks." *Proc. Utah Acad. Sci. Arts Lett.* **40**: 175-184.

Knight, A. W. and A. R. Gaufin. 1964. "Relative importance of varying oxygen concentration, temperature, and water flow on the mechanical activity and survival of the Plecoptera nymph, *Pteronarcys californica* Newport." *Proc. Utah Acad. Sci. Arts Lett.* **41**: 14-28.

Knight, A. W. and A. R. Gaufin. 1966a. "Altitudinal distribution of stoneflies (Plecoptera) in a Rocky Mountain drainage system." *J. Kans. Ent. Soc.* **39**: 668-675.

Knight, A. W. and A. R. Gaufin. 1966b. "Oxygen consumption of several species of stoneflies (Plecoptera)." *J. Insect Physiol.* **12**: 347-355.

Knight, A. W., M. A. Simmons, and C. S. Simmons. 1976. "A phenomenological approach to the growth of the winter stonefly, *Taeniopteryx nivalis* (Fitch) (Plecoptera: Taeniopterygidae)." *Growth* **40**: 343-367.

Knowles, J. N. and W. D. Williams. 1973. "Salinity range and osmoregulatory ability of corixids (Hemiptera: Heteroptera) in south-east Australian inland waters." *Austral. J. Mar. Freshwat. Res.* **24**: 297-302.

Kohshima, S. 1984. "A novel cold-tolerant insect found in a Himalayan glacier." *Nature* **310**: 225-227.

Kokkinn, M. J. 1986. "Osmoregulation, salinity tolerance and the site of ion excretion in the halobiont chironomid, *Tanytarsus barbitarsis* Freeman." *Austral. J. Mar. Freshwat. Res.* **37**: 243-250.

Komnick, H. 1977. "Chloride cells and chloride epithelia of aquatic insects." *Int. Rev. Cytol.* **49**: 285-329.

Konstantinov, A. S. 1958. "The effect of temperature on growth and development of chironomid larvae." *Akad. Nauk. SSSR. Doklady Biol. Sci. Sect.* **20**: 506-509.

Kovalak, W. P. 1976. "Seasonal and diel changes in the positioning of *Glossosoma nigrior* Banks (Trichoptera: Glossosomatidae) on artificial substrates." *Can. J. Zool.* **54**: 1585-1594.

Kownacka, M., and A. Kownacki. 1972. "Vertical distribution of zoocenoses in the streams of the Tatra, Caucasus and Balkans Mts." *Verh. Internat. Verein. Limnol.* **18**: 742-750.

Kownacka, M. and A. Kownacki. 1975. "Gletscherbach-Zuchmücken der Ötztaler Alpen in Tirol (Diptera: Chironomidae: Diamesinae)." *Ent. Germ.* **2**: 35-43.

Kownacki, A. and M. Kownacka. 1973. "The distribution of the bottom fauna in several streams of the Middle Balkan in the summer period." *Acta Hydrobiol.* **15**: 295-310.

Kownacki, A. and R. S. Zosidze. 1980. "Taxocens of Chironomidae (Diptera) in some rivers and streams of the Adzhar ASSR (Little Caucasus Mts)." *Acta Hydrobiol.* **22**: 67-87.

Kozhov, M. 1963. *Lake Baikal and Its Life.* Dr. W. Junk, The Hague.

Krantzberg, G. 1985. "The influence of bioturbation on physical, chemical and biological parameters in aquatic environments: A review." *Envir. Poll. (Series A)* **39**: 99-122.

Krecker, F. H. 1939. "A comparative study of the animal population of certain submerged aquatic plants." *Ecology* **20**: 553-562.

Krecker, F. H. and L. Y. Lancaster. 1933. "Bottom shore fauna of western Lake Erie: A population study to a depth of six feet." *Ecology* **14**: 79-93.

Kristensen, N. P. 1981. "Phylogeny of insect orders." *Ann. Rev. Ent.* **26**: 135-157.

Kroger, R. L. 1973. "Biological effects of fluctuating water levels in the Snake River, Grand Teton National Park, Wyoming." *Am. Midl. Nat.* **89**: 478-481.

Krogh, A. 1939. *Osmotic Regulation in Aquatic Animals.* Cambridge Univ. Press.

Krogh, A. 1943. "Some experiments on the osmoregulation and respiration of *Eristalis* larvae." *Ent. Medd.* **23**: 49-65.

Krull, J. N. 1970. "Aquatic plant-macroinvertebrate associations and waterfowl." *J. Wildlife Mgmt.* **34**: 707-718.

Krumbein, W. C. 1936. "Application of logarithmic moments to size frequency distributions of sediments." *J. Sediment Petrol.* **6**: 35-47.

Kuflikowski, T. 1970. "Fauna in vegetation in carp ponds at Goczalkowice." *Acta Hydrobiol.* **12**: 439-456.

Kuflikowski, T. 1974. "The phytophylous fauna of the dam reservoir at Goczalkowice." *Acta Hydrobiol.* **16**: 189-207.

Kukalová, J. 1968. "Permian mayfly nymphs." *Psyche* **75**: 310-327.

Kukalová-Peck, J. 1978. "Origin and evolution of insect wings and their relation to metamorphosis, as documented by the fossil record." *J. Morphol.* **156**: 53-126.

Kukalová-Peck, J. 1985. "Ephemeroid wing venation based upon new gigantic carboniferous mayflies and basic morphology, phylogeny, and metamorphosis of pterygote insects (Insecta, Ephemerida)." *Can. J. Zool.* **63**: 933-955.

Kukalová-Peck, J. 1987. "New Carboniferous Diplura, Monura, and Thysanura, the hexapod ground plan, and the role of thoracic side lobes in the origin of wings (Insecta)." *Can. J. Zool.* **65**: 2327-2345.

Kureck, A. 1978. "Alternative Schlüpfzeiten der Mücke *Chironomus thummi* und deren okologische Bedeutung." *Verh. Ges. F. Ökologie* (Kiel 1977): 207-210.

Kushlan, J. A. 1979. "Temperature and oxygen in an Everglades alligator pond." *Hydrobiologia* **67**: 267-271.

Labandeira, C. C., B. S. Beall, and F. M. Hueber. 1988. "Early insect diversification: Evidence from a Lower Devonian bristletail from Québec." *Science* **242**: 913-916.

Ladle, M. and J. A. B. Bass. 1981. "The ecology of a small chalk stream and its responses to drying during drought conditions." *Arch. Hydrobiol.* **90**: 448-466.

Laessle, A. M. 1961. "A micro-limnological study of Jamaican bromeliads." *Ecology* **42**: 499-517.

Laird, M. 1988. *The Natural History of Larval Mosquito Habitats.* Academic Press, New York.

Lamberti. G. A. and V. H. Resh. 1983. "Geothermal effects on stream benthos: Separate influences of thermal and chemical components on periphyton and macroinvertebrates." *Can. J. Fish. Aquat. Sci.* **40**: 1995-2009.

Landa, V. 1968. "Developmental cycles of Central European Ephemeroptera and their interrelationships." *Acta Ent. Bohemoslov.* **65**: 276–284.

Landa, V., T. Soldán, and M. Tonner (eds.). 1984. *Proceedings of the Fourth International Conference on Ephemeroptera.* Czechoslovak Acad. Sci., Ceské Budejovice.

Lange, W. H. 1984. "Aquatic and semiaquatic Lepidoptera," pp. 348–360 in R. W. Merritt and K. W. Cummins (eds.), *An Introduction to the Aquatic Insects of North America.* Kendall/Hunt, Dubuque, Iowa.

Langford, T. E. 1972. "A comparative assessment of thermal effects in some British and North American rivers," pp. 319–352 in R. T. Oglesby et al. (eds.). *River Ecology and Man.* Academic Press, New York.

Langford, T. E. 1975. "The emergence of insects from a British River, warmed by Power Station cooling-water. Part II. The emergence patterns of some species of Ephemeroptera, Trichoptera and Megaloptera in relation to water temperature and river flow, upstream and downstream of the cooling-water outfalls." *Hydrobiologia* **47**: 91–133.

Larimore, R. W., W. F. Childers, and C. Heckrotte. 1959. "Destruction and re-establishment of stream fish and invertebrates affected by drought." *Trans. Am. Fish. Soc.* **88**: 261–285.

La Rivers, I. 1956. "Aquatic Orthoptera," pp. 154 in R. L. Usinger (ed.), *Aquatic Insects of California.* Univ. Calif. Press, Berkeley.

LaRow, E. J. 1968. "A persistent diurnal rhythm in *Chaoborus* larvae. I. The nature of the rhythmicity." *Limnol. Oceanogr.* **13**: 250-256.

LaRow, E. J. 1969. "A persistent diurnal rhythm in *Chaoborus* larvae. II. Ecological significance." *Limnol. Oceanogr.* **14**: 213–218.

LaRow, E. J. 1970. "The effect of oxygen tension on the vertical migration of *Chaoborus* larvae." *Limnol. Oceanogr.* **15**: 357–362.

LaRow, E. J. 1971. "Response of *Chaoborus* (Diptera: Chaoboridae) larvae to different wavelengths of light." *Ann. Ent. Soc. Am.* **64**: 461–464.

Lauff, G. A. (ed.). 1967. *Estuaries.* Am. Assoc. Adv. Sci. Publ. No. 83. Washington, D.C.

Lavandier, P. 1974. "Écologie d'un Torrent Pyrénéen de Haute Montagne I. Caractéristiques physiques." *Ann. Limnol.* **10**: 173–219.

Lavandier, P. 1981. "Cycle biologique, croissance et production de *Rhithrogena loyolaea* Navas (Ephemeroptera) dans un torrent pyreneen de haute montagne." *Ann. Limnol.* **17**: 163–179.

Lavandier, P. and J.-Y. Pujol. 1975. "Cycle biologique de *Drusus rectus* (Trichoptera) dans les Pyrénées centrales: Influence de la température et de l'enneigement." *Ann. Limnol.* **11**: 255–262.

Lavery, M. A. and R. P. Costa. 1976. "Life history of *Parargyractis canadensis* Munroe (Lepidoptera: Pyralidae)." *Am. Midl. Nat.* **96**: 407–417.

Lee, R. E. 1989. "Insect cold-hardiness: To freeze or not to freeze." *BioScience* **39**: 308–313.

Leech, H. B. and H. P. Chandler. 1956. "Aquatic Coleoptera," pp. 293-371 in R. L. Usinger (ed.), *Aquatic Insects of California.* Univ. Calif. Press, Berkeley.

Leech, H. B. and M. W. Sanderson. 1959. "Coleoptera," pp. 981–1023 in W. T. Edmondson (ed.), *Freshwater Biology.* Wiley, New York.

Lehmkuhl, D. M. 1974. "Thermal regime alterations and vital environmental physiological signals in aquatic organisms," pp. 216-222 in J. W. Gibbons & R. R. Sharitz (eds.), *Thermal Ecology*. AEC Symp. Ser. (CONF-730505).

Leonard, B. V. and B. V. Timms. 1974. "The littoral rock fauna of three highland lakes in Tasmania." *Proc. Roy. Soc. Tasmania* **108**: 151-156.

Leopold, L. B., M. G. Wolman, and J. P. Miller. 1964. *Fluvial Processes in Geomorphology*. Freeman, San Francisco.

Lepneva, S. G. 1964. *Fauna of the U.S.S.R.; Trichoptera*, Vol. 2, No. 1. *Larvae and Pupae of Annulipalpia*. Zool. Inst. Akad. Nauk SSSR, N.S. 88 (in Russian, translated into English by the Israel Program for Scientific Translations, 1970).

Lepneva, S. G. 1966. *Fauna of the U.S.S.R.; Trichoptera*, Vol. 2, No. 2. *Larvae and Pupae of Integripalpia*. Zool. Inst. Akad. Nauk SSSR, n.s. 95 (in Russian, translated into English by the Israel Program for Scientific Translations, 1971).

LeSage, L. and A. D. Harrison. 1980. "The biology of *Cricotopus* (Chironomidae: Orthocladiinae) in an algal-enriched stream: Part I. Normal Biology." *Arch. Hydrobiol. Suppl.* **57**: 375-418.

Lewis, W. M., Jr. 1975. "Distribution and feeding habits of a tropical *Chaoborus* population." *Verh. Internat. Verein. Limnol.* **19**: 3106-3119.

Lewis, W. M., Jr. 1979. *Zooplankton Community Analysis: Studies on a Tropical System*. Springer-Verlag, New York.

Lewis, W. M., Jr. 1983. "A revised classification of lakes based on mixing." *Can. J. Fish. Aquat. Sci.* **40**: 1779-1787.

Lillehammer, A. 1974. "Norwegian stoneflies. II. Distribution and relationship to the environment." *Norsk Ent. Tidsskr.* **21**: 195-250.

Lillehammer, A. 1975a. "Norwegian stoneflies. III. Field studies on ecological factors influencing distribution." *Norw. J. Ent.* **22**: 71-80.

Lillehammer, A. 1975b. "Norwegian Stoneflies. IV. Laboratory studies on ecological factors influencing distribution." *Norw. J. Ent.* **22**: 99-108.

Lillehammer, A. 1976. "Norwegian stoneflies. V. Variations in morphological characters compared to differences in ecological factors." *Norw. J. Ent.* **23**: 161-172.

Lillehammer, A. 1978a. "The Plecoptera of Øvre Heimdalsvatn." *Holarct. Ecol.* **1**: 232-238.

Lillehammer, A. 1978b. "The Trichoptera of Øvre Heimdalsvatn." *Holarct. Ecol.* **1**: 255-260.

Lillehammer, A. 1988. *Stoneflies (Plecoptera) of Fennoscandia and Denmark*. Brill, Leiden.

Lindberg, H. 1948. *Zur Kenntnis der Insektenfauna im Brackwasser des Baltischen Meeres*. Soc. Sci. Fennica, Comm. Biol. Vol. 10, No. 9.

Lindegaard, C., J. Thorup, and M. Bahn. 1975. "The invertebrate fauna of the moss carpet in the Danish spring Ravnkilde and its seasonal, vertical, and horizontal distribution." *Arch. Hydrobiol.* **75**: 109-139.

Linduska, J. P. 1942. "Bottom type as a factor influencing the local distribution of mayfly nymphs." *Can. Ent.* **74**: 26-30.

Linevich, A. A. 1971. "The Chironomidae of Lake Baikal." *Limnologica* **8**: 51-52.

Lingdell, P.-E. and K. Müller. 1982. "Mayflies (Ins.: Ephemeroptera) in coastal areas of the Gulf of Bothnia," pp. 233-242 in K. Müller (ed.), *Coastal Research in the Gulf of Bothnia.* Dr. W. Junk, The Hague.

Linley, J. R. 1976. "Biting midges of mangrove swamps and saltmarshes (Diptera: Ceratopogonidae)," pp. 335-376 in L. Cheng (ed.), *Marine Insects.* North-Holland Publ. Co., Amsterdam.

Linley, J. R. and D. G. Evans. 1971. "Behavior of *Aedes taeniorhynchus* larvae and pupae in a temperature gradient." *Ent. Exper. Appl.* **14**: 319-332.

Linsenmair, K. E. and R. Jander. 1963. "Das Entspannungsschwimmen von *Velia* und *Stenus.*" *Naturwissenschaften* **50**: 231.

Lloyd, J. T. 1914. "Lepidopterous larvae from rapid streams." *J. N. Y. Ent. Soc.* **22**: 145-152.

Lock, M. A. and D. D. Williams (eds.). 1981. *Perspectives in Running Water Ecology.* Plenum, New York.

Logan, P. and M. P. Brooker. 1983. "The macroinvertebrate faunas of riffles and pools." *Water Res.* **17**: 263-270.

Lubbock, J. 1862. "On two aquatic Hymenoptera, one of which uses its wings in swimming." *Trans. Linn. Soc. Lond.* **24**: 135-142.

Luferov, V. P. 1963. "Epifauna of flooded forest of the Rybinsk Reservoir." *Biol. Aspekty. Vod. Biol. Vrutr. Trudy* **6**: 123-129 (in Russian).

Luferov, V. P. 1966. "Role of light in the distribution of Chironomidae larvae in the lakes of Karelia," pp. 282-301 in B. K. Shtegman (ed.), *Plankton and Benthos of Inland Waters.* Trudy Inst. Biol. Vnutren. Vod, No. 12(15) (in Russian, translated into English by the Israel Program for Scientific Translations, 1969).

Lundbeck, J. 1926. "Die Bodentierweld Norddeutscher Seen." *Arch. Hydrobiol. Suppl.* **7**: 1-473.

Lundbeck, J. 1936. "Untersuchungen über die Bodenbesidlungder Alpenrandseen." *Arch. Hydrobiol. Suppl.* **10**: 208-358.

Lundqvist, G. 1927. "Bodenablagerungen und Entwicklungstypen der seen." *Binnengesässer* **2**: 1-124.

Lutz, P. E. 1968a. "Life-history studies on *Lestes eurinus* Say (Odonata)." *Ecology* **49**: 576-579.

Lutz, P. E. 1968b. "Effects of temperature and photoperiod on larval development in *Lestes eurinus* (Odonata: Lestidae)." *Ecology* **49**: 637-644.

Lutz, P. E. 1974a. "Environmental factors controlling duration of larval instars in *Tetragoneuria cynosura* (Odonata)." *Ecology* **55**: 630-637.

Lutz, P. E. 1974b. "Effects of temperature and photoperiod on larval development in *Tetragoneuria cynosura.*" *Ecology* **55**: 370-377.

Lutz, P. E. and A. R. Pittman. 1970. "Some ecological factors influencing a community of adult Odonata." *Ecology* **51**: 279-284.

Lyman, F. E. 1945. "Reactions of certain nymphs of *Stenomena* (Ephemeroptera) to light as related to habitat preference." *Ann. Ent. Soc. Am.* **38**: 234-236.

Lyman, F. E. 1956. "Environmental factors affecting distribution of mayfly nymphs in Douglas Lake, Michigan." *Ecology* **37**: 568-576.

Macan, T. T. 1938. "Evolution of aquatic habitats with special reference to the distribution of Corixidae." *J. An. Ecol.* **7**: 1-19.

Macan, T. T. 1949. "Corixidae (Hemiptera) of an evolved lake in the English Lake District." *Hydrobiologia* **2**: 1-23.

Macan, T. T. 1951. "Mosquito records from the southern part of the Lake District." *Ent. Gaz.* **2**: 141-147.

Macan, T. T. 1954. "A contribution to the study of the ecology of Corixidae (Hemipt.)." *J. An. Ecol.* **23**: 115-141.

Macan, T. T. 1955. "Littoral Fauna and lake types." *Verh. Internat. Verein. Limnol.* **12**: 608-612.

Macan, T. T. 1957. "The life histories and migrations of the Ephemeroptera in a stony stream." *Trans. Soc. Br. Ent.* **12**: 129-156.

Macan, T. T. 1958. "The temperature of a small stony stream." *Hydrobiologia* **12**: 89-106.

Macan, T. T. 1960. "The occurrence of *Heptagenia lateralis* (Ephem.) in streams of the English Lake District." *Wett. U. Leben* **12**: 231-234.

Macan, T. T. 1961. "Factors that limit the range of freshwater animals." *Biol. Rev.* **36**: 151-198.

Macan, T. T. 1962. "Ecology of aquatic insects." *Ann. Rev. Ent.* **7**: 261-288.

Macan, T. T. 1963. *Freshwater Ecology.* Wiley, New York.

Macan, T. T. 1965. "The fauna in the vegetation of a moorland fishpond." *Arch. Hydrobiol.* **61**: 273-310.

Macan, T. T. 1970. *Biological Studies of the English Lakes.* American Elsevier, New York.

Macan, T. T. 1974a. *Freshwater Ecology*, 2nd ed. Wiley, New York.

Macan, T. T. 1974b. "Running water." *Mitt. Internat. Verein. Limnol.* **20**: 301-321.

Macan, T. T. and R. Maudsley. 1966. "The temperature of a moorland fishpond." *Hydrobiologia* **27**: 1-22.

Macan, T. T. and R. Maudsley. 1968. "The insects of the stony substratum of Lake Windermere." *Trans. Soc. Br. Ent.* **18**: 1-18.

Macan, T. T. and R. Maudsley. 1969. "Fauna of the stony substratum in lakes in the English Lake District." *Verh. Internat. Verein. Limnol.* **17**: 173-180.

MacArthur, R. H. and E. O. Wilson. 1963. "An equilibrium theory of insular zoogeography." *Evolution* **17**: 373-387.

Macdonald, W. W. 1956. "Observations on the biology of chaoborids and chironomids in Lake Victoria and on the feeding habits of the elephant-snout fish *Mormyrus kannum* Forsk." *J. An. Ecol.* **25**: 36-53.

Maciolek, J. A. and P. R. Needham. 1951. "Ecological effects of winter conditions on trout and trout foods in Convict Creek, California, 1951." *Trans. Am. Fish. Soc.* **81**: 202-217.

Mackay, R. J. 1977. "Behavior of *Pycnopsyche* (Trichoptera: Limnephilidae) on mineral substrates in laboratory streams." *Ecology* **58**: 191-195.

Mackay, R. J. 1979. "Life history patterns of some species of *Hydropsyche* (Trichoptera: Hydropsychidae) in southern Ontario." *Can. J. Zool.* **57**: 963-975.

Mackay, R. J. and J. Kalff. 1969. "Seasonal variation in standing crop and species diversity of insect communities in a small Quebec stream." *Ecology* **50**: 101-109.

Mackay, R. J. and G. B. Wiggins. 1979. "Ecological diversity in Trichoptera." *Ann. Rev. Ent.* **24**: 185–208.

Mackerras, I. M. 1950. "Marine insects." *Proc. Roy. Soc. Queensland* **61**: 19–29.

Mackerras, I. M. 1970. "Evolution and classification of the insects," pp. 152–168 in I. M. Mackerras (ed.), *The Insects of Australia.* Melbourne Univ. Press, Australia.

Mackey, A. P., S. F. Ham, D. A. Cooling, and A. D. Berrie. 1982. "An ecological survey of a limestone stream, the River Coln, Gloucestershire, England, in comparison with some chalk streams." *Arch. Hydrobiol. Suppl.* **64**: 307–340.

MacNeill, N. 1960. "A study of the caudal gills of dragonfly larvae of the sub-order Zygoptera." *Proc. Roy. Irish Acad., B* **61**: 115–140.

MacPhee, C. and M. A. Brusven. 1974. *The Effects of River Fluctuations Resulting from Hydroelectric Peaking on Selected Aquatic Invertebrates.* Water Resources Research Inst., Univ. Idaho, Moscow, Idaho.

Madsen, B. L. 1968. "The distribution of nymphs of *Brachyptera risi* Mort. and *Nemoura flexuosa* Aub. (Plecoptera) in relation to oxygen." *Oikos* **19**: 304–310.

Madsen, T. V. and E. Warncke. 1983. "Velocities of currents around and within submerged aquatic vegetation." *Arch. Hydrobiol.* **97**: 389-394.

Maguire, B. 1971. "Phytotelmata: Biota and community structure determination in plant-held waters." *Ann. Rev. Ecol. Syst.* **2**: 439–464.

Maitland, P. S. 1964. "Quantitative studies on the invertebrate fauna of sandy and stony substrates in the River Endrick, Scotland." *Proc. Roy. Soc. Edinb., Sect. B, Biol.* **68**: 277–300.

Maitland, P. S. 1966. *The Fauna of the River Endrick. Studies on Loch Lomond II.* Blackie, Glasgow.

Maitland, P. S. and M. M. Penney. 1967. "The ecology of the Simuliidae in a Scottish river." *J. An. Ecol.* **36**: 179–206.

Malicky, H. 1973. "Trichoptera (Köcherfliegen)." *Handb. Zool.* **4**(2) 2/29: 1–114.

Malicky, H. (ed.). 1976a. *Proceedings of the First International symposium on Trichoptera.* Dr. W. Junk, The Hague.

Malicky, H. 1976b. "Trichopteren—Emergenz in zwei Lunzer Bächen 1972-74. *Arch. Hydrobiol.* **77**: 51–65.

Malmqvist, B., L. M. Nilsson, and B. S. Svensson. 1978. "Dynamics of detritus in a small stream in southern Sweden and its influence on the distribution of the bottom animal communities." *Oikos* **31**: 3-16.

Manton, S. M. 1977. *The Arthropoda. Habits, Functional Morphology, and Evolution.* Clarendon Press, Oxford.

Margalef, R. 1983. *Limnología.* Ediciones Omega, Barcelona.

Marlier, G. 1951. "La biologie d'un ruisseau de plaine. Le Smohain." *Mém. Inst. Roy. Sci. Nat. Belg.* **114**: 1–98.

Marlier, G. 1954. "Recherches hydrobiologiques dans les rivières du Congo oriental II. Etude écologique." *Hydrobiologia* **6**: 225–264.

Marlier, G. 1962. "Genera des Trichoptères de l'Afrique." *Ann. Mus. Roy. Afr. Centr. Sci. Zool.* **109**: 261.

Martin, N. A. 1972. "Temperature fluctuations within English lowland ponds." *Hydrobiologia* **40**: 455–469.

Martynov, A. 1924. "To the knowledge of Baicalinini, a group of endemic baicalian Trichoptera." *Dokl. Acad. Sci. USSR* **1924**: 93-96.

Masner, L. 1968. "A new scelionid wasp from the intertidal zone of South Africa (Hymenoptera: Scelionidae)." *Ann. Natal Mus.* **20**: 195-198.

Mason, C. F. and R. J. Bryant. 1974. "The structure and diversity of the animal communities in a broadleaf reedswamp." *J. Zool. London* **172**: 289-302.

Matta, J. F. 1979. "Aquatic insects of the Dismal Swamp," pp. 200-221 in P. W. Kirk, Jr. (ed.), *The Great Dismal Swamp.* Univ. Virg. Press, Charlottesville, Va.

Mauch, E. 1963. "Untersuchungen über das Benthos der deutschen Mosel unter besonderer Breüchsichtigung der Wassergute." *Mitt. Zool. Mus. Berl.* **39**: 1-172.

Maurer, M. A. and M. A. Brusven. 1983. "Insect abundance and colonization rate in *Fontinalis neo-mexicana* (Bryophyta) in an Idaho batholith stream, U.S.A." *Hydrobiologia* **98**: 9-15.

May, M. L. 1976. "Thermoregulation and adaptation to temperature in dragonflies (Odonata: Anisoptera)." *Ecol. Monogr.* **46**: 1-32.

May, M. L. 1978. "Thermal adaptations of dragonflies." *Odonatologica* **7**: 27-47.

McAlpine, J. F., B. V. Peterson, G. E. Shewell, H. J. Teskey, J. R. Vockeroth, and D. M. Wood (coord.). 1981. *Manual of Nearctic Diptera,* Vol. 1. Monogr. 27, Agric. Canada, Ottawa.

McAlpine, J. F., B. V. Peterson, G. E. Shewell, H. J. Teskey, J. R. Vockeroth, and D. M. Wood (coord.). 1987. *Manual of Nearctic Diptera,* Vol. 2, pp. 675-1332. Monogr. 28. Agric. Canada, Ottawa.

McAuliffe, J. R. 1984. "Competition for space, disturbance, and the structure of a benthic stream community." *Ecology* **65**: 894-908.

McCafferty, W. P. and G. F. Edmunds, Jr. 1979. "The higher classification of the Ephemeroptera and its evolutionary basis." *Ann. Ent. Soc. Am.* **72**: 5-12.

McCafferty, W. P. and C. Pereira. 1984. "Effects of developmental thermal regimes on two mayfly species and their taxonomic interpretation." *Ann. Ent. Soc. Am.* **77**: 69-87.

McCall, P. L. and M. J. S. Tevesz. 1982. "The effects of benthos on physical properties of freshwater sediments," pp. 105-176 in P. L. McCall and M. J. S. Tevesz (eds.), *Animal-Sediment Relations.* Plenum, New York.

McCarraher, D. B. 1972. *A Preliminary Bibliography and Lake Index of the Inland Mineral Waters of the World.* FAO Fish. Circ. No. 146, Food and Agriculture Organization of the United Nations, Rome.

McGaha, Y. J. 1952. "Limnological relations of insects to certain aquatic flowering plants." *Trans. Am. Microsc. Soc.* **71**: 355-381.

McLachlan, A. J. 1969. "The effect of aquatic macrophytes on the variety and abundance of benthic fauna in a newly created lake in the tropics (Lake Kariba)." *Arch. Hydrobiol.* **66**: 212-231.

McLachlan, A. J. 1970. "Submerged trees as a substrate for benthic fauna in the recently created Lake Kariba (Central Africa)." *J. Appl. Ecol.* **7**: 253-266.

McLachlan, A. J. 1974. "Development of some lake ecosystems in tropical Africa, with special reference to the invertebrates." *Biol. Rev.* **49**: 365-397.

McLachlan, A. J. 1979. "Decline and recovery of the benthic invertebrate communities," pp. 143–160 in M. Kalk, A. J. McLachlan, and C. Howard-Williams (eds.), *Lake Chilwa.* Dr. W. Junk, The Hague.

McLachlan, A. J. and S. M. McLachlan. 1975. "The physical environment and bottom fauna of a bog lake." *Arch. Hydrobiol.* **76**: 198–217.

McShaffrey, D. and W. P. McCafferty. 1987. "The behaviour and form of *Psephenus herricki* (DeKay) (Coleoptera: Psephenidae) in relation to water flow." *Freshwat. Biol.* **18**: 319–324.

Mecom, J. O. 1972. "Productivity and distribution of Trichoptera larvae in a Colorado mountain stream." *Hydrobiologia* **40**: 151-176.

Meffe, G. K. and W. L. Minckley. 1987. "Persistence and stability of fish and invertebrate assemblages in a repeatedly disturbed Sonoran Desert stream." *Am. Midl. Nat.* **117**: 177–191.

Mendl, H. and K. Müller. 1982. "Stoneflies (INS: Plecoptera) in the mouth and estuary of the river Ångerån in the northern Bothnian Sea," pp. 243–252 in K. Müller (ed.), *Coastal Research in the Gulf of Bothnia.* Dr. W. Junk, The Hague.

Menke, A. S. (ed.). 1979. *The Semiaquatic and Aquatic Hemiptera of California.* Bull. Calif. Insect Surv. 21.

Menzies, R. J. and M. Selvakumaran. 1974. "The effects of hydrostatic pressure on living aquatic organisms V. Eurybiotic environmental capacity as a factor in high pressure tolerance." *Internat. Revue ges. Hydrobiol.* **59**: 199–205.

Merritt, R. W. and K. W. Cummins (eds.). 1984. *An Introduction to the Aquatic Insects of North America,* 2nd ed. Kendall/Hunt Publ. Co., Dubuque, Iowa.

Messner, B. 1965. "Bemerkungen zur Biologie von *Agriotypus armatus* Walk. (Hymenoptera, Agriotypidae)." *Zool. Anz.* **174**: 354–362.

Mĕstrov, M. and V. Tavčar. 1972. "Hyporheic as a selective biotope for some kinds of larvae of Chironomidae." *Bull. Sci., Sect. A, Yougosl.* **17**: 7–8.

Miall, L. C. 1895. *The Natural History of Aquatic Insects.* Macmillan, London.

Michaelis, F. B. 1977. "Biological features of Pupu Springs." *N. Z. J. Mar. Freshwat. Res.* **11**: 357–373.

Milbrink, G., S. Lundquist, and H. Pramsten. 1974. "On the horizontal distribution of the profundal bottom fauna in a limited area of central Lake Malaren, Sweden. Part 1. Studies on the distribution of the bottom fauna." *Hydrobiologia* **45**: 509–526.

Mill, P. J. 1974. "Respiration: Aquatic Insects," pp. 403–467 in M. Rockstein (ed.), *The Physiology of Insecta,* 2nd ed., Vol. VI. Academic Press, New York.

Mill, P. J. and R. S. Pickard. 1972. "Anal valve movement and normal ventilation in aeshnid dragonfly larvae." *J. Exper. Biol.* **56**: 537–543.

Mill, P. J. and R. S. Pickard. 1975. "Jet-propulsion in anisopteran dragonfly larvae." *J. Comp. Physiol.* **97**: 329–338.

Miller, P. L. 1964. "The possible role of haemoglobin in *Anisops* and *Buenoa* (Hemiptera; Notonectidae)." *Proc. Roy. Ent. Soc. London, A* **39**: 166–175.

Miller, P. L. 1966. "The function of haemoglobin in relation to the maintenance of neutral buoyancy in *Anisops pellucens* (Notonectidae, Hemiptera)." *J. Exp. Biol.* **44**: 529–544.

Miller, P. L. 1974. "Respiration-aerial gas transport," pp. 345–402 in M. Rockstein (ed.), *The Physiology of Insecta,* 2nd ed., Vol VI. Academic Press, New York.

Milliman, J. D. and R. H. Meade. 1983. "World-wide delivery of river sediment to the oceans." *J. Geol.* **91**: 1-21.

Minckley, W. L. 1963. "The ecology of a spring stream, Doe Run, Meade County, Kentucky." *Wildl. Monogr.* **11**: 1-124.

Minshall, G. W. 1967. "Role of allochthonous detritus in the trophic structure of a woodland springbrook community." *Ecology* **48**: 139-149.

Minshall, G. W. 1968. "Community dynamics of the benthic fauna in a woodland springbrook." *Hydrobiologia* **32**: 305-339.

Minshall, G. W. 1978. "Autotrophy in stream ecosystems." *BioScience* **28**: 767-771.

Minshall, G. W. 1984. "Aquatic insect-substratum relationships," pp. 358–400 in V. H. Resh and D. M. Rosenberg (eds.), *The Ecology of Aquatic Insects.* Praeger, New York.

Minshall, G. W. 1988. "Stream ecosystem theory: A global perspective." *J. N. Am. Benthol. Soc.* **7**: 263–388.

Minshall, G. W., D. A. Andrews, and C. Y. Manuel-Faler. 1983a. "Application of island biogeographic theory to streams: Macroinvertebrate recolonization of the Teton River, Idaho," pp. 279-297 in J. R. Barnes and G. W. Minshall (eds.), *Stream Ecology—Application and Testing of General Ecological Theory.* Plenum, New York.

Minshall, G. W., R. C. Petersen, K. W. Cummins, T. L. Bott, J. R. Sedell, C. E. Cushing, and R. L. Vannote. 1983b. "Interbiome comparison of stream ecosystem dynamics." *Ecol. Monogr.* **53**: 1-25.

Minshall, G. W., J. T. Brock, and T. W. LaPoint. 1982. "Characterization and dynamics of benthic organic matter and invertebrate functional feeding group relationships in the Upper Salmon River, Idaho (USA)." *Internat. Revue ges. Hydrobiol.* **67**: 793-820.

Minshall, G. W., K. W. Cummins, R. C. Petersen, C. E. Cushing, D. A. Bruns, J. R. Sedell, and R. L. Vannote. 1985. "Developments in stream ecosystem theory." *Can. J. Fish. Aquat. Sci.* **42**: 1045-1055.

Minshall, G. W. and R. A. Kuehne. 1969. "An ecological study of invertebrates of the Duddon, an English mountain stream." *Arch. Hydrobiol.* **66**: 169–191.

Minshall, G. W. and J. N. Minshall. 1977. "Microdistribution of benthic invertebrates in a Rocky Mountain (U.S.A.) stream." *Hydrobiologia* **55**: 231–249.

Minshall, G. W. and P. V. Winger. 1968. "The effect of reduction in stream flow on invertebrate drift." *Ecology* **49**: 580–582.

Mitchell, R. 1974. "The evolution of thermophily in hot springs." *Rev. Biol.* **49**: 229-242.

Mizuno, T., K. Gose, R. P. Lim, and J. I. Furtado. 1982. "Benthos and attached animals," pp. 286–306 in J. I. Furtado and S. Mori (eds.), *Tasek Bera. The Ecology of a Freshwater Swamp.* Dr. W. Junk, The Hague.

Moeller, J. R., G. W. Minshall, K. W. Cummins, R. C. Petersen, C. E. Cushing, J. R. Sedell, R. A. Larson, and R. L. Vannote. 1979. "Transport of dissolved organic carbon in streams of differing physiographic characteristics." *Org. Geochem.* **1**: 139-150.

Moffett, J. W. 1936. *A Quantitative Study of the Bottom Fauna in Some Utah Streams Variously Affected by Erosion.* Bull. Univ. Utah, Biol. Ser. 26, No. 9.

Monakov, A. V. 1964. "Zooplankton of the Oka River." *Trudy Zool. Instit. Leningrad* **32**: 92-105 (in Russian).

Monakov, A. V. 1969. "The zooplankton and zoobenthos of the White Nile and adjoining waters in the Republic of the Sudan." *Hydrobiologia* **33**: 161-185.

Moniez, F. 1894. "Sur un Hyménopterè Halophile." *Rev. Biol. Nord France* **6**: 439-441.

Moon, H. P. 1934. "An investigation of the littoral region of Windermere." *J. An. Ecol.* **3**: 8-28.

Moon, H. P. 1935. "Flood movements of the littoral fauna of Windermere." *J. An. Ecol.* **4**: 216-228.

Moon, H. P. 1936. "The shallow littoral region of a bay at the north west end of Windermere." *Proc. Zool. Soc. London* **1936**: 490-515.

Moon, H. P. 1939. "Aspects of the ecology of aquatic insects." *Trans. Soc. Brit. Ent.* **6**: 39-49.

Moore, I. and E. F. Legner. 1976. "Intertidal rove beetles (Coleoptera: Staphylinidae)," pp. 521-551 in L. Cheng (ed.), *Marine Insects.* North-Holland Publ. Co., Amsterdam.

Moore, J. W. 1980. "Factors influencing the composition, structure and density of a population of benthic invertebrates." *Arch. Hydrobiol.* **88**: 202-218.

Morduchai-Boltovskoy, Ph.D. and A. I. Shilova. 1955. "On the temporary planktonic modus vivendi of the larvae of *Glyptotendipes.*" *Dokl. Akad. Nauk. SSSR* **105**: 163-165.

Moretti, G. P. (ed.). 1981. *Proceedings of the Third International Symposium on Trichoptera.* Dr. W. Junk, The Hague.

Morgan, A. H. and M. C. Grierson. 1932. "The function of the gills in burrowing may flies *(Hexagenia recurvata).*" *Physiol. Zool.* **5**: 230-245.

Morgan, A. H. and H. D. O'Neil. 1931. "The function of the tracheal gills in larvae of the caddis fly, *Macronema zebratum* Hagen." *Physiol. Zool.* **4**: 361-379.

Morgan, N. C. and A. Waddell. 1961. "Diurnal variation in the emergence of some aquatic insects." *Trans. Roy. Ent. Soc. London* **113**: 123-137.

Morse, J. C. (ed.). 1984. *Proceedings of the Fourth International Symposium on Trichoptera.* Dr. W. Junk, The Hague.

Mortimer, C. H. 1981. "The oxygen content of air-saturated fresh waters over ranges of temperature and atmospheric pressure of limnological interest." *Mitt. Internat. Verein. Limnol.* **22**: 1-23.

Morton, E. S. 1977. "Ecology and behavior of some Panamanian Odonata." *Proc. Ent. Soc. Wash.* **79**: 273.

Müller, K. 1951. "Fische und Fischregionen der Fulda." *Ber. Limnol. Flussstn. Freudenthal* **2**: 18-23.

Müller, K. 1956. "Das produktions-biologische Zusammenspiel zwischen See und Fluss." *Ber. Limnol. Flussstn. Freudenthal* **7**: 1-8.

Müller, K. (ed.). 1982. *Coastal Research in the Gulf of Bothnia.* Dr. W. Junk, The Hague.

Müller, K. and H. Mendl. 1979. "The importance of a brackish water area for the stonefly colonization cycle in a coastal river." *Oikos* **33**: 272-277.

Müller-Liebenau, I. 1956. "Die Besiedlung der *Potamogeton*-Zone ostholsteinischer Seen." *Arch. Hydrobiol.* **52**: 470-606.

Mundie, J. H. 1959. "The diurnal activity of the larger invertebrates at the surface of Lac la Ronge, Saskatchewan." *Can. J. Zool.* **37**: 945-956.

Murphey, R. K. 1971. "Sensory aspects of the control of orientation to prey by the waterstrider *Gerris remigis*." *Z. Vgl. Physiol.* **72**: 168–185.

Mutch, R. A. and G. Pritchard. 1986. "Development rates of eggs of some Canadian stoneflies in relation to temperature." *J. N. Am. Benthol. Soc.* **5**: 272–277.

Muttkowski, R. A. 1918. "The fauna of Lake Mendota." *Trans. Wisc. Acad. Sci. Arts Lett.* **11**: 374–454.

Myers, J. G. 1935. "Aquatic woolly-bear caterpillars from British Guiana." *Proc. Roy. Ent. Soc. London, A* **10**: 65–70.

Nachtigall, W. 1974. "Locomotion: Mechanics and hydrodynamics of swimming in aquatic insects," pp. 381–432 in M. Rockstein (ed.), *The Physiology of Insects,* 2nd ed., Vol. III. Academic Press, New York.

Nachtigall, W. 1985. "Swimming in aquatic insects," pp. 467–490 in G. A. Kerkut and L. I. Gilbert (eds.), *Comprehensive Insect Physiology, Biochemistry and Pharmacology,* Vol. 5. Pergamon Press, Oxford.

Nachtigall, W. and D. Bilo. 1965. "Die Strömungsmechanik des *Dytiscus*—Rumpfes." *Z Vergl. Physiol.* **50**: 371-401.

Nachtigall, W. and D. Bilo. 1975. "Hydrodynamics of the body of *Dytiscus marginalis* (Dytiscidae, Coleoptera)," pp. 585–595 in T. Y. Wu, C. J. Brokaw, and C. Brennen (eds.), *Swimming and Flying in Nature,* Vol. 2. Plenum, New York.

Nagell, B. 1974. "The basic picture and the ecological implications of the oxygen consumption/oxygen concentration curve of some aquatic insect larvae." *Ent. Tidskr. Suppl.* **95**: 182–187.

Nagell, B. 1977. "Phototactic and thermotactic responses facilitating survival of *Cloeon dipterum* (Ephemeroptera) larvae under winter anoxia." *Oikos* **29**: 342–347.

Nagell, B. 1980. "Overwintering strategy of *Cloeon dipterum* (L.) larvae," pp. 259–264 in J. F. Flannagan and K. E. Marshall (eds.), *Advances in Ephemeroptera Biology.* Plenum, New York.

Nagell, B. and J. E. Brittain. 1977. "Winter anoxia—a general feature of ponds in cold temperate regions." *Internat. Revue ges. Hydrobiol.* **62**: 821–824.

Nagell, B. and C.-C. Landahl. 1978. "Resistance to anoxia of *Chironomus plumosus* and *Chironomus anthracinus* (Diptera) larvae." *Holarct. Ecol.* **1**: 333–336.

Naiman, R. J., J. M. Melillo, M. A. Lock, T. E. Ford, and S. R. Reice. 1987. "Longitudinal patterns of ecosystem processes and community structure in a subarctic river continuum." *Ecology* **68**: 1139-1156.

Naiman, R. J. and J. R. Sedell. 1979. "Characterization of particulate organic matter transported by some Cascade Mountain streams." *J. Fish. Res. Board Can.* **36**: 17–31.

Navas, L. 1935. "Monografia de lo Familia de los Sisiridos." *Mem. Acad. Cienc. Zaragoza* **4**: 1–83.

Nebeker, A. V. 1971a. "Effect of water temperature on nymphal feeding rate, emergence, and adult longevity of the stonefly *Pteronarcys dorsata*." *J. Kans. Ent. Soc.* **44**: 21–26.

Nebeker, A. V. 1971b. "Effect of high winter water temperature on adult emergence of aquatic insects." *Water Research* **5**: 777–783.

Nebeker, A. V. 1971c. "Effect of temperature at different altitudes on the emergence of aquatic insects from a single stream." *J. Kans. Ent. Soc.* **44**: 26–35.

Needham, J. G. and P. W. Claassen. 1925. *A Monograph of the Plecoptera or Stoneflies of America North of Mexico.* Thomas Say Foundation, Lafayette, Ind.

Needham, J. G., J. R. Traver, and Yin-Chi Hsu. 1935. *The Biology of Mayflies.* Comstock, Ithaca, N.Y.

Needham, P. R. and R. L. Usinger. 1956. "Variability in the macrofauna of a single riffle in Prosser Creek, California, as indicated by the Surber sampler." *Hilgardia* **24**: 383-409.

Neel, J. K. 1973. "Biotic character as related to stream mineral content." *Trans. Am. Micros. Soc.* **92**: 404-415.

Nelder, K. H. and R. W. Pennak. 1955. "Seasonal faunal variations in a Colorado alpine pond." *Am. Midl. Nat.* **53**: 419-430.

Neumann, D. 1967. "Genetic adaptation in the emergence time of *Clunio* populations to different tidal conditions." *Helgol. Wiss. Meeresunter.* **15**: 163-171.

Neumann, D. 1975. "Lunar and tidal rhythms in the development and reproduction of an intertidal organism," pp. 451-463 in F. J. Vernberg (ed.), *Physiological Adaptation to the Environment.* Intext Educ. Publ., New York.

Neumann, D. 1976. "Adaptations of Chironomids to intertidal environments." *Ann. Rev. Ent.* **21**: 387-414.

Newcomer, E. J. 1918. "Some stoneflies injurious to vegetation." *J. Agric. Res.* **13**: 37-41.

Newell, R. L. and G. W. Minshall. 1978. "Life history of a multivoltine mayfly, *Tricorythodes minutes*: An example of the effect of temperature on the life cycle." *Ann. Ent. Soc. Am.* **71**: 876-881.

Nicola, S. 1968. "Scavenging by *Alloperla* (Plecoptera: Chloroperlidae) on dead pink *Oncorhynchus gorbuscha)* and chum *(Oncorhynchus keta)* salmon embryos." *Can. J. Zool.* **46**: 787-796.

Nielsen, A. 1950a. "The torrential invertebrate fauna." *Oikos* **2**: 176-196.

Nielsen, A. 1950b. "On the zoogeography of springs." *Hydrobiologia* **2**: 313-321.

Nielsen, A. 1976. "Revision of some opinions expressed in my 1942 paper," pp. 163-165 in H. Malicky (ed.), *First International Symposium on Trichoptera.* Dr. W. Junk, The Hague.

Niesiolowski, S. 1980. "Studies on the abundance, biomass and vertical distribution of larvae and pupae of black flies (Simuliidae, Diptera) on plants of the Grabia River, Poland." *Hydrobiologia* **75**: 149-156.

Nilsen, H. C. and R. W. Larimore. 1973. "Establishment of invertebrate communities on log substrates in the Kaskaskia River, Illinois." *Ecology* **54**: 366-374.

Nilsson, A. N. 1982. "Aquatic Coleoptera of the northern Swedish Bothnian coast," pp. 273-283 in K. Müller (ed.), *Coastal Research in the Gulf of Bothnia.* Dr. W. Junk, The Hague.

Nilsson, N.-A. 1955. "Studies on the feeding habits of trout and char in north-Swedish lakes." *Rep. Instit. Freshwat. Res. Drottningholm* **36**: 163-225.

Nimmo, A. P. 1971. "The adult Rhyacophilidae and Limnephilidae (Trichoptera) of Alberta and Eastern British Columbia and their post-glacial origins." *Quaest. Ent.* **7**: 3-234.

Njogu, A. R. and G. K. Kinoti. 1971. "Observations on the breeding sites of mosquitoes in Lake Manyara, a saline lake in the East African Rift Valley." *Bull. Ent. Res.* **60**: 473–479.

Norris, K. R. 1970. "General biology," pp. 107–140 in I. M. Mackerras (ed.), *The Insects of Australia.* Melbourne University Press, Australia.

Nuttall, P. M. and G. H. Bielby. 1973. "The effect of china-clay wastes on stream invertebrates." *Envir. Pollut.* **5**: 77–86.

Oberdorfer, R. Y. and K. W. Stewart. 1977. "The life cycle of *Hydroperla crosbyi* (Plecoptera: Perlodidae)." *Great Basin Nat.* **37**: 260-273.

Odum, E. P. 1971. *Fundamentals of Ecology.* Saunders, Philadelphia.

Odum, H. T. 1957. "Trophic Structure and Productivity of Silver Springs, Florida." *Ecol. Monogr.* **27**: 55–112.

Oemke, M. P. 1987. "The effect of temperature and diet on the larval growth of *Glossosoma nigrior*," pp. 257-262 in M. Bournaud and H. Tachet (eds.), *Proc. Fifth Internat. Symp. Trichoptera.* Dr. W. Junk, Dordrecht.

Okada, Yô K. 1928. "Two Japanese aquatic glowworms." *Trans. Roy. Ent. Soc. London* **76**: 101–108.

Okazawa, T. 1975. "Aquatic insect survey of the River Kaunnai, with special reference to the rocky chute bed fauna." *Kontyû* **43**: 497-512.

Ökland, J. 1964. "The eutrophic lake Borrevann (Norway)—an ecological study on shore and bottom fauna with special reference to gastropods, including a hydrographic survey." *Folia Limnol. Scand.* **13**: 1-337.

Olafsson, J. 1979. "Physical characteristics of Lake Myvatn and River Laxa." *Oikos* **32**: 38–66.

Olander, R. and E. Palmén. 1968. "Taxonomy and ecology of the northern Baltic *Clunio marinus* Halid." *Ann. Zool. Fenn.* **5**: 97–110.

Oliff, W. D. 1960. "Hydrobiological studies on the Tugela River system. Part I. The main Tugela River." *Hydrobiologia* **14**: 281-385.

Oliver, D. R. 1968. "Adaptations of arctic Chironomidae." *Ann. Zool. Fenn.* **5**: 111–118.

Olsson, T. I. 1981. "Overwintering of benthic macroinvertebrates in ice and frozen sediment in a north Swedish river." *Holarct. Ecol.* **4**: 161–166.

Olsson, T. I. 1985. "Stationarity versus migration—alternative behaviours of invertebrates in subarctic rivers." *Aquilo Ser. Zool.* **24**: 71–76.

Olsson, T. and O. Söderström. 1978. "Springtime migration and growth of *Parameletus chelifer* (Ephemeroptera) in a temporary stream in northern Sweden." *Oikos* **31**: 284–289.

O'Meara, G. F. 1976. "Saltmarsh mosquitoes (Diptera: Culicidae)," pp. 303–333 in L. Cheng (ed.), *Marine Insects.* North-Holland Publ. Co., Amsterdam.

Ordish, R. G. 1976. "Two new genera and species of subterranean water beetle from New Zealand." *N. Z. J. Zool.* **3**: 1–10.

Orghidan, T. 1959. "Ein neuer Lebensraum des unterirdischen Wassers: Der hyporheische Biotop." *Arch. Hydrobiol.* **55**: 392–414.

Orth, D. J. and O. E. Maughan. 1983. "Microhabitat preferences of benthic fauna in a woodland stream." *Hydrobiologia* **106**: 157-168.

Osborne, L. L. and E. E. Herricks. 1987. "Microhabitat characteristics of *Hydropsyche* (Trichoptera: Hydropsychidae) and the importance of body size." *J. N. Am. Benthol. Soc.* **6**: 115–124.

Otto, C. 1982. "Habitat, size and distribution of Scandinavian limnephilid caddisflies." *Oikos* **38**: 355–360.

Otto, C. 1983. "Adaptations to benthic freshwater herbivory," pp. 199-205 in R. G. Wetzel (ed.), *Periphyton of Freshwater Ecosystems*. Dr. W. Junk, The Hague.

Otto, C. and B. S. Svensson. 1980. "The significance of case material selection for the survival of caddis larvae." *J. An. Ecol.* **49**: 855–865.

Otto, C. and B. S. Svensson. 1983. "Properties of acid brown water streams in south Sweden." *Arch. Hydrobiol.* **99**: 15–36.

Palmer, M. 1973. "A survey of the animal community of the main pond at Castor Hanglands National Nature Reserve, near Peterborough." *Freshwat. Biol.* **3**: 397–407.

Pandian, T. J., S. Mathavan, and C. P. Jeyagopal. 1979. "Influence of temperature and body weight on mosquito predation by the dragonfly nymph *Mesogomphus lineatus*." *Hydrobiologia* **62**: 99–104.

Parfin, S. I. and A. B. Gurney. 1956. "The spongilla-flies, with special reference to those of the western Hemisphere." *Proc. U. S. Nat. Mus.* **105**: 421–529.

Parma, S. 1971. "*Chaoborus flavicans* (Meigen) (Diptera, Chaoboridae): An autecological study." Doctoral thesis, University of Groningen.

Parsons, M. C. 1966. "Labial skeleton and musculature of the Hydrocorisae." *Can. J. Zool.* **44**: 1051–1084.

Parsons, M. C. and R. J. Hewson. 1975. "Plastral respiratory devices in adult *Cryphocricos*." *Psyche* **81**: 510–527.

Pasternak, K. and R. Sowa (eds.). 1979. *Proceedings of the Second International Conference on Ephemeroptera*. Panstwowe Wydawnictwo Naukow, Warsaw.

Paterson, C. G. and C. H. Fernando. 1969. "The effect of winter drainage on reservoir benthic fauna." *Can. J. Zool.* **47**: 589-595.

Patterson, C. G. 1971. "Overwintering ecology of the aquatic fauna associated with the pitcher plant *Sarracenia purpurea* L." *Can. J. Zool.* **49**: 1455–1459.

Pearson, R. G., L. J. Benson, and R. E. W. Smith. 1986. "Diversity and abundance of the fauna in Yuccabine Creek, a tropical rainforest stream," pp. 329–342 in P. DeDekker and W. D. Williams (eds.), *Limnology in Australia*. Dr. W. Junk, Dordrecht.

Peckarsky, B. L. 1983. "Biotic interactions or abiotic limitations? A model of lotic community structure," pp. 303–324 in T. D. Fontaine and S. M. Bartell (eds.), *Dynamics of Lotic Ecosystems*. Ann Arbor Science Publ., Ann Arbor, Mich..

Penáz, M., F. Kubícek, P. Marvan, and M. Zelinka. 1968. "Influence of the Vir River Valley Reservoir on the hydrobiological and ichthyological conditions in the River Svratka." *Acta sc. nat. Brno* **2**(1): 1-60.

Pennak, R. W. 1968. "Historical origins and ramifications of interstitial investigations." *Trans. Am. Microsc. Soc.* **87**: 214-218.

Pennak, R. W. 1978. *Fresh-Water Invertebrates of the United States*, 2nd ed. Wiley, New York.

Pennak, R. W. and C. M. McColl. 1944. "An experimental study of oxygen absorption in some damselfly naiads." *J. Cell. Comp. Physiol.* **23**: 1–10.

Pennak, R. W. and J. V. Ward. 1986. "Interstitial faunal communities of the hyporheic and adjacent groundwater biotopes of a Colorado mountain stream." *Arch. Hydrobiol. Suppl.* **74**: 356–396.

Percival, E. and H. Whitehead. 1926. "Observations on the biology of the mayfly *Ephemera danica* Müll." *Proc. Leeds Phil. Soc. (Sci. Sect.)* **1**: 136–148.

Percival, E. and H. Whitehead. 1929. "A quantitative study of some types of stream-bed." *J. Ecol.* **17**: 282–314.

Perkins, D. J. 1974. *The Biology of Estuaries and Coastal Waters.* Academic Press, London.

Perkins, P. D. 1976. "Psammophilous aquatic beetles in Southern California: A study of microhabitat preferences with notes on responses to stream alteration (Coleoptera: Hydraenidae and Hydrophilidae)." *Coleopterists Bull.* **30**: 309–324.

Pescador, M. L. and W. L. Peters. 1974. "The life history and ecology of *Baetisca rogersi* Berner (Ephemeroptera: Baetiscidae)." *Bull. Florida State Mus., Biol. Sci.* **17**: 151–209.

Peschet, R. 1932. *Description d'un Bidessus nouveau hypogé de l'Afrique occidentale Francaise (Coleoptera Dytiscidae),* pp. 571-574. Soc. Ent. France, Livre du Centenaire.

Pesta, O. 1933. "Beiträge zur Kenntnis der Limnologischen Beschaffenheit ostalpiner Tumpelgewässer." *Arch. Hydrobiol.* **25**: 68–80.

Peters, J. G., W. L. Peters, and T. J. Fink. 1987. "Seasonal synchronization of emergence in *Dolania americana* (Ephemeroptera: Behningiidae)." *Can. J. Zool.* **65**: 3177–3185.

Peters, W. L. and J. G. Peters (eds.). 1973. *Proceedings of the First International Conference on Ephemeroptera.* E. J. Brill, Leiden.

Peters, W. L. and J. G. Peters. 1977. "Adult life and emergence of *Dolania americana* in northwestern Florida (Ephemeroptera: Behningiidae)." *Internat. Revue ges. Hydrobiol.* **62**: 409–438.

Petersen, R. C. and K. W. Cummins. 1974. "Leaf processing in a woodland stream." *Freshwat. Biol.* **4**: 343–368.

Petit, G. and D. Schachter. 1954. "La Camarque. Étude écologique et faunistique." *Ann. Biol.* **30**(5–6): 193–253.

Petr, T. 1970a. "The bottom fauna of the rapids of the Black Volta River in Ghana." *Hydrobiologia* **36**: 399–418.

Petr, T. 1970b. "Macroinvertebrates of flooded trees in the man-made Volta Lake (Ghana) with special reference to the burrowing mayfly *Povilla adusta* Navás." *Hydrobiologia* **36**: 373–398.

Petr, T. 1973. "Some factors limiting the distribution of *Povilla adusta* Navas (Ephemeroptera, Polymitarcidae) in African lakes," pp. 223–230 in W. L. Peters and J. G. Peters (eds.), *Proc. First Internat. Conf. Ephemeroptera.* E. J. Brill, Leiden.

Peyton, E. L., J. F. Reinert, and N. E. Peterson. 1964. "The occurrence of *Deinocerites pseudes* Dyar and Knab in the United States, with additional notes on the biology of *Deinocerites* Species of Texas." *Mosquito News* **24**: 449–458.

Philipson, G. N. 1954. "The effect of water flow and oxygen concentration on six species of caddis fly (Trichoptera) larvae." *Proc. Zool. Soc. London* **124**: 547–564.

Philipson, G. N. 1978. "The undulatory behaviour of larvae of *Hydropsyche pellucidula* Curtis and *Hydropsyche* siltalai Döhler," pp. 241-247 in M. I. Crichton (ed.), *Proc. Second Internat. Symp. Trichoptera.* Dr. W. Junk, The Hague.

Philipson, G. N. and B. H. S. Moorhouse. 1974. "Observations on ventilatory and net-spinning activities of larvae of the genus *Hydropsyche* Pictet (Trichoptera, Hydropsychidae) under experimental conditions." *Freshwat. Biol.* **4**: 524–533.

Phillips, J. E. and J. Meredith. 1969. "Active sodium and chloride transport by anal papillae of a salt water mosquito larva *(Aedes campestris)*." *Nature* **222**: 168–169.

Phillipson, J. 1956. "A study of factors determining the distribution of the blackfly *Simulium ornatum* Mg." *Bull. Ent. Res.* **47**: 227-238.

Pictet, F. J. 1843. *Histoire Naturelle, générale et particulière des Insectes Néuropterès. Famille des Éphémèrines.* Baillière édit., Paris.

Pleskot, G. 1951. "Wassertemperatur und Leben im Bach." *Wett. U. Leben* **3**: 129–143.

Pleskot, G. 1953. "Zur Ökologie der Leptophlebiiden (Ins. Ephemeroptera)." *Österreich. Zool. Z.* **4**: 45–107.

Pleskot, G. 1958. "Die Periodizität einiger Ephemeropteren der Schwechat." *Wasser Abwasser* **1958**: 1-32.

Pleskot, G. 1961. "Die Periodizität der Ephemeropteren-Fauna einiger österreichischer Fliessgewässer." *Verh. Internat. Verein. Limnol.* **14**: 410-416.

Plummer, G. and J. B. Kethley. 1964. "Foliar absorption of amino acids, peptides, and other nutrients by the pitcher plant, *Sarracenia flava*." *Bot. Gaz.* **125**: 245–260.

Poff, N. L. and J. V. Ward. 1989. "Implications of streamflow variability and predictability for lotic community structure: A regional analysis of streamflow patterns." *Can. J. Fish. Aquat. Sci.* **46**: 1805–1818.

Poff, N. L. and J. V. Ward. 1990. "Physical habitat template of lotic systems: Recovery in the context of historical pattern of spatiotemporal heterogeneity." *Envir. Manag.* **14**: 629–645.

Polhemus, J. T. 1976. "Shore bugs (Hemiptera: Saldidae, etc.)," pp. 225–262 in L. Cheng (ed.), *Marine Insects.* North-Holland Publ. Co., Amsterdam.

Polhemus, J. T. 1984. "Aquatic and semiaquatic Hemiptera," pp. 231-260 in R. W. Merritt and K. W. Cummins (eds.), *An Introduction to the Aquatic Insects of North America*, 2nd ed. Kendall/Hunt, Dubuque, Iowa.

Pomeisl, E. 1953. "Der Mauerbach." *Wett. Leben Sonderh.* **2**: 103–121.

Pomeisl, E. 1961. "Ökologische-biologische untersuchungen an Plecopteren in Gebiet der niederösterreichischen kalkalpen und des diesem vorgelagerten Flyschgebietes." *Verh. Internat. Verein. Limnol.* **14**: 351-354.

Pomeroy, L. R. and R. G. Wiegert (eds.). 1981. *The Ecology of a Salt Marsh.* Springer-Verlag, Berlin.

Poole, W. C. and K. W. Stewart. 1976. "The vertical distribution of macrobenthos within the substratum of the Brazos River, Texas." *Hydrobiologia* **50**: 151–160.

Pope, G. E., J. C. H. Carter, and G. Power. 1973. "The influence of fish on the distribution of *Chaoborus* spp. (Diptera) and density of larvae in the Matamek River system, Québec." *Trans. Am. Fish. Soc.* **102**: 707–714.

Popham, E. J. 1943. "Ecological studies of the commoner species of British Corixidae." *J. An. Ecol.* **12**: 124–136.

Popham, E. J. 1961. "The function of the paleal pegs of Corixidae (Hemiptera: Heteroptera)." *Nature* **190**: 742.

Popov, Y. A. 1971. "Historical development of the hemipterous infraorder Nepomorpha." *Tr. Paleontol. Inst. Akad. Nauk SSSR* **129**: 1-228 (in Russian).

Potts, W. T. W. and G. Parry. 1964. *Osmotic and Ionic Regulation in Animals.* Macmillan, New York.

Prosser, C. L. (ed.). 1973. *Comparative Animal Physiology.* Saunders, Philadelphia.

Pugsley, C. W. and H. B. N. Hynes. 1983. "A modified freeze-core technique to quantify the depth distribution of fauna in stony streambeds." *Can. J. Fish. Aquat. Sci.* **40**: 637-743.

Puthz, V. 1978. "Ephemeroptera," pp. 256-263 in J. Illies (ed.), *Limnofauna Europaea.* Fischer, Stuttgart.

Quennerstedt, N. 1958. "Effect of water level fluctuation on lake vegetation." *Verh. Internat. Verein. Limnol.* **13**: 901-906.

Rabe, W. 1953. "Beitrage zum Orientierungsproblem der Wasserwanzen." *Z. vergl. Physiol.* **35**: 300-320.

Rabeni, C. F. and G. W. Minshall. 1977. "Factors affecting microdistribution of stream benthic insects." *Oikos* **29**: 33-43.

Rader, R. B. and J. V. Ward. 1989. "Influence of impoundments on mayfly diets, life histories, and production." *J. N. Am. Benthol. Soc.* **8**: 64-73.

Rader, R. B. and J. V. Ward. 1990. "Mayfly growth and population density in constant and variable temperature regimes." *Great Basin Nat.* **50**: 97-106.

Radford, D. S. and R. Hartland-Rowe. 1971. "Subsurface and surface sampling of benthic inverts in two streams." *Limnol. Oceanogr.* **16**: 114-119.

Radovanovic, M. 1935. "Überdie gegenwärtige kenntnis der balkanischen Trichopteren." *Verh. Internat. Verein. Limnol.* **7**: 100-105.

Rankin, J. C. and J. Davenport. 1981. *Animal Osmoregulation.* Wiley, New York.

Rapoport, E. H. and L. Sánchez. 1963. "On the epineuston or the super-aquatic fauna." *Oikos* **14**: 96-109.

Ratte, H. T. 1979. "Tagesperiodische Vertikalwanderung in thermisch geschichteten Gewässern: Einfluss von Temperatur-und Photoperiode-Zyklen auf *Chaoborus crystallinus* de Geer (Diptera: Chaoboridae)." *Arch. Hydrobiol. Suppl.* **57**: 1-37.

Rawson, D. S. 1930. "The bottom fauna of Lake Simcoe and its role in the ecology of the lake." *Univ. Toronto Studies Biology, Publ. Ont. Fish. Res. Lab.* **34**: 1-183.

Rawson, D. S. and J. E. Moore. 1944. "The saline lakes of Saskatchewan." *Can. J. Res, D* **22**: 141-201.

Réaumur, R. A. 1734-1742. *Mémoires pour servir a l'histoire des Insectes.* (6 vols.), N.p., Paris.

Redfield, A. C. 1972. "Development of a New England salt marsh." *Ecol. Monogr.* **42**: 201-237.

Reice, S. R. 1980. "The role of substratum in benthic macroinvertebrate microdistribution and litter decomposition." *Ecology* **61**: 580-590.

Remane, A. and C. Schlieper. 1971. *Biology of Brackish Water*, 2nd ed. Die Binnengewässer **25**: 1-372.

Remmert, H. 1955. "Ökologische Untersuchungen über die Dipteren der Nord-und Ostsee." *Arch. Hydrobiol.* **51**: 2-53.

Resh, V. H., A. V. Brown, A. P. Covich, M. E. Gurtz, H. W. Li, G. W. Minshall, S. R. Reice, A. L. Sheldon, J. B. Wallace, and R. Wissmar. 1988. "The role of disturbance in stream ecology." *J. N. Am. Benthol. Soc.* **7**: 433-455.

Resh, V. H. and J. O. Solem. 1984. "Phylogenetic relationships and evolutionary adaptations of aquatic insects," pp. 66–75 in R. W. Merritt and K. W. Cummins (eds.), *An Introduction to the Aquatic Insects of North America.* Kendall/Hunt, Dubuque, Iowa.

Resh, V. H. and D. M. Rosenberg (eds.). 1984. *The Ecology of Aquatic Insects.* Praeger, New York.

Reynoldson, T. B. and H. R. Hamilton. 1982. "Spatial heterogeneity in whole lake sediments—towards a loading estimate." *Hydrobiologia* **91**: 235–240.

Richardson, R. E. 1921. "The small bottom and shore fauna of the middle and lower Illinois River and its connecting lakes, Chillicothe to Grafton." *Bull. Illinois Nat. Hist. Surv.* **13**: 363–522.

Ricker, W. E. 1934. "An ecological classification of certain Ontario streams." *Univ. Toronto Studies, Biol. Series* **37**: 1–114.

Riek, E. F. 1971. "The origin of insects." *Proc. Internat. Congr. Ent.* **13**: 292–293.

Riek, E. F. 1973. "The classification of the Ephemeroptera," pp. 160-178 in W. L. Peters and J. G. Peters (eds.), *Proc. First Internat. Conf. Ephemeroptera.* E. J. Brill, Leiden.

Riek, E. F. 1976. "The marine caddisfly family Chathamiidae (Trichoptera)." *J. Austral. Ent. Soc.* **15**: 405–419.

Rigler, F. H. 1978. "Limnology in the high Arctic: A case study of Char Lake." *Verh. Internat. Verein. Limnol.* **20**: 127–140.

Riley, C. F. C. 1921. "Distribution of the large water-strider *Gerris remigis* Say throughout a river system." *Ecology* **11**: 32–36.

Robert, P. A. 1958. *Les libellules (odonates).* Delachaux et Niestlé A. A., Neuchâtel, Switzerland.

Roberts, M. J. 1970. "The structure of the mouthparts of syrphid larvae (Diptera) in relation to feeding habits." Acta Zool. **51**: 43–65.

Robbins, J. A. 1982. "Stratigraphic and dynamic effects of sediment reworking by Great Lakes zoobenthos." *Hydrobiologia* **92**: 611-622.

Rodhe, W. 1974. "Limnology turns to warm lakes." *Arch. Hydrobiol.* **73**: 537–546.

Rohdendorf, B. 1964. *The Historical Development of Diptera.* Nauke, Trans. Instit. Paleontology, Vol. 100, Acad. Sci. USSR, Moscow (in Russian, translated into English by Univ. Alberta Press, Edmonton, Canada, 1974).

Rohnert, U. 1950. "Wasserfüllte Baumhöhlen und ihre Besiedlung. Ein Beitrag zur Fauna Dendrolimnetica." *Arch. Hydrobiol.* **44**: 472-516.

Rooke, B. 1984. "The invertebrate fauna of four macrophytes in a lotic system." *Freshwat. Biol.* **14**: 507–513.

Rooke, J. B. 1986. "Macroinvertebrates associated with macrophytes and plastic imitations in the Eramosa River, Ontario, Canada." *Arch. Hydrobiol.* **106**: 307–325.

Rosenberg, D. M. and A. P. Wiens. 1975. "Experimental sediment addition studies on the Harris River, N. W. T., Canada: The effect on macroinvertebrate drift." *Verh. Internat. Verein. Limnol.* **19**: 1568–1574.

Rosenberg, D. M. and A. P. Wiens. 1978. "Effects of sediment addition on macrobenthic invertebrates in a northern Canadian river." *Water Research* **12**: 753-763.

Ross, D. H. and R. W. Merritt. 1978. "The larval instars and population dynamics of five species of black flies (Diptera: Simuliidae) and their responses to selected environmental factors." *Can. J. Zool.* **56**: 1633–1642.

Ross, D. H. and J. B. Wallace. 1982. "Factors influencing the longitudinal distribution of larval Hydropsychidae (Trichoptera) in a southern Appalachian stream system (U.S.A.)." *Hydrobiologia* **96**: 185–199.

Ross, H. H. 1956. *Evolution and Classification of the Mountain Caddisflies*. Univ. Ill. Press, Urbana.

Ross, H. H. 1958. "Affinities and origins of the northern and montane insects of Western North America." *Zoogeography, Am. Assoc. Adv. Sci. Publ. No. 51*, pp. 231–252.

Ross, H. H. 1963. "Stream communities and terrestrial biomes." *Arch. Hydrobiol.* **59**: 235–242.

Ross, H. H. 1967. "The evolution and past dispersal of the Trichoptera." *Ann. Rev. Ent.* **12**: 169–206.

Roth, J. C. 1968. "Benthic and limnetic distribution of three *Chaoborus* species in a southern Michigan lake (Diptera: Chaoboridae)." *Limnol. Oceanogr.* **13**: 242–249.

Rouch, R. and D. L. Danielopol. 1987. "L'origine de la faune aquatique souterraine, entre le paradigme du refuge et le modèle de la colonisation active." *Stygologia* **3**: 345–372.

Rousseau, E. 1921. *Larves et Nymphes Aquatiques des Insectes d'Europe (morphologie, biologie, systematique)*. Vol. I. Lebeque, Brussels.

Rubtsov, I. A. 1956. *Fauna of the USSR. Insects Diptera*. Vol. VI, Part 6. *Midges (Family Simuliidae)*. Zool. Inst. Akad. Nauk SSSR, n.s., No.64 (in Russian).

Rupprecht, R. 1975. "The dependence of emergence period in insect larvae on water temperature." *Verh. Internat. Verein. Limnol.* **19**: 3057–3063.

Russev, B. 1974. "Das Zoobenthos der Donau zwischen dem 845ten und dem 375ten Stromkilometer. III. Dichte und Biomass." *Acad. Bulg. Sci., Bull. Inst. Zool. Mus.* **40**: 175–194 (in Bulgarian).

Rust, B. R. 1982. "Sedimentation in fluvial and lacustrine environments." *Hydrobiologia* **91**: 59–70.

Ruttner, F. 1926. "Bermerkungen über den Sauerstaffgehalt der Gewässer und dessen respiratorischen." *Wert. Naturwissenschaften* **14**: 1237–1239.

Ruttner, F. 1963. *Fundamentals of Limnology*. University of Toronto Press, Toronto.

Rzóska, J. 1961. "Observations on tropical rainpools and general remarks on temporary waters." *Hydrobiologia* **17**: 265–286.

Rzóska, J. 1974. "The Upper Nile swamps, a tropical wetland study." *Freshwat. Biol.* **4**: 1–30.

Rzóska, J. 1976. "Zooplankton of the Nile System," pp. 333–343 in J. Róska (ed.), *The Nile, Biology of an Ancient River*. Dr. W. Junk, The Hague.

Saether, O. A. 1968. "Chironomids of the Finse area, Norway, with special reference to their distribution in a glacier brook." *Arch. Hydrobiol.* **64**: 426–483.

Saether, O. A. 1972. "Chaoboridae. Das Zooplankton der Binnengewässer. 1. Teil." *Binnengewässer* **26**: 257–280.

Saether, O. A. 1977. "Taxonomic studies on Chironomidae: *Nanocladius*, *Pseudochironomus*, and the *Harnischia* complex." *Bull. Fish. Res. Board Can.* **196**: 1–43.

Saether, O. A. 1980. "The influence of eutrophication on deep lake benthic invertebrate communities." *Prog. Wat. Tech.* **12**: 161-180.

Sanfillipo, N. 1958. "Viaggio in Venezuela di Nino Sanfilippo, V: Descrizione de

Trogloguignotus concii N. Gen. N. Sp. di Dytiscidae Freatobio." *An. Mus. Civ. Stor. Natur. Genoa* **70**: 159-164.

Sattler, W. 1963. "Über den Körperbau und Ethologie der Larve und Puppe von *Macronema* Pict. (Hydropsychidae) ein als Larve sich von "Mikro-Drift" ernährendes Trichopter aus dem. *Amazongebiet." Arch. Hydrobiol.* **59**: 26-60.

Sauramo, M. 1958. "Die Geschichte der Ostsee." *Ann. Acad. Sci. Fennicae* **51**: 1-522.

Savage, A. A. 1982. "Use of water boatmen (Corixidae) in the classification of lakes." *Biol. Conserv.* **23**: 55-70.

Sawchyn, W. W. and N. S. Church. 1973. "The effects of temperature and photoperiod on diapause development in the eggs of four species of *Lestes* (Odonata: Zygoptera)." *Can. J. Zool.* **51**: 1257-1265.

Schaller, F. 1968. "Action de la température sur la diapause et le développement de l'embryon d'*Aeschna mixta* (Odonata)." *J. Insect Physiol.* **14**: 1477-1483.

Schenke, G. 1965. "Die Ruderbewegungen bie *Corixa punctata* Illig. (Cryptocerata)." *Internat. Revue ges. Hydrobiol.* **50**: 73-84.

Scherer, E. 1965. "Zur Methodik experimenteller Fliesswasser—Ökologie." *Arch. Hydrobiol.* **61**: 242-248.

Schmassmann, W. 1920 "Die Bodenfauna hochalpiner Seen." *Arch. Hydrobiol. Suppl.* **3**: 1-106.

Schmid, F. 1955. "Contribution à l'étude des Limnephilidae (Trichoptera)." *Mitt. Schweiz. Ent. Ges.* **28**: 1-245.

Schmidt, H.-H. 1984. "Einfluss der Temperatur auf die Entwicklung von *Baetis vernus* Curtis." *Arch. Hydrobiol. Suppl.* **69**: 364-410.

Schmitz, E. H. 1959. "Seasonal biotic events in two Colorado alpine tundra ponds." *Am. Midl. Nat.* **61**: 424-446.

Schmitz, W. 1961. "Fliesswässerforschung-Hydrographie und Botanik." *Verh. Internat. Verein. Limnol.* **14**: 541-586.

Schoonbee, H. J. 1973. "The role of ecology in the species evaluation of the genus *Afronurus* Lestage (Heptageniidae) in South Africa," pp. 88-113 in W. L. Peters and J. G. Peters (eds.), *Proc. First Internat. Conf. Ephemeroptera.* E. J. Brill, Leiden.

Schott, R. J. and M. A. Brusven. 1980. "The ecology and electrophoretic analysis of the damselfy, *Argia vivida* Hagen, living in a geothermal gradient." *Hydrobiologia* **69**: 261-265.

Schwabe, G. H. 1936 "Beiträge zur kenntnis isländischer Thermalbiotope." *Arch. Hydrobiol. Suppl.* **6**: 161-352.

Schwarz, E. A. 1914. "Aquatic beetles, especially *Hydroscapha*, in hot springs, in Arizona." *Proc. Ent. Soc. Wash.* **16**: 163-168.

Schwarz, P. 1973. "Tages und jahresperiodische Imaginalhauting subarktischer und mitteleuropäischer Populationen von *Diura bicaudata.*" *Oikos* **24**: 151-154.

Schwoerbel, J. 1961. "Über die Lebensbedingungen und die Besiedlung des hyporheischen Lebensraumes." *Arch. Hydrobiol.* **25**: 182-214.

Schwoerbel, J. 1967. "Das hyporheische Interstitial als Grenzbiotop zwischen oberirdischem und subterranem Ökosystem und seine Bedeutung für die Primär-Evolution von Kleinsthöhlenbewohnern." *Arch. Hydrobiol. Suppl.* **33**: 1-62.

Scott, D. 1958. "Ecological studies on the Trichoptera of the River Dean, Cheshire." *Arch. Hydrobiol.* **54**: 340-392.

Scott, K. M. F., A. D. Harrison, and W. Macnae. 1952. "The ecology of South African estuaries. Part II. The Klein River Estuary, Hermanus, Cape." *Trans. Roy. Soc. S. Afr.* **33**: 283–332.

Scudder, G. G. E. 1976. "Water-Boatmen of saline waters (Hemiptera: Corixidae)," pp. 263–289 in L. Cheng (ed.), *Marine Insects.* North-Holland Publ. Co., Amsterdam.

Scudder, G. G. E. 1983. "A review of factors governing the distribution of two closely related corixids in the saline lakes of British Columbia." *Hydrobiologia* **105**: 143–154.

Scullion, J. and A. Sinton. 1983. "Effects of artificial freshets on substratum composition, benthic invertebrate fauna and invertebrate drift in two impounded rivers in mid-Wales." *Hydrobiologia* **107**: 261-269.

Sedell, J. R., J. E. Richey, and F. J. Swanson. 1989. "The river continuum concept: A basis for the expected ecosystem behavior of very large rivers." *Can. Spec. Publ. Fish. Aquat. Sci.* **106**: 49-55.

Seeger, W. 1971. "Die Biotopwahl bei Halipliden, zugleich ein Beitrag zum Problem der syntopischen (sympatrschen s. str.) Arten (Haliplidae; Coleoptera)." *Arch. Hydrobiol.* **69**: 155–199.

Segerstråle, S. G. 1949. "The brackish water fauna of Finland." *Oikos* **1**: 127–141.

Segerstråle, S. G. 1957. "Baltic Sea." *Geol. Soc. Am. Mem.* **67**: 751-800.

Segerstråle, S. G. 1959. "Brackishwater classification—historical survey." *Arch. Oceanogr. Limnol.* **11**(Suppl.): 7–33.

Seifert, R. P. 1975. "Clumps of *Heliconia* inflorescences as ecological islands." *Ecology* **56**: 1416–1422.

Seifert, R. P. 1982. "Neotropical *Heliconia* insect communities." *Quart. Rev. Biol.* **57**: 1–28.

Seifert, R. P. and F. H. Seifert. 1976. "A community matrix analysis of *Heliconia* insect communities." *Am. Nat.* **110**: 461–483.

Selgeby, J. H. 1974. "Immature insects (Plecoptera, Trichoptera, and Ephemeroptera) collected from deep water in western Lake Superior." *J. Fish. Res. Board Can.* **31**: 109–111.

Serruya, C. (ed.). 1978. *Lake Kinneret (Lake of Tiberias, Sea of Galilee).* Dr. W. Junk, The Hague.

Shadin, V. I. 1956. "Life in rivers." *Juzni preznih vod SSSR* **3**: 113-256 (in Russian).

Sheldon, A. L. 1984. "Colonization dynamics of aquatic insects," pp. 401–429 in V. H. Resh and D. M. Rosenberg (eds.), *The Ecology of Aquatic Insects.* Praeger, New York.

Shepard, L. J. and P. E. Lutz. 1976. "Larval Responses of *Plathermis lydia* Drury to experimental photoperiods and temperatures (Odonata: Anisoptera)." *Am. Midl. Nat.* **95**: 120–130.

Shepard, W. D. and K. W. Stewart. 1983. "Comparative study of nymphal gills in North American stonefly (Plecoptera) genera and a new, proposed paradigm of Plecoptera gill evolution." *Misc. Publ. Ent. Soc. Am. 13* **55**: 1–57.

Shepherd, B. G., G. F. Hartman, and W. J. Wilson. 1986. "Relationships between stream and intragravel temperatures in coastal drainages, and some implications for fisheries workers." *Can. J. Fish. Aquat. Sci.* **43**: 1818–1822.

Short, R. A., S. P. Canton, and J. V. Ward. 1980. "Detrital processing and associated macroinvertebrates in a Colorado mountain stream." *Ecology* **61**: 727-732.

Short, R. A. and J. V. Ward. 1980a. "Macroinvertebrates of a Colorado high mountain stream." *Southwest. Nat.* **25**: 23-32.

Short, R. A. and J. V. Ward. 1980b. "Leaf litter processing in a regulated Rocky Mountain stream." *J. Fish. Aquat. Sci.* **37**: 123-127.

Short, R. A. and J. V. Ward. 1981. "Benthic detritus dynamics in a mountain stream." *Holarct. Ecol.* **4**: 32-35.

Shreve, R. L. 1966. "Statistical law of stream numbers." *J. Geol.* **74**: 17-37.

Siegfried, C. A. and A. W. Knight. 1977. "The effects of washout in a Sierra foothills stream." *Am. Midl. Nat.* **98**: 200-207.

Siltala (Silfvenius), A. J. 1906. "Zur Trichopterenfauna des finnischen Meerbusens." *Acta Soc. Fauna Flora Fenn.* **28**: 1-21.

Simpson, K. W. 1976. "Shore flies and brine flies (Diptera: Ephydridae)," pp. 465-495 in L. Cheng (ed.), *Marine Insects.* North-Holland Publ. Co., Amsterdam.

Slack, H. D. 1965. "The profundal fauna of Loch Lomond, Scotland." *Proc. Roy. Soc. Edinb.* **69**: 272-297.

Sloan, W. C. 1956. "The distribution of aquatic insects in two Florida springs." *Ecology* **37**: 81-98.

Sly, P. G. (ed.). 1982. "Sediment/freshwater interaction." *Hydrobiologia* **91**: 1-700.

Smith, B. P. 1977. "Water mite parasitism of water boatmen (Hemiptera: Corixidae)." M.S. thesis, Univ. British Columbia, Vancouver, B.C.

Smith, B. D., P. S. Maitland, M. R. Young, and M. J. Carr. 1981. "The littoral zoobenthos," pp. 155-203 in P. S. Maitland (ed.), *The Ecology of Scotland's Largest Lochs.* Dr. W. Junk, The Hague.

Smith, D. 1979. "The larval stage of *Hydropsyche separata* Banks (Trichoptera: Hydropsychidae)." *Pan-Pac. Ent.* **55**: 10-20.

Smith, J. A. and A. J. Dartnall. 1980. "Boundary layer control by water pennies (Coleoptera: Psephenidae)." *Aquatic Insects* **2**: 65-72.

Smith, K. 1972. River water temperatures—an environmental review. *Scott. Geogr. Mag.* **88**: 211-220.

Smith, K. and M. E. Lavis. 1975. "Environmental influences on the temperature of a small upland stream." *Oikos* **26**: 228-236.

Smith, R. E. W. and R. G. Pearson. 1985. "Survival of *Sclerocyphon bicolor* Carter (Coleoptera: Psephenidae) in an intermittent stream in north Queensland." *J. Austral. Ent. Soc.* **24**: 101-102.

Smith, R. E. W. and R. G. Pearson. 1987. "The macro-invertebrate communities of temporary pools in an intermittent stream in tropical Queensland." *Hydrobiologia* **150**: 45-61.

Smith, R. L. 1976. "Male brooding behavior of the water bug *Abedus herberti.*" *Ann. Ent. Soc. Am.* **69**: 740-747.

Smock, L. A., D. L. Stoneburner, and D. R. Lenat. 1981. "Littoral and profundal macroinvertebrate communities of a coastal brown-water lake." *Arch. Hydrobiol.* **92**: 306-320

Snellen, R. K. and K. W. Stewart. 1979a. "The life cycle and drumming behavior of

Zealeuctra claasseni (Frison) and *Zealeuctra hitei* Ricker and Ross (Plecoptera: Leuctridae) in Texas, USA." *Aquatic Insects* **1**: 65-89.

Snellen, R. K. and K. W. Stewart. 1979b. "The life cycle of *Perlesta placida* (Plecoptera: Perlidae) in an intermittent stream in northern Texas." *Ann. Ent. Soc. Am.* **72**: 659-666.

Snodgrass, R. E. 1954. "The dragonfly larva." *Smithsonian Misc. Coll.* **123**(2): 1-38.

Solem, J. O. 1976. "Studies on the behaviour of adults of *Phryganea bipunctata* and *Agrypnia obsoleta* (Trichoptera)." *Norw. J. Ent.* **23**: 23-28.

Solem, J. O. 1983. "Temporary pools in the Dovre Mountains, Norway, and their fauna of Trichoptera." *Acta Ent. Fenn.* **42**: 82-85.

Soluk, D. A. 1985. "Macroinvertebrate abundance and production of psammophilous Chironomidae in shifting sand areas of a lowland river." *Can. J. Fish. Aquat. Sci.* **42**: 1296-1302.

Soluk, D. A. and H. F. Clifford. 1985. "Microhabitat shifts and substrate selection by the psammophilous predator *Pseudiron centralis* McDunnough (Ephemeroptera: Heptageniidae)." *Can. J. Zool.* **63**: 1539-1543.

Somero, G. N. 1978. "Temperature adaptation of enzymes: Biological optimization through structure-function compromises." *Ann. Rev. Ecol. Syst.* **9**: 1-29.

Soós, A. and L. Papp (eds.). 1984 et seq. *Catalogue of Palaearctic Diptera.* Elsevier, Amsterdam (an unfinished multiple-volume series).

Sorensen, D. L., M. M. McCarthy, E. J. Middlebrooks, and D. B. Porcella. 1977. *Suspended and Dissolved Solids Effects on Freshwater Biota: A Review.* EPA-600/3-77-042, Ecol. Res. Series, U.S. Environmental Protection Agency, Corvallis, Oregon.

Soszka, G. J. 1975a. "Ecological relationships between invertebrates and submerged macrophytes in the lake littoral." *Ekol. Pol.* **23**: 393-416.

Soszka, G. J. 1975b. "The invertebrates on submerged macrophytes in three Masurian lakes." *Ekol. Pol.* **23**: 371-391.

Sourie, R. 1962. "Les tufs à chironomides dans quelques ruisseaux de la region de Campon (Hautes-Pyrénées)." *Vie Milieu* **13**: 741-746.

Spangler, P. J. 1972. "A new genus and two new species of madicolous beetles from Venezuela (Coleoptera: Hydrophilidae)." *Proc. Biol. Soc. Wash.* **85**: 139-146.

Spangler, P. J. 1981a. "A new water beetle, *Troglochores ashmolei*, n. gen., n. sp., from Ecuador; the first known eyeless cavernicolous hydrophilid beetle (Coleoptera: Hydrophilidae)." *Proc. Ent. Soc. Wash.* **83**: 316-323.

Spangler, P. J. 1981b. "Two new genera of phreatic elmid beetles from Haiti; one eyeless and one with reduced eyes (Coleoptera, Elmidae)." *Bijdr. Dierk.* **51**: 375-387.

Spangler, P. J. 1986. "Insecta: Coleoptera," pp. 622-631 in L. Botosaneanu (ed.), *Stygofauna Mundi.* E. J. Brill, Leiden.

Spence, D. H. N. 1982. "The zonation of plants in freshwater lakes." *Adv. Ecol. Res.* **12**: 37-125.

Spence, J. R., D. H. Spence, and G. G. E. Scudder. 1980a. "The effects of temperature on growth and development of water strider species (Heteroptera: Gerridae) of central British Columbia and implications for species packing." *Can. J. Zool.* **58**: 1813-1820.

Spence, J. R., D. H. Spence, and G. G. E. Scudder. 1980b. "Submergence behavior in *Gerris*: Underwater basking." *Am. Midl. Nat.* **103**: 385-391.

Sprules, W. M. 1947. "An ecological investigation of stream insects in Algonquin Park, Ontario." *Publ. Ont. Fish. Res. Lab.* **69**: 1-81.

St. Quentin, D. and M. Beier. 1968. "Odonata (Libellen)." *Handb. Zool.* **4**(2) 2/6: 1-39.

Stanford, J. A. and J. V. Ward. 1979. "Stream regulation in North America," pp. 215-236 in J. V. Ward and J. A. Stanford (eds.), *The Ecology of Regulated Streams* Plenum, New York.

Stanford, J. A. and J. V. Ward. 1983. "Insect species diversity as a function of environmental variability and disturbance in stream systems," pp. 265-278 in J. R. Barnes and G. W. Minshall (eds.), *Stream Ecology—Applications and Testing of General Ecological Theory.* Plenum, New York.

Stanford, J. A. and J. V. Ward. 1988. "The hyporheic habitat of river ecosystems." *Nature* **335**: 64-66.

Stankovíc, S. 1960. *The Balkan Lake Ohrid and Its Living World.* Dr. W. Junk, The Hague.

Stark, J. D., R. E. Fordyce, and M. J. Winterbourn. 1976. "An ecological survey of the hot springs area, Hurunui River, Canterbury, New Zealand." *Mauri Ora* **4**: 35-52.

Starmühlner, F. 1984. "Checklist and longitudinal distribution of the meso-and macrofauna of mountain streams of Sri Lanka (Ceylon)." *Arch. Hydrobiol.* **101**: 303-325.

Statzner, B. 1975. "Zur longitudinalzonierung eines zentralafrikanischen Fliessgewässersystems unter besonderer Berücksichtigung der Köcherfliegen (Trichoptera, Insecta)." *Arch. Hydrobiol.* **76**: 153-180.

Statzner, B. 1981a. "A progress report on Hydropsychidae from the Ivory Coast: Characters for the specific identification of larvae and population dynamics of four abundant species," pp. 329-335 in G. P. Moretti (ed.), *Proc. Third Internat. Symp. Trichoptera.* Dr. W. Junk, The Hague.

Statzner, B. 1981b. "The relation between hydraulic stress and microdistribution of benthic macroinvertebrates in a lowland running water system, the Schierenseebrooks (North Germany)." *Arch. Hydrobiol.* **91**: 192-218.

Statzner, B. 1988. "Growth and Reynolds number of lotic macroinvertebrates: A problem for adaptation of shape to drag." *Oikos* **51**: 84-87.

Statzner, B., J. A. Gore, and V. H. Resh. 1988. "Hydraulic stream ecology: Observed patterns and potential applications." *J. N. Am. Benthol. Soc.* **7**: 307-360.

Statzner, B. and B. Higler. 1985. "Questions and comments on the river continuum concept." *Can. J. Fish. Aquat. Sci.* **42**: 1038-1044.

Statzner, B. and T. F. Holm. 1982. "Morphological adaptations of benthic invertebrates—an old question studied by means of a new technique (laser doppler anemometry)." *Oecologia* **53**: 290-292.

Statzner, B. and T. F. Holm. 1989. "Morphological adaptation of shape to flow: Microcurrents around lotic macroinvertebrates with known Reynolds numbers at quasi-natural flow conditions." *Oecologia* **78**: 145-157.

Steffan, A. W. 1971. "Chironomid (Diptera) biocoenoses in Scandinavian glacier brooks." *Can. Ent.* **103**: 477-486.

Steinly, B. A. 1986. "Violent wave action and the exclusion of Ephydridae (Diptera) from marine temperate intertidal and freshwater beach habitats." *Proc. Ent. Soc. Wash.* **88**: 427-437.

Steinmann, P. 1907. "Die Tierwelt der Gebirgsbäche, eine faunistischbiologische Studie." *Ann. Biol. Lac.* **2**: 30–162.

Stephens, D. W. 1974. "A summary of biological investigations concerning the Great Salt Lake, Utah (1861–1973)." *Great Basin Nat.* **34**: 221–229.

Stewart, K. W. and B. P. Stark. 1988. *Nymphs of North American Stonefly Genera (Plecoptera)*, Vol. XII. Thomas Say Found.

Stobbart, R. H. and J. Shaw. 1974. "Salt and water balance; excretion," pp. 361–446 in M. Rockstein (ed.), *The Physiology of Insecta*, 2nd ed., Vol V. Academic Press, New York.

Stockner, J. G. 1971. "Ecological energetics and natural history of *Hedriodiscus truquii* (Diptera) in two thermal spring communities." *J. Fish. Res. Board Can.* **28**: 73–94.

Stortenbeker, C. W. 1967. *Observations on the Population Dynamics of the Red Locust, Nomadacris septemfasciata* (Serville) *in Its Outbreak Areas.* Agric Res. Rep., Pudoc, Wageningen No. 694.

Strahler, A. N. 1957. "Quantitative analysis of watershed geomorphology." *Trans. Am. Geophys. Union* **38**: 913–920.

Stube, M. 1958. "The fauna of a regulated lake." *Rep. Instit. Freshwat. Res. Drottningholm* **39**: 162–224.

Stumm, W. and J. J. Morgan. 1970. *Aquatic Chemistry; an Introduction Emphasizing Chemical Equilibria in Natural Waters.* Wiley, New York.

Sublette, J. E. 1957. "The ecology of the macroscopic bottom fauna in Lake Texoma (Denison Reservoir), Oklahoma and Texas." *Am. Midl. Nat.* **57**: 371–402.

Suren, A. M. 1988. "The ecological role of bryophytes in high alpine streams of New Zealand." *Verh. Internat. Verein. Limnol.* **23**: 1412–1416.

Sutcliffe, D. W. 1961. Studies on salt and water balance in caddis larvae (Trichoptera): 1. Osmotic and ionic regulation of body fluids in *Limnephilus affinis* Curtis." *J. Exp. Biol.* **38**: 501-519.

Sutcliffe, D. W. 1962. "Studies on salt and water balance in caddis larvae (Trichoptera). III. Drinking and excretion." *J. Exp. Biol.* **39**: 141–160.

Sutcliffe, D. W. and T. R. Carrick. 1973. "Studies on mountain streams in the English Lake District. I. pH, calcium and the distribution of invertebrates in the River Duddon." *Freshwat. Biol.* **3**: 437–462.

Swammerdam, J. 1675. *Ephemeri Via, of Afbeeldingh van's Menschen leven Vertoont in de Wonderbaarelijcke en nooyt gehoorde Historie van het vliegent ende een-dagh-levent Halft of Oever-Aas . . . (avec préface manuscrite de l'éditeur A. Wolfgang, Amsterdam, et en préface imprimée, une lettre d'Antoinette Bourignon incitant Swammerdam à publier ses observations et lui souhaitant un bon nouvel an, en ce 5 janvier 1675).* Bibliothèque nationale, Paris.

Sweeney, B. W. 1978. "Bioenergetic and developmental response of a mayfly to thermal variation." *Limnol. Oceanogr.* **23**: 461–477.

Sweeney, B. W. 1984. "Factors influencing life-history patterns of aquatic insects," pp. 56–100 in V. H. Resh and D. M. Rosenberg (eds.), *The Ecology of Aquatic Insects.* Praeger, New York.

Sweeney, B. W. and J. A. Schnack. 1977. "Egg development, growth, and metabolism of *Sigara alternata* (Say) (Hemiptera: Corixidae) in fluctuating thermal environments." *Ecology* **58**: 265–277.

Sweeney, B. W. and R. L. Vannote. 1986. "Growth and production of a stream stonefly: influence of diet and temperature." *Ecology* **67**: 1396–1410.

Swift, M. C. and R. B. Forward, Jr. 1980. "Photoresponses of *Chaoborus* larvae." *J. Insect Physiol.* **26**: 365–372.

Symoens, J. J. 1955. "Oécouverte de tufs à chironomides dans la région mosane." *Verh. Internat. Verein. Limnol.* **12**: 604–607.

Takemon, Y. 1990. "Timing and synchronicity of the emergence of *Ephemera strigata*," pp. 61–70 in I. C. Campbell (ed.), *Mayflies and Stoneflies.* Kluwer, Dordrecht.

Tarzwell, C. M. 1936. "Experimental evidence on the value of trout stream improvement in Michigan." *Trans. Am. Fish. Soc.* **66**: 177-187.

Taylor, T. P. and D. C. Erman. 1980. "The littoral bottom fauna of high elevation lakes in Kings Canyon National Park, California, USA." *Calif. Fish Game* **66**: 112–119.

Tebo, L. B., Jr. 1955. "Effects of siltation, resulting from improper logging, on the bottom fauna of a small trout stream in the southern Appalachians." *Prog. Fish Cult.* **17**: 64–70.

Teraguchi, M. and T. C. Northcote. 1966. "Vertical distribution and migration of *Chaoborus flavicans* larvae in Corbett Lake, British Columbia." *Limnol. Oceanogr.* **11**: 164–176.

Terry, F. W. 1913. "On a new genus of Hawaiian chironomids." *Proc. Hawaiian Ent. Soc.* **2**: 291–295.

Thibault, M. 1971. "Écologie d'un ruisseau à truites des Pyrénées-Atlantiques, le Lissuraga II. Les fluctuations thermiques de l'eau; répercussion sur les périodes de sortie et la taille de quelques Éphéméroptères, Plécoptères et Trichoptères." *Ann. Hydrobiol.* **2**: 241–274.

Thienemann, A. 1912. "Der Bergbach des Sauerlands. Faunistisch-biologische Untersuchungen." *Internat. Revue ges. Hydrobiol. Hydrogr. Suppl.* **4**: 1–125.

Thienemann, A. 1918. "Untersuchungen über die Bezeihungen zwischen dem sauerstoffgehalt des Wassers und der Zusammensetzung der fauna in Norddeutschen Seen." *Arch. Hydrobiol.* **12**: 1–65.

Thienemann, A. 1925. "Die Binnengewässer Mitteleuropas." *Binnengewässer* **1**: 1–255.

Thienemann, A. 1932. "Tropische seen und seetypenlehre." *Arch. Hydrobiol. Suppl.* **9**: 205–231.

Thienemann, A. 1934. "Die Tierwelt der tropischen Pflanzengewässer." *Arch. Hydrobiol. Suppl.* **13**: 1–91.

Thienemann, A. 1950. "Verbreitungsgeschichte der Süsswassertierwelt Europas." *Binnengewässer* **18**: 1–809.

Thienemann, A. 1954. "Chironomus." *Binnengewässer* **20**: 1–834.

Thompson, D. J. 1978. "Towards a realistic predator-prey model: the effect of temperature on the functional response and life history of larvae of the damselfly, *Ischnura elegans*." *J. An. Ecol.* **47**: 757–767.

Thorpe, W. H. 1950. "Plastron respiration in aquatic insects." *Biol. Rev.* **25**: 344–390.

Thorpe, W. H. and D. J. Crisp. 1947a. "Studies on plastron respiration. I. The biology of *Aphelocheirus* [Hemiptera, Aphelocheiridae (Naucoridae)] and the mechanism of plastron retention." *J. Exp. Biol.* **24**: 227–269.

Thorpe, W. H. and D. J. Crisp. 1947b. "Studies on plastron respiration. II. The respiratory efficiency of the plastron in *Aphelocheirus*." *J. Exp. Biol.* **24**: 270–303.

Thorpe, W. H. and D. J. Crisp. 1947c. "Studies on plastron respiration. III. The orientation responses of *Aphelocheirus* (Hemiptera, Aphelocheiridae (Naucoridae) in relation to plastron respiration: Together with an account of specialized pressure receptors in aquatic insects." *J. Exp. Biol.* **24**: 310-328.

Thorpe, W. H. and D. J. Crisp. 1949. "Studies on plastron respiration. IV. Plastron respiration in the Coleoptera." *J. Exp. Biol.* **26**: 219-260.

Thorup, J. 1966. "Substrate-type and its value as a basis for the delimitation of bottom fauna communities in running waters." *Pymatuning Lab. Ecol.* **4**: 59-74.

Thorup, J. 1970. "The influence of a short-termed flood on a springbrook community." *Arch. Hydrobiol.* **66**: 447-457.

Thorup, J. and C. Lindegaard. 1977. "Studies on Danish springs." *Folia Limnol. Scand.* **17**: 7-15.

Thut, R. N. 1969. "A study of the profundal bottom fauna of Lake Washington." *Ecol. Monogr.* **39**: 79-100.

Tiegs, O.W. and S. M. Manton. 1958. "The evolution of the Arthropoda." *Biol. Rev.* **33**: 255-337.

Tilly, L. J. 1968. "The structure and Dynamics of Cone Spring." *Ecol. Monogr.* **38**: 169-197.

Tillyard, R. J. 1917. *The Biology of Dragonflies (Odonata or Paraneuroptera).* Cambridge Univ. Press, Cambridge.

Timms, B. V. 1981. "Animal communities in three Victorian lakes of differing salinity." *Hydrobiologia* **81**: 181-193.

Timms, B. V. 1983. "A study of benthic communities in some shallow saline lakes of western Victoria, Australia." *Hydrobiologia* **105**: 165-177.

Tjønneland, A. 1962. "The nocturnal flight activity and the lunar rhythm of emergence in the African midge, *Conochironomus acutistilus* (Freeman)." *Contr. Facult. Sci., Univ. Coll. Addis Ababa, C* **4**: 1-21.

Tokunaga, M. 1932. *Morphological and Biological Studies on a New Marine Chironomid Fly, Pontomyia pacifica, from Japan.* Me. College Agric., Kyoto Imperial Univ., No. 19 (Ent. Series No. 3).

Tokunaga, M. 1935. "Chironomidae from Japan (Diptera) 4. The early stages of a marine midge *Telmatogeton japonicus* Tokunaga." *Philippine J. Sci.* **57**: 491-511.

Tolkamp, H. H. and J. C. Both. 1978. "Organism-substrate relationship in a small Dutch lowland stream. Preliminary results." *Verh. Internat. Verein. Limnol.* **20**: 1509-1515.

Towns, D. R. 1981. "Effects of artificial shading on periphyton and invertebrates in a New Zealand stream." *N. Z. J. Mar. Freshwat. Res.* **15**: 185-192.

Townsend, C. R. and A. G. Hildrew. 1976. "Field experiments on the drifting, colonization and continuous redistribution of stream benthos." *J. An. Ecol.* **45**: 759-772.

Townsend, C. R., A. G. Hildrew, and J. Francis. 1983. "Community structure in some southern English streams: The influence of physicochemical factors." *Freshwat. Biol.* **13**: 521-544.

Tozer, W. 1979. "Underwater behavioural thermoregulation in the adult stonefly, *Zapada cinctipes.*" *Nature* **281**: 566-567.

Trapp, K. E. and A. C. Hendricks. 1984. "Modifications in the life history of *Glossosoma nigrior* exposed to three different thermal regimes," pp. 397-406 in J. C. Morse (ed.), *Proc. Fourth Internat. Symp. Trichoptera.* Dr. W. Junk Publ., The Hague.

Traver, J. R. 1939. "Himalayan mayflies (Ephemeroptera)." *Ann. Magazine Nat. Hist. Ser. 11* **IV**: 32-56.

Trottier, R. 1973. "Influence of temperature and humidity on the emergence behaviour of *Anax junius* (Odonata: Aeshnidae)." *Can. Ent.* **105**: 975-984.

Trotzsky, H. M. and R. W. Gregory. 1974. "The effects of water flow manipulation below a hydroelectric power dam on the bottom fauna of the Upper Kennebec River, Maine." *Trans. Am. Fish. Soc.* **103**: 318-324.

Tshernova, O. A. 1970. "On the classification of fossil and Recent Ephemeroptera." *Ent. Rev. Wash.* **49**: 71-81 (transl. of Entomologicheskii Obozhrenie).

Tucker, V. A. 1969. "Wave making by whirligig beetles (Gyrinidae)." *Science* **166**: 897-899.

Turcotte, P. and P. P. Harper. 1982. "The macro-invertebrate fauna of a small Andean stream." *Freshwat. Biol.* **12**: 411-419.

Tuxen, S. L. 1944. "The Hot Springs, their animal communities and their zoogeographical significance." *Zool. Iceland* **I**(11): 1-206.

Uéno, S. 1957. "Blind aquatic beetles of Japan, with some accounts of the fauna of Japanese subterranean waters." *Arch. Hydrobiol.* **53**: 250-296.

Ulfstrand, S. 1967. "Microdistribution of benthic species in Lapland stream." *Oikos* **18**: 293-310.

Ulfstrand, S. 1968. "Life cycles of benthic insects in Lapland streams (Ephemeroptera, Plecoptera, Trichoptera, Diptera Simuliidae)." *Oikos* **19**: 167-190.

Usinger, R. L. 1956. "Aquatic Hemiptera," pp. 182-228 in R. L. Usinger (ed.), *Aquatic Insects of California.* Univ. Calif. Press, Berkeley.

Usinger, R. L. 1957. "Marine insects." *Geol. Soc. Am. Mem.* **67**: 1177-1182.

Vaillant, F. 1956. "Recherches sur la faune madicole (hygropétrique s.l.) de France, de Corse et d'Afrique de Nord." *Mém. Mus. Hist. Nat. Paris* **11**: 1-258.

Vaillant, F. 1961. "Fluctuations d'une population madicole au cours d'une année." *Verh. Internat. Verein. Limnol.* **14**: 513-516.

Vandel, A. 1965. *Biospeleology.* Pergamon Press, Oxford.

Vannote, R. L. 1978. "A geometric model describing a quasi-equilibrium of energy flow in populations of stream insects." *Proc. Nat. Acad. Sci.* **75**: 381-384.

Vannote, R. L., G. W. Minshall, K. W. Cummins, J. R. Sedell, and C. E. Cushing. 1980. "The river continuum concept." *Can. J. Fish. Aquat. Sci.* **37**: 130-137.

Vannote, R. L. and B. W. Sweeney. 1980. "Geographic analysis of thermal equilibria: A conceptual model for evaluating the effect of natural and modified thermal regimes on aquatic insect communities." *Am. Nat.* **115**: 667-695.

Varga, L. 1928. "Ein interessanter Biotop der Biocönose von Wasserorganismen." *Biol. Zbl.* **48**: 143-162.

Varley, G. C. 1937. "Aquatic insect larvae which obtain oxygen from the roots of plants." *Proc. Roy. Ent. Soc. London, A* **12**: 55-60.

Varley, G. C. 1939. "On the structure and function of the hind spiracles of the larva of the beetle *Donacia* (Coleoptera, Chrysomelidae)." *Proc. Roy. Ent. Soc. London, A* **14**: 115-123.

Ventner, G. E. 1961. "A new ephemeropteran record from Africa." *Hydrobiologia* **18**: 327-331.

Vepsäläinen, K. 1974. "The life cycles and wing lengths of Finnish *Gerris* Fabr. species (Heteroptera, Gerridae)." *Acta Zool. Fenni. 141*: 1-73.

Verbeke, J. 1957. *Recherches écologiques sur la faune des grands lacs de'L'Est Congo Belge. Explor. Hydrobiol. Lacs Kivu, Edouard et Albert (1952–54)*, Vol. 3, No. 1. Inst. Roy. Sci. Nat. Belg., Brusseles.

Verrier, M.-L. 1956. *Biologie des Éphémères.* Collection Armand Colin, Paris, No. 306.

Vogel, S. 1981. *Life in Moving Fluids.* Willard Grant Press, Boston.

von Ende, C. N. 1979. "Fish predation, interspecific predation, and the distribution of two *Chaoborus* species." *Ecology* **60**: 119-128.

Walker, E. M. 1953. *The Odonata of Canada and Alaska,* Vol. I. Univ. Toronto Press.

Walker, I. R. and R. W. Mathewes. 1989. "Chironomidae (Diptera) remains in surficial lake sediments from the Canadian Cordillera: analysis of the fauna across an altitudinal gradient." *J. Paleolimnol.* **2**: 61–80.

Walker, K. F. 1973. "Studies on a saline lake ecosystem." *Austral. J. Mar. Freshwat. Res.* **24**: 21–71.

Wallace, J. B. and R. W. Merritt. 1980. "Filter-feeding ecology of aquatic insects." *Ann. Rev. Ent.* **25**: 103–132.

Walsh, G. B. 1925. "The coast Coleoptera of the British Isles." *Ent. Monthly Mat.* **6**: 137–151.

Walshe, B. M. 1947. "On the function of haemoglobin in *Chironomus* after oxygen lack." *J. Exp. Biol.* **24**: 329–342.

Walshe, B. M. 1948. "The oxygen requirements and thermal resistance of chironomid larvae from flowing and from still waters." *J. Exp. Biol.* **25**: 25–44.

Walshe, B. M. 1950. "The function of haemoglobin in *Chironomus plumosus* under natural conditions." *J. Exp. Biol.* **27**: 73–95.

Walshe, B. M. 1951. "The function of haemoglobin in relation to filter feeding in leaf-mining chironomid larvae." *J. Exp. Biol.* **38**: 57-61.

Waltz, R. D. and W. P. McCafferty. 1979. *Freshwater Springtails (Hexapoda: Collembola) of North America.* Purdue Univ. Agric. Exp. Station Res. Bull. 960.

Ward, G. M. and K. W. Cummins. 1979. "Effects of food quality on growth of a stream detritivore, *Paratendipes albimanus* (Meigen) (Diptera: Chironomidae)." *Ecology* **60**: 57–64.

Ward, J. V. 1975. "Bottom fauna-substrate relationships in a northern Colorado trout stream: 1945 and 1974." *Ecology* **56**: 1429–1434.

Ward, J. V. 1976. "Effects of thermal constancy and seasonal temperature displacement on community structure of stream macroinvertebrates," pp. 302–307 in G. W. Esch and R. W. McFarlane (eds.), *Thermal Ecology II.* National Technical Information Service, Springfield, VA.

Ward, J. V. 1981. "Altitudinal distribution and abundance of Trichoptera in a Rocky Mountain stream," pp. 377–383 in G. P. Moretti (ed.), *Proc. Third Internat. Symp. Trichoptera.* Dr. W. Junk, The Hague.

Ward, J. V. 1982. "Altitudinal zonation of Plecoptera in a Rocky Mountain stream." *Aquatic Insects* **4**: 105–110.

Ward, J. V. 1985. "Thermal characteristics of running waters." *Hydrobiologia* **125**: 31–46.

Ward, J. V. 1986. "Altitudinal zonation in a Rocky Mountain stream." *Arch. Hydrobiol. Suppl.* **74**: 133-199.

Ward, J. V. 1989a. "The four-dimensional nature of lotic ecosystems." *J. N. Am. Benthol. Soc.* **8**: 2-8.

Ward, J. V. 1989b. "Riverine-wetland interactions," pp. 385-400 in R. R. Sharitz and J. W. Gibbons (eds.), *Freshwater Wetlands and Wildlife.* DOE Sympos. Ser. 61, Oak Ridge, Tenn.

Ward, J. V. and L. Berner. 1980. "Abundance and altitudinal distribution of Ephemeroptera in a Rocky Mountain stream," pp. 169-177 in J. F. Flannagan and K. E. Marshall (eds.), *Advances in Ephemeroptera Biology.* Plenum, New York.

Ward, J. V. and R. G. Dufford. 1979. "Longitudinal and seasonal distribution of macroinvertebrates and epilithic algae in a Colorado springbrook-pond system." *Arch. Hydrobiol.* **86**: 284-321.

Ward, J. V. and J. R. Holsinger. 1981. "Distribution and habitat diversity of subterranean amphipods in the Rocky mountains of Colorado." *Internat. J. Speleol.* **11**: 63-70.

Ward, J. V. and J. A. Stanford. 1982. "Thermal responses in the evolutionary ecology of aquatic insects." *Ann. Rev. Ent.* **27**: 97-117.

Ward, J. V. and J. A. Stanford. 1983. "The intermediate disturbance hypothesis: An explanation for biotic diversity patterns in lotic ecosystems," pp. 347-356 in T. D. Fontaine and S. M. Bartell (eds.), *Dynamics of Lotic Ecosystems.* Ann Arbor Science Publ., Ann Arbor, Mich.

Ward, J. V., H. J. Zimmermann, and L. D. Cline. 1986. "Lotic zoobenthos of the Colorado System," pp. 403-423 in B. R. Davies and K. F. Walker (eds.), *The Ecology of River Systems,* Dr. W. Junk, Dordrecht (The Netherlands).

Waringer, J. A. and U. H. Humpesch. 1984. "Embryonic development, larval growth and life cycle of *Coenagrion puella* (Odonata: Zygoptera) from an Austrian pond." *Freshwat. Biol.* **14**: 385-399.

Wartinbee, D. C. 1979. "Diel emergence patterns of lotic Chironomidae." *Freshwat. Biol.* **9**: 147-156.

Waters, T. F. 1968. "Diurnal periodicity in the drift of a day-active stream invertebrate." *Ecology* **49**: 152-153.

Waters, T. F. 1972. "The drift of stream insects." *Ann. Rev. Ent.* **17**: 253-272.

Waters, T. F. 1981. "Drift of stream invertebrates below a cave source." *Hydrobiologia* **78**: 169-176.

Wautier, J. and E. Pattée. 1955. "Expérience physiologique et expérience écologique. L'influence du substrat sur la consommation d'oxygène chez les larves d'Ephémeroptères." *Bull. Mens. Soc. Linn. Lyon* **24**: 178-183.

Weber, R. E. 1980. "Functions of invertebrate hemoglobins with special reference to adaptations to environmental hypoxia." *Am. Zool.* **20**: 79-101.

Webster, D. A. and P. C. Webster. 1943. "Influence of water current on case weight of the caddisfly *Goera calcarata* Banks." *Can. Ent.* **75**: 105-108.

Weele, H. W. Van der. 1910. "Megaloptera, monographic revision." *Coll. Zool. Selys Longch.* **5**: 1-93.

Wegener, W., V. Williams, and T. D. McCall. 1975. "Aquatic macroinvertebrate responses to an extreme drawdown." *Proc. Southeast. Assoc. Game Fish Commiss.* **28**: 126-144.

Welch, H. E. 1976. "Ecology of chironomidae (Diptera) in a polar lake." *J. Fish. Res. Board Can.* **33**: 227–247.

Welch, P. S. 1916. "Contribution to the biology of certain aquatic lepidoptera." *Ann. Ent. Soc. Am.* **9**: 159–187.

Welch, P. S. 1952. *Limnology*. McGraw-Hill, New York.

Welcomme, R. L. 1979. *Fisheries Ecology of Floodplain Rivers.* Longman, London.

Wells, R. M. G., M. J. Hudson, and T. Brittain. 1981. "Function of the hemoglobin and the gas bubble in the backswimmer *Anisops assimilis* (Hemiptera: Notonectidae)." *J. Comp. Physiol.* **142**: 515–522.

Wentworth, C. K. 1922. "A scale of grade and class terms for clastic sediments." *J. Geol.* **30**: 377–392.

Wesenberg-Lund, C. 1908. "Die Littoralen Tiergesellschaften unserer grösseren Seen." *Internat. Revue ges. Hydrobiol. Hydrogr.* **1**: 574–607.

Wesenberg-Lund, C. 1913. "Odonaten-Studien." *Internat. Revue ges. Hydrobiol. Hydrogr.* **6**: 155–422.

Wesenberg-Lund, C. 1943. *Biologie der Süsswasserinsekten.* Springer-Verlag, Berlin.

Wetmore, S. H., R. J. Mackay, and R. W. Newbury. 1990. "Characterization of the hydraulic habitat of *Brachycentrus occidentalis*, a filter-feeding caddisfly." *J. N. Am. Benthol. Soc.* **9**: 157–169.

Wetzel, R. G. 1983. *Limnology*. Saunders, Philadelphia.

White, D. S. 1978. "Life cycle of the riffle beetle, *Stenelmis sexlineata* (Elmidae)." *Ann. Ent. Soc. Am.* **71**: 121–125.

White, D. S., C. H. Elzinga, and S. P. Hendricks. 1987. "Temperature patterns within the hyporheic zone of a northern Michigan river." *J. N. Am. Benthol. Soc.* **6**: 85–91.

Whitton, B. A. (ed.). 1975. *River Ecology.* Blackwell, Oxford.

Wichard, W. 1978. "Structure and function of the tracheal gills of *Molanna angustata* Curt.," pp. 293–296 in M. I. Crichton (ed.), *Proc. Second Internat. Symp. Trichoptera.* Dr. W. Junk, The Hague.

Wichard, W., P. T. P. Tsui, and H. Komnick. 1973. "Effect of different salinities on the coniform chloride cells of mayfly larvae." *J. Insect Physiol.* **19**: 1825–1835.

Wiederholm, T. and L. Eriksson. 1977. "Benthos of an acid lake." *Oikos* **29**: 261–267.

Wiggins, G. B. 1973. "A contribution to the biology of caddisflies (Trichoptera) in temporary pools." *Life Sci. Contrib. Roy. Ontario Mus.* **88**: 1–28.

Wiggins, G. B. 1977. *Larvae of the North American Caddisfly Genera.* Univ. Toronto Press, Ontario.

Wiggins, G. B. 1984. "Trichoptera," pp. 271–311 in R. W. Merritt and K. W. Cummins (eds.), *An Introduction to the Aquatic Insects of North America,* 2nd ed. Kendall/Hunt, Dubuque, Iowa.

Wiggins, G. B. and R. J. Mackay. 1978. "Some relationships between systematics and trophic ecology in Nearctic aquatic insects, with special reference to Trichoptera." *Ecology* **59**: 1211–1220.

Wiggins, G. B., R. J. Mackay, and I. M. Smith. 1980. "Evolutionary and ecological strategies of animals in annual temporary pools." *Arch. Hydrobiol. Suppl.* **58**: 97–206.

Wiggins, G. B. and W. Wichard. 1989. "Phylogeny and pupation in Trichoptera, with proposals on the origin and higher classification of the order." *J. N. Am. Benthol. Soc.* **8**: 260-276.

Wigglesworth, V. B. 1963. "Origin of wings in insects." *Nature* **197**: 97–98.

Wigglesworth, V. B. 1972. *The Principles of Insect Physiology,* 7th ed. Chapman and Hall, London.

Wigglesworth, V. B. 1973. "Evolution of insect wings and flight." *Nature* **246**: 127–129.

Wigglesworth, V. B. 1976. "The evolution of insect flight," pp. 255-269 in R. C. Rainey (ed.), *Insect Flight.* Blackwell, Oxford.

Wilber, C. G. 1983. *Turbidity in the Aquatic Environment.* C. C. Thomas, Springfield, Ill.

Wilcox, R. S. 1972. "Communication by surface waves; mating behavior of a water strider (Gerridae)." *J. Comp. Physiol.* **80**: 255–266.

Wiley, M. J. and S. L. Kohler. 1980. "Positioning changes of mayfly nymphs due to behavioral regulation of oxygen consumption." *Can. J. Zool.* **58**: 618–622.

Wiley, M. and S. L. Kohler. 1984. "Behavioral adaptations of aquatic insects," pp. 101–133 in V. H. Resh and D. M. Rosenberg (eds.), *The Ecology of Aquatic Insects.* Praeger, New York.

Williams, D. D. 1977. "Movements of benthos during the recolonization of temporary streams." *Oikos* **29**: 306–312.

Williams, D. D. 1983. "The natural history of a Nearctic temporary pond in Ontario with remarks on continental variation in such habitats." *Internat. Revue ges. Hydrobiol.* **68**: 239–253.

Williams, D. D. 1984. "The hyporheic zone as a habitat for aquatic insects and associated arthropods," pp. 430–455 in V. H. Resh and D. R. Rosenberg (eds.), *The Ecology of Aquatic Insects.* Praeger, New York.

Williams, D. D. 1987. *The Ecology of Temporary Waters.* Timber Press, Portland, Oregon.

Williams, D. D. and I. D. Hogg. 1988. "Ecology and production of invertebrates in a Canadian coldwater spring-springbrook system." *Holarct. Ecol.* **11**: 41–54.

Williams, D. D. and H. B. N. Hynes. 1974. "The occurrence of benthos deep in the substratum of a stream." *Freshwat. Biol.* **4**: 233-256.

Williams, D. D. and H. B. N. Hynes. 1976a. "The ecology of temporary streams I. The faunas of two Canadian streams." *Internat. Revue ges. Hydrobiol.* **61**: 761–787.

Williams, D. D. and H. B. N. Hynes. 1976b. "The recolonization mechanisms of stream benthos." *Oikos* **27**: 265–272.

Williams, D. D. and H. B. N. Hynes. 1977. "The ecology of temporary streams. II. General remarks on temporary streams." *Internat. Revue ges. Hydrobiol.* **62**: 53–61.

Williams, D. D., A. F. Tavares, and E. Bryant. 1987. "Respiratory device or camouflage? A case for the caddisfly." *Oikos* **50**: 42-52.

Williams, T. R. and H. B. N. Hynes. 1971. "A survey of the fauna of streams on Mount Elgon, East Africa, with special reference to the Simuliidae (Diptera)." *Freshwat. Biol.* **1**: 227–248.

Williams, W. D. 1978. "Limnology of Victoria salt lakes, Australia." *Verh. Internat. Verein. Limnol.* **20**: 1165–1174.

Williams, W. D. 1980. *Australian Freshwater Life*, 2nd ed. Macmillan, Melbourne.

Williams, W. D. 1981a. "Inland salt lakes: An introduction." *Hydrobiologia* **81**: 1-14.

Williams, W. D. (ed.). 1981b. *Salt Lakes*. Dr. W. Junk, The Hague.

Williams, W. D. 1981c. "The limnology of saline lakes in western Victoria. A review of some recent studies." *Hydrobiologia* **82**: 233-259.

Williams, W. D. 1985. "Biotic adaptations in temporary lentic waters, with special reference to those in semi-arid and arid regions." *Hydrobiologia* **125**: 85-110.

Willoughby, L. G. and R. G. Mappin. 1988. "The distribution of *Ephemerella ignita* (Ephemeroptera) in streams: The role of pH and food resources." *Freshwat. Biol.* **19**: 145-155.

Wingfield, C. A. 1939. "The function of the gills of mayfly nymphs from different habitats." *J. Exp. Biol.* **16**: 363-373.

Winterbourn, M. J. 1968. "The faunas of thermal waters in New Zealand." *Tuatara* **16**: 111-122.

Winterbourn, M. J. 1969. "The distribution of algae and insects in hot spring thermal gradients at Waimangu, New Zealand." *N.Z. J. Mar. Freshwat. Res.* **3**: 459-465.

Winterbourn, M. J. 1976. "Fluxes of litter falling into a small beech forest stream." *N.Z. J. Mar. Freshwat. Res.* **10**: 399-416.

Winterbourn, M. J. 1982. "The invertebrate fauna of a forest stream and its association with fine particulate matter." *N.Z. J. Mar. Freshwat. Res.* **16**: 271-281.

Winterbourn, M. J. and N. H. Anderson. 1980. "The life history of *Philanisus plebeius* Walker (Trichoptera: Chathamiidae), a caddisfly whose eggs were found in a starfish." *Ecol. Ent.* **5**: 293-303.

Winterbourn, M. J. and T. J. Brown. 1967. "Observations on the faunas of two warm streams in the Taupo thermal region." *N.Z. J. Mar. Freshwat. Res.* **1**: 38-50.

Winterbourn, M. J. and K. J. Collier. 1987. "Distribution of benthic invertebrates in acid, brown water streams in the South Island of New Zealand." *Hydrobiologia* **153**: 277-286.

Winterbourn, M. J. and S. F. Davis. 1976. "Ecological role of *Zelandopsyche ingens* (Trichoptera: Oeconesidae) in a beech forest stream ecosystem." *Austral. J. Mar. Freshwat. Res.* **27**: 197-215.

Winterbourn, M. J., J. S. Rounick, and B. Cowie. 1981. "Are New Zealand streams really different?" *N.Z. J. Mar. Freshwat. Res.* **15**: 321-328.

Wirth, W. W. 1949. "A revision of the clunionine midges with descriptions of a new genus and four new species (Diptera: Tendipedidae)." *Univ. Calif. Publ. Ent.* **8**: 151-182.

Wirth, W. W. 1951. "A revision of the dipterous family Canaceidae." *Bishop Mus. Occas. Pap.* **20**: 245-275.

Wirth, W. W. and A. Stone. 1956. "Aquatic Diptera," pp. 372-482 in R. L. Usinger (ed.), *Aquatic Insects of California*. Univ. Calif. Press, Berkeley.

Wise, E. J. 1976. "Studies on the Ephemeroptera of a Northumbrian river system I. Serial distribution and relative abundance." *Freshwat. Biol.* **6**: 363-372.

Wise, E. J. 1980. "Seasonal distribution and life histories of Ephemeroptera in a Northumbrian river." *Freshwat. Biol.* **10**: 101-111.

Wisely, B. 1962. "Studies on Ephemeroptera. II. *Coloburiscus humeralis* (Walker); ecology and distribution of the nymphs." *Trans. Roy. Soc. N.Z.* **2**: 209-220.

Wisseman, R. W. and N. H. Anderson. 1987. "The life history of *Cryptochia pilosa* (Trichoptera: Limnephilidae) in an Oregon Coast Range watershed," pp. 243-246 in M. Bournaud and H. Tachet (eds.), *Proc. Fifth Internat. Symp. Trichoptera.* Dr. W. Junk, Dordrecht.

Wolvekamp, H. P. 1955. "Die physikalische kieme der Wasserinsekten." *Experientia* **11**: 294-301.

Wood, K. G. 1956. "Ecology of *Chaoborus* (Diptera: Culicidae) in an Ontario Lake." *Ecology* **37**: 639-643.

Woodmansee, R. A. and B. J. Grantham. 1961. "Diel vertical migrations of two zooplankters *(Mesocyclops* and *Chaoborus)* in a Mississippi lake." *Ecology* **42**: 619-628.

Wootton, R. J. 1972. "The evolution of insects in freshwater ecosystems," pp. 69-82 in R. B. Clark and R. J. Wootton (eds.), *Essays in Hydrobiology.* Univ. Exeter, Exeter, England.

Wootton, R. J. 1976. "The fossil record and insect flight," pp. 235-254 in R. C. Rainey (ed.), *Insect Flight.* Blackwell, Oxford.

Wotton, R. S. 1986. "The use of silk life-lines by larvae of *Simulium noelleri* (Diptera)." *Aquatic Insects* **8**: 255-261.

Wright, M. and C. S. Shoup. 1945. "Dragonfly nymphs from the Obey River and adjacent streams in Tennessee." *J. Tenn. Acad. Sci.* **20**: 266-278.

Wu, Y. F. 1931. "A contribution to the biology of *Simulium.*" *Pap. Mich. Acad. Sci.* **13**: 543-599.

Wurtz, C. B. 1969. "The effects of heated discharges on freshwater benthos," pp. 199-213 in P. A. Krenkel and F. L. Parker (eds.), *Biological Aspects of Thermal Pollution.* Vanderbuilt Univ. Press, Nashville, Tenn.

Yang, C. T. 1971. "Formation of riffles and pools." *Water Resources Res.* **7**: 1567-1574.

Young, F. N. and G. Longley. 1976. "A new subterranean aquatic beetle from Texas (Coleoptera: Dytiscidae-Hydroporinae)." *Ann. Ent. Soc. Am.* **69**: 787-792.

Young, J. O. 1974. "Life-cycles of some invertebrate taxa in a small pond together with changes in their numbers over a period of three years." *Hydrobiologia* **45**: 63-90.

Young, J. O. 1975. "Seasonal and diurnal changes in the water temperature of a temperate pond (England) and a tropical pond (Kenya)." *Hydrobiologia* **47**: 513-526.

Zahar, A. R. 1951. "The ecology and distribution of black-flies (Simuliidae) in south-east Scotland." *J. An. Ecol.* **20**: 33-62.

Zahner, R. 1959. "Über die Bindung der mitteleuropäischen *Calopteryx*-Arten (Odonata, Zygoptera) an den lebensraum des strömenden Wassers I. Der Anteil der Larven an der Biotopbindung." *Internat. Revue ges. Hydrobiol. Hydrogr.* **44**: 51-130.

Zalom, F. G., A. A. Grigarick, and M. O. Way. 1980. "Diel flight periodicities of some Dytiscidae (Coleoptera) associated with California rice paddies." *Ecol. Ent.* **5**: 183-187.

Zapekina-Dulkeyt, Yu. I. and L. A. Zhiltsova. 1973. "A new genus of stoneflies (Plecoptera) from Lake Baikal." *Ent. Obozv.* **52**: 340-345.

Zhadin, V. I. and S. V. Gerd. 1961. *Fauna and Flora of the Rivers, Lakes and Reservoirs of the U.S.S.R.* (in Russian, translated into English by the Israel Program for Scientific Translations, 1963). Gosudarstvennoe Uchebno-Pedagogicheskoe Izdatel'stvo Ministerstva Prosveshcheniya RSFSR, Moscow.

Ziegler, D. D. and K. W. Stewart. 1977. "Drumming behavior of eleven Nearctic stonefly (Plecoptera) species." *Ann. Ent. Soc. Am.* **70**: 495-505.

Ziemann, H. 1975. "Über den Einfluss der Wasserstoffionenkonzentration und des Hydrogenkarbonatgehaltes auf die Ausbildung von Bergbachbiozönosen." *Internat. Revue ges. Hydrobiol.* **60**: 523-555.

Zimmerman, M. C. and T. E. Wissing. 1978. "Effects of temperature on gut-loading and gut-clearing times of the burrowing mayfly, *Hexagenia limbata.*" *Freshwat. Biol.* **8**: 269-277.

Zschokke, F. 1894. "Die Tierwelt der Juraseen." *Rev. Suisse Zool.* **2**: 349-376.

Zschokke, F. 1900. "Die Tierwelt der Hochgebirgsseen." *Neue Deukschr. allg. Schweiz. Ges. Naturwissenschaften.* **37**: 1-400.

Zwick, P. 1973. "Insecta: Plecoptera. Phylogenetisches system und katalog." *Das Tierreich* **94**: 1-465.

Zwick, P. 1980. "Plecoptera (Steinfliegen)." *Handb. Zool.* **4**(2) 2/7: 1-115.

INDEX